Design and Management of Manufacturing Systems

Design and Management of Manufacturing Systems

Editor

Arkadiusz Gola

MDPI • Basel • Beijing • Wuhan • Barcelona • Belgrade • Manchester • Tokyo • Cluj • Tianjin

Editor
Arkadiusz Gola
Lublin University of Technology
Poland

Editorial Office
MDPI
St. Alban-Anlage 66
4052 Basel, Switzerland

This is a reprint of articles from the Special Issue published online in the open access journal *Applied Sciences* (ISSN 2076-3417) (available at: https://www.mdpi.com/journal/applsci/special_issues/Manufacturing_System).

For citation purposes, cite each article independently as indicated on the article page online and as indicated below:

LastName, A.A.; LastName, B.B.; LastName, C.C. Article Title. *Journal Name* **Year**, *Volume Number*, Page Range.

ISBN 978-3-0365-1146-7 (Hbk)
ISBN 978-3-0365-1147-4 (PDF)

Contents

About the Editor

Arkadiusz Gola received the D.Sc. degree in mechanical engineering from Krakow University of Technology (2019) and Ph.D. in industrial engineering from Lublin University of Technology (2011). Currently, he works as a Professor of Lublin University of Technology (Poland) and he is a Head of the Division of Production Robotisation and Organisation in the Department of Production Computerisation and Robotisation (Faculty of Mechanical Engineering, Lublin University of Technology). He is an expert at the Polish National Center for Research and Development and he was also employed as an external evaluator of the international project ILA-Lean (2016–2018). In fact, he leads a R&D project of about 30 mln PLN that aims to develop an innovative warehouse logistics process through a synergistic combination of dispersed workstations for various purposes. Arkadiusz Gola has supervised more than 90 master's and engineering theses, and has published over 130 manuscripts, including textbooks and refereed papers in international journals and conferences, with about 500 quotations due to Science Citation Index Expanded. His major fields of research include various aspects of manufacturing systems design, production logistics and operations management.

itorial

)esign and Management of Manufacturing Systems

kadiusz Gola

Department of Production Computerization and Robotization, Faculty of Mechanical Engineering, Lublin University of Technology, ul. Nadbystrzycka 36, 20-618 Lublin, Poland; a.gola@pollub.pl

tion: Gola, A. Design and
nagement of Manufacturing
tems. *Appl. Sci.* **2021**, *11*, 2216.
s://doi.org/10.3390/app11052216

eived: 1 March 2021
epted: 2 March 2021
lished: 3 March 2021

lisher's Note: MDPI stays neutral
n regard to jurisdictional claims in
lished maps and institutional affil-
ons.

Although the design and management of manufacturing systems have been explored in the literature for many years now, they still remain topical problems in current scientific research. Changing market trends, globalization, the constant pressure to reduce production costs, and technical and technological progress make it necessary to search for new manufacturing methods and ways of organizing them, and to modify manufacturing-system design paradigms.

Even though the very concept of a manufacturing system emerged at the beginning of the 19th century, production continued to be an artisanal activity until the beginning of the 20th century. An important milestone (referred to today as the Second Industrial Revolution) was Henry Ford's introduction of the moving assembly line, which radically increased the efficiency of manufacturing processes. The following years, which, on the one hand, brought unprecedented progress in the development of manufacturing techniques, mechanization, and methods of controlling production devices, and, on the other hand, saw the evolution of customer expectations, necessitating the individualization of products, completely changed the paradigms of designing manufacturing systems at that time. To remain competitive, companies had to design manufacturing systems that not only produced high-quality products at low costs, but also allowed for producing a wide range of different products using the same system. As a consequence, research at the end of the 20th century was focused on the optimal design of flexible manufacturing systems (FMSs) capable of producing a variety of goods belonging to a defined family of a specific class of products. Unfortunately, FMSs turned out to be costly, most particularly because the equipment that possessed features enabling general flexibility was expensive to build and maintain. Those systems were also expensive because the machines that they used had more functionality than what they really needed, and this additional flexibility and functionality in many cases caused a waste of resources, since the added cost paid for this general functionality equalled unrealized capital investment until the extra functionality was actually used.

To eliminate the negative characteristics of both dedicated manufacturing lines (DMLs) and FMSs, and to combine the two opposing goals of reducing production costs and ensuring high system flexibility, new paradigms had to be defined, and new solutions for the design and management of manufacturing processes had to be found. The slogan "exactly the capacity and functionality needed, exactly when needed" became the keynote and main challenge of the process of designing manufacturing systems. Accordingly, in the past several years, research on the development of manufacturing systems has revolved around three main concepts that meet the assumptions of focused flexibility and the challenges of the Industry 4.0 philosophy: focused-flexibility manufacturing systems (FFMSs), reconfigurable manufacturing systems (RMSs), and smart manufacturing systems (SMSs).

Overall, designing manufacturing processes and systems is a complex multilevel procedure influenced by a large number of factors. Designing requires the indepth analysis of market targets, and possible ways of preparing and implementing usually automated and robotized manufacturing systems, assessing the impact of crucial factors, as well as integrating the knowledge of many branches of science and individual divisions. The

1

target of each design is to optimally provide the design processes while maintaining the required quality and minimizing costs.

This Special Issue presents the current research in different areas connected with the design and management of manufacturing systems. In particular, papers published in the volume cover the following subject areas:

- methods supporting the design of manufacturing systems [1–6],
- methods of improving maintenance processes in companies [7–9],
- the design and improvement of manufacturing processes [10–14],
- the control of production processes in modern manufacturing systems [15,16],
- production methods and techniques used in modern manufacturing systems [17], and
- environmental aspects of production and their impact on the design and management of manufacturing systems [18–20].

The wide range of research findings reported in this Special Issue confirms that the design of manufacturing systems is a complex problem, and the achievement of goals set for modern manufacturing systems requires interdisciplinary knowledge and simultaneous design of product, process, and system, as well as the knowledge of modern manufacturing and organizational methods and techniques. The need for and ability to reduce the negative impact of manufacturing processes on the natural environment are also of importance, as signaled in this introductory article and this volume. I wish to thank all the authors for the effort that they expended in preparing their papers. I hope that this Special Issue will be of wide interest to readers and inspire further research, leading to the development of new effective solutions supporting the processes of designing and managing manufacturing systems.

Funding: This research received no external funding.

Institutional Review Board Statement: Not applicable.

Conflicts of Interest: The authors declare no conflict of interest.

References

1. Won, Y. On Solving Large-Size Generalized Cell Formation Problems via a Hard Computing Approach Using the PMP. *Appl. Sci.* **2020**, *10*, 3478. [CrossRef]
2. Relich, M.; Świć, A. Parametric Estimation and Constraint Programming-Based Planning and Simulation of Production Cost of a New Product. *Appl. Sci.* **2020**, *10*, 6330. [CrossRef]
3. Nagy, L.; Ruppert, T.; Abonyi, J. Analytic Hierarchy Process and Multilayer Network-Based Method for Assembly Line Balancing. *Appl. Sci.* **2020**, *10*, 3932. [CrossRef]
4. Kowalski, A.; Waszkowski, R. Layout Guidelines for 3D Printing Devices. *Appl. Sci.* **2020**, *10*, 6333. [CrossRef]
5. Matuszek, J.; Seneta, T.; Moczała, A. Assessment of the Design for Manufacturability Using Fuzzy Logic. *Appl. Sci.* **2020**, *10*, 3935. [CrossRef]
6. Calderón-Andrade, R.; Hernández-Gress, E.S.; Montufar Benítez, M.A. Productivity Improvement through Reengineering and Simulation: A Case Study in a Footwear-Industry. *Appl. Sci.* **2020**, *10*, 5590. [CrossRef]
7. Borucka, A.; Grzelak, M. Application of Logistic Regression for Production Machinery Efficiency Evaluation. *Appl. Sci.* **2019**, *9*, 4770. [CrossRef]
8. Świderski, A.; Borucka, A.; Grzelak, M.; Gil, L. Evaluation of Machinery Readiness Using Semi-Markov Processes. *Appl. Sci.* **2020**, *10*, 1541. [CrossRef]
9. Antosz, K.; Paśko, Ł.; Gola, A. The Use of Artificial Intelligence Methods to Assess the Effectiveness of Lean Maintenance Concept Implementation in Manufacturing Enterprises. *Appl. Sci.* **2020**, *10*, 7922. [CrossRef]
10. Realyvásquez-Vargas, A.; Arredondo-Soto, K.C.; García-Alcaraz, J.L.; Macías, E.J. Improving a Manufacturing Process Using the 8Ds Method. A Case Study in Manufacturing Company. *Appl. Sci.* **2020**, *10*, 2433. [CrossRef]
11. Park, K.; Kim, G.; No, H.; Jeon, H.W.; Kremer, G.E.O. Identification of Optimal Process Parameter Settings Based on Manufacturing Performance for Fused Filament Fabrication of CFR-PEEK. *Appl. Sci.* **2020**, *10*, 4630. [CrossRef]
12. Stoma, P.; Stoma, M.; Dudziak, A.; Caban, J. Bootstrap Analysis of the Production Processes Capability Assessment. *Appl. Sci.* **2019**, *9*, 5360. [CrossRef]
13. Grznar, P.; Gregor, M.; Krajcovic, M.; Mozol, S.; Schickerle, M.; Vavrik, V.; Durica, L.; Marschall, M.; Bielik, T. Modeling and Simulation of Processes in a Factory of Future. *Appl. Sci.* **2020**, *10*, 4503. [CrossRef]

Sobaszek, Ł.; Gola, A.; Kozłowski, E. Predictive Scheduling with Markov Chains and ARIMA Models. *Appl. Sci.* **2020**, *10*, 6121. [CrossRef]

Koo, P.-H.; Ruiz, R. Simulation-Based Analysis on Operational Control of Batch Processors in Wafer Fabrication. *Appl. Sci.* **2020**, *10*, 5936. [CrossRef]

Kaid, H.; Al-Ahmari, A.; Li, Z.; Davidrajuh, R. Automatic Supervisory Controller for Deadlock Control in Reconfigurable Manufacturing Systems with Dynamic Changes. *Appl. Sci.* **2020**, *10*, 5270. [CrossRef]

Patalas-Maliszewska, J.; Topczak, M.; Kłos, S. The Level of the Additive Manufacturing Technology Use in Polish Metal and Automotive Manufacturing Enterprises. *Appl. Sci.* **2020**, *10*, 735. [CrossRef]

Ratner, S.; Gomonov, K.; Revinova, S.; Lazanyuk, I. Eco-Design of Energy Production Systems: The Problem of Renewable Energy Capacity Recycling. *Appl. Sci.* **2020**, *10*, 4339. [CrossRef]

Piotrowska, K.; Piasecka, I.; Bałdowska-Witos, P.; Kruszelnicka, W. LCA as a Tool for the Environmental Management of Car Tire Manufacturing. *Appl. Sci.* **2020**, *10*, 7015.

Nielsen, I.E.; Majumder, S.; Saha, S. Game-Theoretic Analysis to Examine How Government Subsidy Policies Affect a Closed-Loop Supply Chain Decision. *Appl. Sci.* **2020**, *10*, 145. [CrossRef]

Article

Parametric Estimation and Constraint Programming-Based Planning and Simulation of Production Cost of a New Product [†]

Marcin Relich [1,*] **and Antoni Świć** [2]

[1]   Faculty of Economics and Management, University of Zielona Gora, 65-417 Zielona Gora, Poland
[2]   Department of Production Computerisation and Robotisation, Faculty of Mechanical Engineering, Lublin University of Technology, 20-618 Lublin, Poland; a.swic@pollub.pl
*   Correspondence: m.relich@wez.uz.zgora.pl
†   This work is an extended version of paper published in proceedings of the 12th Asian Conference on Intelligent Information and Database Systems (ACIIDS), held in Phuket, Thailand, 23–26 March 2020.

Received: 23 July 2020; Accepted: 9 September 2020; Published: 11 September 2020

Abstract: Currently-used decision support solutions allow decision makers to estimate the cost of developing a new product, its production, and promotion, and compare the estimated cost to the target cost. However, these solutions are inadequate for supporting simulations of identifying conditions, by which the specific cost is reached. The proposed approach provides a framework for searching for possible variants towards reaching the target production cost. This paper is concerned with a prototyping problem of product development described in terms of a constraint satisfaction problem. The proposed method uses parametric estimation to identify relationships between variables, and constraint programming to search for project completion variants within the company's resources and project requirements. The results of an experiment indicate that constraint programming provides effective search strategies for finding admissible solutions. Consequently, the proposed approach allows decision makers to obtain alternative scenarios within the limits imposed by the production process. In this, it outperforms current methods dedicated to the support of evaluating the total cost of a new product. The declarative approach presented in this paper is used to model the production cost; however, it can be effortlessly extended to other aspects of product development (e.g., product reliability).

Keywords: constraint programming; constraint satisfaction problem; cost estimation; decision support systems; multicriteria optimization; production planning; project management

1. Introduction

The new product development (NPD) process belongs to crucial processes in contemporary companies, which have to compete in a saturated market and by short product life cycles. Successful implementation of NPD projects affects the company's profitability and survival. Unfortunately, many NPD projects either fail completely, miss the deadline, or exceed the budget. According to a Project Management Institute (PMI) study, some of the most common causes of project failure include changes in priorities within a company and limited resources. There are a few reasons why project duration and cost may be exceeded [1,2]: use of static and inflexible methods, use of low-end software solutions that lack features related to the entire NPD process, failure to align with the company's goals and strategies, weak executive support, and poor communication. The last three reasons depend on organizational issues, whereas the first two refer to technical support for improving project performance. The goal of this study was to design a method for supporting decision makers in searching for alternative project completion scenarios.

Currently-used software solutions dedicated to project planning provide information on the estimated cost of a project [3]. If the company's resources (e.g., financial, human) are not sufficient to develop an NPD project according to schedule, then the decision makers may be interested in obtaining information on alternative project completion variants. Additionally, they may find the estimated cost of NPD unacceptable. Current software solutions, however, fail to support project managers in reviewing the possibilities of project performance so that the preferred NPD cost is achieved given the company's resources and project requirements. The proposed approach fills in this gap and is geared towards specifying foundations for developing a decision support system dedicated to solving project prototyping problems. The aim of this paper is to elaborate an approach for planning and simulating NPD project completion, in which factors related to research and development (R&D), production, and sales promotion are involved. As the production cost usually constitutes the majority of the total cost in the process of product placement in the market, this study mainly focuses on the field of production. The proposed approach aims to identify the cost of R&D, production, and promotion at the early phase of an NPD project in order to select the most promising NPD project.

The current project planning methods represent the procedural approach, in which an NPD model is built for a specific problem, and the process of designing the model ends when the structure of the model is sufficient to solve the given problem. By contrast, in the declarative approach a single NPD model is developed, which can be used to formulate various NPD-related decision problems. A declarative representation of an NPD model allows to use effective techniques for reducing the search space of admissible solutions. This is particularly important when many decision variables with large domains have to be selected for simulations. There exist declarative simulation modeling methods that can be used to increase the efficiency of identifying alternative NPD project performance variants. By using a declarative representation of an NPD model, a decision maker can perform simulations for an entire set of admissible solutions. Consequently, the decision maker obtains more alternative NPD project performance scenarios than they would using a traditional scenario analysis, which includes the basic, optimistic, and pessimistic variants. Moreover, the decision maker may obtain variants of NPD project performance that they would never come up with themselves. This is particularly important in multi-project environments, in which resources are shared.

In this study, a project prototyping problem is formulated in terms of a constraint satisfaction problem and implemented using constraint programming techniques. This study develops previous research [4] towards using the declarative approach to search for variants of project completion within production cost constraints. The novelty of this research is twofold: (1) specifying an NPD project model and company's resources as a set of variables and constraints; (2) designing a method for solving the project prototyping method. The proposed method uses parametric estimation to identify complex relationships among data, and constraint programming to effectively search for possible solutions. Consequently, the proposed approach is more adaptable to new conditions related to project performance than other currently used methods.

The paper is organized as follows: Section 2 provides a literature review on new product development, cost estimation techniques and constraint programming. In Section 3 a prototyping problem of product development is formulated as a constraint satisfaction problem (CSP). A method for searching variants that meet the desirable production cost is presented in Section 4. An illustrative example of the proposed approach is presented in Section 5. Finally, a conclusion is drawn and directions of future research are indicated in Section 6.

2. Literature Review

2.1. New Product Development

New product development begins with the identification of market needs and ends with introducing a new product on the market. Intermediate phases of the NPD process are variously distinguished in the literature. Ulrich and Eppinger [5] proposed the following phases of the NPD process: planning, concept development, system-level design, detail design, tests and refinement, and production ramp-up. In turn, Crawford and Benedetto [6] distinguished phases in the NPD model such as opportunity identification and selection, concept generation, concept evaluation, development of a new product, and launch. After introducing a product in the market, the next phase of product life cycle is sales and accompanying activities related to production and promotion. If a new product has been successfully launched in the market, the production phase generates the majority of costs in the total cost of the product life cycle. Figure 1 illustrates cumulative cash flows (sales revenue, costs, and corresponding profit/loss).

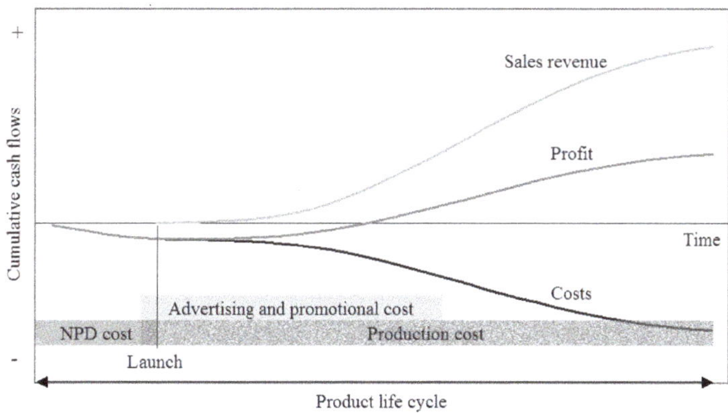

Figure 1. Product development cash flow.

The potential of a new product can be measured as the comparison of predicted cash inflows with outflows within a given period of time. Consequently, there is a need to predict sales and costs of product development, production, and promotion. Sales forecasting and cost estimation depend on available data regarding similar previous NPD projects. If a new product belongs to the existing product line, then analogical models can be used to cost estimation. In turn, if a company develops an entirely new product, then analytical models are more useful for evaluating sales or costs of the new product.

Cumulative costs include the cost of product development, production, and promotion. The cost of NPD occurs before launching a new product on the market. The production cost appears in the NPD phase by manufacturing prototypes of a new product, and it lasts until the end of product life cycle. In turn, the advertising and promotional cost can begin before launching a new product, and it usually takes place in the first phase of sales. The comparison of alternative new products requires the prediction of product lifespan, sales, and costs. The evaluation of the potential of new products can also include the discount rate (as in the NPV method) towards reflecting opportunity cost of capital and inflation in the investment period.

2.2. Cost Estimation Techniques

Empirical studies indicate that the size of a project team and the project budget are key structural variables affecting the quality of the NPD process and success of the project [7]. An investigation of new product forecasting practices in industrial companies shows that the most popular techniques include customer/market research, looks-like analysis, trend line analysis, moving average, scenario analysis, and multi-attribute models [8,9]. To a lesser extent, companies use forecasting techniques related to nonlinear regression, expert systems, neural, and neuro-fuzzy networks [8,10].

Cost estimation techniques can be divided into the following groups [11,12]: intuitive, analogical, parametric, and analytical. Intuitive methods use past experience of an estimator. Analogical methods estimate the cost of new products using similarity to previous products. Parametric methods estimate the cost of a new product from parameters that significantly influence the cost. In turn, analytical methods estimate the cost of a product by decomposing product development into elementary tasks with the known cost.

Parametric estimation techniques may be based on regression analysis [13,14], artificial neural networks [15,16], fuzzy logic systems [17,18], or hybrid systems such as neuro-fuzzy systems [16,19] and genetic fuzzy systems [20,21]. Engineering approaches are predicated on a detailed analysis of product features and the manufacturing process. For example, the cost of a new product is calculated in this approach as the sum of the resources used to design and produce each component of the product (e.g., raw materials, labor, equipment). As a result, the engineering approach is best used in the final phases of product development, in which the product and the manufacturing process are well defined.

2.3. Constraint Programming Techniques

A project prototyping problem can be formulated as a CSP by specifying constraints and variables. CSPs, which are combinatorial problems, are solved with the use of constraint programming (CP) [22,23]. CP includes consistency techniques and systematic search strategies that are crucial to improving the search efficiency in solving the CSPs [23,24]. Consequently, CP provides an appropriate framework for developing decision-making software to support identification of project completion alternatives.

Constraint programming consists of two phases: first, the problem is specified in terms of constraints, and then it is solved. The specification of a problem by means of constraints is very flexible because constraints can be added, removed, or modified [24]. The goal of CP is to develop efficient domain-specific methods to be used instead of general methods. They can be employed to develop more efficient constraint solvers, constraint propagation algorithms, and search algorithms. Consequently, CP is a powerful paradigm for solving combinatorial search problems, in which the user declaratively states the constraints on feasible solutions for a set of decision variables [25].

The advantages of using CP refer to declarative problem modeling, propagation of the effects of decisions by means of efficient algorithms, and the search for optimal solutions. The specific search methods and constraint propagation algorithms used in CP allow to significantly reduce the search space. Consequently, CP is suitable for modeling complex problems related to scheduling [26,27], manufacturing [28,29], resource allocation [30,31], supply chain problems [32–34], and others. In the context of an NPD project, the CP paradigm has been used in areas such as design and product configuration [35,36], project scheduling [37,38], and project prototyping [39]. Although the use of CP to project selection and scheduling problems has been widely discussed in the literature, the NPD project prototyping problem has not been considered thus far.

3. Problem Formulation

The project prototyping problem is a problem in which alternative NPD project completion scenarios are searched for, taking into account the adopted constraints. This study is concerned with searching for project completion variants that meet the desired NPD cost. In the traditional approach to project evaluation, when the decision maker finds a specific cost unacceptable the project is rejected. However, if the project is important from a strategic point of view, the decision maker is interested in prerequisites that must be met to achieve the desired cost threshold. The proposed approach consists of identifying all possible project performance scenarios (variants) that meet constraints related to project objectives, project budget, human resources, machines, etc. Figure 2 compares the traditional approach to project evaluation (Figure 2a) with the proposed approach in which variants within the target project performance are searched for (Figure 2b). The traditional approach may be considered as a project prototyping problem stated in a forward form, whereas the approach presented in this study can be viewed as the same type of problem stated in an inverse form.

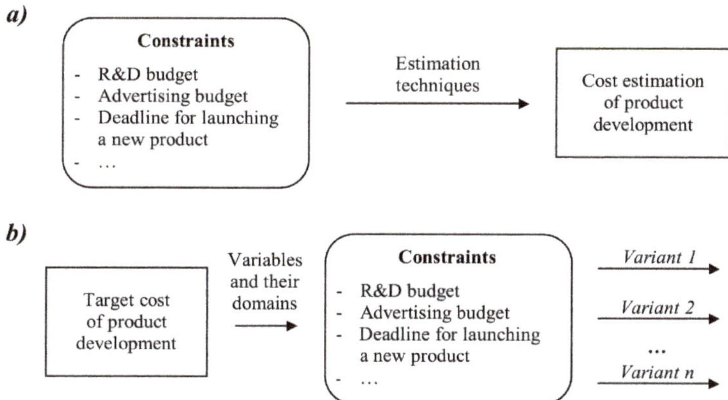

Figure 2. The project prototyping problem stated in a forward (**a**) and inverse form (**b**).

The proposed approach allows the decision maker to identify the prerequisites that must be met for the project to attain the desired performance level given the specified constraints, variables, and relationships between these variables. The number of possible project performance variants depends on what constraints and variable domains are considered and the granularity of decision variables. Relationships between variables can be identified on the basis of previous experiences with similar projects and represented as if-then rules. Then, they may be used to predict the potential of a project (the traditional approach), or to test whether there exist alternative project completion scenarios that could deliver the desired project performance (the proposed approach). The use of CSP to the formulation of the project prototyping problem offers significant flexibility of the proposed approach. The addition/removal of variables or constraints to/from CSP causes a simultaneous recalculation and a new set of possible solutions.

The present project prototyping approach requires the specification of variables, their domains, and constraints. This allows to identify all available solutions, if there are any. The prototyping problem can be easily expressed as a CSP which can be described in the following form [28]:

$$(V, D, C) \tag{1}$$

where V is a finite set of n variables $\{V_1, V_2, \dots, V_n\}$, D is a finite set of discrete domains $\{D_1, D_2, \dots, D_n\}$ related to variables V, and C is a finite set of constraints $\{C_1, C_2, \dots, C_m\}$ that restrict the values of the variables and link them.

Each constraint is treated as a predicate that can be seen as an n-relation defined by Cartesian product $D_1 \times D_2 \times \dots \times D_n$. The solution to the CSP is a vector $(D_{1i}, D_{2k}, \dots, D_{nj})$ related to the assessment of the values of each variable that satisfy all constraints C. Generally, constraints may be specified using analytical and/or logical formulas.

The variables are associated with the company's resources and the NPD project. Characteristically, the variables $(V_4, \dots, V_{16})$, which are purported to affect costs $(V_1, \dots, V_3)$, are controllable by a company and can be simulated to identify a set of their values that satisfy all constraints and ensure the desired level of a specific cost. The following set of variables for estimating the NPD, production, and promotional cost were proposed:

V_1—NPD cost (in thousand €),

V_2—unit production cost (in €),

V_3—advertising and promotional cost (in thousand €),

V_4—number of R&D employees involved in product design,

V_5—number of R&D employees involved in prototype tests,

V_6—duration of product design (in months),
V_7—duration of prototype tests (in months),
V_8—number of prototype tests (in hundreds),
V_9—number of product components,
V_{10}—amount of materials needed to manufacture a unit of a new product,
V_{11}—amount of energy needed to manufacture a unit of a new product,
V_{12}—assembly time for a unit of a new product,
V_{13}—processing time for a unit of a new product,
V_{14}—number of workplace units needed for assembling and processing a new product,
V_{15}—duration of advertising and promotional campaign of a new product,
V_{16}—number of potential receivers of advertising and promotional campaign,
V_{17}—sales volume,
V_{18}—production cost (in thousand €),
V_{19}—desirable margin (in €),
V_{20}—product price (in €),
V_{21}—sales revenue of a new product (in thousand €).

There are the following constraints regarding the available quantity of resources in a company and technical parameters of a new product: NPD project budget (*PB*, in thousand €), total number of R&D employees involved in an NPD project (*TE*), deadline for launching the new product into the market (*LD*, in months), time needed to manufacture a unit of a new product (*MT*), and advertising and promotional budget (*AB*, in thousand €). The set of constraints and relationships is as follows:

$$V_1 \leq PB \tag{2}$$

$$V_4 + V_5 \leq TE \tag{3}$$

$$V_6 + V_7 \leq LD \tag{4}$$

$$V_{12} + V_{13} \leq MT \tag{5}$$

$$V_3 \leq AB \tag{6}$$

$$V_2 \cdot V_{17} = V_{18} \tag{7}$$

$$V_2 + V_{19} \leq V_{20} \tag{8}$$

$$V_1 + V_3 + V_{18} \leq V_{21} \tag{9}$$

The model formulated as an CSP incorporates the technical parameters of the new product that refer to the planned project performance and the available resources. The problem is solved by searching for answers to the following questions:

- What is the cost of product development (including unit production cost)?
- What values should the variables have to reach the desirable level of the cost?

A project prototyping problem can be expressed as a CSP and then be solved with the use of specific techniques such as constraint propagation and variable distribution. Constraint propagation applies constraints to prune the search space. Propagation techniques aim to reach a certain level of consistency, and accelerate the search procedures to reduce the size of the search tree [28]. The values of the variables excluded by constraints are removed from their domains. A CSP may be effectively solved with the use of CP techniques. The declarative nature of CP is particularly useful in applications where it is enough to state what has to be solved without saying how to solve it [28]. As CP uses specific search methods and constraint propagation algorithms, it allows to considerably reduce the search space. Consequently, CP is suitable for modeling and solving complex problems.

4. The Proposed Method for Planning and Simulation of Production Cost

The proposed method consists of the following phases: (1) collecting data from previous projects that are similar to the new project, (2) identifying relationships between variables, (3) estimating the production cost, and, if needed, (4) searching for variants that allow to obtain the desired cost. Figure 3 shows a framework for the proposed decision support system that uses parametric estimation techniques to identify relationships and constraint programming to reduce the search space and test the possibility of reaching the desired production cost.

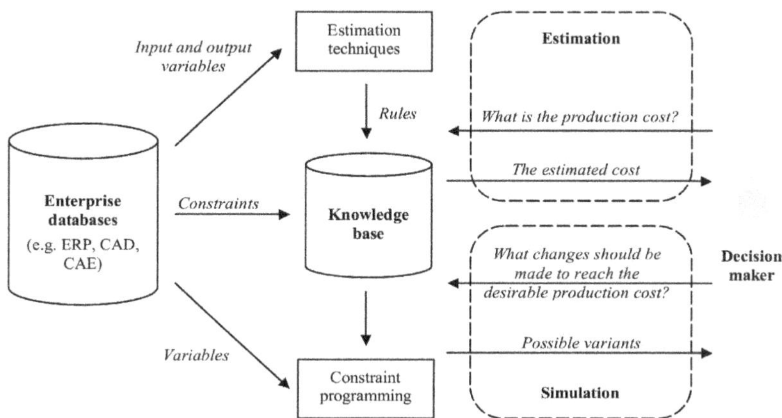

Figure 3. A framework for the proposed decision support system.

In the first phase, the data is collected from enterprise databases, for example related to information systems such as enterprise resource planning (ERP), computer-aided design (CAD), and computer-aided engineering (CAE). This requires the use of some project management standards, including project performance planning, monitoring, control, and appropriate project planning and execution techniques. The use of the proposed approach requires the access to the sufficient amount of data related to the past NPD projects that can be outdated, which causes the need to consider the delay aspect. The applicability of the proposed method depends on whether or not the following data management procedures are used: the enterprise adjusts the common project management standards to its needs, distinguishes phases in the NPD process, uses standards for specifying tasks in an NPD project and for project portfolio management, measures the success of new products on corporate financial performance, uses the results of a financial performance analysis to improve the effectiveness of NPD projects, registers performance and metrics of NPD projects, uses a primary schedule for monitoring performance in NPD projects, and uses the defined procedure to allocate employees to NPD projects.

In the second phase of the proposed method, cause-and-effect relationships are identified and then used to estimate the cost of product development, and search for a desirable outcome of an NPD project. The input variables should impact the cost and be controlled by a company, such as the number of project team members, product components, and prototype tests. A set of variables, their domains, and constraints constitute a CSP, which provides a framework for finding the value of the NPD, production and promotional cost (the third phase), and, if that is unacceptable, the values of variables at which the desired cost of a new product can be achieved (the fourth phase).

In the third phase, the cost is estimated using the parametric models based on linear regression and artificial neural networks. The quality of the obtained results is compared using the root mean square errors. The data set is divided into learning and testing sets to verify the quality of the learned neural network. The small errors in the testing set exhibit good predictive abilities of the learned network.

The fourth phase of the proposed method refers to the search for possible solutions to achieve the desired cost. The size of the search space depends on the number of variables analyzed, the range of decision variable domains, and the constraints that link the variables and limit the set of possible solutions. An exhaustive search always finds a solution if one exists, but its performance is proportional to the number of admissible solutions. Therefore, an exhaustive search tends to grow very quickly as the size of the problem increases, which limits its usage in solving many practical problems. Consequently, more effective methods for searching the space and finding possible solutions are needed. This study proposes a CP that can be used to efficiently solve a project prototyping problem modelled as a CSP. Figure 4 shows a framework for solving this problem in the inverse form.

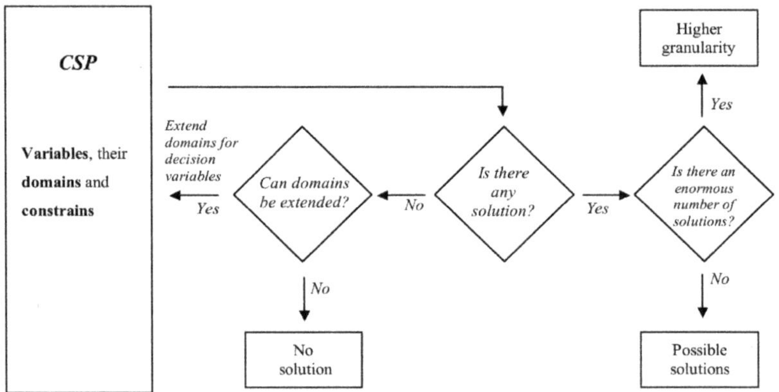

Figure 4. A framework for solving the inverse form of the project prototyping problem.

The proposed approach includes three stop/go conditions. The first stop/go condition checks whether there exists a solution for a set of decision variables, given the specified input and output variables, their domains, and constraints. If such a solution exists, then it is verified whether a number of solutions is acceptable for the user or not. If not, the granularity of the solution is increased. Furthermore, changes in granularity are related to scaling domains and significantly affect the effectiveness of the CP application (e.g., the production cost specified in 1000 euros provides a smaller number of solutions than when specified in single euros). If there is any solution, the possibility of extending the domains related to the selected decision variables is tested. Finally, if the domains of the selected decision variables cannot be changed, the last stop condition leads to the empty solution set.

5. An Example of Using the Proposed Method

5.1. Cost Estimation of a New Product

The relationships between the input and output variables (V_1, V_2, V_3) have been identified with the use of parametric estimation techniques such as the artificial neuron networks (ANNs), linear regression (LR), and compared with the average. The use of ANNs offers advantages over ability to learn and identification of complex nonlinear relationships. The dataset used in the analysis included 25 completed NPD projects that featured products in the same line as the investigated project. The data were divided into two sets—learning (20 cases) and testing (five cases)—to evaluate the quality of the estimating model. The relationships between the input and output variables are determined using the data related to the previous NPD projects to estimate the cost of product development. The set of input variables is obtained taking into account the significant impact of an input variable on an output variable, as well as the possibility of their controllability by a company and the access to the data through the past specifications of NPD projects. There are the following relationships between input variables and costs: the NPD cost (V_1) depends on $\{V_4, \dots , V_9\}$, the unit production cost (V_2) depends on $\{V_9, \dots , V_{14}\}$, and the advertising and promotional cost (V_3) depends on $\{V_{15}, V_{16}\}$.

In this study, a multilayer feed-forward ANN has been trained according to the back-propagation algorithm and weights optimized according to the Levenberg-Marquardt algorithm (LM) and gradient descent momentum with adaptive learning rate algorithm (GDX). The neural network structure has been determined in an experimental way, by the comparison of a learning set and testing set for the different number of layers and hidden neurons. The root mean square errors (RMSEs) have been calculated as the average of 50 iterations for each structure of a neural network with a number to the extent of 30 hidden neurons. Table 1 presents the RMSEs calculated in the learning and testing set using the different parametric models for estimating the cost of NPD (V_1), production (V_2), and promotion (V_3).

Table 1. A comparison of root mean square errors (RMSEs) for estimation models.

Data Set	Model	V_1	V_2	V_3
	ANN LM	0.005	0.001	0.003
Learning	ANN GDX	0.521	0.051	0.118
	LR	0.637	0.072	0.142
	Average	1.312	0.134	0.326
	ANN LM	0.496	0.067	0.129
Testing	ANN GDX	0.602	0.049	0.085
	LR	0.913	0.106	0.138
	Average	1.826	0.180	0.548

The ANNs trained according to the LM algorithm obtained the least RMSEs in the learning set. However, the least RMSEs in the testing set was calculated using the ANNs trained according to the GDX algorithm (for V_2 and V_3) and LM algorithm (for V_1). The trained ANN is used to estimate the unit production cost for the following values of input variables: $V_9 = 45$, $V_{10} = 830$, $V_{11} = 55$, $V_{12} = 25$, $V_{13} = 15$, $V_{14} = 7$. The estimated unit production cost reaches 24.5 €, and the total production cost 970 thousand € in the two-year projected period of product life cycle. In turn, the NPD cost reaches 190 thousand €, and the advertising and promotional cost 150 thousand €.

Let us assume that the estimated unit production cost does not satisfy the decision maker who is interested in reducing this cost to 24 €. To check whether these expectations can be fulfilled, the problem is reformulated into an inverse problem in which such values of input variables are sought to ensure the desired level of the unit production cost.

5.2. Simulation for Identifying the Desired Level of Costs

The example consists of two steps presented in Figure 4: (1) a basic variant for the originally selected decision variables, their domains, and constraints; and (2) an extension of the domains of the selected decision variables.

5.2.1. The Basic Variant

Let us assume that the decision maker is interested in reducing the unit production cost to 24 €. To test whether there exist solutions, the problem under consideration is reformulated into an inverse problem. A solution to the inverse problem is sought using constraint programming, which requires that decision variables, their domains, and constraints, including relationships between variables (e.g., their mutual impact on one another), are specified. The domains for the considered variables are as follows: $D_9 = \{45\}$, $D_{10} = \{830\}$, $D_{11} = \{54, 55, 56\}$, $D_{12} = \{24, 25\}$, $D_{13} = \{14, 15\}$, and $D_{14} = \{7\}$. The domains include only integer numbers and the simulation step is related to the increase of these numbers by one for each variable separately. As a result, all available solutions, referring to all numbers in the domains, are sought.

The inverse problem is implemented in Mozart/Oz software, which is a multiparadigm programming language. Mozart/Oz contains most of the major programming paradigms, including logic, functional, imperative, object-oriented, concurrent, constraint, and distributed programming. The major strengths of Mozart/Oz are related to its constraint and distributed programming components, which are able to effectively solve many practical problems, for example, timetabling and scheduling problems [40].

Table 2 presents eight possible solutions (variants of the production process) that take into account the specified constraints and variables. The minimal unit production cost appears for the values of variables in the fifth variant.

Table 2. A set of possible solutions.

Variant	Values of Variables	V_2
1	$V_9 = 45$, $V_{10} = 830$, $V_{11} = 55$, $V_{12} = 25$, $V_{13} = 15$, $V_{14} = 7$	24.00
2	$V_9 = 45$, $V_{10} = 830$, $V_{11} = 54$, $V_{12} = 25$, $V_{13} = 15$, $V_{14} = 7$	23.97
3	$V_9 = 45$, $V_{10} = 830$, $V_{11} = 56$, $V_{12} = 24$, $V_{13} = 15$, $V_{14} = 7$	23.98
4	$V_9 = 45$, $V_{10} = 830$, $V_{11} = 55$, $V_{12} = 24$, $V_{13} = 15$, $V_{14} = 7$	23.95
5	$V_9 = 45$, $V_{10} = 830$, $V_{11} = 54$, $V_{12} = 24$, $V_{13} = 15$, $V_{14} = 7$	23.82
6	$V_9 = 45$, $V_{10} = 830$, $V_{11} = 56$, $V_{12} = 25$, $V_{13} = 14$, $V_{14} = 7$	23.99
7	$V_9 = 45$, $V_{10} = 830$, $V_{11} = 55$, $V_{12} = 25$, $V_{13} = 14$, $V_{14} = 7$	23.96
8	$V_9 = 45$, $V_{10} = 830$, $V_{11} = 54$, $V_{12} = 25$, $V_{13} = 14$, $V_{14} = 7$	23.93

The changes presented in Table 2 regard three variables V_{11}, V_{12}, and V_{13}. The set of possible solutions informs the decision maker what changes may be incorporated into the production process to reduce the unit production cost.

Let us assume that the decision maker's expectations regarding the unit production cost decrease from 24 to 22 €. There is no solution by the indicated domains that satisfies this new value of the preferred cost. This triggers the next step in the problem solving procedure, i.e., testing whether it is possible to extend the domains related to the selected variables.

5.2.2. Extension of Variable Domains

This step of the procedure of solving the inverse form of the problem involves selecting variables whose domains can be extended. As a result of domain extension, the problem is solved again for new domains assigned to some variables, all other conditions being equal. The set of domains is as follows: $D_9 = \{45\}$, $D_{10} = \{800, \dots, 850\}$, $D_{11} = \{50, \dots, 60\}$, $D_{12} = \{20, \dots, 30\}$, $D_{13} = \{10, \dots, 20\}$, and $D_{14} = \{7\}$. The remaining constraints are the same as in the basic variant. An extension of domains for four variables causes an increase in the number of choice nodes in the explored search tree. Consequently, there is a need to use an effective technique to search the space of admissible solutions.

Table 3 presents the results of searching for an admissible solution for different strategies of variable distribution. Different strategies of variable distribution in the constraint programming with exhaustive search (ES) are compared with regard to the number of nodes checked, depth, and the time needed to find solutions. The calculations were tested on an IntelCore (tm) i5-8300H 2.3-4GHz, RAM 8 GB platform.

Table 3. A comparison of strategies for variable distribution.

Distribution Strategy	Number of Nodes Checked	Depth	Time [s]
ES	67,880	65	5.25
CP Naïve	32,852	48	1.41
CP First-fail	32,852	48	1.16
CP Split	32,852	48	1.07

The results show that the application of the constraint programming reduces the computational time, which is especially important when there is a larger number of possible solutions. The user can obtain the entire set of solutions or one optimal solution. The extension of variable domains can generate many very similar solutions. The number of solutions can be reduced through considering only the minimum and maximum values of domains. Then, the decision maker can develop the scenario analysis towards the preferable direction to recognize all scenarios ensuring the desired cost. Constraint programming techniques allow to use strategies related to constraint propagation and variable distribution, significantly reducing the set of admissible solutions and the average computational time, improving in this way the interactive properties of a decision support system.

6. Conclusions

The declarative approach proposed in the present paper for NPD project prototyping is an alternative to current methods, which only estimate the cost of NPD, production, and promotion. In the proposed project prototyping method, all possible project completion scenarios are sought, if any. These variants inform the decision maker whether an NPD project can be completed based on the company's resources and within project requirements. Businesses with limited resources need to invest extra effort in managing NPD projects. This approach is especially useful to this type of firms (e.g., ones that have a limited project budget), as it allows them to check whether a project can be completed under the specified constraints (e.g., the desired level of the unit production cost). Consequently, there is a need to develop a decision support system for searching possible variants of the cost reduction. The proposed model encompasses areas related to a product and a company's resources. These areas were described in terms of a CSP that includes sets of decision variables, their domains, and constraints. The project prototyping problem is a problem in which answers to queries about the estimated cost and the values of input variables that ensure the desired level of cost are searched for. The results of this research include not only the possibility of verification to reach the desired level of the cost, but it can also be developed towards identifying the possible variants of project completion. The presented method can be used to verify the possibility of completing an NPD project at the target time, and to specify the resources needed to the reduction of project duration. This is one of the most important issues related to the new product development process. The problem presented in this paper refers to the project prototyping problem that is stated in an inverse form, i.e., the possible variants are sought to ensure the desirable level of production cost. However, it is also possible to use the proposed approach in other areas, for example, in project scheduling, product configuration, resource allocation, and supply chain problems.

The results show that the application of a CP improves the search efficiency in the context of the project prototyping problem, especially when there are a larger number of admissible solutions. Moreover, this study presents the use of artificial neural networks to identify the relationships for estimating costs within a product life cycle. The identified relationships are stored in a knowledge base and used to generate alternative variants of manufacturing a new product. If decision makers find that the cost related to product development is unacceptable, they can use the identified variants as a support tool in identifying the impact of input variables on an output variable (e.g., the unit production cost) under the specified constraints. The drawback of the proposed approach is that sufficient amounts of data on similar past NPD projects need to be collected, and several parameters must be specified to build and train an artificial neural network. Moreover, the limitation of the presented study is the selection of input variables that significantly affect the cost of a new product and are controllable by a company, as well as the time needed to analyze all solutions by the decision maker. In the case of an enormous number of admissible solutions, the granularity of domain can be increased or the minimum and maximum values of domain can be taken into account in the calculations. In our future work, we would like to verify the proposed approach in project-oriented companies in different business sectors. We also plan to extend the application of the presented approach by incorporating the warranty cost in the total product development cost.

Author Contributions: Conceptualization, M.R.; methodology, M.R.; validation, M.R. and A.Ś.; formal analysis, M.R.; investigation, M.R. and A.Ś.; resources, M.R. and A.Ś.; writing—original draft preparation, M.R. and A.Ś.; writing—review and editing, M.R. and A.Ś.; visualization, M.R. and A.Ś.; project administration, M.R. All authors have read and agreed to the published version of the manuscript.

Funding: This research received no external funding.

Conflicts of Interest: The authors declare no conflict of interest.

References

1. The 6 Not-So-Obvious Reasons A Project Plan Fails. Available online: www.microsoft.com/en-us/microsoft-365/business-insights-ideas/resources/the-6-not-so-obvious-reasons-a-project-plan-fails (accessed on 31 August 2020).
2. Spalek, S. *Data Analytics in Project Management*; CRC Press: Boca Raton, FL, USA, 2018.
3. Kuster, J.; Huber, E.; Lippmann, R.; Schmid, A.; Schneider, E.; Witschi, U.; Wüst, R. *Project Management Handbook*; Springer: Berlin/Heidelberg, Germany, 2015.
4. Relich, M.; Nielsen, I.E.; Bocewicz, G.; Banaszak, Z. Constraint Programming for New Product Development Project Prototyping. In *Proceedings of 12th Asian Conference on Intelligent Information and Database Systems*; Springer: Cham, Switzerland, 2020; pp. 26–37.
5. Ulrich, K.T.; Eppinger, S.D. *Product Design and Development*, 5th ed.; MacGraw-Hill: New York, NY, USA, 2012.
6. Crawford, M.; Benedetto, A.D. *New Products Management*, 10th ed.; McGraw-Hill Education: New York, NY, USA, 2011.
7. Ernst, H.; Hoyer, W.; Rubsaamen, C. Sales, marketing, and research-and-development cooperation across new product development stages: Implications for success. *J. Mark.* **2010**, *74*, 80–92. [CrossRef]
8. Derbyshire, J.; Giovannetti, E.G. Understanding the failure to understand New Product Development failures: Mitigating the uncertainty associated with innovating new products by combining scenario planning and forecasting. *Technol. Forecast. Soc. Chang.* **2017**, *125*, 334–344. [CrossRef]
9. Hird, A.; Mendibil, K.; Duffy, A.; Whitfield, R.I. New product development resource forecasting. *R D Manag.* **2015**, *46*, 857–871. [CrossRef]
10. Relich, M. Computational Intelligence for Estimating Cost of New Product Development. *Found. Manag.* **2016**, *8*, 21–34. [CrossRef]
11. Voltolini, R.; Vasconcelos, K.; Borsato, M.; Peruzzini, M. Product development cost estimation through ontological models—A literature review. *J. Manag. Anal.* **2019**, *6*, 209–229. [CrossRef]
12. Więcek, D.; Więcek, D.; Kuric, I. Cost Estimation Methods of Machine Elements at the Design Stage in Unit and Small Lot Production Conditions. *Manag. Syst. Prod. Eng.* **2019**, *27*, 12–17. [CrossRef]
13. Liu, H.; Gopalkrishnan, V.; Quynh, K.T.N.; Ng, W.-K. Regression models for estimating product life cycle cost. *J. Intell. Manuf.* **2008**, *20*, 401–408. [CrossRef]
14. Świć, A.; Gola, A. Economic analysis of casing parts production in a flexible manufacturing system. *Actual Probl. Econ.* **2013**, *141*, 526–533.
15. Kumar, P.S.; Behera, H.; Kumari K, A.; Nayak, J.; Naik, B. Advancement from neural networks to deep learning in software effort estimation: Perspective of two decades. *Comput. Sci. Rev.* **2020**, *38*, 100288. [CrossRef]
16. Relich, M. Portfolio selection of new product projects: A product reliability perspective. *Ekspolatacja i Niezawodn. Maint. Reliab.* **2016**, *18*, 613–620. [CrossRef]
17. Habibi, F.; Birgani, O.T.; Koppelaar, H.; Radenovic, S. Using fuzzy logic to improve the project time and cost estimation based on Project Evaluation and Review Technique (PERT). *J. Proj. Manag.* **2018**, *3*, 183–196. [CrossRef]
18. Kłosowski, G.; Gola, A. Risk-Based Estimation of Manufacturing Order Costs with Artificial Intelligence. In Proceedings of the 2016 Federated Conference on Computer Science and Information Systems, Gdansk, Poland, 11–14 September 2016; IEEE: Gdansk, Poland, 2016; pp. 729–732.
19. Rajab, S.; Sharma, V. A review on the applications of neuro-fuzzy systems in business. *Artif. Intell. Rev.* **2017**, *49*, 481–510. [CrossRef]
20. Yassine, A.A.; Mostafa, O.; Browning, T.R. Scheduling multiple, resource-constrained, iterative, product development projects with genetic algorithms. *Comput. Ind. Eng.* **2017**, *107*, 39–56. [CrossRef]

21. Labbi, O.; Ouzizi, L.; Douimi, M.; Ahmadi, A. Genetic algorithm combined with Taguchi method for optimisation of supply chain configuration considering new product design. *Int. J. Logist. Syst. Manag.* **2018**, *31*, 531–561. [CrossRef]
22. Fruhwirth, T.; Abdennadher, S. *Essentials of Constraint Programming*; Springer: Berlin, Germany, 2003.
23. Liu, S.; Wang, C.-J. Optimizing project selection and scheduling problems with time-dependent resource constraints. *Autom. Constr.* **2011**, *20*, 1110–1119. [CrossRef]
24. Apt, K. *Principles of Constraint Programming*; Cambridge University Press: Cambridge, UK, 2003.
25. Rossi, F.; Van Beek, P.; Walsh, T. *Handbook of Constraint Programming*; Elsevier Science: Amsterdam, The Netherlands, 2006.
26. Liu, J.; Lu, M. Constraint Programming Approach to Optimizing Project Schedules under Material Logistics and Crew Availability Constraints. *J. Constr. Eng. Manag.* **2018**, *144*, 04018049. [CrossRef]
27. Nielsen, I.E.; Dang, Q.-V.; Nielsen, P.; Pawlewski, P. Scheduling of Mobile Robots with Preemptive Tasks. *Adv. Intell. Syst. Comput.* **2014**, *290*, 19–27. [CrossRef]
28. Banaszak, Z. CP-Based Decision Support for Project Driven Manufacturing. In *Perspectives in Modern Project Scheduling*; Springer: Boston, MA, USA, 2006; pp. 409–437.
29. Soto, R.; Kjellerstrand, H.; Gutierrez, J.; Lopez, A.; Crawford, B.; Monfroy, É. Solving Manufacturing Cell Design Problems Using Constraint Programming. *Comput. Vis.* **2012**, *7345*, 400–406. [CrossRef]
30. Kreter, S.; Schütt, A.; Stuckey, P.J.; Zimmermann, J. Mixed-integer linear programming and constraint programming formulations for solving resource availability cost problems. *Eur. J. Oper. Res.* **2018**, *266*, 472–486. [CrossRef]
31. Laborie, P. An Update on the Comparison of MIP, CP and Hybrid Approaches for Mixed Resource Allocation and Scheduling. In *Proceedings of International Conference on the Integration of Constraint Programming, Artificial Intelligence, and Operations Research*; Springer: Cham, Switzerland, 2018; pp. 403–411.
32. Grzybowska, K.; Kovács, G. Sustainable Supply Chain—Supporting Tools. In Proceedings of the Federated Conference on Computer Science and Information Systems, Warsaw, Poland, 7–10 September 2014; IEEE: Warsaw, Poland, 2014; pp. 1321–1329.
33. Janardhanan, M.N.; Li, Z.; Bocewicz, G.; Banaszak, Z.; Nielsen, P. Metaheuristic algorithms for balancing robotic assembly lines with sequence-dependent robot setup times. *Appl. Math. Model.* **2019**, *65*, 256–270. [CrossRef]
34. Sitek, P.; Wikarek, J. A multi-level approach to ubiquitous modeling and solving constraints in combinatorial optimization problems in production and distribution. *Appl. Intell.* **2017**, *48*, 1–24. [CrossRef]
35. Ochoa, L.; González-Rojas, O.; Cardozo, N.; González, A.; Chavarriaga, J.; Casallas, R.; Díaz, J.F. Constraint programming heuristics for configuring optimal products in multi product lines. *Inf. Sci.* **2019**, *474*, 33–47. [CrossRef]
36. Yang, D.; Dong, M. A constraint satisfaction approach to resolving product configuration conflicts. *Adv. Eng. Inform.* **2012**, *26*, 592–602. [CrossRef]
37. Trojet, M.; H'Mida, F.; Lopez, P. Project scheduling under resource constraints: Application of the cumulative global constraint in a decision support framework. *Comput. Ind. Eng.* **2011**, *61*, 357–363. [CrossRef]
38. Szeredi, R.; Schütt, A. Modelling and Solving Multi-mode Resource-Constrained Project Scheduling. In *Proceedings of the International Conference on Principles and Practice of Constraint Programming*; Springer: Cham, Switzerland, 2016; pp. 483–492.
39. Relich, M. Identifying Project Alternatives with the Use of Constraint Programming. *Adv. Intell. Syst. Comput.* **2016**, *521*, 3–13. [CrossRef]
40. Van Roy, P. *Multiparadigm Programming in Mozart/Oz*; Springer: Berlin/Heidelberg, Germany, 2005; p. 3389.

Article

On Solving Large-Size Generalized Cell Formation Problems via a Hard Computing Approach Using the PMP

Youkyung Won

Department of Business Administration, Kunsan National University, 558 Daehak-ro, Gunsan-si 54150, Korea; ykwon@kunsan.ac.kr

Received: 5 April 2020; Accepted: 15 May 2020; Published: 18 May 2020

Abstract: In this paper, we show that the hard computing approach using the p-median problem (PMP) is a very effective strategy for optimally solving large-size generalized cell formation (GCF) problems. The soft computing approach, relying on heuristic or metaheuristic search algorithms, has been the prevailing strategy for solving large-size GCF problems with a short computation time at the cost of the global optimum in large instances of GCF problems; however, due to recent advances in computing technology, using hard computing techniques to solve large-sized GCF problems optimally is not time-prohibitive if an appropriate mathematical model is built. We show that the hard computing approach using the PMP-type model can even solve large 0–1 GCF instances optimally in a very short computation time with a powerful mixed integer linear programming (MILP) solver adopting an exact search algorithm such as the branch-and-bound algorithm.

Keywords: hard computing approach; p-median problem; generalized cell formation

---

1. Introduction

The cell formation (CF) problem has attracted researchers in academia as well as practitioners in the field since it was introduced as a part of group technology (GT) [1]. The initial step of CF is to create machine cells and their associated part families. A machine cell is a collection of functionally dissimilar machines which are grouped together and dedicated to process its associated part family, which is a collection of parts which are similar with regard to their geometric shape and size or processing requirements. By creating efficient cells, the maximum operations of the machines within cells (intra-cell operations) and minimum transfers of parts from one cell to another (inter-cell operations) are achieved. This leads numerous operational benefits, such as a reduction in setup time, work-in-process inventories, improvement in quality and a high degree of flexibilities to product demand changes [2].

Since the CF is an NP-hard problem [3,4], a number of approaches and methods have been proposed to solve the CF problem effectively. Papaioannou and Wilson [5] provided a recent review of the CF solution methodologies. The CF problems are classified into two categories: the standard CF (SCF) problem, considering only one process plan for each part, and the generalized CF (GCF) problem, considering alternative process plans for each part. Both problems can include replicate machines; i.e., extra copies for a machine type. GCF is more complicated than SCF since SCF is a special case of GCF. When a part has alternative process plans, operations can be performed on different types of machines or extra copies of a machine type. By considering alternative process plans and replicate machines, more independent cells and higher machine utilization due to reduced inter-cell flows can be achieved [6].

The first step in solving CF problems is to construct the mathematical model which is best suited to achieving the objectives of a specific CF. However, this usually leads to a huge model that has many integer and continuous variables, constraints, and/or nonlinear functions. As the number of machines and parts directly influencing the size of CF problem increases, the optimal solution methodology fails to solve large CF instances [7]. More specifically, if a mathematical model of CF contains nonlinear and/or multi-objective functions over multiple periods, it is very difficult to optimally solve that model, although those nonlinear functions can be linearized. Therefore, a number of soft computing approaches relying on local search methodologies such as artificial intelligence, heuristic/meta-heuristic, or hybrid algorithms have been proposed. Soft computing approaches attempt to find good or acceptable solutions to the proposed mathematical model of CF in a short computation time at the expense of the global optimum. Most soft computing approaches use specific mathematical models to set up the CF problem rather than solving them optimally. In this regard, almost all soft computing approaches for CF are heuristic in nature. However, the cell design experience with industrial experts shows that designers would rather spend more time to achieve an optimal or near optimal solution than use a heuristic approach to get an inferior solution [8].

On the contrary, hard computing approaches can use exact search algorithms such as branch-and-bound to solve the mathematical models of CF optimally if reasonable computation time is allowed. The application of hard computing approaches for CF significantly has relied on recent advances in computer hardware and commercially available mixed integer linear programming (MILP) solvers, such as CPLEX, LINGO, or Gurobi. Borrero et al. [9] stated that hard computing approaches can yield optimal solutions to large MILP problems with a reasonable running time if appropriate mathematical models are constructed.

Recently, two hard computing approaches for solving the mathematical models of CF have been mentioned in the CF-related literature. The first is an exact method that attempts to find the best cell configuration by directly maximizing the objective function of CF. The grouping efficacy (GE) measure [10] has been widely used as an objective function of CF. Since the GE takes a fractional function, the CF problem with the GE objective function results in a 0–1 nonlinear fractional programming problem. Thus, maximizing the GE directly has attracted many researchers since the early 2010s [11–18]. In order to evaluate the performance of their exact methods, 35 small to intermediate-size benchmark incidences [19] have been widely used for benchmark testing, and their solutions have been compared. However, some instances were not solved optimally even under the time limit of 100,000 s using the CPLEX MILP solver.

The other approach aims to indirectly maximize the GE or other alternative performance measures by using the classic or modified p-median problem (PMP). Since Hakimi [20,21] first introduced the PMP on a network of nodes and arcs, the PMP has been widely studied and extended to many practical situations including the location of plants, warehouses, distribution centers, hubs, and public service facilities [22]. Revelle and Swain [23] used Balinski-type constraints [24] to present an integer linear programming (ILP) formulation of the PMP. Unfortunately, since the original Revelle and Swain model (ORSM) defined on an n-node network contains n^2 binary variables and $n^2 + 1$ constraints, it is computationally infeasible to exactly solve the ORSM even for moderately sized networks. Therefore, many attempts have been made to formulate equivalent PMP models including fewer binary variables and constraints than the original ORSM [25–31]. Those reduced PMP models have been solved using hard computing techniques on MILP solvers, and their performances have been compared to those of past PMP models.

Kusiak [32,33] first proposed using the PMP-type model as an alternative mathematical programming model for the CF, replacing exact methods. However, the PMP itself does not explicitly optimize the objective of CF in the same way as the GE. Nevertheless, the PMP grasps the clustering nature of CF and presents a flexible framework by allowing additional constraints reflecting realistic aspects to be introduced [34]. In this context, the PMP matches the CF problem well and shows good solution performance for small to intermediate-size SCF/GCF instances [35–49]. Recently, Goldengorin et al. [34] proposed a flexible PMP-based approach for solving large-sized 0–1

SCF problems and used the Xpress MILP solver to optimally solve most of the SCF instances available in the literature within one second.

However, few studies reporting successful applications of the hard computing approach of the PMP-type model to large-sized GCF instances have appeared in the literature. There are two main reasons for this:

- First, with regard to SCF, the 35 standard incidences available in the literature have been widely used for benchmark testing for the last 20 years, and a recent study adopting a hybrid algorithm [50] has reported the best optimal solutions with huge time-savings. However, there are few open large GCF data sets available in the literature regarding the standard instances for benchmark testing and the performance comparison of the solution algorithms used. As far as the present author knows, the largest example of an open GCF available in the literature has at most 55 machines, 60 part types, and a total of 124 process plans [51].

- Second, execution strategies for CF and complicated aspects inherent in the GCF problem itself make the solution quality of CF methods very sensitive to subsequent part assignment or improvement procedures that are necessary to follow after machine cells are obtained. Three different strategies have been used to execute the CF algorithm [52]: the part family identification (PGI) strategy forms part families first and then groups machines, the machine group identification (MGI) strategy creates machine cells first and then allocate parts to cells, and the part family/machine grouping (PF/MG) strategy forms machine cells and part families simultaneously. Most soft and hard computing approaches for CF use the MGI strategy to execute CF algorithms since it usually takes enormous computation time to implement the PF/MG strategy even for intermediate-size incidences. The PMP-based approach for CF also uses the MGI strategy to create cells. Therefore, once machine cells are obtained from the PMP solution, part families need to be formed by allocating parts to the best cells. Danilovic and Ilic [50] and Li et al. [53] have established sufficient conditions for the optimal assignment of parts to machine cells given a partition of machines of a SCF problem. However, it should be noted that their sufficient conditions may not guarantee the optimal assignment of parts maximizing the GE in the GCF problem due to the existence of alternative process plans and/or replicate machines.

Motivated by the drawbacks of extant studies attacking the GCF problem, this paper proposes an effective hard computing approach using the PMP-type model to solve large-sized GCF problems. Our hard computing approach has the following distinctive features compared to previous hard computing approaches dealing with the GCF problem:

- Two new linear 0–1 mathematical models of GCF are formulated: an exact model that directly maximizes the GE and a PMP-type model that indirectly maximizes the GE. Because the exact model contains too many binary variables and constraints, the PMP-type model is used to solve large-sized GCF instances optimally. According to the computational experiments applied to large GCF instances with over 10,000 binary variables, our PMP-type model solves those large GCF instances optimally within one second using the LINGO MILP solver.

- Since the PMP-type approach uses MGI strategy to form machine cells first, a subsequent part allocating step is needed to form the corresponding part families. In this paper, a systematic heuristic part assignment procedure based on a new classification scheme of part types with alternative process plans is used to assign the best process plan of each part to its best cell. A subsequent refinement procedure is the used to further improve the block diagonal solution by reassigning improperly assigned exceptional machines (EMs) in such a way that the GE is maximized. The computational burden of implementing these extra procedures is negligibly small since they accomplish a high-quality CF within 0.2 s, even for the largest GCF instances tested in our computational experiments.

- Unlike many comparative studies of SCF using the standard data set provided in Goncalves and Resende [19], studies of GCF lack the standard data set. Our computational experiment has been conducted over the widest range of GCF incidences that have ever appeared in the CF-related literature. Our collection of the GCF incidences can be used as a standard data set for subsequent benchmark tests in the future.

2. Materials and Methods

2.1. Basic Input

Different approaches for the GCF take different approaches to the definition of alternative process plans or routings. Vin and Delchambre [54] classify the approaches by sorting the processes or routings into six categories: (i) fixed routing (process route), (ii) routing with replicate machines, (iii) routing with alternate machines for some operations, (iv) several fixed routings, (v) fixed process plan, and (vi) alternative process plans. In this paper, we take the combination of manners (ii) and (iv) to define alternative process plans to model the GCF problem.

To model the GCF problem, the binary part–machine incidence matrix (PMIM), which represents the association between the process plans of parts and machines, will be used as a basic input. Given m part types with a total of t process plans and n different machine types, the $t \times n$ binary PMIM $A\left(= \left[a_{irj}\right]\right)$ is defined as follows:

$$a_{irj} = \begin{cases} 1 & \text{if process plan } r \text{ of part } i \text{ is processed on machine } j \\ 0 & \text{otherwise.} \end{cases}$$

In addition, the following indices, notation, and decision variables will be used throughout the paper:

Indices

i = part index;
r = process plan index;
j = machine or cell index;
k = copy index of replicate machine;
c = machine cell/part family index.

Parameters

m = number of part types;
R_i = set of process plans of part type i;
t = total number of process plans;
n = number of different machine types;
p = number of cells;
U = upper limit on the cell size;
DM = set of replicate machine types;
M_j = set of copies of replicate machine type $j \in DM$;
$q = n - |DM| + \sum_{j \in DM} |M_j|$ = total number of machines including copies of replicate machine types (the symbol $|X|$ denotes the cardinality of the set X);
gs_{j_1, j_2} = generalized similarity coefficient between machine types j_1 and j_2;
c_{j_1, j_2} = similarity coefficient between machines j_1 and j_2;
MC_c = set of machines in machine cell c;
PF_c = set of parts in part family c;
e = total number of 1s in the block diagonal solution matrix;
e_0 = number of exceptional elements(EEs) in the block diagonal solution matrix;
e_v = number of voids in the block diagonal solution matrix.;

Decision variables

$$x_{irc} = \begin{cases} 1 & \text{if process plan } r \text{ of part } i \text{ is assigned to cell; } c \\ 0 & \text{otherwise.} \end{cases}$$

$$y_{jkc} = \begin{cases} 1 & \text{if copy } k \text{ of replicate machine } j \text{ is assigned to cell; } c \\ 0 & \text{otherwise.} \end{cases}$$

$$z_{j_1,j_2} = \begin{cases} 1 & \text{if machine } j_1 \text{ belongs to cell; } j_2 (j_1, j_2 = 1, \cdots, q) \\ 0 & \text{otherwise.} \end{cases}$$

Then, the 0–1 GCF problem can be illustrated with a generalized 0–1 PMIM as shown in Figure 1a, in which each row indicates a part and each column a machine. The manufacturing system shown in Figure 1a has eight part types with a total of 19 process plans and five different machine types with an extra copy for machine type 3. An entry of "1" indicates that a part is processed by its associated machine, and an entry of "0", which is not shown for visual convenience, indicates that it is not processed by its associated machine. The alphabetical letters after the part numbers indicate alternative process plans. Rearranging the rows and columns in such a way that only one process plan is selected for each part and only one copy is allowed for each different machine type in each cell results in a block diagonal solution matrix, as shown in Figure 1b. The solution of Figure 1b shows two machine cells and two part families. Machine cell 1 (MC$_1$), consisting of machines 1, 3, and 5 (MC$_1$ = {1, 3, 5}), processes plans 2a, 3c, 5a, 6b, and 9a of part family 1 (PF$_1$) (PF$_1$ = {2a, 3c, 5a, 6b, 9a}). MC$_2$, consisting of machines 2, 4 and an extra copy of machine type 3 (MC$_2$ = {2, 4, 3}), processes plans 1b, 4a, 7b, and 8c of PF$_2$ (PF$_2$ = {1b, 4a, 7b, 8c}). As a result, the solution of Figure 1b yields no EEs and four voids.

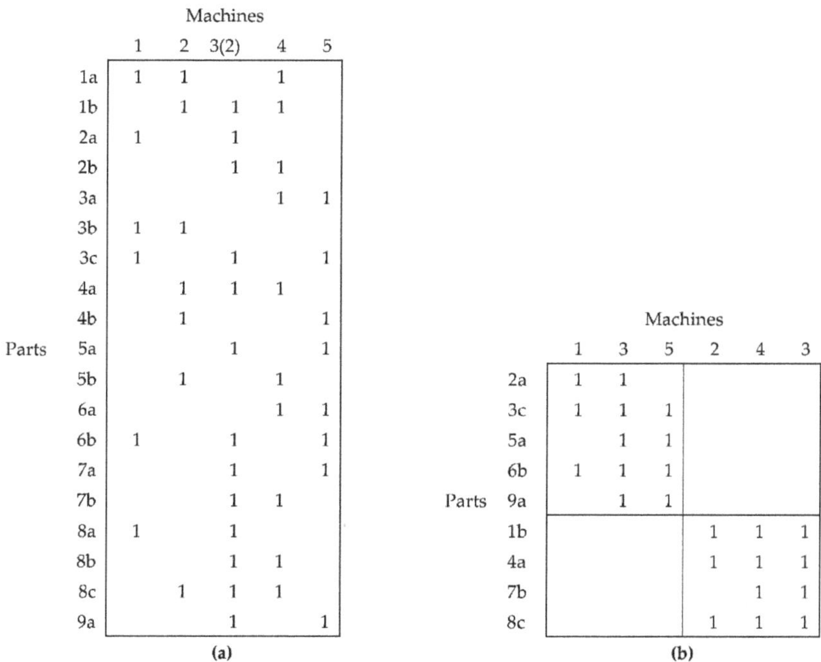

(a) Initial generalized part–machine incidence matrix

Parts	1	2	3(2)	4	5
1a	1	1		1	
1b		1	1	1	
2a	1		1		
2b			1	1	
3a				1	1
3b	1	1			
3c	1		1		1
4a		1	1	1	
4b		1			1
5a			1		1
5b		1		1	
6a				1	1
6b	1		1		1
7a			1		1
7b			1	1	
8a	1		1		
8b			1	1	
8c		1	1	1	
9a			1		1

(b) Block diagonal solution matrix

Parts	1	3	5	2	4	3
2a	1	1				
3c	1	1	1			
5a		1	1			
6b	1	1	1			
9a		1	1			
1b				1	1	1
4a				1	1	1
7b					1	1
8c				1	1	1

Figure 1. (a) Initial generalized part–machine incidence matrix (PMIM=; (b) block diagonal solution matrix.

2.2. Performance Measure

Several comprehensive grouping efficiency measures considering both EEs and voids have been proposed to evaluate the quality of 0–1 block diagonal solutions and are reviewed critically [55,56]. Of those measures, the grouping efficacy (GE) [10] has been most widely used to evaluate the performance of the 0–1 GCF problem as well as the 0–1 SCF problem. The GE, Γ, is defined as

$$\Gamma = \frac{\text{total no. of 1s in the block diagonal matrix} - \text{EEs}}{\text{total no. of 1s in the block diagonal matrix} + \text{voids}} = \frac{e - e_0}{e + e_v}. \tag{1}$$

According to the above definition, the block diagonal solution of Figure 1b gives a GE of 85.19%. A GE of 100% indicates a perfect CF without EEs and zeros.

2.3. Mathematical Models

2.3.1. Exact Model

Several authors have provided exact formulations maximizing the grouping efficacy in the 0–1 SCF problem with a single copy of each machine type [11–18]. In this subsection, we present an exact formulation extended for the 0–1 GCF problem with multiple copies of replicate machine types.

To construct the exact formulation which directly maximizes the GE, the values of e, e_0, and e_v should be calculated. They are given by the following equations:

$$e = \sum_{c=1}^{p} \sum_{i=1}^{m} \sum_{r \in R_i} \sum_{j=1}^{n} \sum_{k \in M_j} a_{irj} x_{irc} y_{jkc}. \tag{2}$$

$$\begin{aligned}
e_0 &= \sum_{c=1}^{p} \sum_{i=1}^{m} \sum_{r \in R_i} \sum_{j=1}^{n} a_{irj} x_{irc} \left[\sum_{k \in M_j} \left(1 - y_{jkc}\right) - \left(\left|M_j\right| - 1\right) \right] \\
&= \sum_{c=1}^{p} \sum_{i=1}^{m} \sum_{r \in R_i} \sum_{j=1}^{n} a_{irj} x_{irc} \left[1 - \sum_{k \in M_j} y_{jkc} \right] \\
&= \sum_{c=1}^{p} \sum_{i=1}^{m} \sum_{r \in R_i} \sum_{j=1}^{n} a_{irj} x_{irc} - \sum_{c=1}^{p} \sum_{i=1}^{m} \sum_{r \in R_i} \sum_{j=1}^{n} \sum_{k \in M_j} a_{irj} x_{irc} y_{jkc}
\end{aligned} \tag{3}$$

$$e_v = \sum_{c=1}^{p} \sum_{i=1}^{m} \sum_{r \in R_i} \sum_{j=1}^{n} \sum_{k \in M_j} \left(1 - a_{irj}\right) x_{irc} y_{jkc}. \tag{4}$$

Note that once a process plan r of part i is processed on a copy k of replicate machine type j in cell c, the remaining $\left(\left|M_j\right| - 1\right)$ elements with $a_{irj} = 1$ processed by different copies of that replicate machine type in other cells are not the EEs.

Then, the mathematical model directly maximizing the GE is formulated as follows:
(Model 1)

$$\text{Maximize } \Gamma = \frac{\sum_{c=1}^{p} \sum_{i=1}^{m} \sum_{r \in R_i} \sum_{j=1}^{n} \sum_{k \in M_j} a_{irj} x_{irc} y_{jkc}}{\sum_{c=1}^{p} \sum_{i=1}^{m} \sum_{r \in R_i} \sum_{j=1}^{n} \sum_{k \in M_j} a_{irj} x_{irc} + \sum_{c=1}^{p} \sum_{i=1}^{m} \sum_{r \in R_i} \sum_{j=1}^{n} \sum_{k \in M_j} \left(1 - a_{irj}\right) x_{irc} y_{jkc}} \tag{5}$$

Subject to

$$\sum_{c=1}^{p} \sum_{r \in R_i} x_{irc} = 1, \ i = 1, \cdots, m \tag{6}$$

$$\sum_{c=1}^{p} \sum_{k \in M_j} y_{jkc} = 1, \ j = 1, \cdots, m \tag{7}$$

$$\sum_{k \in M_j} y_{jkc} \leq 1, \; j \in DM, \; j = 1, \cdots, n; c = 1, \cdots, p \tag{8}$$

$$\sum_{j=1}^{n} \sum_{k \in M_j} y_{jkc} \leq U, c = 1, \cdots, p \tag{9}$$

$$x_{irc}, y_{jkc} = 0 \text{ or } 1, r \in R_i, i = 1, \cdots, m; k \in M_j, j = 1, \cdots, n; c = 1, \cdots, p. \tag{10}$$

The objective function in Equation (5) maximizes the GE. Constraint (6) ensures that only one process plan of each part is assigned to only one cell. Constraint (7) ensures that each machine belongs to exactly one machine cell. Constraint (8) ensures that at most a single copy of a replicate machine type is assigned to each cell. Constraint (9) ensures that the number of machines in each cell does not exceed U machines. Constraint (10) ensures the binary restriction of variables.

Model 1, which is a non-linear 0–1 fractional programming model, can be linearized by introducing the following auxiliary binary variables:

$$w_{irjkc} = x_{irc} y_{jkc}, \; r \in R_i, i = 1, \cdots, m; k \in M_j, j = 1, \cdots, n; c = 1, \cdots, p. \tag{11}$$

To linearize the variable w_{irjkc}, the following extra constraints should be added [57,58]:

$$w_{irjkc} - x_{irc} - y_{jkc} \geq -1.5, r \in R_i, i = 1, \cdots, m; k \in M_j, j = 1, \cdots, n; c = 1, \cdots, p. \tag{12}$$

$$1.5w_{irjkc} - x_{irc} - y_{jkc} \leq 0, r \in R_i, i = 1, \cdots, m; k \in M_j, j = 1, \cdots, n; c = 1, \cdots, p. \tag{13}$$

Then, we have the linear 0–1 fractional programming model 2 which is equivalent to model 1 as follows:

(Model 2)

$$\text{Maximize } \Gamma = \frac{\sum_{c=1}^{p} \sum_{i=1}^{m} \sum_{r \in R_i} \sum_{j=1}^{n} \sum_{k \in M_j} a_{irj} w_{irjkc}}{\sum_{c=1}^{p} \sum_{i=1}^{m} \sum_{r \in R_i} \sum_{j=1}^{n} \sum_{k \in M_j} a_{irj} x_{irc} + \sum_{c=1}^{p} \sum_{i=1}^{m} \sum_{r \in R_i} \sum_{j=1}^{n} \sum_{k \in M_j} \left(1 - a_{irj}\right) w_{irjkc}} \tag{14}$$

Subject to Equations (6)–(13) and

$$w_{irjkc} = 0 \text{ or } 1, r \in R_i, i = 1, \cdots, m; k \in M_j, j = 1, \cdots, n; c = 1, \cdots, p. \tag{15}$$

Model 2 can then be solved by using a solver such as LINGO adopting the branch-and-bound algorithm. However, the computational burden of optimally solving model 2 seems still to be heavy even if a powerful solver is used due to the presence of too many binary variables and constraints. The total number of binary variables of model 2 is

$$\left[\sum_{j=1}^{n} |M_j| + \sum_{i=1}^{m} |R_i| + \left(\sum_{j=1}^{n} |M_j| \right) \left(\sum_{i=1}^{m} |R_i| \right) \right] p$$

and the number of constraints is

$$n + m + (|DM| + 1)p + 2p \left[\sum_{j=1}^{n} |M_j| + \sum_{i=1}^{m} |R_i| + \left(\sum_{j=1}^{n} |M_j| \right) \left(\sum_{i=1}^{m} |R_i| \right) \right].$$

For example, to solve a large-sized 0–1 GCF problem containing 110 machines, 120 part types, and 248 process plans—which will be tested in Section 4—331,656 binary variables and 663,554 constraints are needed. Therefore, relying on an alternative model with much fewer binary variables and constraints that leads to the maximization of the GE can be a better strategy. Thus, we use the PMP-type model to solve a large-sized GCF problem by maximizing the GE indirectly.

2.3.2. PMP-Type Model

Since the exact model 2 has too many binary variables and constraints to optimally solve large-sized GCF incidences within a reasonably short computation time, we use the PMP-type model as a better alternative formulation to maximize the GE indirectly. However, to develop the PMP-type model of GCF, we need the definition of similarity coefficients incorporating both alternative process plans and replicate machines. In this paper, we use a modified version of Won and Kim's similarity coefficient [59] defined between pairs of machine types to formulate the mathematical model of the GCF. Won and Kim's similarity coefficient based on the binary PMIM is a generalization of the Jaccard similarity coefficient used for the SCF problem. Their similarity coefficient gs_{j_1,j_2} between two machine types j_1 and j_2 is defined by

$$gs_{j_1,j_2} = \frac{\sum_{i=1}^{m} \beta(i, j_1, j_2)}{\sum_{i=1}^{m} \alpha(i, j_1) + \sum_{i=1}^{m} \alpha(i, j_2) - \sum_{i=1}^{m} \beta(i, j_1, j_2)} \tag{16}$$

where

$$\alpha(i, j_1) = \begin{cases} 1 & \text{if } a_{irj_1} = 1 \text{ for some } r \in R_i \\ 0 & \text{otherwise,} \end{cases}$$

$$(1)\beta(i, j_1, j_2) = \begin{cases} 1 & \text{if } a_{irj_1} = a_{irj_2} = 1 \text{ for some } r \in R_i \\ 0 & \text{otherwise.} \end{cases}$$

However, since the above similarity coefficient does not consider the replicate machines, we use Won and Logendran's similarity coefficient [48] to formulate the PMP-type model of GCF. The similarity coefficient gs_{j_1,j_2} between two machines j_1 and j_2 is defined as follows:

$$c_{j_1,j_2} = \begin{cases} -\infty & \text{if both } j_1 \text{ and } j_2 \text{ belong to the same replicate machine type} \\ gs_{j_1,j_2} & \text{otherwise.} \end{cases} \tag{17}$$

Based on the similarity coefficient defined in Equation (13), the PMP-type model of GCF can be formulated as follows:

(Model 3)

$$\text{Maximize} \sum_{j_1=1}^{q} \sum_{j_2=1}^{q} c_{j_1,j_2} z_{j_1,j_2} \tag{18}$$

Subject to

$$\sum_{j_2=1}^{q} z_{j_1,j_2} = 1, \ j_1 = 1, \cdots, q \tag{19}$$

$$\sum_{j_1=1}^{q} z_{j_1,j_1} = p \tag{20}$$

$$\sum_{j_2=1}^{q} z_{j_1,j_2} \leq U z_{j_1,j_2} = 1, \ j_2 = 1, \cdots, q \tag{21}$$

$$\sum_{j_1 \in M_{j_3}} z_{j_1,j_2} \leq 1, \ j_2 = 1, \cdots, q; \ j_3 \in DM \tag{22}$$

$$z_{j_1,j_2} = 0 \text{ or } 1, \ j_1, j_2 = 1, \cdots, q. \tag{23}$$

The objective function in Equation (18) maximizes the sum of similarities among all pairs of machines including replicate machines. Constraint (19) ensures that each machine belongs to exactly one machine cell. Constraint (20) specifies the required number of machine cells. Constraint (21) ensures that the number of machines in each cell does not exceed U machines. Constraint (22) ensures that at most a single copy of a replicate machine type is assigned to each cell. Constraint (23) ensures the binary restriction of variables.

Since model 3 is a linear model and the number of machine types is usually much lower than the number pf process plans, moderately large-sized instances can be solved optimally within a reasonable computation time regardless of the total number of process plans. On the contrary, the number of process plans critically affects the model size in model 2. To solve the example incidence addressed in Section 2.3.1, 12,100 binary variables and 497 constraints are needed. Clearly, model 3 is very economical since it contains much fewer binary variables and/or constraints than model 2. We will show that even a large GCF incidence can be solved optimally within only one second using the LINGO solver.

2.3.3. Part Assignment to Cells

The solution of PMP-type model 3 identifies only the machine cells. Once the machine cells are formed, parts need to be assigned to their best associated cells in such a way that the objective of CF is optimized. The classic part assignment rule that has been widely used is the maximum density rule, assigning a part to the cell in which it has most operations. Since different part assignment rules affect the solution quality of CF, several modified part assignment rules have been proposed [50,59–64]. A critical drawback of these rules is that the solution quality due to the assignment of a part depends on the assignment of other parts since they are heuristic rules. Some authors [50,53] have established sufficient conditions for optimal part assignment that do not depend on the assignment of other parts for the SCF problem. Several heuristic part assignment procedures for the GCF problem considering the number of EEs and voids have been proposed in the literature [59,61–63]; however, due to the existence of alternative process plans and replicate machines, no optimal part assignment rules for the GCF problem have been proposed.

Won [63] used the nonbinary generalized PMIM to develop a heuristic part assignment procedure for the GCF problem. In this paper, we use a modified heuristic part assignment procedure based on the binary generalized PMIM to classify parts into eight categories as follows:

- A type I strongly nonexceptional part (SNEP) for which a unique process plan has the most 1s in a unique machine cell without EEs;
- A type II SNEP for which multiple process plans have the most 1s in a unique machine cell without EEs;
- A type I neutrally nonexceptional part (NNEP) for which a unique process plan has the most 1s in more than one machine cell without EEs;
- A type II NNEP for which multiple process plans have the most 1s in more than one machine cell without EEs;
- A type I weakly exceptional part (WEP) for which a unique process plan has the most 1s in a unique machine cell with EEs;
- A type II WEP for which multiple process plans have the most 1s in a unique machine cell with EEs;
- A type I neutrally exceptional part (NEP) for which a unique process plan has the most 1s in more than one machine cell with EEs;
- A type II NEP for which multiple process plans have the most 1s in more than one machine cell with EEs.

To determine the specific type of a part and assign the best process plan to its best associated cell in such a way that the GE is maximized, the following measures need to be calculated for all process plans of parts:

1. The numbers of EEs due to the assignment of each process plan to each cell;
2. The number of voids due to the assignment of each process plan to each cell;
3. The number of cells which have the most 1s for completing the required operations due to the assignment of each process plan to each cell;
4. The total number of operations (1s) contained in each cell with all the parts assigned until the current stage;
5. The number of parts assigned to each cell until the current stage; and
6. The number of operations processed by the machines in each cell.

Criteria 1 and 3 contribute to the independent cell configuration with the least inter-cell moves. Criteria 2 and 6 contribute to the compact cell configuration with high machine utilization. Criteria 4 and 5 contribute to the balanced cell configuration with an even workload among cells. A specific type of a part can be determined from criteria 1 and 3. Figure 2 presents a flow chart showing the procedure which identifies the type of a part.

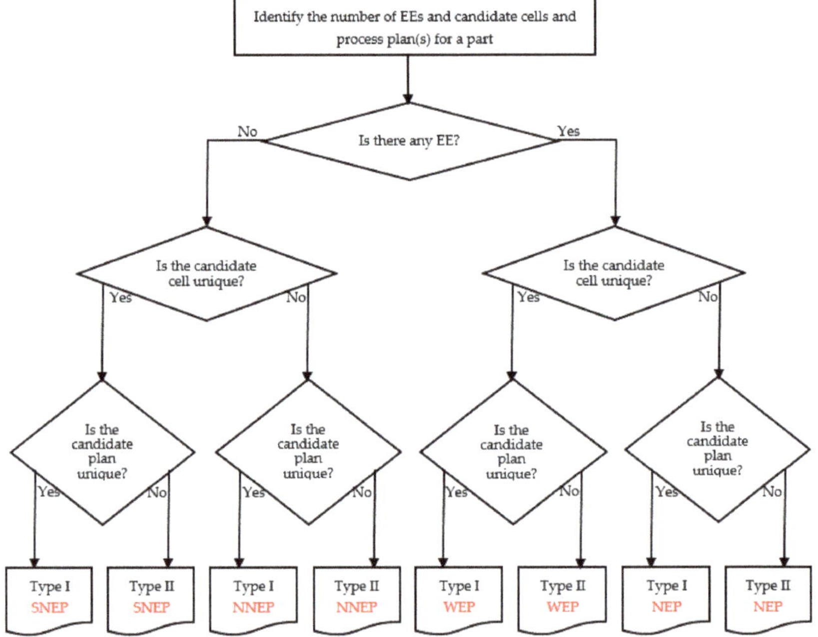

Figure 2. Flow chart identifying the category of part. SNEP: strongly nonexceptional part; NNEP: neutrally nonexceptional part; WEP: weakly exceptionally part; NEP: neutrally exceptional part.

The assignment of a type I SNEP or type I WEP is straightforward: its unique candidate process plan is assigned to its associated unique cell. However, the assignment of remaining part types requires tie-breaking rules to select the best candidate process plan and its best associated cell. Given the information of criteria 1 to 6 for each part, assignment rules for the remaining part types are stated as follows:

Assignment rule for Type II SNEP or Type II WEP

There is a unique candidate cell for each of these part types. The process plan with the least number of EEs from 1 is selected. If ties occur, the plan with the maximum number of operations from 6 is selected. If ties occur again, the smallest-numbered cell is selected.

Assignment rule for Type I NNEP or Type I NEP

There is a unique candidate plan for each of these part types. The cell with the lowest number of EEs from 1 is selected. If ties occur, the cell with the least number of voids from 2 is selected. If ties occur again, the cell assigned with the least number of 1s from 4 is selected. If ties occur again, the cell assigned with the least number of part types from 5 is selected. If ties occur again, the cell with the maximum number of operations in a cell from 6 is selected. If ties occur again, the smallest-numbered cell is selected.

Assignment rule for Type II NNEP or Type II NEP

There are multiple candidate plans and cells for each of these part types. The plan and the associated cell with the least number of EEs from 1 are selected. If ties occur, the process plan and the associated cell with the least number of voids from 2 are selected. If ties occur again, the process plan and the associated cell assigned with the least number of 1s from 4 are selected. If ties occur again, the process plan and the associated cell assigned with the least number of part types from 5 are selected. If ties occur again, the process plan and the associated cell with the maximum number of operations in a cell from 6 are selected. If ties occur again, the smallest-numbered process plan and cell are selected.

2.3.4. Reassigning Improperly Assigned Exceptional Machines (EMs) and Redundant Machines (RMs)

The CF based on the solution from model 3 and the part family formation may result in an unsatisfactory block diagonal solution due to improperly assigned Ems, which process most parts in other cells, or RMs, which process no parts in their parent cell. Therefore, a subsequent refinement procedure is used to improve the quality of incumbent block diagonal solutions through an inspection by reassigning them to their most appropriate cells if there are any improperly assigned EMs. The reassignment of improperly assigned EMs/RMs can lead to higher GE by decreasing EEs and voids.

The improperly assigned EMs/RMs are categorized as follows:

- An absolute RM which processes no parts in any cell;
- A type I RM which processes most parts in other unique cells except for its parent cell;
- A type II RM which processes most parts in two or more other cells except for its parent cell;
- A type I EM which processes some parts in its parent cell but processes most parts in other unique cells except for its parent cell;
- A type II EM which processes most parts in two or more cells including its parent cell.

To illustrate the types of improperly assigned EMs/RMs, consider a block diagonal matrix as shown in Figure 3. From this matrix, we can observe the set of machine cells, $MC_1 = \{1, 2, 3\}$, $MC_2 = \{4, 5, 6\}$, and $MC_3 = \{7, 8\}$; and the set of part families, $PF_1 = \{1a, 2b, 3c\}$, $PF_2 = \{4a, 5b, 6a\}$, and $PF_3 = \{7b, 8c, 9b\}$. According to the above definition, machines 1, 2, and 3 are type I RM, type II RM, and absolute RM, respectively, and machines 6 and 8 are type I EM and type II EM, respectively.

		Machines							
		1	2	3	4	5	6	7	8
	1a						1	1	
	2b						1		1
	3c						1		1
	4a	1	1		1	1	1		
Parts	5b	1	1		1	1	1		
	6a	1			1				
	7b	1	1		1			1	1
	8c		1					1	
	9b							1	1

Figure 3. A block diagonal matrix for identifying the type of exceptional machines (EMs)/redundant machines (RMs).

To determine the type of EM and RM, the following elements need to be evaluated for each machine:

- Its parent cell;
- Cells processing most parts;
- The total number of parts processed; and
- The number of parts processed in its parent cell.

Figure 4 is a flow chart showing the determination of the type of EM or RM for a machine. The reassignment procedure for improperly assigned EMs/RMs is then stated as follows:

- Reassign a type I RM or type I EM to another unique candidate cell;
- Reassign an absolute RM, type II RM or type II EM to the cell with the fewest 1s. If ties occur, reassign them to the cell with the lowest number of machines.
- Remove an absolute RM from the current cell since it does not process any parts under the incumbent cell configuration.

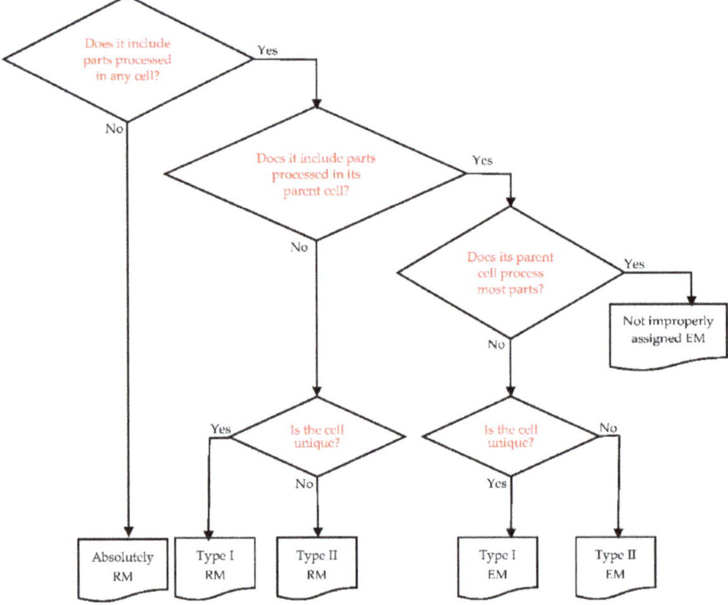

Figure 4. Flow chart identifying the type of EM and RM.

2.3.5. Illustrative Example

The proposed procedures for part assignment and improperly assigned EM/RM reassignment are illustrated with a hypothetical GCF incidence which includes seven machine types, 15 part types, and 35 process plans as shown in Table 1, which lists the routing information of parts. Machine types 3 and 4 have two and one extra copies, respectively. After model 3 is solved under the condition of $p = 3$ and $U = 4$, the following machine cells are obtained: $MC_1 = \{1, 3, 6\}$, $MC_2 = \{2,4,3\}$, and $MC_3 = \{5,7,3,4\}$.

Once these machine cells are determined, partial assignment steps of parts 1, 2, and 3 are presented in Table 2, where the column (1) indicates the number of EEs generated by the process plan assigned to a specific cell. The symbol * indicates the entries corresponding to the candidate process plan and cells selected by the criterion of that step. The symbol ** indicates the entries corresponding to the best process plan and the cell selected by the criterion of the final step. If the best candidate plan is determined, the associated best process plan and cell can be identified by scanning the columns from left to right. After all parts are assigned, no RMs and improperly assigned EMs are found. Figure 5 is the resulting block diagonal solution matrix with a GE of 77.19%.

Table 1. Routing information of parts.

Part No.	Process Plan	Machines	Part No.	Process Plan	Machines
1	a	1, 5	9	a	5, 7
	b	1, 3, 4, 6		b	2, 4
2	a	3, 5, 7	10	a	1, 3, 4
	b	2, 3		b	1, 2, 5
3	a	2, 3, 4	11	a	2, 3, 4, 6
4	a	2, 3, 4, 5		b	1, 2, 5
	b	1, 2, 4, 5		c	5, 6, 7
	c	2, 5, 6		d	2, 3, 4, 5, 7
	d	2, 3, 4, 7	12	a	3, 5
5	a	3, 4, 5, 6		b	2, 6
	b	2, 7		c	2, 3, 4
	c	1, 3, 4, 5, 7	13	a	1, 2, 3, 6
6	a	2, 4	14	a	2, 3, 4, 6
	b	1, 3, 6		b	1, 2, 3, 4
7	a	3, 5	15	a	2, 7
	b	4, 5, 7		b	3, 4
8	a	2, 3, 5			
	b	4, 6, 7			
	c	1, 2, 3, 6			

Table 2. Assignment of parts 1, 2, and 3 to cells.

Part No.	Process Plan	Cells Assigned	Criteria		Part Type	Criteria		
			1	2		4	5	6
1	a *	1 *	1 *	2				
		2	2	3				
		3 *	1 *	3	Type II NEP			
	b **	1 **	1 *	0 *				
		2	2	1				
		3	2	2				
2	a **	1	2	2				
		2	2	2				
		3 **	0 *	1 *		0 *	0 *	3 **
	b *	1	1	2	Type II NNEP			
		2 *	0 *	1 *		0 *	0 *	2
		3	1	3				
3	a **	1	2	2				
		2 **	0 *	0 *	Type I SNEP			
		3	1	2				

* indicates the candidate process plans and cells and ** indicates the best process plan and cell.

Parts \ Machines	1	3	6	2	4	3	5	7	3	4
1b	1	1	1		1					
6b	1	1	1							
8c	1	1	1	1						
13a	1	1	1	1						
3a				1	1	1				
4a				1	1	1	1			
9b				1	1					
10a	1				1	1				
12c				1	1	1				
14a			1	1	1	1				
15b					1	1				
2a							1	1	1	
5c	1						1	1	1	1
7b							1	1		1
11d				1			1	1	1	1

Figure 5. Block diagonal solution to the hypothetical example.

3. Results

The purpose of the computational experiment performed in this section was two-fold. One aim was to show the effectiveness of our approach compared to the solutions already reported in the literature under the same restrictions as the reference approaches with regard to the number of cells p and cell size U; the other was to provide solutions that can be used as benchmarks for comparative testing with different CF solution approaches by finding better alternative solutions that have not been reported in the literature under different values of p and U.

The proposed PMP-based hard computing approach has been implemented and tested with two groups of GCF incidences: 26 small to intermediate-sized problem sets available in the literature and three expanded large-sized ones. The large-sized incidences are expanded with values of p which are different from the original ones. Tables 3 and 4 show a total of 39 test incidences for 26 small to intermediate-sized original problems and a total of five expanded test incidences for three large-sized original problems, respectively.

Table 3. Computational results with small to intermediate-sized original incidences. GE: grouping efficacy.

Problem	Problem Size					Reference Algorithm				Proposed Approach			
Source	$n(q)$	m	t	p	U	e	EEs	Voids	GE (%)	e	EEs	Voids	GE (%)	CPU (s)
1. [33]	4	5	11	2	2	9	0	1	**90.00**	9	0	1	**90.00**	0.05 [a] + 0.044 [b]
2. [35]	6	10	20	2	4	28	2	8	**72.22**	28	2	8	**72.22**	0.05 + 0.047
	6(7)	10	20	2	4	27	0	10	72.97	31	0	8	**79.49**	0.09 + 0.054
3. [65]	20	20	51	5	5	66	1	17	**78.31**	67	6	16	73.49	0.09 + 0.189
4. [66]	6	6	13	2	3	15	0	3	**83.33**	15	0	3	**83.33**	0.05 + 0.044
5. [67]	10	10	27	3	5	47	10	17	57.81	49	12	12	**60.66**	0.08 + 0.062
	10(13)	15	27	3	5	49	5	20	63.77 [c]	49	4	21	**64.29**	0.08 + 0.050
6. [68]	7	14	32	3	3	31	5	6	**70.27** [d]	29	5	6	68.57	0.06 + 0.048
7. [69]	10	10	24	3	5	31	1	3	**88.24**	31	1	3	**88.24**	0.08 + 0.050
	10(11)	10	24	3	5	33	3	10	69.77	33	1	3	**88.89**	0.06 + 0.049
8. [70]	8	13	26	3	3	33	2	5	**81.58**	33	3	4	81.08	0.09 + 0.049
9. [70]	30	30	89	6	7	151	1	48	75.38	150	1	46	**76.02**	0.11 + 0.065
10. [59]	4	4	8	2	2	8	0	0	**100.00**	8	0	0	**100.00**	0.05 + 0.040
	7	10	23	3	3	25	3	2	**81.48**	23	4	5	67.86	0.06 + 0.045
	11	10	22	4	3	28	3	3	**80.65**	27	3	4	77.42	0.08 + 0.041
	26	28	71	6	7	124	16	25	**72.48**	121	15	27	71.62	0.10 + 0.059

Table 3. *Cont.*

Problem	Problem Size									Reference Algorithm			Proposed Approach				
Source	$n(q)$	m	t	p	U	e	EEs	Voids	GE (%)	e	EEs	Voids	GE (%)	CPU (s)
11. [71]	8	15	46	2	4	30	1	31	47.54	35	1	26	**55.74**	0.08 + 0.049
	8	15	46	2	5	NA	NA	NA	NA	33	0	22	**60.00**	0.07 + 0.048
	8	15	46	3	3	32	1	12	70.45	33	1	9	**76.19**	0.09 + 0.047
12. [72]	12	20	26	3	5	NA	29	NA	47.06 [e]	85	31	26	**48.65**	0.08 + 0.047
13. [72]	14	20	45	3	5	85	24	35	**50.83** [e]	85	31	27	48.21	0.09 + 0.051
14. [72]	18	18	59	3	6	116	26	108	40.18 [e]	116	32	84	**42.00**	0.09 + 0.054
15. [73]	10(13)	7	13	3	5	22	0	9	**70.97**	23	1	11	64.71	0.08 + 0.053
16. [74]	6(15)	20	34	3	5	49	0	52	48.51 [e]	49	0	51	**49.00**	0.08 + 0.048
17. [75]	6	8	14	2	3	21	1	4	80.00	22	1	3	**84.00**	0.05 + 0.046
18. [75]	5	7	11	2	3	15	0	3	83.33	17	1	1	**88.89**	0.05 + 0.047
19. [75]	10(14)	16	31	2	7	68	8	52	50.00	68	1	45	**59.29**	0.08 + 0.050
20. [76]	10(16)	20	35	2	10	77	0	87	**46.95**	77	4	79	46.79	0.09 + 0.049
21. [77]	9	8	20	2	6	28	2	13	63.41	28	2	10	**68.42**	0.08 + 0.046
22. [78]	30	70	149	2	15	NA	NA	NA	NA	477	132	717	28.89	0.15 + 0.079
	30	70	149	3	10	NA	NA	NA	NA	478	193	415	31.91	0.28 + 0.082
	30	70	149	4	8	NA	NA	NA	NA	477	219	290	33.64	0.17 + 0.081
	30	70	149	5	6	NA	NA	NA	NA	478	249	192	34.18	0.16 + 0.081
23. [51]	30	35	82	4	8	224	52	92	54.43	229	52	86	**56.19**	0.18 + 0.067
	30	35	100	5	7	NA	NA	NA	NA	233	58	38	64.58	0.13 + 0.066
24. [51]	40	45	100	6	8	263	39	86	64.18	263	37	85	**64.94**	0.19 + 0.076
	40	45	100	7	7	NA	NA	NA	NA	266	42	36	**74.17**	0.26 + 0.069
25. [51]	55	60	124	8	9	402	50	62	75.86	411	50	53	**77.80**	0.27 + 0.076
26. [62]	8	20	27	3	4	60	10	2	**80.65**	60	10	2	**80.65**	0.07 + 0.051

[a] Running time implementing model 3. [b] Running time assigning parts and reassigning improperly assigned EMs/RMs. [c] The best solution reported in [69]. [d] The best solution reported in [79]. [e] The best solution reported in [61].

Table 4. Computational results with double-expanded large-size incidences.

Original Problem	Expanded Problem Size									Proposed Approach		
	$n(q)$	m	t	p	U	e	EEs	Voids	GE (%)	CPU (s)
1. [70]	60	60	178	12	7	302	2	90	**76.53**	0.28 + 0.085
2. [78]	60	140	298	10	6	956	502	386	33.83	0.89 + 0.129
3. [51]	60	70	164	10	7	466	116	76	**64.58**	0.58 + 0.095
4. [51]	80	90	200	14	7	533	81	70	**74.96**	0.77 + 0.101
5. [51]	110	120	248	16	9	822	100	106	**77.80**	0.63 + 0.121

The former incidences are used for comparative purposes along with the results obtained by the reference approaches; the latter incidences are used to show the effectiveness of our approach in solving large-sized GCF incidences. The double-expanded large-size incidences have been produced by following Adil et al.'s data expansion scheme [80], which duplicates rows and columns of an existing intermediate-sized binary PMIM instead of using randomly generated data sets.

For the original problems 6, 11, 22, 23, and 24 in Table 3, extra incidences have been tested for values of p and U different from those used the in the original problems. Some of the original data sets include information on the operation sequences and/or production volumes of parts. Those data sets are slightly changed so that the resulting routing data or PMIMs are suited for the 0–1 GCF problem format. For problem 5, process plans b and c of part type 2 are merged into an identical process plan since they require the same tools. For problems 17 and 19, process plan c of part type 1 has been deleted since it requires the same machines as process plan b. All the GCF incidences and block diagonal solution matrices are available upon request.

Model 3 has been solved using the LINGO 16.0 solver on an ASUS laptop computer including an Intel Core i7-9750H processor running at 2.6 GHz with 16 GB RAM under the Windows 10 operating system. The procedures for part assignment and improperly assigned EM/RM reassignment were coded in C + + and implemented using the Visual Studio 2015. The execution times taken to implement all the reference CF methods selected for comparative purposes will not be reported since each incidence selected in our experiment was solved using different machines and compilers; only the execution time taken to implement our approach is reported for future comparison.

4. Discussion

In Table 3, the bold-faced GEs indicate the best GEs obtained from the corresponding reference approach and proposed PMP hard computing approach. Of these 39 test incidences, the best published values of GE under the corresponding conditions of p and U for four incidences of problem 22 and three extra test incidences of problems 11, 23, and 24 have not been reported in the literature. Therefore, these instances are not used for performance comparison between the reference approaches and the proposed PMP approach and are reported only for the purpose of future comparison.

According to the computational result presented in Table 3, for six incidences (6/32 = 18.8%) of problems 1, 2, 4, 7, 10, and 26, both approaches yield the same GEs. For nine incidences (9/32 = 28.1%) of problems 3, 6, 8, 10, 13, 15, and 20, the reference approaches yield better GEs than the PMP approach. For 17 incidences (17/32 = 53.1%) of problems 2, 5, 7, 9, 11, 12, 14, 16, 17, 18, 19, 21, 23, 24, and 25, the PMP approach yields better GEs than the reference approaches. In summary, the proposed PMP hard computing approach absolutely outperforms the existing soft or hard computing approaches by 53.1% for the small to intermediate GCF incidences available in the literature. Furthermore, the proposed PMP hard computing approach finds better alternative solutions which were not found by the reference approaches with values of p and U which were different from the original problems 11, 23, and 24.

Regarding the computing time required to implement model 3, the proposed PMP hard computing approach reached the global optimum within 0.3 s even for a large incidence of 24 including 3025 binary variables. The execution times required to implement subsequent procedures for part assignment and improperly assigned EM/RM reassignment were also negligibly small since those procedures were terminated within 0.1 s for all incidences.

The proposed PMP hard computing approach has been applied to larger GCF incidences in order to show how efficiently the PMP hard computing approach solves large-sized GCF incidences. To the best of our knowledge, problem 24 is the largest open GCF incidence that is available in the literature and therefore can be used for comparative purposes with different CF solution approaches. Some authors [61,81] tested the efficiency of their CF solution approaches with randomly generated incidences in order to show the efficiency of their methods for large-sized incidences. However, they did not provide solutions that show an explicit configuration of machine cells and associated part families. As a result, the best published values of GE for those incidences are not available for benchmark testing with such randomly generated incidences.

To show how efficiently the proposed PMP hard computing approach solves even larger GCF incidences, we have chosen to double-expand some original large incidences instead of using randomly generated incidences. For this purpose, problems 8 and 21 to 24 were selected and expanded. A typical strategy used to randomly generate large-sized matrices is to create an ideal block diagonal structure first and then destroy it gradually by using random flips. The random flipping of the original GCF matrices is performed to change 1s in the diagonal blocks into zeros and zeros in the off-diagonal blocks into 1s [34]. The more flipped the expanded matrices, the less random they become. However, the large GCF incidences generated through double-expansion in our computational experiment are not randomly flipped, meaning that the resulting matrices do not approach completely random ones. By avoiding random flips over the expanded matrices, we can test both the efficiency and robustness of the proposed PMP approach. The only element randomly scrambled is the order of part numbers.

Table 4 shows the double-expanded incidences and computational results with those incidences. The numbers of machines of expanded incidences vary from 60 to 110, and the numbers of binary variables therefore vary from 3600 to 12,100. As far as the author knows, few mathematical models have been implemented to use a hard computing approach to optimally solve large-sized GCF incidences with such a large number of binary variables. According to Table 4, model 3 reaches the global optimum for all the expanded incidences within one second. Subsequent procedures for part assignment and improperly assigned EM/RM reassignment were terminated within 0.2 s. Clearly, this shows the computational efficiency of the proposed PMP hard computing approach for large-sized GCF problems. Furthermore, except for incidence 2, which was double-expanded for problem 21, the remaining double-expanded incidences even yielded better GEs than the original incidence before double-expansion. This reveals the robustness of the proposed PMP approach when applied to large-sized GCF problems.

5. Conclusions

In this paper, we have proposed an effective hard computing technique to solve large-sized GCF incidences to the global optimum within a short computation time. The distinctive contributions of this paper are summarized as follows:

- Two new linear 0–1 mathematical models have been formulated to solve the GCF problem: an exact model to directly maximize the GE and a PMP-type model to indirectly maximize the GE;
- It seems still to be very difficult to use a hard computing approach to optimally solve large-sized GCF incidences of an exact mathematical model optimizing the objective function directly.
- The PMP-type model, as an alternative to optimizing the objective function of GCF indirectly, can use hard computing techniques to solve large-sized GCF incidences optimally in a very short computation time. Even a large GCF incidence containing 12,100 binary variables can be solved within only one second to the global optimum, and the machine cells can be identified.
- The computation time for subsequent part assignment procedures finding the associated part families and refinement steps and reassigning improperly assigned EMs/RMs is also negligibly small, even if some soft computing heuristics can work faster, as shown by Goldengorin et al. [34].
- The solution quality based on the proposed hard computing approach compares favorably to existing GCF solution approaches.
- The GCF incidences collected in our computation experiment can be used as a standard data set for subsequent benchmark tests in the future.

A limitation of the paper should be addressed. Unlike Danilovic & Ilic's part assignment rule [50], the part assignment procedure proposed in this paper does not guarantee the optimal part assignment since it is heuristic. The solution quality due to the assignment of a part depends on the assignment of other parts. Therefore, establishing sufficient conditions for optimal part assignment in the GCF problem will be a very attractive future research issue.

Funding: This research received no external funding.

Acknowledgments: The author deeply appreciates the valuable comments and suggestions of three anonymous reviewers who led to the improvement of the earlier manuscript.

Conflicts of Interest: The author declares no conflict of interest.

References

1. Burbidge, J.L. The new approach to production. *Prod. Eng.* **1961**, *40*, 769–794. [CrossRef]
2. Wemmerlöv, U.; Johnson, D.J. Empirical findings on manufacturing cell design. *Int. J. Prod. Res.* **2000**, *38*, 481–507. [CrossRef]
3. Ballakur, A.; Steudel, H.J. A within–cell utilization based heuristic for designing cellular manufacturing systems. *Int. J. Prod. Res.* **1987**, *25*, 639–655. [CrossRef]
4. Batsyn, M.V.; Batsyna, E.K.; Bychkov, I.S. On NP-completeness of the cell formation problem. *Int. J. Prod. Res.* **2019**, in press. [CrossRef]
5. Papaioannou, G.; Wilson, J.M. The evolution of cell formation problem methodologies based on recent studies (1997–2008): Review and directions for future research. *Eur. J. Oper. Res.* **2010**, *206*, 509–521. [CrossRef]
6. Lozano, S.; Guerrero, F.; Eguia, I.; Onieva, L. Cell design and loading in the presence of alternative routing. *Int. J. Prod. Res.* **1999**, *37*, 3289–3304. [CrossRef]
7. Islam, K.M.S.; Sarker, B.R. A similarity coefficient measure and machine-parts grouping in cellular manufacturing systems. *Int. J. Prod. Res.* **2000**, *38*, 699–720. [CrossRef]
8. Heragu, S.S.; Chen, J.S. Optimal solution of cellular manufacturing system design: Benders' decomposition approach. *Eur. J. Oper. Res.* **1998**, *107*, 175–192. [CrossRef]
9. Borrero, J.S.; Gillen, C.; Prokopyev, O.A. Fractional 0–1 programming: Applications and algorithms. *J. Glob. Optim.* **2017**, *69*, 255–282. [CrossRef]
10. Kumar, C.S.; Chandrasekharan, M.P. Grouping efficacy: A quantitative criterion for goodness of block diagonal forms of binary matrices in group technology. *Int. J. Prod. Res.* **1990**, *28*, 233–243. [CrossRef]
11. Elbenani, B.; Ferland, J.A. Cell formation problem solved exactly with the Dinkelbach algorithm. *CIRRET* **2012**, *7*, 1–14.
12. Elbenani, B.; Ferland, J.A. An exact method for solving the manufacturing cell formation problem. *Int. J. Prod. Res.* **2012**, *50*, 4038–4045. [CrossRef]
13. Bychkov, I.; Batsyn, M.; Pardalos, P.M. Exact model for the cell formation problem. *Optim. Lett.* **2014**, *8*, 2203–2210. [CrossRef]
14. Brusco, M.J. An exact algorithm for maximizing grouping efficacy in part-machine clustering. *IIE Trans.* **2015**, *47*, 653–671. [CrossRef]
15. Pinheiro, R.G.S.; Martins, I.C.; Protti, F.; Ochi, L.S.; Simonetti, L.G.; Subramanian, A. On solving manufacturing cell formation via bicluster editing. *Eur. J. Oper. Res.* **2016**, *254*, 769–779. [CrossRef]
16. Utkina, I.; Batsyn, M.V.; Batsyna, E.K. A branch and bound algorithm for a fractional 0-1 programming problem. In *Discrete Optimization and Operations Research*; Kochetov, Y., Khachay, M., Beresnev, V., Nurminski, E., Pardalos, P., Eds.; Lecture Notes in Computer Science; Springer: New York, NY, USA, 2016; Volume 9869, pp. 244–255.
17. Bychkov, I.; Batsyn, M. An efficient exact model for the cell formation problem with a variable number of production cells. *Comput. Oper. Res.* **2018**, *91*, 112–120. [CrossRef]
18. Utkina, I.; Batsyn, M.V.; Batsyna, E.K. A branch-and-bound algorithm for the cell formation problem. *Int. J. Prod. Res.* **2018**, *56*, 3262–3273. [CrossRef]
19. Goncalves, J.F.; Resende, M.G.C. An evolutionary algorithm for manufacturing cell formation. *Comput. Ind. Eng.* **2004**, *47*, 247–273. [CrossRef]
20. Hakimi, S.L. Optimum location of switching centers and the absolute centers and medians of a graph. *Oper. Res.* **1964**, *12*, 450–459. [CrossRef]
21. Hakimi, S.L. Optimum distribution of switching centers and some graph related theoretic problems. *Oper. Res.* **1965**, *13*, 462–475. [CrossRef]
22. Shi, J.; Zheng, X.; Jiao, B.; Wang, R. Multi-scenario cooperative evolutionary algorithm for the β-Robust p-median problem with demand uncertainty. *Appl. Sci.* **2019**, *9*, 4174. [CrossRef]
23. ReVelle, C.S.; Swain, R.W. Central Facilities location. *Geogr. Anal.* **1970**, *2*, 30–42. [CrossRef]
24. Balinski, M. Integer programming: Methods, uses, computations. *Manag. Sci.* **1965**, *12*, 253–313. [CrossRef]
25. Efroymson, M.A.; Ray, T.L. A branch-bound algorithm for plant location. *Oper. Res.* **1966**, *14*, 361–368. [CrossRef]
26. Church, R.L. COBRA: A new formulation of the classic p-median location problem. *Ann. Oper. Res.* **2003**, *122*, 103–120. [CrossRef]

27. Goldengorin, B.; Ghosh, D.; Sierksma, G. Branch and peg algorithms for the simple plant location problem. *Comput. Oper. Res.* **2003**, *30*, 967–981. [CrossRef]

28. Goldengorin, B.; Tijssen, G.A.; Ghosh, D.; Sierksma, G. Solving the simple plant location problems using a data correcting approach. *J. Global Optim.* **2003**, *25*, 377–406. [CrossRef]

29. Church, R.L. BEAMR: An exact and approximate model for the p-median problem. *Comput. Oper. Res.* **2008**, *35*, 417–426. [CrossRef]

30. Elloumi, S. A tighter formulation of the p-median problem. *J. Glob. Optim.* **2010**, *19*, 69–83. [CrossRef]

31. García, S.; Labbé, M.; Marín, A. Solving large p-median problems with a radius formulation. *INFORMS J. Comput.* **2011**, *23*, 546–556. [CrossRef]

32. Kusiak, A. The part families problem in flexible manufacturing systems. *Ann. Oper. Res.* **1985**, *3*, 279–300. [CrossRef]

33. Kusiak, A. The Generalized group technology concept. *Int. J. Prod. Res.* **1987**, *25*, 561–569. [CrossRef]

34. Goldengorin, B.; Krushinsky, D.; Slomp, J. Flexible PMP approach for large-size cell formation. *Oper. Res.* **2012**, *60*, 1157–1166. [CrossRef]

35. Sankran, S.; Kasilingam, R.G. An integrated approach to cell formation and part routing in group technology. *Eng. Optim.* **1990**, *16*, 235–245. [CrossRef]

36. Kaparthi, S.; Suresh, N.C. Performance of selected part-machine grouping techniques for data sets of wide ranging sizes and imperfection. *Decis. Sci.* **1994**, *25*, 515–539. [CrossRef]

37. Lee, H.; Garcia-Diaz, A. Network flow procedures for the analysis of cellular manufacturing systems. *IIE Trans.* **1996**, *28*, 333–345. [CrossRef]

38. Viswanathan, S. A new approach for solving the p-median problem in group technology. *Int. J. Prod. Res.* **1996**, *34*, 2691–2700. [CrossRef]

39. Deutsch, S.J.; Freeman, S.F.; Helander, M. Manufacturing cell formation using an improved *p*-median model. *Comput. Ind. Eng.* **1998**, *34*, 135–146. [CrossRef]

40. Won, Y. New *p*-median approach to cell formation with alternative process plans. *Int. J. Prod. Res.* **2000**, *38*, 229–240. [CrossRef]

41. Won, Y. Two-phase approach to GT cell formation using efficient *p*-median formulations. *Int. J. Prod. Res.* **2000**, *38*, 1601–1613. [CrossRef]

42. Won, Y.; Lee, K.C. Modified *p*-median approach for efficient GT cell formation. *Comput. Ind. Eng.* **2004**, *46*, 495–510. [CrossRef]

43. Ashayeri, J.; Heuts, R.; Tammel, B. A modified simple heuristic for the p-median problem, with facilities design applications. *Robot. CIM Int. Manuf.* **2005**, *21*, 451–464. [CrossRef]

44. Won, Y.; Currie, K.R. An effective *p*-median model considering production factors in machine cell/part family formation. *J. Manuf. Syst.* **2006**, *25*, 58–64. [CrossRef]

45. Süer, G.A.; Huang, J.H.; Maddisetty, S. Design of dedicated, shared and remainder cells in a probabilistic demand environment. *Int. J. Prod. Res.* **2010**, *48*, 5613–5646. [CrossRef]

46. Egilmez, G.; Süer, G.A.; Huang, J. Stochastic cellular manufacturing system design subject to maximum acceptable risk level. *Comput. Ind. Eng.* **2012**, *63*, 842–854. [CrossRef]

47. Egilmez, G.; Süer, G.A. The impact of risk on the integrated cellular design and control. *Int. J. Prod. Res.* **2014**, *52*, 1455–1478. [CrossRef]

48. Won, Y.; Logendran, R. Effective two-phase *p*-median approach for the balanced cell formation in the design of cellular manufacturing system. *Int. J. Prod. Res.* **2015**, *53*, 2730–2750. [CrossRef]

49. Alhawari, O.; Süer, G. Modified p-median model with minimum threshold for average family similarity. *Procedia Manuf.* **2019**, *39*, 1048–1056. [CrossRef]

50. Danilovic, M.; Ilic, O. A novel hybrid algorithm for manufacturing cell formation problem. *Expert Syst. Appl.* **2019**, *135*, 327–350. [CrossRef]

51. Kao, Y.; Chen, C.C. Automatic clustering for generalised cell formation using a hybrid particle swarm optimisation. *Int. J. Prod. Res.* **2014**, *52*, 3466–3484. [CrossRef]

52. Riccardo, M.; Riccardo, A.; Marco, B. Similarity-based cluster analysis for the cell formation problem. In *Operations Management Research and Cellular Manufacturing Systems: Innovative Methods and Approaches*; Modrák, V., Pandian, R.S., Eds.; IGI Global: Hershey, PA, USA, 2012; pp. 140–163.

53. Li, X.; Baki, M.; Aneja, Y. An ant colony optimization metaheuristic for machine-part cell formation problems. *Comput. Oper. Res.* **2010**, *37*, 2071–2081. [CrossRef]

54. Vin, E.; Delchambre, A. Generalized cell formation: Iterative versus simultaneous resolution with grouping genetic algorithm. *J. Intell. Manuf.* **2014**, *25*, 1113–1124. [CrossRef]
55. Sarker, B.R. Measures of grouping efficiency in cellular manufacturing systems. *Eur. J. Oper. Res.* **2001**, *130*, 588–611. [CrossRef]
56. Sarker, B.R.; Khan, M. A comparison of existing grouping efficiency measures and a new weighted grouping efficiency measure. *IIE Trans.* **2001**, *33*, 11–27. [CrossRef]
57. Mahdavi, I.; Javadi, B.; Fallah-Alipour, K.; Slomp, J. Designing a new mathematical model for cellular manufacturing system based on cell utilization. *Appl. Math. Comput.* **2007**, *190*, 662–670. [CrossRef]
58. Paydar, M.M.; Saidi-Mehrabad, M. A hybrid genetic-variable neighborhood search algorithm for the cell formation problem based on grouping efficacy. *Comput. Oper. Res.* **2013**, *40*, 980–990. [CrossRef]
59. Won, Y.; Kim, S. Multiple criteria clustering algorithm for solving the group technology problem with multiple process routings. *Comput. Ind. Eng.* **1997**, *32*, 207–220. [CrossRef]
60. Mukattash, A.M.; Adil, M.B.; Tahboub, K.K. Heuristic approaches for part assignment in cell formation. *Comput. Ind. Eng.* **2002**, *42*, 327–341. [CrossRef]
61. Wu, T.H.; Chung, S.H.; Chang, C.C. Hybrid simulated annealing algorithm with mutation operator to the cell formation problem with alternative process routings. *Expert Syst. Appl.* **2009**, *36*, 3652–3661. [CrossRef]
62. Shiyasa, C.R.; Pillaia, V.M. Cellular manufacturing system design using grouping efficacy-based genetic algorithm. *Int. J. Prod. Res.* **2014**, *52*, 3504–3517. [CrossRef]
63. Won, Y. P-median approach for the large-size multi-objective generalized cell formation. *Korean Manag. Sci. Rev.* **2018**, *35*, 35–55. [CrossRef]
64. Al-Zawahreha, A.; Dahmanib, N.; Alethem, K.A.; Mukattash, A. Sensitivity analysis of the impact of part assignment in cellular manufacturing systems. *Decis. Sci. Lett.* **2019**, *8*, 109–120. [CrossRef]
65. Nagi, R.; Harhalakis, G.; Proth, J. Multiple routings and capacity consideration in group technology applications. *Int. J. Prod. Res.* **1990**, *28*, 2243–2257. [CrossRef]
66. Moon, Y.B.; Chi, S.C. Generalized part family formation using neural network techniques. *J. Manuf. Syst.* **1992**, *11*, 149–159. [CrossRef]
67. Kasilingam, R.G.; Lashkari, R.S. Cell formation in the presence of alternate process plans in flexible manufacturing systems. *Prod. Plan. Control.* **1991**, *2*, 135–141. [CrossRef]
68. Logendran, R.; Ramakrishna, P.; Sriskandarajah, C. Tabu search-based heuristics for cellular manufacturing systems in the presence of alternative process plans. *Int. J. Prod. Res.* **1994**, *32*, 273–297. [CrossRef]
69. Adil, G.K.; Rajamani, D.; Strong, D. Cell formation considering alternate routeings. *Int. J. Prod. Res.* **1996**, *34*, 1361–1380. [CrossRef]
70. Lee, M.K.; Luong, H.S.; Abhary, K. A genetic algorithm based cell design considering alternative routing. *Comput. Integr. Manuf.* **1997**, *10*, 93–107. [CrossRef]
71. Han, J. Formation of Part and Machine Cells with Consideration of Alternative Machines. Master's Thesis, Ohio University, Athens, OH, USA, 1998.
72. Sofianopoulou, S. Manufacturing cells design with alternative process plans and/or replicate machines. *Int. J. Prod. Res.* **1999**, *37*, 707–720. [CrossRef]
73. Gen, M.; Cheng, R. Manufacturing cell design. In *Genetic Algorithms and Engineering Optimization*; John Wiley & Sons: New York, NY, USA, 2000; pp. 390–450.
74. Aktürk, M.S.; Turkcan, A. Cellular manufacturing system design using a holonistic approach. *Int. J. Prod. Res.* **2000**, *38*, 2327–2347. [CrossRef]
75. Yin, Y.; Yasuda, K. Manufacturing cells' design in consideration of various production factors. *Int. J. Prod. Res.* **2002**, *40*, 885–906. [CrossRef]
76. Solimanpur, M.; Vrat, P.; Shankar, R. A multi-objective genetic algorithm approach to the design of cellular manufacturing systems. *Int. J. Prod. Res.* **2004**, *42*, 1419–1441. [CrossRef]
77. Bhide, P.; Bhandwale, A.; Kesavadas, T. Cell formation using multiple process plans. *J. Intell. Manuf.* **2005**, *16*, 53–65. [CrossRef]
78. Hu, L.; Yasuda, K. Minimising material handling cost in cell formation with alternative processing routes by grouping genetic algorithm. *Int. J. Prod. Res.* **2006**, *44*, 2133–2167. [CrossRef]
79. Hwang, H.; Ree, P. Routes selection for the cell formation problem with alternative part process plans. *Comput. Ind. Eng.* **1996**, *30*, 423–431. [CrossRef]

80. Adil, G.K.; Rajamani, D.; Strong, D. Assignment allocation and simulated annealing algorithms for cell formation. *IIE Trans.* **1997**, *29*, 53–67. [CrossRef]
81. Wu, T.H.; Chen, J.F.; Yeh, J.Y. A decomposition approach to the cell formation problem with alternative process plans. *Int. J. Adv. Manuf. Technol.* **2004**, *24*, 834–840. [CrossRef]

 *applied sciences*

Article

Analytic Hierarchy Process and Multilayer Network-Based Method for Assembly Line Balancing

László Nagy, Tamás Ruppert and János Abonyi *

MTA-PE "Lendület" Complex Systems Monitoring Research Group, University of Pannonia, P.O. Box 158, H-8200 Veszprém, Hungary; laszlo.nagy@fmt.uni-pannon.hu (L.N.); ruppert@abonyilab.com (T.R.)
* Correspondence: janos@abonyilab.com

Received: 30 April 2020; Accepted: 2 June 2020; Published: 5 June 2020

Abstract: Assembly line balancing improves the efficiency of production systems by the optimal assignment of tasks to operators. The optimisation of this assignment requires models that provide information about the activity times, constraints and costs of the assignments. A multilayer network-based representation of the assembly line-balancing problem is proposed, in which the layers of the network represent the skills of the operators, the tools required for their activities and the precedence constraints of their activities. The activity–operator network layer is designed by a multi-objective optimisation algorithm in which the training and equipment costs as well as the precedence of the activities are also taken into account. As these costs are difficult to evaluate, the analytic hierarchy process (AHP) technique is used to quantify the importance of the criteria. The optimisation problem is solved by a multi-level simulated annealing algorithm (SA) that efficiently handles the precedence constraints. The efficiency of the method is demonstrated by a case study from wire harness manufacturing.

Keywords: assembly-line balancing; multi-objective optimization; simulated annealing; multilayer network

---

1. Introduction

Production line-based assembly lines are still the most widely applied manufacturing systems [1]. Assembly-Line Balancing (ALB) [2] deals with the balanced assignment of tasks to the workstations, resulting in the optimisation of a given objective function without violating precedence constraints [3]. The efficiency of these optimisation tasks is mostly determined by the model of the manufacturing process represented [4].

The concept of Industry 4.0 has already had a significant influence on how production and assembly lines are designed [5] and managed [6]. The requirement of practical design at a high automation level ensures that sensors and equipment can be integrated in a fast, secure, and reliable way. In our research, we study how the interoperability capabilities of Industry 4.0 solutions can be improved and how the efficiency of solution development can be increased.

As Internet of Things-based products and processes are rapidly developing in the industry, there is a need for solutions that can support their fast and cost-effective implementation. There is a need for further standardization to achieve more flexible connectivity, interoperability, and fast application-oriented development; furthermore, advanced model-based control and optimisation functions require a better understanding of sensory and process data [7].

Usually, production systems include multiple subsystems and layers of connectivity. Thus, although research-based solutions for classical operations typically use a graph-based representation of problems and flow-based optimisation algorithms, conventional single-layer networks quickly become incapable of representing the complexity and connectivity of all the details of the production line. With the overlapping data in Industry 4.0 solutions, it should be highlighted that multilayer networks are expected to be the most suitable options for representing modern production lines. The concept of a multilayer network was developed to represent multiple types of relationships [8], and these models have been proven to be applicable to the representation of complex connected systems [9]. Network-based models can also represent how products, resources and operators are connected [10], which is beneficial in terms of solving manufacturing cell formation problems [11]. This work demonstrates how the multilayer network representation of production lines can be utilised in line balancing.

In the proposed novel network model, the layers represent the skills of the operators, the tools required for the activities, and the precedence constraints of the activities. At the same time, a multi-objective optimisation algorithm designs the assignment of activities and operators to network layers. The proposed multilayer network approach supports the intuitive formulation of multi-objective line balancing optimisation tasks. Besides the utilisation of operators, the utilisation of the tools and the number of skills of an operator are also taken into account. The main advantage of the proposed network-based representation is that the latter two objectives are directly related to the structural properties of the optimised network.

Line balancing is a non-deterministic polynomial-time hard (NP-hard) optimisation problem, which means that the computational complexity of the optimisation problem increases exponentially as the dimensions of the problem increase. This challenge explains why numerous meta heuristic approaches such as simulated annealing (SA) [12,13], hybrid heuristic optimisation [13,14], chance-constrained integer programming [15], recursive and dynamic programming [16], as well as tabu search [17] have been utilised in the field of production management. Fuzzy set theory provides a transparent and interpretable framework to represent the uncertainty of information and solve the ALB problem [18,19]. Among the wide range of heuristic methods capable of achieving reasonable solutions [20], SA is the most widely used search algorithm [21], so it has already been applied to solve mixed and multi-model line-balancing problems [22].

To deal with the complexity of this problem, an SA algorithm was also developed. The proposed algorithm utilises a unique problem-oriented sequential representation of the assignment problem and applies a neighbourhood-search strategy that generates feasible task sequences for every iteration. Since the algorithm has to handle multiple aspects of line balancing, the analytic hierarchy process (AHP) technique is used to quantify the importance of the objectives, also known as Saaty's method [23]. AHP is a method for multi-criteria decision-making which is used to evaluate complex multiple criteria alternatives involving subjective judgments [24]. This method is a useful and practical approach to solving complex and unstructured decision-making problems by calculating the relative importance of the criteria based on the pairwise comparison of different alternatives [25]. The method has been widely applied thanks to its effectiveness and interpretability. Two papers were found in which it has already been applied to determine the cost function of multi-objective SA optimisation problems. In the first case study of supplier selection, AHP was applied to calculate the weight of every objective by applying the Taguchi method [26] (Adaptive Tabu Search Algorithm—ATSA [27]). In contrast, in the second report, this concept was applied to the maintenance of road infrastructure [28].

The novelties of the work are the following:

- In Section 2, the main problem formulation is introduced, including the multilayer network representation of multiple aspects concerning the balancing of production lines, and the details of the objective function.
- In Section 3, an SA algorithm will be introduced based on a novel sequential representation of the line-balancing problem and the search algorithm that guarantees the fulfilment of the precedence constraints.
- Section 4 demonstrates how AHP can be used to aggregate the multi layer network-represented objectives of the line-balancing problem for SA.

2. Problem Formulation

In this section, the problem formulation is presented. First, the representation of production line modelling with the multilayer network is introduced in Section 2.1. The details of the minimised function and its AHP-based aggregation are given in Section 2.2.

2.1. Multilayer Network-Based Representation of Production Lines

The proposed network model of the production line consists of a set of bipartite graphs that represent connections between operators, $\mathbf{o} = \{o_1, \ldots, o_{N_o}\}$; skills of the operators needed to perform the given activity, $\mathbf{s} = \{s_1, \ldots, s_{N_s}\}$; equipment, $\mathbf{e} = \{e_1, \ldots, e_{N_e}\}$; activities (operations), $\mathbf{a} = \{a_1, \ldots, a_{N_a}\}$; and the precedence constraints between activities, $\mathbf{a}' = \{a_1', \ldots, a_{N_a}'\}$. The relationships between these sets are defined by bipartite graphs $G_{i,j} = (O_i, O_j, E_{i,j})$ represented by $\mathbf{A}[O_i, O_j]$ biadjacency matrices, where O_i and O_j denote a general representation of the sets of objects, such that $O_i, O_j \in \{\mathbf{s}, \mathbf{e}, \mathbf{a}', \mathbf{a}, \mathbf{o}\}$.

The edges of these bipartite networks represent structural relationships; e.g., the biadjacency matrix $\mathbf{A}[\mathbf{a}, \mathbf{a}']$ represents the precedence constraints or $\mathbf{A}[\mathbf{a}, \mathbf{o}]$ represents the assignments of activities to operators. Moreover, the edge weights can be proportional to the number of shared components/resources or time/cost (see Table 1) [10].

Table 1. Definition of the biadjacency matrices of the bipartite networks used to illustrate how a multidimensional network can represent a production line.

	Nodes	Description
W	Activity (**a**)–operator (**o**)	Operator assigned to the activity
S	Activity (**a**)–skill (**s**)	Skill/education required for a category of activities
E	Activity (**a**)–equipment (**e**)	Equipment which is in use in an activity
A′	Activity (**a**)–activity (**a′**)	Precedence constraint between activities

As can be seen in Figure 1, these bipartite networks are strongly connected. The proposed model can be considered as an interacting or interconnected network [8], where bipartite networks define the layers. Since different types of connections are defined, the model can also be handled as a multidimensional network. As illustrated in Figure 2, when relationships between the sets O_i and O_j are not directly defined, it is possible to evaluate the relationship between their elements $o_{i,k}$ and $o_{j,l}$ in terms of the number of possible paths or the length of the shortest path between these nodes [10].

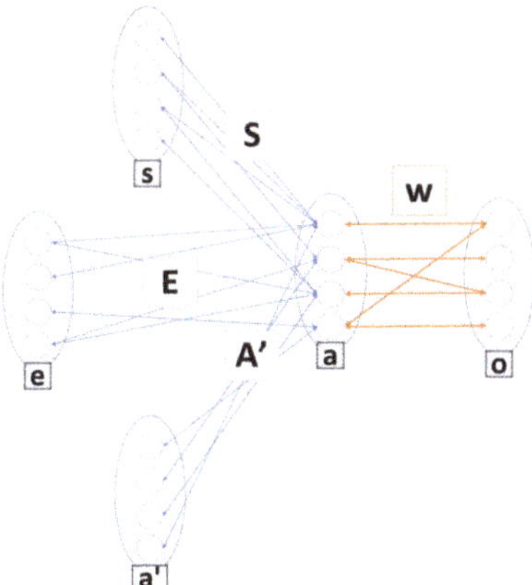

Figure 1. Illustrative network representation of a production line. The definitions of the symbols are given in Table 1.

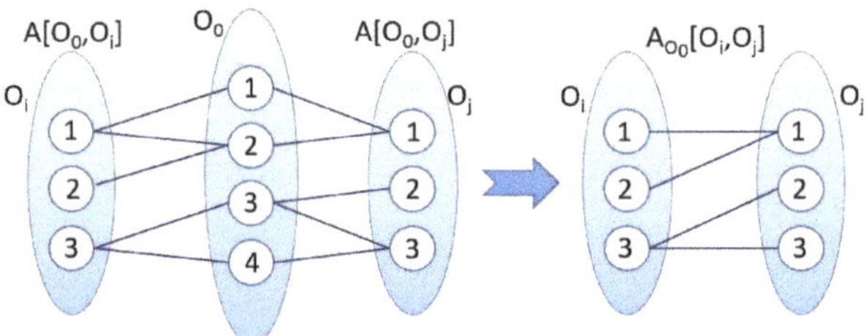

Figure 2. Projection of a property connection.

In the case of connected unweighted multipartite graphs, the number of paths intersecting the set O_0 can be easily calculated based on the connected pairs of bipartite graphs as follows:

$$\mathbf{A}_{O_0}[O_i, O_j] = \mathbf{A}[O_0, O_i]^T \times \mathbf{A}[O_0, O_j]. \tag{1}$$

In the proposed network model, the optimisation problem is defined by the allocation of tasks that require the allocation of different skills and tools to an operator that might necessitate extra training, labour and investment costs. The main benefit of the proposed network representation is that these costs can be directly evaluated based on the products of the biadjacency matrices **S** and **W**:

$$\mathbf{A}[\mathbf{s}, \mathbf{o}] = \mathbf{A}[\mathbf{s}, \mathbf{a}]\mathbf{A}[\mathbf{a}, \mathbf{o}] = \mathbf{SW}. \tag{2}$$

The resultant network $\mathbf{A}[\mathbf{s}, \mathbf{o}]$ represents how many times a given skill should be utilised by an operator, while its unweighted version $\mathbf{A}^u[\mathbf{s}, \mathbf{o}]$ models which skills the operators should have.

The design of the presented network model is based on the analysis of the semantically standardized models of production lines [29], and the experience gained in the development project connected to the proposed case study. The details of the multilayer network-based modelling of a wire-harness production process can be found in [10].

2.2. The Objective Function

A simple assembly line balancing problem (SALBP) assigns N_a tasks/activities to N_o workstations/operators. Each activity is assigned to precisely one operator, and the sum of task times of workstation should be less or equal to the cycle time T_c [30]. Precedence relations between activities must not be violated [31]. There are two important variants of this problem [32]: SALBP-1 aims to minimise N_o for a given T_c, while the goal of SALBP-2 is to minimise T_c for a predefined N_o [1,33,34]. In this paper, the SALBP-2 problem was investigated and extended to include the following skill and equipment-related objective functions:

Station-time-related objective: The main objective of line balancing is to minimise the cycle time T_c, which is equal to the sum of the maximum of the station times T_j. The utilisation of the whole assembly line can be calculated as follows:

$$T_c = \arg\max_j T_j = \sum_{i=1}^{N_a} w_{i,j} t_i, \tag{3}$$

where t_i represents the elementary activity times of the a_i-th activity.

As the theoretical minimum of T_c is

$$T_c^* = \frac{\sum_{i=1}^{N_a} t_i}{N_o}, \tag{4}$$

the following ratio evaluates the efficiency of the balancing of the activity times:

$$Q_T(\pi) = \frac{T_c^*}{T_c} = \frac{\frac{\sum_{i=1}^{N_a} t_i}{N_o}}{\sum_{i=1}^{N_a} w_{i,j} t_i}. \tag{5}$$

Skill-related (training) objective: The training cost is calculated with the node degree between skill-operator elements $s - o$. The number of skills N_s is divided by the sum of the node degrees k_i between sub-networks s and o in the multilayer representation:

$$Q_S(\pi) = \frac{N_s}{\sum_i^{s-o,o} k_i}. \tag{6}$$

Equipment-related objective function: The equipment cost is calculated with the node degree between equipment-operator elements $e - o$. The number of pieces of equipment N_e is divided by the sum of the node degrees k_i between sub-networks e and o in the multilayer representation:

$$Q_E(\pi) = \frac{N_e}{\sum_i^{e-o,o} k_i}. \tag{7}$$

Since the importance of these objectives is difficult to quantify, a pairwise comparison is used to evaluate their relative importance, and the analytic hierarchy process (AHP) is used to determine the weights λ in the objective function:

$$Q(\pi) = \lambda_1 Q_T(\pi) + \lambda_2 Q_S(\pi) + \lambda_3 Q_E(\pi), \tag{8}$$

where $Q_T(\pi) \in [0, 1]$ represents the balance of the production line, and $Q_S(\pi) \in [0, 1]$ and $Q_E(\pi) \in [0, 1]$ measure the efficiency of how the skills and tools are utilised, respectively.

The application of AHP-based weighting is beneficial to integrate the normalised values of the easy to evaluate station-time and equipment-related objectives, and the less specific training-related costs. Although the pairwise comparison of the importance of these objectives and cost-items is subjective, the consistency of the comparisons can be evaluated based on the numerical analysis of the resulted comparison matrices (which will be shown in the next section), which clarifies the reason for our choice of AHP as an ideal tool to extract expert knowledge for the formalisation of the cost function.

3. Simulated Annealing-Based Line-Balancing Optimization

This section presents the proposed optimisation algorithm. The representation of the SA problem is introduced in Section 3.1. Section 3.2 discusses how the precedence constraints of the activities are represented, while Section 3.3 presents how the assignment of activities to operators is formulated by a sequencing problem that can be efficiently solved by the proposed simulated annealing algorithm.

3.1. Representation of the Problem

In the proposed network representation (Figure 1), the assignment of activities to operators is defined by the elements $w_{i,j}$ of the matrix **W** that represent the ith activity assigned to the jth operator. Instead of the direct optimisation of these $N_a \times N_o$ elements, a sequence $N_\pi = N_a + N_o - 1$ is optimised, where N_a represents the number of activities and N_o denotes the number of operators.

The concept of sequence-based allocation is illustrated in Figure 3, where the horizontal axis represents the fixed order of the operators o_j and the vertical axis stands for the activities a_i, where $\pi(i)$ represents the index of the activity by the ith sequence number. The ordered activities are assigned to the operators by $N_o - 1$ boundary elements, represented as $a_{\pi(i)} = *$, which ensure that the next activity in the sequence is assigned to the following operator.

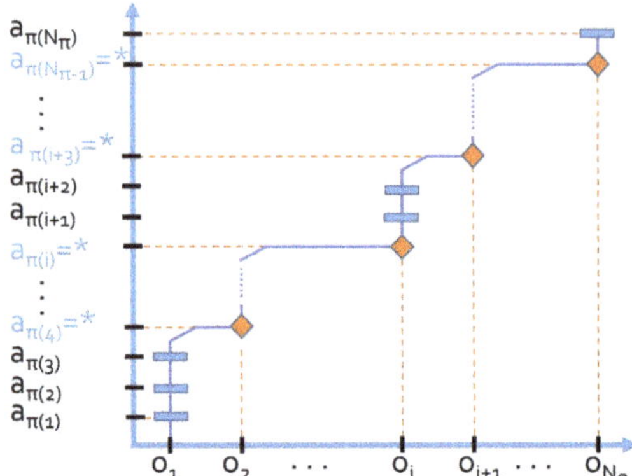

Figure 3. Illustration of the sequencing method. The activities are separated into the different groups of activities that are assigned to different operators.

3.2. Handling Precedence Constraints

In addition to these three objectives of the simulated production line, a so-called soft limit is also defined, which is the amount of the unaccomplished precedence of the activities (A'). This limitation of the order with regard to the activities is stored in the multilayer network.

The completion of a task is a precondition for the start of another because tasks depend on other tasks. The π sequence has some constraining condition and cannot be entirely arbitrary. The precedence graph is used to represent these dependencies in SALBP [1,35,36]. Figure 4 shows a problem from a well-known example by Jackson [37] with $N_a = 11$ tasks, where task 7 requires tasks 3–5 to be completed directly (direct predecessor) and task 1 indirectly (indirect predecessor). The precedence graph can be described by matrix $A'(i,j)$, $i,j = 1,2,\ldots,N_a$, where $A'(i,j) = 1$ if task i is the direct predecessor of task j, otherwise, it is 0 [32]. The precedence graph is partially ordered if tasks cannot be performed in parallel. It must be determined whether a permutation $\pi = (\pi_1, \pi_2, \ldots, \pi_{N_a})$ is feasible or not according to the precedence constraint.

Based on the transitive closure A^* of A', π is feasible if $A^*(p_j, p_i) = 0, \forall i,j, i < j$; otherwise, π is infeasible [32]. A sub-sequence $(\pi_i, \pi_{i+1}, \ldots, \pi_j)$, where $i < j$ of π, can be defined by $\pi_{(i:j)}$. For example, a feasible sequence π of the precedence graph in Figure 4 is $\pi = (1,4,3,2,5,7,6,8,9,10,11)$ and $\pi_{(2:4)} = (4,3,2)$ is a sub-sequence of π.

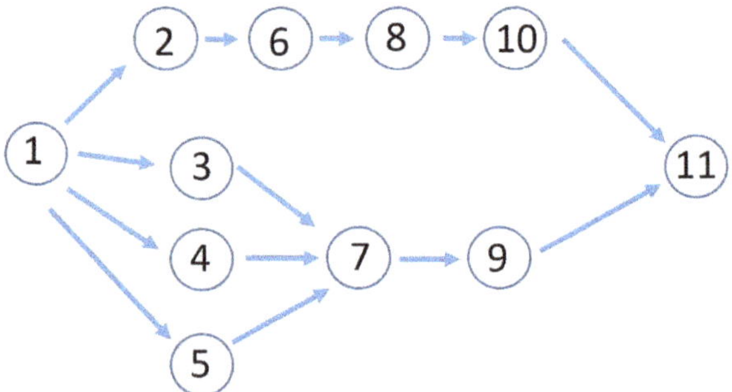

Figure 4. Precedence graph of the example problem taken from Jackson [32,37].

As will be presented in the next subsection, the key idea of the algorithm is that it determines the interchangeable sets of activity pairs and uses these in the guided simulated annealing optimisation.

3.3. Sequence-Based Activity Grouping and Operator Assignment

The optimization algorithm is shown in Algorithm 1 and consists of the following steps:

- Generating the initial feasible sequence.
- SA I: Optimization of the sequences of the activities.

  - SA II (embedded in SA I): in the case of a specific sequence, the activities are assigned to the operators by optimizing the location of the boundary elements in sequence π as has been presented in Figure 3, so SA I uses a cost function that relates to the optimal assignment.

Algorithm 1: Pseudocode of the proposed SA-ALB algorithm

Input: $s, e, Time, Precedence$

Output: $\pi, Q(\pi)$

Annealing: $maxiter, T, T_{max}, T_{min}, m_{max}, m_{min}$

1   $\alpha = (\frac{T_{min}}{T_{max}})^{1/maxiter}, T^1 = T_{max}$

2   $\alpha_m = (\frac{m_{min}}{m_{max}})^{1/maxiter}, m^1 = m_{max}$

3   **Begin**

4   $T = T_{max}; T_{max} =$ maximum value of temperature

5   while $T < T_{min}; T_{min} =$ minimum value of temperature

6

7   Generate initial sequence π, which satisfies the constraints

8   Generate initial placement of the boundary elements

9   Evaluate the cost function $Q(\pi)$, as functions (5), (6) and (7)

10   **for** $i = 1$ *to maxiter* **do**

11

12   Select randomly one interchangeable activity pair

13   Interchange the activities and evaluate the new solution by implementing SA II that optimizes the placement of the boundary elements in this sequence

14   `// SA II is working with the same principle as this main SA I`

15

16     $NewQ(\pi_{new}) = Q(\pi_{new})$

17     $\Delta = Q(\pi_{new}) - Q(\pi)$

18   **if** $\Delta < 0$ **then**

19       $\pi = \pi_{new}$

20       $Q(\pi) = Q(\pi_{new})$

21   **else**

22   **if** $random() < exp(\frac{-\Delta}{T^i})$ **then**

23          $\pi = \pi_{new}$

24          $Q(\pi) = Q(\pi_{new})$

25     $T^{i+1} = \alpha T^i, m^{i+1} = \alpha_i m^i$

26   **End**

---

4. Case Study

This study was inspired by an industrial case study of wire harness manufacturing, where operators work with several tools that perform different activities at workstations to manufacture cables. The problem assumes that it is possible to improve the manufacturing efficiency if the resources, activities, skills and precedence are better designed.

The development of the proposed line-balancing algorithm is motivated by a development project which was defined to improve the efficiency of an industrial wire harness manufacturing process [38].

In this work, a subset of this model is used which consists of 24 activities, five operators, six skills and eight pieces of equipment.

The elementary activity times that influence the line balance were determined based on expert knowledge [39] (see Table 2).

A more detailed description of the activities, pieces of equipment and skills can be found in Tables 2–6.

Table 2. List of the elementary activities that should be allocated in the line balancing problem.

Activity ID	Description	Time
A1	Connector handling	4 s
A2	Connector handling	3 s
A3	Connector handling	2 s
A4	Connector handling	3 s
A5	Insert 1st end + routing	10 s
A6	Insert 2nd end	5 s
A7	Insert 1st end + routing	10 s
A8	Insert 2nd end	5 s
A9	Insert 1st end + routing	10 s
A10	Insert 2nd end	5 s
A11	Insert 1st end + routing	10 s
A12	Insert 2nd end	5 s
A13	Insert 1st end + routing	10 s
A14	Insert 2nd end	5 s
A15	Insert 1st end + routing	10 s
A16	Insert 2nd end	5 s
A17	Insert 1st end + routing	10 s
A18	Insert 2nd end	5 s
A19	Taping	15 s
A20	Taping	13 s
A21	Taping	11 s
A22	Taping	17 s
A23	Taping	15 s
A24	Quality check	10 s

The following tables give a more detailed description of the activities, equipment (Table 3) and skills (Table 4) which are involved in the proposed case study. Furthermore, the activity–equipment (Table 5) and activity–skill (Table 6) connectivity matrices show the requirements of the given base activity.

Table 3. List of equipment that should be allocated in the line balancing problem.

Equipment ID	Description
E1	Connector fixture
E2	Connector fixture
E3	Routing tool
E4	Insertion tool
E5	Taping tool (expert)
E6	Taping tool (normal)
E7	Taping tool (normal)
E8	Repair tool

Table 4. Description of skills that should be used in the studied production process.

Skill ID	Description
S1	Connector handling skill
S2	Insertion (normal) and routing skills
S3	Insertion (expert) skill
S4	Taping (normal) skill
S5	Taping (expert) skill
S6	Quality (expert) skill

Table 5. Activity–equipment matrix that defines which equipment are required to perform a given activity.

	E1	E2	E3	E4	E5	E6	E7	E8
A1	1	1						
A2	1	1						
A3	1	1						
A4	1	1						
A5			1					
A6				1				
A7			1					
A8				1				
A9			1					
A10				1				
A11			1					
A12				1				
A13			1					
A14				1				
A15			1					
A16				1				
A17			1					
A18				1				
A19					1			
A20					1			
A21						1	1	
A22						1	1	
A23						1	1	
A24								1

Table 6. Activity–skill matrix that defines which skills are required to perform a given activity.

	S1	S2	S3	S4	S5	S6
A1	1					
A2	1					
A3	1					
A4	1					
A5		1				
A6			1			
A8			1			
A9		1				
A10			1			
A11		1				
A12			1			
A13		1				
A14			1			
A15		1				

Table 6. *Cont.*

	S1	S2	S3	S4	S5	S6
A7		1				
A16			1			
A17		1				
A18			1			
A19				1		
A20				1		
A21					1	
A22					1	
A23					1	
A24						1

The tables illustrate that the practical implementation of line balancing problems is also influenced by how much equipment is needed for the designed production line and how many skills should be learnt by the operators.

All the collected information is transformed into network layers, as shown in Figure 5. The top of the figure shows the bipartite networks that represent the details of the assignments, while the bottom of the figure represents the tree layers of the network that define the activity–operator, skill–operator, and equipment–operator assignments. As can be seen, this representation is beneficial as it shows how similar operators, skills and equipment can be grouped into clusters.

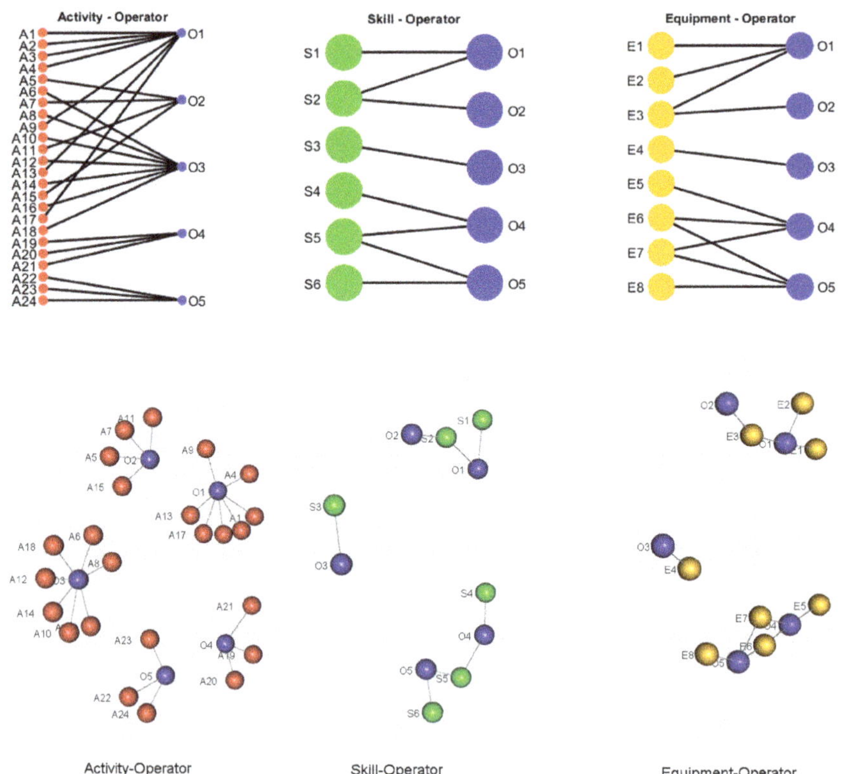

Figure 5. Illustration of the skill–operator and equipment–operator assignments after line balancing.

Although this is not shown in the figure, the weights of the edges represent the costs or benefits of the assignments. The final form of the network is formed based on a multi-objective optimisation of the sets of active edges.

As some of these objectives are difficult to measure, we utilised the proposed AHP-based method to convert the pairwise comparisons of the experts into weights of criteria. The structure of the decision problem is represented in Figure 6. As this figure illustrates, the AHP is used to compare difficult to evaluate equipment and skill assignment costs and the importance of the objectives. The pairwise comparison was performed by a process engineer, and the resulting comparison matrices can be found in Tables 7–9. Based on the analysis of the the eigenvalues of these matrices [25], we found that the evaluations were consistent.

Since the activities cannot be performed in parallel, a precedence graph defines the most crucial question, namely whether a permutation of sequence π is feasible. Based on the transitive closure of the adjacency matrix of the graph, the interchangeable sets of activities can be defined as depicted in Figure 7.

The result of the optimization is shown in Figure 5, which illustrates that the five operators assigned to different skills and pieces of equipment.

The reliability and the robustness of the proposed method are evaluated by ten independent runs of the optimisation algorithm to highlight how the stochastic nature of the proposed method influences the result, as well as showing the effect of the number of operators on the solutions. The aim of the analysis of the independent runs was to estimate the variance of the solutions caused by the stochastic nature of the process and the optimisation algorithm. The sample size of such repeat studies can be determined based on the statistical tests of the estimated variance. In our analysis, we found that ten experiments were sufficient to get a proper estimation of the variance (which is in line with the widely applied ten-fold cross-validation concept).

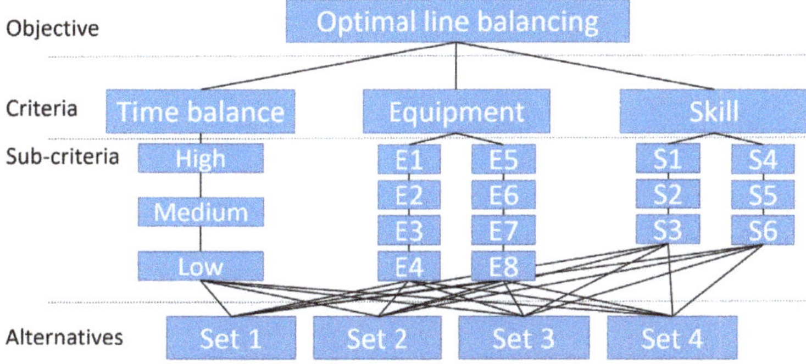

Figure 6. Analytic hierarchy process (AHP) used to solve a decision problem.

Table 7. AHP TOP matrix that shows the relative importance of the objectives. It can be seen that, in this pair-wise comparison, the skill-related cost is evaluated as being twice as important as the equipment-related costs.

	Balancing	Equipment	Skill
Balancing		2.00	4.00
Equipment	0.50		2.00
Skill	0.25	0.50	

Table 8. AHP equipment matrix that shows the relative importance of the equipment.

	E1	E2	E3	E4	E5	E6	E7	E8
E1		1	0.33	0.50	2	3	3	2
E2	1		0.33	0.50	2	3	3	2
E3	3	3		2	5	7	7	5
E4	2	2	0.50		3	5	5	3
E5	0.50	0.50	0.20	0.33		2	2	1
E6	0.33	0.33	0.14	0.20	0.50		1	0.50
E7	0.33	0.33	0.14	0.20	0.50	1		0.50
E8	0.50	0.50	0.20	0.33	1	2	2	

Table 9. AHP skill matrix that shows the relative importance of the skills.

	S1	S2	S3	S4	S5	S6
S1		0.20	0.30	0.50	0.50	0.14
S2	5		2	3	3	0.50
S3	3	0.50		2	2	0.33
S4	2	0.30	0.50		1	0.20
S5	2	0.30	0.50	1		0.20
S6	7	2	3	5	5	

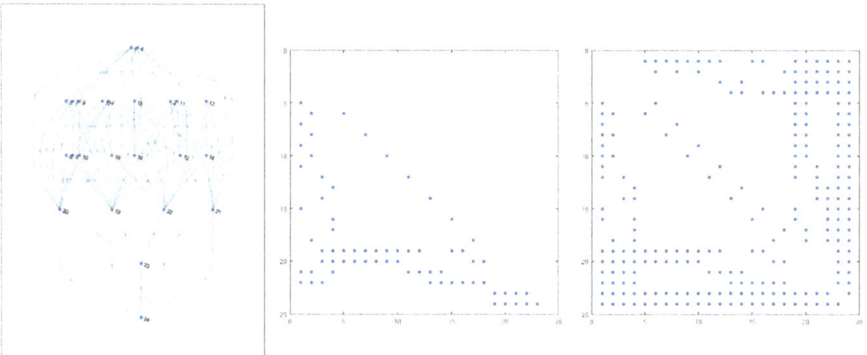

Figure 7. Possible path (**left**), precedence (**middle**) and transitive closure (**right**) of the activities (the unmarked pairs are interchangeable).

Figure 8 presents the different time, skill and equipment-related objectives in the case of different operators. As the results show, the increase in the number of operators decreases the efficiency of the utilisation of the tools and skills (this trend is the main driving force for forming manufacturing cells). The process can be well balanced in the case of 3–5 operators; e.g., in the case of five operators, in one of the best solutions, the balancing objectives are a time cost of 94.3%, training cost of 75.0% and equipment cost of 72.8%. Figure 9 shows the different total activity times of each operator during the simulation. In this case, the station times do not differ greatly, and the result is optimal [10]. The proposed algorithm was implemented in MATLAB and is available on the website of the authors (www.abonyilab.com/about-us/software-and-data); interested readers can make further comparisons, and the proposed problem can serve as a benchmark for constrained multi-objective line balancing.

Based on the simple modification of the code, the algorithm can be compared to classical simulated annealing-based line balancing; this comparison demonstrates that the main benefit of the proposed constrained handling is the acceleration of the optimisation. At the same time, the application of the inner-loop-based assignment significantly reduces the variance and increases the chance of obtaining improved line balancing results.

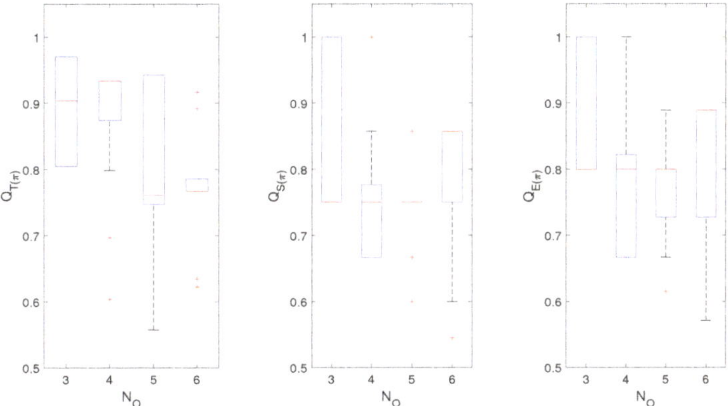

Figure 8. Boxplot of time, skill and equipment-related objectives for different independent runs of the algorithm and with different numbers of operators.

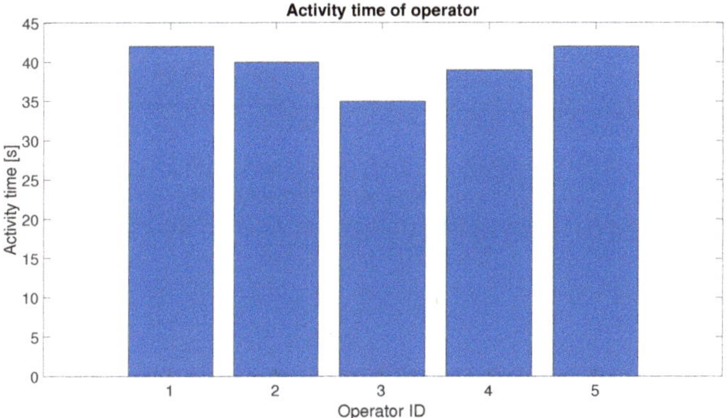

Figure 9. Comparison of the operators' activity times.

5. Conclusions

We proposed an assembly line balancing algorithm to improve the efficiency of production systems by the multiobjective assignment of tasks to operators. The optimisation of this assignment is based on a multilayer network model that provides information about the activity times, constraints and benefits (objectives) of the assignments, where the layers of the network represent the skills of the operators, the tools required for their activities and the precedence constraints of their activities.

The training and equipment costs as well as the precedence of the activities are also taken into account in the activity–operator layer of the network. As these costs and benefits are difficult to evaluate, the analytic hierarchy process (AHP) technique is used to quantify the importance of the criteria. The optimisation problem is solved by a multi-level simulated annealing algorithm (SA) that efficiently handles the precedence constraints thanks to the proposed problem-specific representation.

The proposed algorithm was implemented in MATLAB and the applicability of the method demonstrated with an industrial case study of wire harness manufacturing. The results confirm that multilayer network-based representations of optimisation problems in manufacturing seem to be potential promising solutions in the future.

The main contribution of the work is that it presents tools that can be used for the efficient representation of expert knowledge that should be utilised in complex production management problems. The proposed multilayer network-based representation of the production line supports the incorporation of advanced (ontology-based) models of production systems and provides an interpretable and flexible representation of all the objectives of the line balancing problem.

The AHP-based pairwise comparison of the importance of the nodes, edges and complex paths of this network can be used to evaluate the objectives of the optimisation problems. The integration of the network-based knowledge representation and the AHP-based knowledge extraction makes the application of the proposed methodology attractive in complex optimisation problems.

Author Contributions: L.N., T.R. and J.A. developed the methodology and software; L.N. prepared the original draft; J.A. reviewed and edited the manuscript. All authors have read and agreed to the published version of the manuscript.

Funding: This research was supported from the Higher Educational Institutional Excellence Program 2019 the grant of the Hungarian Ministry for Innovation and Technology (Grant Number: NKFIH-1158-6/2019). Tamás Ruppert was supported by the project 2018-1.3.1-VKE-2018-00048–Development of intelligent Industry 4.0 solutions of production optimization in existing plants and the ÚNKP-19-3 New National Excellence Program of the Ministry of Human Capacities.

Acknowledgments: The authors are grateful for the valuable comments and suggestions offered by the anonymous reviewers.

Conflicts of Interest: The authors declare no conflicts of interest.

Abbreviations

The following abbreviations are used in this manuscript:

SA	Simulated annealing
ALB	Assembly line balancing
$G_{i,j}$	Bipartite graphs between the ith and jth sets of objects
O_j, O_j	General representation of a set of objects as $O_i, O_j \in \left\{ \mathbf{s}, \mathbf{e}, \mathbf{a}', \mathbf{a}, \mathbf{w} \right\}$
$\mathbf{a} = a_1, \ldots, a_{N_a}$	Index of activities
$\mathbf{o} = o_1, \ldots, o_{N_o}$	Index of operators
$\mathbf{s} = s_1, \ldots, s_{N_s}$	Index of skills
$\mathbf{e} = e_1, \ldots, e_{N_e}$	Index of equipment
$\mathbf{w} = w_1, \ldots, w_{N_w}$	Index of workstations
$\mathbf{W}$	Workstation assigned for the activity, $N_a \times N_w$
$\mathbf{O}$	Operators assigned for the activity, $N_a \times N_o$
$\mathbf{S}$	Skills assigned for the activity, $N_a \times N_s$
$\mathbf{E}$	Equipment assigned for the activity, $N_a \times N_e$
$\mathbf{A}'$	Precedence constraint between activities, $N_a \times N_a$
$\mathbf{T} = t_1, \ldots, t_{N_a}$	Activity time
c_1	Station-time-related cost
c_2	Skill-related (training) cost
c_3	Equipment-related cost
T_c	Cycle time
N_w	Number of workstations
N_o	Number of operators
N_s	Number of skills
N_e	Number of pieces of equipment
N_π	Number of sequence elements

References

1. Becker, C.; Scholl, A. A survey on problems and methods in generalized assembly line balancing. *Eur. J. Oper. Res.* **2006**, *168*, 694–715. [CrossRef]
2. Boysen, N.; Fliedner, M.; Scholl, A. Assembly line balancing: Which model to use when? *Int. J. Prod. Econ.* **2008**, *111*, 509–528. [CrossRef]

3.   Nilakantan, J.M.; Ponnambalam, S.; Nielsen, P. Application of Particle Swarm Optimization to Solve Robotic Assembly Line Balancing Problems. In *Handbook of Neural Computation*; Elsevier: Amsterdam, The Netherlands, 2017; pp. 239–267.

4.   Hazır, Ö.; Delorme, X.; Dolgui, A. A survey on cost and profit oriented assembly line balancing. *IFAC Proc. Vol.* **2014**, *47*, 6159–6167. [CrossRef]

5.   Rüßmann, M.; Lorenz, M.; Gerbert, P.; Waldner, M.; Justus, J.; Engel, P.; Harnisch, M. *Industry 4.0: The Future of Productivity and Growth in Manufacturing Industries*; Technical Report 9; Boston Consulting Group: Boston, MA, USA, 2015.

6.   Gola, A. Reliability analysis of reconfigurable manufacturing system structures using computer simulation methods. *Eksploat. Niezawodn.* **2019**, *21*, 90–102. [CrossRef]

7.   Honti, G.M.; Abonyi, J. A review of semantic sensor technologies in internet of things architectures. *Complexity* **2019**, *2019*. [CrossRef]

8.   Boccaletti, S.; Bianconi, G.; Criado, R.; Del Genio, C.I.; Gómez-Gardenes, J.; Romance, M.; Sendina-Nadal, I.; Wang, Z.; Zanin, M. The structure and dynamics of multilayer networks. *Phys. Rep.* **2014**, *544*, 1–122. [CrossRef]

9.   Kivelä, M.; Arenas, A.; Barthelemy, M.; Gleeson, J.P.; Moreno, Y.; Porter, M.A. Multilayer networks. *J. Complex Netw.* **2014**, *2*, 203–271. [CrossRef]

10. Ruppert, T.; Honti, G.; Abonyi, J. Multilayer network-based production flow analysis. *Complexity* **2018**, *2018*, 1–15. [CrossRef]

11. Pigler, C.; Fogarassy-Vathy, Á.; Abonyi, J. Scalable co-Clustering using a Crossing Minimization–Application to Production Flow Analysis. *Acta Polytech. Hung.* **2016**, *13*, 209–228.

12. Romeijn, H.E.; Smith, R.L. Simulated annealing for constrained global optimization. *J. Glob. Optim.* **1994**, *5*, 101–126. [CrossRef]

13. Suresh, G.; Sahu, S. Stochastic assembly line balancing using simulated annealing. *Int. J. Prod. Res.* **1994**, *32*, 1801–1810. [CrossRef]

14. Chiang, W.C.; Urban, T.L. The stochastic U-line balancing problem: A heuristic procedure. *Eur. J. Oper. Res.* **2006**, *175*, 1767–1781. [CrossRef]

15. Ağpak, K.; Gökçen, H. A chance-constrained approach to stochastic line balancing problem. *Eur. J. Oper. Res.* **2007**, *180*, 1098–1115. [CrossRef]

16. Guerriero, F.; Miltenburg, J. The stochastic U-line balancing problem. *Nav. Res. Logist. (NRL)* **2003**, *50*, 31–57. [CrossRef]

17. Chiang, W.C. The application of a tabu search metaheuristic to the assembly line balancing problem. *Ann. Oper. Res.* **1998**, *77*, 209–227. [CrossRef]

18. Tsujimura, Y.; Gen, M.; Kubota, E. Solving fuzzy assembly-line balancing problem with genetic algorithms. *Comput. Ind. Eng.* **1995**, *29*, 543–547. [CrossRef]

19. Özcan, U.; Toklu, B. Multiple-criteria decision-making in two-sided assembly line balancing: A goal programming and a fuzzy goal programming models. *Comput. Oper. Res.* **2009**, *36*, 1955–1965. [CrossRef]

20. Baykasoglu, A. Multi-rule multi-objective simulated annealing algorithm for straight and U type assembly line balancing problems. *J. Intell. Manuf.* **2006**, *17*, 217–232. [CrossRef]

21. McMullen, P.R.; Frazier, G. Using simulated annealing to solve a multiobjective assembly line balancing problem with parallel workstations. *Int. J. Prod. Res.* **1998**, *36*, 2717–2741. [CrossRef]

22. Güden, H.; Meral, S. An adaptive simulated annealing algorithm-based approach for assembly line balancing and a real-life case study. *Int. J. Adv. Manuf. Technol.* **2016**, *84*, 1539–1559. [CrossRef]

23. de Jong, P. A statistical approach to Saaty's scaling method for priorities. *J. Math. Psychol.* **1984**, *28*, 467–478. [CrossRef]

24. Ho, W.; Ma, X. The state-of-the-art integrations and applications of the analytic hierarchy process. *Eur. J. Oper. Res.* **2018**, *267*, 399–414. [CrossRef]

25. Saaty, T.L. Decision making—The analytic hierarchy and network processes (AHP/ANP). *J. Syst. Sci. Syst. Eng.* **2004**, *13*, 1–35. [CrossRef]

26. Stoma, P.; Stoma, M.; Dudziak, A.; Caban, J. Bootstrap Analysis of the Production Processes Capability Assessment. *Appl. Sci.* **2019**, *9*, 5360. [CrossRef]

27. Che, Z. Clustering and selecting suppliers based on simulated annealing algorithms. *Comput. Math. Appl.* **2012**, *63*, 228–238. [CrossRef]

28. Coulter, E.D.; Sessions, J.; Wing, M.G. Scheduling forest road maintenance using the analytic hierarchy process and heuristics. *Silva Fenn.* **2006**, *40*, 143–160. [CrossRef]
29. Lu, Y.; Morris, K.C.; Frechette, S. Current standards landscape for smart manufacturing systems. *Natl. Inst. Stand. Technol. NISTIR* **2016**, *8107*, 39.
30. Bryton, B. Balancing of a Continuous Production Line. Ph.D. Thesis, Northwestern University, Evanston, IL, USA, 1954.
31. Erel, E.; Sarin, S.C. A survey of the assembly line balancing procedures. *Prod. Plan. Control* **1998**, *9*, 414–434. [CrossRef]
32. Leitold, D.; Vathy-Fogarassy, A.; Abonyi, J. Empirical working time distribution-based line balancing with integrated simulated annealing and dynamic programming. *Cent. Eur. J. Oper. Res.* **2019**, *27*, 455–473. [CrossRef]
33. Hackman, S.T.; Magazine, M.J.; Wee, T. Fast, effective algorithms for simple assembly line balancing problems. *Oper. Res.* **1989**, *37*, 916–924. [CrossRef]
34. Scholl, A.; Voß, S. Simple assembly line balancing—Heuristic approaches. *J. Heuristics* **1997**, *2*, 217–244. [CrossRef]
35. Hoffmann, T.R. Assembly line balancing with a precedence matrix. *Manag. Sci.* **1963**, *9*, 551–562. [CrossRef]
36. Sacerdoti, E.D. *A Structure for Plans and Behavior*; Technical Report; SRI International's Artificial Intelligence Center: Menlo Park, CA, USA, 1975.
37. Jackson, J.R. A computing procedure for a line balancing problem. *Manag. Sci.* **1956**, *2*, 261–271. [CrossRef]
38. Ruppert, T.; Abonyi, J. Software sensor for activity-time monitoring and fault detection in production lines. *Sensors* **2018**, *18*, 2346. [CrossRef] [PubMed]
39. Ong, N.; Boothroyd, G. Assembly times for electrical connections and wire harnesses. *Int. J. Adv. Manuf. Technol.* **1991**, *6*, 155–179. [CrossRef]

Article

Layout Guidelines for 3D Printing Devices

Arkadiusz Kowalski [1] and Robert Waszkowski [2,*]

[1] Faculty of Mechanical Engineering, Wrocław University of Science and Technology, 50-371 Wrocław, Poland;
   arkadiusz.kowalski@pwr.edu.pl
[2] Cybernetics Faculty, Military University of Technology, 00-908 Warszawa, Poland
[*] Correspondence: robert.waszkowski@wat.edu.pl

Received: 14 August 2020; Accepted: 8 September 2020; Published: 11 September 2020

Abstract: 3D printing methods are constantly gaining in popularity among investors, allowing for the production of products with a complex structure, and also used in the production of products with increasingly longer production series. It is planned to build factories (or those already under construction) in which 3D printing devices are the basic production devices. It is therefore important to develop guidelines and recommendations for layout design principles for additive technologies' devices. A question should be asked: will the development of a layout for additive technology machines in future factories differ from the preparation of a layout plan, for example, removal machining devices? Is it safe to assume that the mathematical methods of optimizing the layout of Computerized Relative Allocation of Facilities Technique (CRAFT), Computerized Relationship Layout Planning (CORELAP), and Modified Spanning Tree (MST) workstations or Schmigalla triangles will also work for 3D printing machines and devices? In search of answers to these questions, the article will attempt to apply a selected mathematical method to optimize the layout of workstations when machines and devices of additive technology are deployed for the assumed technological process and implemented according to frequently used Selective Laser Sintering (SLS) and Selective Laser Melting (SLM) technologies. A sample layout will be prepared for the assumed production plan and selected 3D printing technologies. Requirements and guidelines relevant to the development of layout plans will be collected regarding the necessary space, installations, and connections.

Keywords: layout; 3D printing devices; methods of optimizing the arrangement of workstations

---

1. Introduction

With the development of 3D printing methods, factories equipped with this type of equipment are increasingly being built. It is important from the investor's point of view that the developed layout plan takes into account the specific requirements of 3D printing machines. When developing the above-mentioned layout plans, the optimization of workstation layout is an important step. The question has arisen as to whether and how the known methods (e.g., CRAFT, CORELAP, MST, or Schmigalla triangles) will work well with 3D printing machines.

The concept of layout is very broad and there is no strict definition of it. This is presented and interpreted in various ways; one of the most popular definitions states [1] that "A facility layout is an arrangement of everything needed for the production of goods or delivery of services. A facility is an entity that facilitates the performance of any job." [2–5]. On the other hand, the factors that should be taken into account when designing the layout are clearly specified, ensuring adjustment to the requirements of the production system [6]:

- the type, variety, quality, and quantity of raw materials, work-in-process, and finished goods;
- the technological specification at each stage and the process of conversion as depicted in a flow process chart;

- the type of machinery and manufacturing aids necessary for the conversion process as stated above;
- the human factor of skill level at each stage, the number of persons required, the safety requirements and their ergonomics;
- the statutory and regulatory requirements as per provisions of the factories act must be considered for the plant layout;
- the line balancing and waiting factor decides the space requirement for the material storage;
- the material handling system is an integral part of the plant layout. It is one of the major cost areas. It is decided based on the extent of automation required and associated cost, speed, volume to be handled and type of material;
- the change factors in terms of future expansion, product modification, product range, flexibility and versatility of the organization, its products and technology; and
- service factors for machines and men have to be provided for. The space for machine maintenance, safety requirements for men and machines, provision of canteen, urinals, and toilets for workers, etc. is to be provided in the plant layout.

The variety of layout requirements, their large number and scale (from the location of the production hall on the industrial plot to the design of a single workstation), and the need to combine knowledge from various areas (from health and safety regulations, requirements regarding the order of arrangement of workstations, access between them, necessary connections, to their ergonomics) makes the implementation of a comprehensive layout project long and costly.

With the growing popularity of 3D printing techniques [7,8], layout designers will increasingly face the task of designing the layout of workstations taking into account the requirements for this type of device.

2. 3D Printing Methods—Information Essential for Designing Layout Plans

Additive manufacturing techniques offer great opportunities compared to traditional manufacturing methods such as, for example, foundry, plastic working, or plastic processing, allowing for the production of objects with complex geometries [9,10]. The model produced with the use of additive techniques was built by adding material layer by layer. The successively applied layers ultimately create the finished product. Various 3D printing methods can be used to apply the material. The following main types of additive technologies can be distinguished [11]:

- FDM (3D printing with the use of thermoplastics);
- SLA, DLP, MJP (using light-curing resins);
- SLS and MJF (3D printing with the use of powdered plastics);
- SLM, DMP, DMLS and EBM (3D printing with the use of powdered metals);
- CJP (3D printing with the use of gypsum powder); and
- LOM (3D printing with the use of foil or paper).

The most widespread and known techniques include SLA, SLS, and FDM. For further analysis, SLS and SLM devices were selected as basic devices for "factory building", based on the assumption that the offered products will use various materials (i.e., powdered plastics and metals).

2.1. Description of Selected Popular 3D Printing Methods

The SLA method (i.e., stereolithography) was the first of the developed rapid prototyping methods. The company 3D Systems patented this method in 1986 in the United States. In 1987, it began to manufacture machines using the technique of stereolithography [12]. The SLA method is based on layered polymerization of epoxy or acrylic resin with a laser beam. It is necessary to build structures that support the model, which take the form of thin rods and are usually removed mechanically [13]. The process begins by building a model in a 3D CAD system, which is then converted to the STL format. The prepared material in the form of a liquid resin is placed in the tub of the device. Before each layer is hardened, the scraper smooths the sheets of liquid and removes air bubbles. The laser beam scans the areas that show the current cross-section of the created element, which causes polymerization. Then, the working plate is lowered by the thickness of the layer. The whole process is repeated until the final geometry of the manufactured item is obtained. The finished element must be cleaned of unbound resin. This is done by rinsing the product in isopropanol or acetone. Then, the supporting structures must be removed mechanically. The last step is supplementary UV irradiation so that the polymerization process is completed in the entire volume of the model [14]. The element prepared in this way can be subjected to additional finishing:

- grinding/smoothing, and
- varnishing.

Various types of resins are the materials used in the SLA method. Standard resin is mainly used to create concept models and prototypes [15,16]. ABS resin is used for elements that must exhibit high strength and significant elongation. The products that are to be elastically compliant are made of elastic resin. There are also resins such as high-temperature, foundry, or medical [17]. When designing a layout for rapid prototyping machines using the SLA method, it is necessary to take into account that apart from the printer itself, other devices are also used. The entire instrumentation also includes:

- UV irradiation chamber, and
- model rinsing bathtub.

The full name of the SLS method is selective laser sintering. This was developed and patented by the founders of one of the first 3D printing companies—Desk Top Manufacturing Corporation [18]. The SLS method is based on the layered solidification of materials that are used in the form of a powder. The individual layers are joined by a laser beam that affects the powder surface. The whole process begins with loading a 3D spatial CAD model, which is then converted by the software to the STL format, thanks to which the model is divided into individual layers, then control instructions for the machine are generated [19]. The prepared plastic powder is applied to the work area with a scraper or a roller. There, it is heated and then laser sintered. The areas that show the current cross-section of the model are scanned by the laser beam, which causes their fusion. Then, the working plate is lowered by the thickness of the layer and another dose of powder is applied. The whole process is repeated until the final geometry of the manufactured item is obtained [20]. The final product, which we receive after printing, should be cleaned of any residual powder. Additionally, depending on the requirements and needs, it can be subject to finishing [21]:

- smooth grinding (sandblasting),
- surface sealing, and
- varnishing.

The materials used in the SLS method are plastic powders. The most popular is the PA12 polyamide due to its high flexibility and high mechanical properties. Other materials used in this method include PA12 polyamide with the addition of glass balls, thanks to which the manufactured elements will show greater strength and stiffness to compression as well as slightly increased thermal resistance of the material and alumide (a mixture of polyamide powder and aluminum filings). The SLS method allows one to produce a finished product using one device, although other machines and devices are also necessary for the entire process, which should be taken into account when building the layout plan. Powder material must be properly prepared before it can be used in the process and its remains after printing can be reused after prior preparation. The elements of the instrumentation equipment include:

- control panel,
- control cabinet,

- cooler,
- unpacking and screening station,
- unpacking device, and
- cabin sandblaster.

Another increasingly popular technique of 3D printing is the SLM method—Selective Laser Melting. This technology is protected by several patents [22,23]. Using this method, elements from metal powders are produced. However, they exhibit lower strength and durability than parts made with traditional shaping techniques. The SLM method ensures repeatability of mechanical properties, which allows the use of parts produced in this way as elements of machines and devices. The SLM method consists in melting metal powder with a metal beam. The process begins with applying a layer of powder and then levelling it. Selected areas are melted by the laser beam. The next step in the process is to lower the working plate by the layer thickness and apply the powder layer again. The whole process is repeated until the finished model is obtained. The produced model should be cleaned of unmelted powder and most often subjected to finishing treatments, which include:

- CNC machining and milling, and
- varnishing.

The materials most often used in the SLM method include stainless steels, pure titanium, and its alloys as well as low-melting, zinc, copper, tool steel, silicon carbide, or aluminum oxide alloys. In machines used for SLM printing, it is necessary to use a protective gas in the working chamber and the type depends on the powder that will be used to print the element. The melting parameters of the powder must be selected in such a way that the overheating of the area that is irradiated with the laser is as little as possible, since too much heating could distort the resulting product [12]. When designing the layout plan, one has to take into account that in the SLM method, apart from the main machine that produces the finished model, other devices are also necessary and without them, the entire process could not take place. Such elements include, for example:

- control cabinet,
- cooler,
- unpacking and screening station,
- lift truck,
- cabin sandblaster, and
- unpacking device.

To test the methods of optimizing the placement of 3D printing devices, we planned to choose one of the several available methods. The selection was based on the popularity of the methods and their adequacy to the problem being solved.

2.2. Characteristics of Methods for Optimizing the Arrangement of Workstations

The classification of methods for optimizing the arrangement of workstations can be implemented, among others, according to the following criteria [24,25]:

- type of solution (exact and approximate);
- restriction of the choice of place (with or without restrictions);
- the way of presenting the shapes and dimensions of the stands (point methods—all devices have the same dimensions and modular methods—the size of square or triangular modules corresponds to the size of the devices);
- method of positioning workstations (stepwise and iterative);
- other (e.g., methods taking into account the outline of the shape of machines and devices; and genetic algorithms, expert systems, or solutions offered by producers of CAD programs).

When choosing a method to optimize the placement of stations for the planned research on the placement of 3D printing machines and devices, it was decided to take into account the most commonly used methods that simultaneously differed from each other in terms of the operation algorithm. The length of transport routes, the number of transport operations, transport costs, weight of transported materials, or the volume of transports can be used for optimization criteria. For the purposes of research, a total of five methods were selected for further analysis:

- CRAFT,
- Bloch-Schmigalla,
- ROC,
- MST, and
- CORELAP.

The CRAFT (computerized relative allocation of facilities technique) method was proposed in 1964 by E. Buff, G. Armour, and T. Vollmann. The CRAFT algorithm changes the positions of the stations in the initial layout to determine improved solutions based on the flow of materials; subsequent changes lead to the layout with the lowest total cost. CRAFT uses the material flow cost as a criterion to consider the best layout. The best design is the one that has a minimum total cost. CRAFT does not guarantee the least costly solution, as not all possible exchanges are considered. The quality of the final solution depends on the initial solution. Therefore, it is common practice to identify several different initial designs and try all the exchange combinations and then select the best solution generated [26–28].

In the CRAFT method, an existing layout or a completely new plan is considered as a block arrangement. The algorithm calculates the allocations of workstations and estimates the costs that will be incurred in the initial phase of the project. The impact on the cost measure is computed for two or more different position settings in the layout plan. The main goal of this method is to minimize the total cost of the TC function [16], which is defined by the formula:

$$TC = \sum_{i=1}^{n} \sum_{j=1}^{n} D_{ij} \cdot W_{ij} \cdot C_{ij}, \tag{1}$$

where D_{ij} is the distance between station i and station j; W_{ij} is the volume of flow between station i and station j; C_{ij} is the cost of transport between station i and station j; and n is the number of stations.

The key steps of the algorithm are the calculation of the total cost of the TC function for the distance matrix describing the examined layout and the part flow matrix built on the basis of the sequence of technological operations and the demand for parts at different time periods.

The Bloch-Schmigalla method was initiated in the 1950s by W. Bloch. It uses a mesh of equilateral triangles in order to find the correct distribution of positions. The method was further developed and modified by H. Schmigalla [29]. The Schmigalla method of triangles begins by determining the order in which the workstations will be located and their distribution at individual nodes of a mesh composed of equilateral triangles. One should start with setting up a pair of stations with the highest flow intensity. It was assumed that the optimal system will be obtained when the W value of the objective function is the smallest possible. The function W is expressed as follows:

$$W = \sum_{i=1}^{n} \sum_{j=1}^{n} S_{ij} \cdot L_{ij} \rightarrow minimum, \tag{2}$$

where S_{ij} is the amount of part flow between the stations i and j; and L_{ij} is the distance between the stations, which is always equal to the side length of the equilateral triangle of the mesh.

The next steps of the algorithm provide for the construction of a matrix of the sequence of technological operations to select later, from the matrix of connections, the objects connected with the highest flow intensity; from these objects, we start arranging the positions on the mesh of equilateral triangles. A table of the intensity of connections between sites that have already been deployed and sites that have not yet been located helps in determining the location of subsequent objects.

The description of the ROC (rank order clustering) method was published in 1980 by J. King. ROC enables the arrangement of workstations by grouping them into manufacturing cells, taking into account families of parts classified according to technological similarity [30,31]. Its algorithm provides for the construction of a matrix of transport links and then, for each column of this matrix, its so-called binary weight is determined, according to the formula:

$$BW_j = 2^{m-j},\tag{3}$$

where m is the number of machines and j is the machine number. For the value of each row, the decimal binary equivalent is then calculated using the formula:

$$DE_i = \sum_{j=1}^{m} 2^{m-j} \cdot a_{ij},\tag{4}$$

where a_{ij} is the relationship between element i and position j, according to the matrix of transport connections. The effect of the ROC method is obtaining the arrangement of workstations with a job shop structure.

The modified spanning tree (MST) algorithm is based on the selection of adjacent pairs of workstations using a designated adjacency weight matrix. This method is similar to the spanning tree algorithm, which relies on a set of vertices and their connecting edges, where each edge with a separate weight connects two vertices. The vertices correspond to the workstations and the edges to the transport routes [32]. In the MST method, the adjacency weight matrix f'_{ij} is built based on Equation (5) [1]:

$$f'_{ij} = \left(f_{ij}\right) \cdot \left(d_{ij} + 0,5 \cdot \left(l_i + l_j\right)\right),\tag{5}$$

where f'_{ij} is the adjacency weight matrix; f_{ij} is the flow matrix; d_{ij} is the recommended distances between i and j machines; l_i is the i machine dimension; and l_j is the j machine dimension.

Equation (5) uses the product of the number of transport connections between workstations and the distance to be covered during the implementation of transport activities to determine the weight of the adjacency. It should be noted that the orientation of the machines is known in advance, or possibly presumed, which may be considered as a limitation of this method. As a consequence, it may also be necessary to prepare several variants of potential solutions that differ only in the orientation of machines and devices, and re-evaluate them by the MST algorithm.

From the adjacency weight matrix f'_{ij}, a pair of workstations connected by the highest weight are selected—they will be placed next to each other on the layout plan—a row and a column with a pair of devices already placed are plotted from the matrix. The next device is selected on the same principle, building a one-dimensional matrix of the order of arrangement of workstations.

The CORELAP (computerized relationship layout planning) method is one of the approximate methods. The use of this method is beneficial if there are various relationships between the objects being arranged that cannot be represented by one quantity [33].

The CORELAP method is based on determining the adjacency weight. The method consists of three stages: planning surfaces, their size and connections between them; calculating and designing CORELAP; and the final stage is drawing the layout plan. In order to define the relationship between the individual surfaces, each pair is assigned a symbol A, E, I, O, U, or X with a specific value, starting with "absolutely necessary" and ending with "undesirable". Relationships between positions recorded in the matrix are assessed by calculating the value of the proximity indicator (TCR). On this basis, the order of placing the positions is determined based on the six rules proposed in the CORELAP method. Depending on the adjacency method (fully contiguous, point contact, and non-contiguous, for which the adhesion coefficient is 1, 0.5, and 0, respectively), subsequent stations are arranged. The method was first used as an algorithm for arranging rooms or departments in industrial plants.

The presented characteristics of the methods of optimizing the arrangement of workstations show that their application for 3D printing machines is virtually problem-free with one exception. A significant limitation is related to the need to take into account auxiliary devices, often not connected to the main 3D printing device by means of transport. In this case, most of the methods based on material flow analysis simply omit these devices, and this may result in the lack of space for their placement on the layout plan. The CORELAP method does not have this disadvantage.

2.3. Environmental Conditions for Machines and Devices for 3D Printing, Important for the Layout Plan

Machines and devices for 3D printing must be provided with appropriate environmental conditions, which should be taken into account when designing layout plans. In addition, access to various media is necessary: electricity, compressed air, cooling water, or protective gas. In addition to the main machines that print a given element, a number of devices are also used that support the entire process so additional space should be allocated for them. For research work, it was assumed that the 3D printing devices used in the newly built plant—along with appropriate auxiliary machines—will be the devices offered by EOS GmbH:

- FORMIGA P 110, using the SLS method [34], and
- EOS M 290, using the SLM method [35].

3D printing machines and devices have strictly defined environmental conditions during operation, and the requirements relate to the permissible temperature in the room and relative air humidity. Separate requirements apply to the storage environment for the powder used in the SLS method. The specification of the mains connection includes the values of voltage, its fluctuations and frequency, the necessary mains protection, and the type of connection. The specification of the compressed air connection includes its consumption, operating pressure, minimum, and maximum pressure, compressed air temperature, its quality (including water and oil content), and the type of connection. All these information and requirements must be considered by the layout plan designer [36]. The key information in developing layout plans, however, are the dimensions of the machines as well as their weight. The main dimensions of an exemplary SLS device are shown in Figure 1a,b.

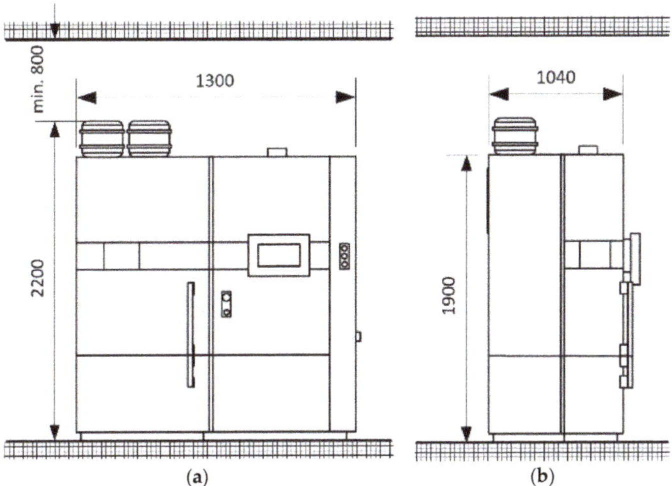

Figure 1. Main dimensions of an exemplary SLS device, dimensions in [mm]: (**a**) Front view of machine; (**b**) Side view of machine [34].

Manufacturers of 3D printing machines and devices also provide detailed information on the recommended distances from walls and other objects and the location of individual types of media, which is important for designers of layout plans (Figure 2).

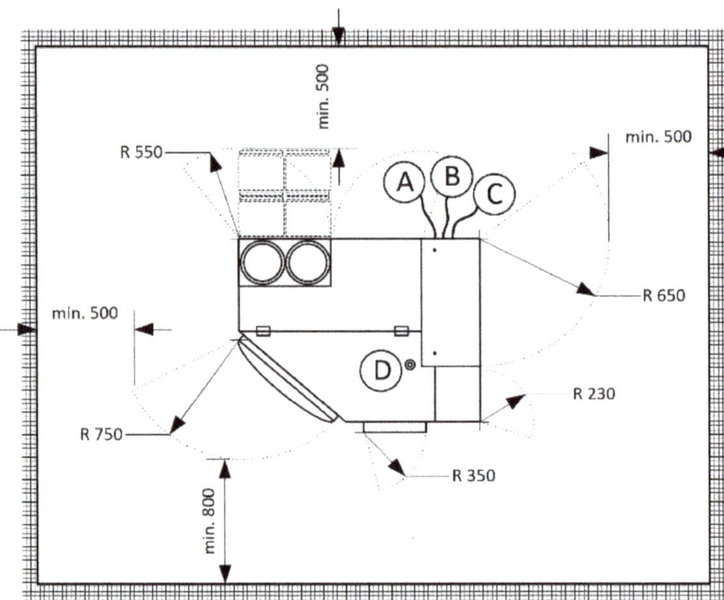

Figure 2. Requirements regarding space and location of connections for an exemplary SLS device [34].

3D printing devices that use a laser to sinter metal powder have specific requirements that must be met in order to ensure the correct operation of these machines. These requirements relate to the permissible temperature and humidity of the ambient air. During the SLM process, heat is released, which must be collected by a cooler operating in the system (e.g., air–water), hence, the requirements for the quality of the cooling water (e.g., pH value, chloride content, temperature, and minimum flow). An important role is also played by the shielding gas circuit filtering system (most often this gas is argon), which effectively collects the smelting contamination; a suitable place for it should be provided on the layout plan. During the operation of the SLM device, the noise emission at the level of approx. 70 dB (A) should be taken into account, which is also a factor that the designer of the layout plan cannot underestimate as the vicinity of a working 3D printing device can be troublesome in this respect. The power supply connection is particularly important for this type of 3D printing machine, due to the high energy demand of the laser sintering metal powder, the rated powers reach significant values so it is necessary to check these requirements with regard to the parameters of the power connection each time.

2.4. Manufacturing Processes Implemented in SLS and SLM Technologies, Determining the Necessary Number of Machines

For computational purposes, a hypothetical production plan was assumed; the production processes of two types of products will be carried out on 3D machines: product X manufactured in the SLS technique and product Y manufactured in the SLM technique. The production plan assumes the production of 20,000 pieces of product X and 23,000 pieces of product Y per year. Additional assumptions:

- number of products simultaneously printed on the working plate: Eight (for the SLS method) and 10 (for the SLM method);

- number of working days in a year: 250;
- number of production shifts: 2;
- the annual production plan for part X is 20,000 pcs, for part Y it is 22,800 pcs; and
- working day utilization factor (depends on the length of the work breaks, one break of 30 min is assumed): 0.9375.

To calculate the number of necessary devices in order to be able to implement the assumed production plan within the prescribed time, a formula was used that considered the ratio of the time necessary for technological operations to the available number of man-hours per year:

$$i_0 = \frac{t_{pz} + n \cdot t_j}{\psi_d \cdot D \cdot I \cdot 8} \tag{6}$$

where t_{pz} is the unit time of a technological operation; t_j is the preparation and completion time of a technological operation; n is the number of manufactured items; ψ_d is the working day utilization factor; D is the number of working days per year; and I is the number of shifts.

The form of the technological process for product X and Y is recorded in Tables 1 and 2, these tables present the designated number of machines according to Equation (6):

Table 1. Sequence of technological operations for product X with the calculated i_0 number of machines and devices necessary to implement the production plan.

Operation No.	Device Name	Unit Time t_j [min]	Preparation and Completion Time t_{pz} [min]	Value i_0	Number of Devices
10	Preparatory station	10	20	0.889	1
20	Mixing station	15	20	1.333	2
30	Sifter	5	10	0.444	1
40	SLS device	320	30	3.556	4
50	Cabin sandblaster	5	10	0.444	1
60	Grinder	8	15	0.711	2 [1]
70	Varnishing station	5	30	0.445	1 [1]
80	Packing	7	10	0.622	2 [1]

[1] Devices shared for the manufacturing process of products X and Y, i_0 values are then added up before rounding up.

Table 2. Sequence of technological operations for product Y along with the calculated i_0 number of machines and devices necessary to implement the production plan.

Operation No.	Device Name	Unit Time t_j [min]	Preparation and Completion Time t_{pz} [min]	Value i_0	Number of Devices
10	Feeding module	15	10	1.520	2
20	Filling module	20	20	2.027	3
30	SLM device	430	35	4.375	5
40	Microsand blaster	10	10	1.013	2
50	Grinder	8	15	0.811	2 [1]
60	Varnishing station	5	30	0.507	1 [1]
70	Packing	7	10	0.709	2 [1]

[1] Devices shared for the manufacturing process of products X and Y, i_0 values are then added up before rounding up.

The calculated number of 3D printing machines and devices, along with their dimensions, will be the key information at the next stage of developing layout plans.

3. Results—The Effects of the Arrangement Optimization of 3D Printing Stations

To optimize the arrangement of workstations, it was decided to use the MST method, based on the adjacency weight matrix, built on the basis of material flows. With this algorithm of procedure, auxiliary stations, characteristic for 3D printing machines and devices, are not taken into account. Thanks to this, it will be possible to collect information on whether this limitation is a significant disadvantage of this type of optimization method in the case of 3D printing devices.

3.1. Results of Optimization of the Arrangement of Workstations

For the purposes of the calculations, the MST method assumed that the size of the transport batch would correspond to the number of parts on the platforms of SLS and SLM devices. Machines will be located in a way providing, apart from the minimum recommended distances between them, access to the transport road for each of them. It was assumed that a hand truck would be used to transport a batch of elements. It will also be necessary to provide additional space for buffers (intermediate storage areas). The minimum recommended distances depend on the requirements of the manufacturers or on the size of the machines, the type of adjacency (side of the device, rear of the device, side of the machine where the operator works in relation to the sides of another machine, wall or transport road, etc.), and the dimensions of the machines.

Table 3 contains the information matrix of the part flow for the assumed technological process of products X and Y. The sequence of technological operations to be implemented is recorded in this matrix. It also includes the calculated number of transport activities, based on the number of manufactured products and the number of pieces in the transport batch.

The next step in the MST method is the development of the flow matrix (Table 4), containing information on the number of transport activities connecting workstations into subsequent pairs. Certainly, these calculations are based on the part flow information matrix. The numbering of workstations from Table 3 was also retained in the following tables (Tables 4–8).

Table 5 contains the selected distances between the stations connected by transport activities, depending on the method of the assumed adjacency (in this case "side to side") or the requirements taken from the technical and commissioning documentation provided by the manufacturer of the device.

The selected distances between the devices in Table 5 also depended on the dimensions of the machines (Table 6), where the larger the dimensions of the machines, the greater the recommended distances between them.

The values in the adjacency weight matrix f′$_{ij}$, presented in Table 7, were calculated on the basis of the previously presented Equation (5) from Section 2.2.

The result of the MST method is the sequence of workstations, read from the adjacency weight matrix f′$_{ij}$, written in the characteristic form of a single-line matrix. For the analyzed case, the results are presented in Table 8.

The pair of Output Warehouse (14) and Packing (13) stations showed the greatest weight of adjacency, therefore the pair from which the reading of the sequence of the arrangement of workstations began. The Output Warehouse (14) was the last station in the chain, therefore subsequent stations were selected from the adjacency weight matrix preceding the Packing (13) station, guided by the calculated weight for transport activities.

Table 3. Part flow information matrix.

	Machine														Part Demand	Batch Size	Number of Batches
	Input Warehouse	Preparatory Station	Mixing Station	Sifter	SLS Device	Cabin Sandblaster	Feeding Module	Filling Module	SLM Device	Microsand blaster	Grinder	Varnishing Station	Packing	Output Warehouse			
	1	2	3	4	5	6	7	8	9	10	11	12	13	14			
Part X	1	2	3	4	5	6					7	8	9	10	20,000	5	4000
Part Y	1						2	3	4	5	6	7	8	9	22,800	6	3800

Table 4. Flow matrix f_{ij}.

Machine	1	2	3	4	5	6	7	8	9	10	11	12	13	14
Machine 1		4000					3800							
2			4000											
3				4000										
4					4000									
5						4000								
6											4000			
7								3800						
8									3800					
9										3000				
10											3800			
11												7800		
12													7800	
13														7800
14														

Table 5. Clearance matrix d_{ij} for the analyzed case.

Machine	1	2	3	4	5	6	7	8	9	10	11	12	13	14
Machine 1		0.90					0.90							
2			0.40											
3				0.40										
4					0.50									
5						0.50								
6											0.50			
7								0.40						
8									0.50					
9										0.50				
10											0.50			
11												0.40		
12													0.40	
13														0.90 [1]
14														

[1] All selected distances between devices are expressed in meters.

Table 6. Machine lengths *l* for the analyzed case.

Machine Lengths *l* [m]													
1	**2**	**3**	**4**	**5**	**6**	**7**	**8**	**9**	**10**	**11**	**12**	**13**	**14**
5.00	1.13	0.68	0.65	1.75	1.26	0.80	0.80	1.57	0.75	1.55	1.40	1.30	5.00

Appl. Sci. **2020**, 10, 6333

Table 7. Adjacency weight matrix f'_{ij}.

Machine														
	1	2	3	4	5	6	7	8	9	10	11	12	13	14
1		15,850					14,440							
2			5210											
3				4260										
4					6800									
5						8020								
6											7,620			
7								4560						
8									6403					
9										6308				
10											6270			
11												14,625		
12													13,650	
13														31,590
14														

Table 8. Designated sequence of workstations for the analyzed case.

Computational Sequence of Workstations													
10	9	8	7	1	2	3	4	5	6	11	12	13	14

3.2. Developed Layout Plan

When developing the layout plan, the information collected thus far regarding workstations, the required and recommended distances between them, their number and dimensions, and the location determined according to the MST method were used. In addition, it was assumed that the shape of the plot on which the production hall is located enables the delivery of the necessary raw materials and other consumables on one side of the production hall and the receipt of finished products on the other. With this location of the Input and Output Warehouses, it was easier to ensure non-intersecting material flows. Access to the transport road was ensured via inter-operational buffers for batch-based transport operations. The developed layout plan, taking into account the above requirements, is shown in Figure 3.

A certain difficulty in the design process was the transfer of the designated layout of workstations from the MST method to a specific layout plan, due to the need to consider technological limitations related to:

- taking into account the required/recommended minimum distances between devices, not only "side to side", but also clearances (e.g., from the transport road or walls);
- the need to adjust to the shape and dimensions of the production hall;
- limited availability of the required connections for machines and devices in the production hall; and
- the need to minimize routes for operators between devices and buffers, in particular in cases where there are several identical devices.

The social part for employees was not included in the designed production hall in order to limit the size and detail of the drawing and thus increase its readability. For the same reason, the dimensions of machines and devices or the assumed distances between them are not shown.

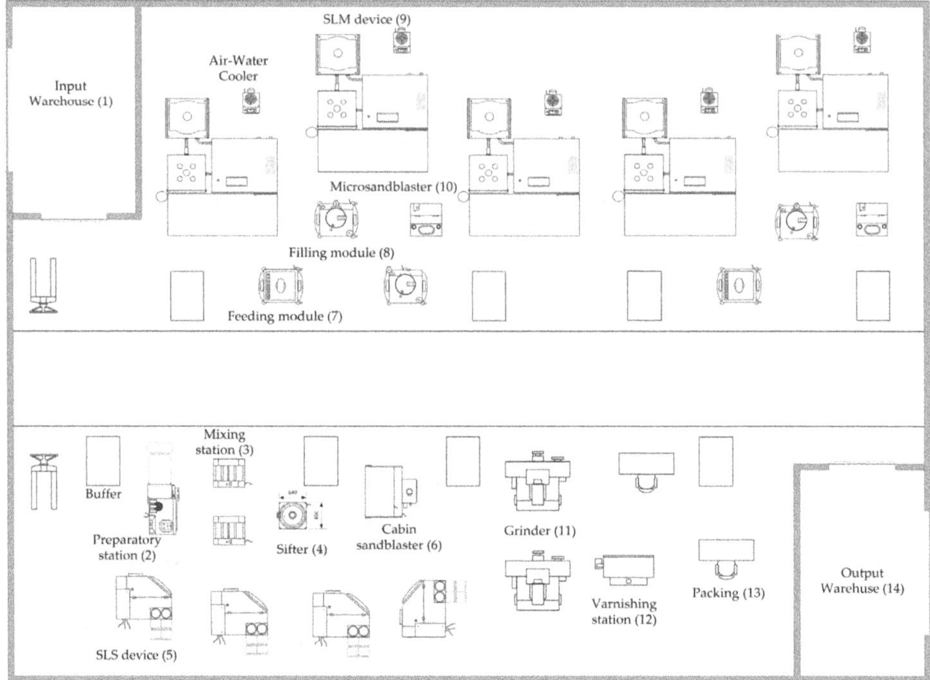

Figure 3. Sample layout plan design for 3D printing machines and devices.

4. Discussion—Guidelines for the Placement of 3D Printing Devices

In general, mathematical methods of optimizing the arrangement of workstations work well when using 3D printing machines; special attention should be paid to the following aspects when designing layout plans:

- there is a problem with the arrangement of auxiliary devices for 3D printing machines since some of them are not covered by materials or product transport, and therefore they are not included in the methods of optimizing the arrangement based on the number of transport activities. Certainly, they should be placed on the layout plan, which often distorts the optimal—under given conditions—solution. Such devices include control cabinets, coolers of various types or filtration systems;
- the calculated load of 3D printing machines shows that a significant part of the auxiliary devices is used to a small extent, the solution is to share these devices, which complicates the task of the layout plan designer;
- special attention should be paid to the connections as they differ depending on the 3D printing method;
- intermediate storage areas are most often not included in mathematical methods of arranging workstations, but they are always placed as close to machines and devices as possible in order to shorten the distance covered by operators as much as possible; and
- recommended distances between machines, available in the literature on the subject are usually much smaller than the requirements of manufacturers of specific 3D printing devices and the limitations resulting from the need to maintain access to the service doors.

A significant difficulty is also the process of transforming the solution from a mathematical method of optimizing the arrangement of workstations to a specific layout since the limited space, dimensions of workstations, and the required distance to the transport route and the need to connect utilities must be taken into account. This design stage is only partially supported by CAD software, which provides ready-made machines and devices as parameterized objects, so it relies on the knowledge and experience of the layout designer.

Author Contributions: Conceptualization, A.K. and R.W.; Methodology, A.K.; Validation, R.W.; Formal analysis, A.K.; Resources, A.K. and R.W.; Data curation, A.K.; Writing—original draft preparation, A.K. and R.W.; Writing—review and editing, A.K.; Visualization, A.K. and R.W.; Funding acquisition, R.W. All authors have read and agreed to the published version of the manuscript.

Funding: This research received no external funding.

Conflicts of Interest: The authors declare no conflict of interest.

References

1. Heragu, S.S. *Facilities Design*; CRC Press: Boca Raton, FL, USA, 2008; ISBN 978-1-4200-6627-2.
2. Ashraf, M.; Hasan, F.; Murtaza, Q. Optimum Machines Allocation in a Serial Production Line Using NSGA-II and TOPSIS. In *Precision Product-Process Design and Optimization*; Pande, S.S., Dixit, U.S., Eds.; Springer: Singapore, 2018; pp. 411–434.
3. Plinta, D.; Krajčovič, M. Production System Designing with the Use of Digital Factory and Augmented Reality Technologies. In *Progress in Automation, Robotics and Measuring Techniques*; Szewczyk, R., Zieliński, C., Kaliczyńska, M., Eds.; Springer International Publishing: Berlin/Heidelberg, Germany, 2015; pp. 187–196.
4. Gola, A.; Kłosowski, G. Development of computer-controlled material handling model by means of fuzzy logic and genetic algorithms. *Neurocomputing* **2019**, *338*, 381–392. [CrossRef]
5. Anjos, M.F.; Vieira, M.V.C. Mathematical optimization approaches for facility layout problems: The state-of-the-art and future research directions. *Eur. J. Oper. Res.* **2017**, *261*, 1–16. [CrossRef]
6. Kachwala, T.T.; Mukherjee, P.N. *Operations Management and Productivity Techniques*; PHI Learning Pvt. Ltd.: New Delhi, India, 2009; ISBN 978-81-203-3602-5.
7. Plocher, J.; Panesar, A. Review on design and structural optimisation in additive manufacturing: Towards next-generation lightweight structures. *Mater. Des.* **2019**, *183*, 108164. [CrossRef]
8. Alcácer, V.; Cruz-Machado, V. Scanning the Industry 4.0: A Literature Review on Technologies for Manufacturing Systems. *Eng. Sci. Technol. Int. J.* **2019**, *22*, 899–919. [CrossRef]
9. Jasiulewicz-Kaczmarek, M.; Legutko, S.; Kluk, P. Maintenance 4.0 technologies—New opportunities for sustainability driven maintenance. *Manag. Prod. Eng. Rev.* **2020**, *11*, 74–87. [CrossRef]
10. Wang, C.; Tan, X.P.; Tor, S.B.; Lim, C.S. Machine learning in additive manufacturing: State-of-the-art and perspectives. *Addit. Manuf.* **2020**, *36*, 101538. [CrossRef]
11. Ngo, T.D.; Kashani, A.; Imbalzano, G.; Nguyen, K.T.Q.; Hui, D. Additive manufacturing (3D printing): A review of materials, methods, applications and challenges. *Compos. Part B Eng.* **2018**, *143*, 172–196. [CrossRef]
12. Kruth, J.-P.; Leu, M.C.; Nakagawa, T. Progress in Additive Manufacturing and Rapid Prototyping. *CIRP Ann.* **1998**, *47*, 525–540. [CrossRef]
13. Kataria, A.; Rosen, D.W. Building around inserts: Methods for fabricating complex devices in stereolithography. *Rapid Prototyp. J.* **2001**, *7*, 253–262. [CrossRef]
14. Mankovich, N.J.; Cheeseman, A.M.; Stoker, N.G. The display of three-dimensional anatomy with stereolithographic models. *J. Digit. Imaging* **1990**, *3*, 200–203. [CrossRef]
15. Miedzińska, D.; Małek, E.; Popławski, A. Numerical modelling of resins used in stereolitography rapid prototyping. *Appl. Comput. Sci.* **2019**, 74–83. [CrossRef]
16. Korga, S.; Barszcz, M.; Dziedzic, K. Development of software for identification of filaments used in 3d printing technology. *Appl. Comput. Sci.* **2019**, 74–83. [CrossRef]
17. Wang, X.; Jiang, M.; Zhou, Z.; Gou, J.; Hui, D. 3D printing of polymer matrix composites: A review and prospective. *Compos. Part B Eng.* **2017**, *110*, 442–458. [CrossRef]
18. Deckard, C.R. Method and Apparatus for Producing Parts by Selective Sintering. U.S. Patent 4,863,538, 5 September 1989.

19. Kai, C.C.; Jacob, G.G.K.; Mei, T. Interface between CAD and Rapid Prototyping systems. Part 2: LMI? An improved interface. *Int. J. Adv. Manuf. Technol.* **1997**, *13*, 571–576. [CrossRef]
20. Petrovic, V.; Gonzalez, J.V.H.; Ferrando, O.J.; Gordillo, J.D.; Puchades, J.R.B.; Griñan, L.P. Additive layered manufacturing: Sectors of industrial application shown through case studies. *Int. J. Prod. Res.* **2011**, *49*, 1061–1079. [CrossRef]
21. Stansbury, J.W.; Idacavage, M.J. 3D printing with polymers: Challenges among expanding options and opportunities. *Dent. Mater.* **2016**, *32*, 54–64. [CrossRef]
22. Meiners, W.; Wissenbach, K.D.; Gasser, A.D. Shaped Body Especially Prototype or Replacement Part Production. DE Patent 19649865C1, 19 September 1998.
23. Meiners, W.; Wissenbach, K.; Gasser, A. Selective Laser Sintering at Melting Temperature. U.S. Patent 6,215,093, 10 April 2001.
24. Liggett, R.S. Automated facilities layout: Past, present and future. *Autom. Constr.* **2000**, *9*, 197–215. [CrossRef]
25. Montreuil, B. Requirements for representation of domain knowledge in intelligent environments for layout design. *Comput. Aided Des.* **1990**, *22*, 97–108. [CrossRef]
26. Buffa, E.S.; Armour, G.C.; Vollmann, T.E. *Allocating Facilities with CRAFT*; Harvard University: Boston, MA, USA, 1964.
27. Prasad, N.H.; Rajyalakshmi, G.; Reddy, A.S. A Typical Manufacturing Plant Layout Design Using CRAFT Algorithm. *Procedia Eng.* **2014**, *97*, 1808–1814. [CrossRef]
28. Gay, R.; Iyer, S.A.; Winsor, J. *Computer Integrated Manufacturing—Proceedings of the 3rd International Conference (In 2 Volumes)*; World Scientific: Singapore, 1995; ISBN 978-981-4548-79-3.
29. Schmigalla, H. *Fabrikplanung: Begriffe und Zusammenhänge*; Hanser: München, Germany, 1995; ISBN 978-3-446-18572-2.
30. King, J.R. Machine-component group formation in group technology. *Omega* **1980**, *8*, 193–199. [CrossRef]
31. Mahadevan, B. *Operation Management: Theory and Practice*; Pearson Education India: New Delhi, India, 2009; ISBN 978-81-7758-564-3.
32. Heragu, S.S.; Kusiak, A. Machine Layout Problem in Flexible Manufacturing Systems. *Oper. Res.* **1988**, *36*, 258–268. [CrossRef]
33. Lee, R.C. *Computerized Relationship Layout Planning (CORELAP)*; Northeastern University: Boston, MA, USA, 1966.
34. EOS. *Installation Conditions FORMIGA P 110. Laser-Sintering System for Plastics*; EOS: Krailling, Germany, 2013.
35. EOS. *EOS M 290. Produces Highest Quality Metal Parts in Additive Manufacturing*; EOS: Krailling, Germany, 2016.
36. Kowalski, A.; Chlebus, T.; Serwatka, K. Technical aspects of relocation of production and assembly lines in automotive industry. *Top. Intell. Comput. Ind. Des.* **2017**, *1*, 111–117.

Article

Assessment of the Design for Manufacturability Using Fuzzy Logic

Józef Matuszek *, Tomasz Seneta and Aleksander Moczała

Faculty of Mechanical Engineering and Computer Science, University of Bielsko-Biala, Willowa 2, 43-309 Bielsko-Biała, Poland; tomasz.seneta@zf.com (T.S.); amoczala@ath.bielsko.pl (A.M.)
* Correspondence: jmatuszek@ath.bielsko.pl; Tel.: +48-338-279-253

Received: 19 April 2020; Accepted: 31 May 2020; Published: 5 June 2020

Featured Application: New method to assess design for manufacturability based on fuzzy variables.

Abstract: The study proposes a procedure for assessing the designed manufacturing process for a new products. The purpose of the developed procedure is to evaluate the production process from the point of view of product design manufacturability of a unit and the small-lot production process. Evaluation of the design for the production process of a new product is based on criteria like process performance efficiency. Fuzzy logic-based methods were used to assess the designed process at different stages of its implementation—processing, assembly and organization of production. The developed method was illustrated by an example. The method presented in the study may be used by designers of production processes and employees of companies involved in the rationalization of already implemented production processes. The proposed method applies specifically to small-lot and unit production.

Keywords: production process design; unit and small-lot production design for manufacturability; fuzzy logic

---

1. Introduction

It can be assumed that the first practical examples of designing the structural form of the product components like the "design for assembly at that time were associated with the concept (PDM—product design merit)" activities can be seen in the early days of H. Ford around 1920. The plants began to produce at high-volumes, different products in several variants without any significant difficulties in the sales markets. In this period, the focus was mainly on the external appearance and functionality of the products rather than on the properties of their features in the technological and production processes. Development departments did not feel much pressure to apply appropriate activities related to the concept of "design for assembly—PDM" [1]. In the 1960s, a growing discrepancy between the obtained product quality parameters and growing customer requirements was noted in the United States [1]. An attempt was made to solve the problem by introducing additional design solutions. A temporary effect was obtained, the quality improved, but a significant increase in the production costs of the products resulted [2]. In the 1970s, global competition between enterprises grew significantly and increasing emphasis was placed on improving the competitiveness of production. High costs of designing and making the product were no longer acceptable. Much emphasis was put on the effectiveness of project management for the implementation of new products due to the significant impact of the designed manufacturing processes on the production costs [3].

2. Literature Review

Various methods of assembly support called DFX—design for X—has been developed and spread across industry methods such as QFD (quality function deployment) [4–6] used in the processes of implementing product customer requirements, FMEA (failure mode and effect analysis) [7]—related to the prediction and prevention of problems at the product design stage, DFA (design for assembly) [8–10]—e.g., design for manufacturing (DFM) regarding the shaping of the design process of components and the product itself [9,11–15]. Decisions made at the product design stage have a significant impact on production costs, efficiency and quality of production.

In the process of implementing the product, the impact of design on the cost of its implementation is very significant. The share of design costs varies around 5% of the costs of starting production of a new product but affects about 70% of the cost of the product after its implementation into production. While the rationalization activities of the production process at the production stage of the product (direct labor costs and indirect production costs often account for around 40% of the cost of the final product) affect only about 10% on the cost of production of the product [1,14,16]. This information is based on the US market cost structure. This means that the actions related to the changes in the production project (with relatively low costs incurred) in the right time have the greatest real impact on the production costs of the final product [1,15,17]. Swift [15] and others [1,7,18] analyzed the percentage of problems that occur in companies that have not performed and DFA activity during the product design development. For example, 35.9% had problems with the assembly of individual parts of the evaluated design. DFA methods have been selected to evaluate how their effects provide the greatest impact on manufacturing cost [1,14,15,19,20]. The DFA methods described in the literature and used in manufacturing practice were aimed at serial and mass product production [8–10,12,15,17,21]. There has been described in the latest scientific studies some of new DFA methods connected with CAD/CAM systems or Life Cycle assessment, some attempts to use fuzzy logic were also done however complete fuzzy DFA method for different kinds of production hasn't been recognized in literature [18,21–25] thus the proposed method also opens to small-lot and unit production.

Considering the methods described in the literature, we chose to focus on the analysis of unit and small-series production. We also point out that a method is needed to evaluate the production process in a comprehensive way that takes care of machining, assembly and production organizations. The use of fuzzy logic is justified by the need to estimate data due to the lower availability of complex analytical tools in small businesses, development budgets and product testing are limited there and there may be no accurate data for the reasons listed above. The method developed is open and other or additional criteria may be considered according to the production conditions of the company concerned. It may also have been interesting to develop DFA methods in medical procedures and the medical industry [26–30]. Another area of interest is the processes of electrical and electronic components that have been rapidly developing in recent years [28,31,32].

3. Design Manufacturability Assessment in Terms of Unit and Low-Volume Production

3.1. Assumptions for the New Method Design for Manufacturability

The justification for the emergence of a new fuzzy method to assess the technology of the structure resulted from the observed lack of flexibility of the described methods of Boothroyd-Dewhurst and Lucas. These methods were created in the 1980s where there was demand in the economy and was focused on serial and mass production. The current development of the economy and technology means that the modern economic system is characterized by a much greater need for flexibility in terms of production methods: high volume, low volume and in units. The need to create a more flexible method which is adaptable to the type of production was noticeable [33].

The design process should be determined from the point of view of various usability criteria—Figure 1. The assessment should consider many other various factors, sales, service, spare parts availability, production series, types of equipment, available assembly techniques, level of automation, cooperative services, possibilities of application commercial components, crew technical culture, etc. In small-lot and serial production conditions, the design process for new product production was based on simplified production documentation. Due to the low production series, production data results from the project were rarely verified at the production stage, while the experience gained from this stage was used in the production projects of new products. Concerning mass production, particular attention from the point of view of cost criterion was paid to the possibility of using unified and standardized elements included in the final product, the use of work stations and workshop aids for processing and assembly of various elements included in the products making up the program production and introduction of group machining processes, process phases, group operations for various elements [34–36]. The newly proposed method using fuzzy inference was characterized by such flexibility. In the literature cited in the study [15,19,37–40] there has been lack of studies enabling in the absence or uncertain data to estimate the times of assembly operations. The developed method has the features of novelty and meets the needs of production practice.

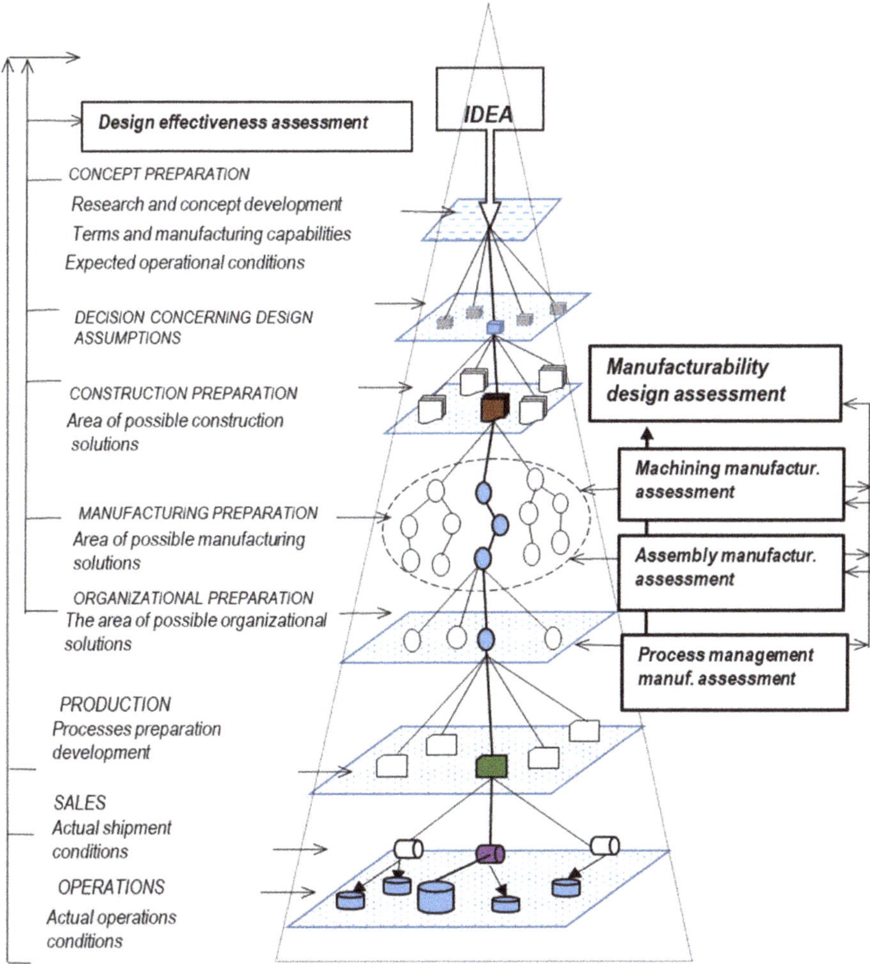

Figure 1. Modified design and development process to produce a new product.

3.2. The Course of the New Method Design for Manufacturability

In this study, there was a proposed new model of the product design analysis process which was carried out by experts representing: product design, machining process design, assembly process design, quality assurance, product cost analysis, OHS and environmental protection. Their inputs were assessed with the help of fuzzy sets methods following Figure 2. The assessment of the design manufacturability from the point of view of the assembly process was the first step followed by the machining process and production organization. According to experts, the order of assessment results from the size of the impact of the assessed design manufacturability on production efficiency. The feedback in the assessment activities results from the impact of decisions made in one stage on the other stage assessments.

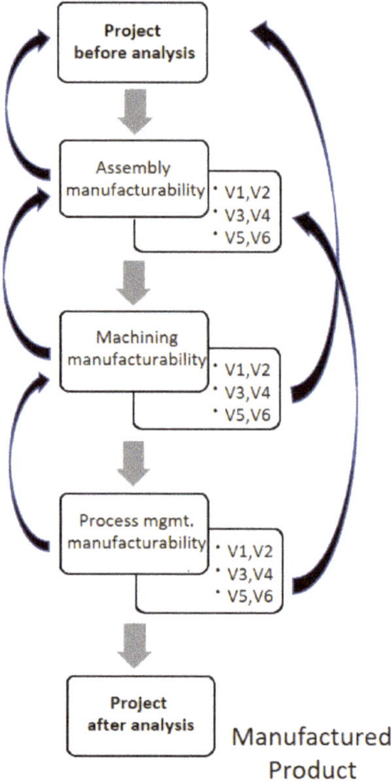

Figure 2. Structural analysis of the structure's technology in the proposed method.

Due to the costs of accurate analyses, there was less possibility to determine the performance parameters of the designed process in a unit, small-lot production due to unique, unstable and non-rhythmic production in the form of values, in the form of deterministic assessments. Therefore, fuzzy logic may be useful in such production conditions. Experts determine the fuzzy marks based on their own experience in the order given in Figure 2. The assessment was made on a scale of 0 to 100. Triangular symmetrical distributions were used for the assessments. The assessment method was presented below: when assessing, experts can be guided by their own production experience, they can also use data tables in the Boothroyd & Dewhurst and Lucas methods.

The assessment was related to the set of linguistic variables $V_i = \{V_1, \ldots, V_n\}$ and $\in N-\{0\}$, defining the input and output criteria of technology. The linguistic variable V_i was described by a quadruple:

$$[L_i, T_i(L), \Omega_i, M_i] \tag{1}$$

where: $L_i = \{L_1, \ldots, L_n\}$, $i \in N-\{0\}$—set of linguistic variable names, $T_i(L_i) = \{T_1(L_1), \ldots, T_n(L_n)\}$, $i \in N-\{0\}$—set of countable determinations of linguistic variables, $t_{ij} = \{t_{11}, t_{12}, \ldots, t_{nm}\}$, $i, j \in N-\{0\}$,

$t_{ij} \in T_i (L_i)$—set of linguistic values of linguistic variables, $\Omega_i = \{\Omega_1, \ldots, \Omega_n\}$, $i \in N$—{0}—set of linguistic ranges of variables V_i, $M_i = \{M_i, \ldots, M_n\}$, $i \in N$—{0}—set of semantic rules, $m_{ij} = \{m_{11}, m_{12}, \ldots, m_{mn}\}$, and, $j \in N$—{0}, $m_{ij} \in M_i$—range of variation in linguistic value t_{ij} with an assessment of belonging from 0 to 1 [41].

The assessment of the assembly process capability followed by the assessment of assembly technology and production organization corresponds to the stage of developing the project documentation of the product design. The applied variables V_1, V_2, V_3, V_4, V_5, V_6 in the scope of machining technologies, assembly, production organization are shown in Figure 2. The assessment, depending on the scope of information obtained, can be carried out for individual components of the product, groups of elements, its assemblies or also in a holistic way [41]. Sets of V_i variables can be modified and changed depending on the nature of the target process for which we design the product. This gives the fuzzy method a significant advantage in terms of flexibility. In the example presented, the set of variables V_i was prepared for medium-sized plant and small-lot production. It is illustrated by an example of one stage of the developed method to better illustrate the course of proceedings.

Variables that, in addition to deterministic values, can assume imprecise values—fuzzy. The triangular membership function can be defined using the following formula.

$$\mu_A(x) = \begin{cases} 0 & dla & x \leq a \; lub \; x \geq c \\ \frac{x-a}{b-a} & dla & a \leq x \leq b \\ \frac{c-x}{c-b} & dla & b \leq x \leq c \end{cases} \tag{2}$$

where a, b and c are parameters meeting the condition $a < b < c$.

Figure 3: presents a graph of the membership function of a given Formula (1).

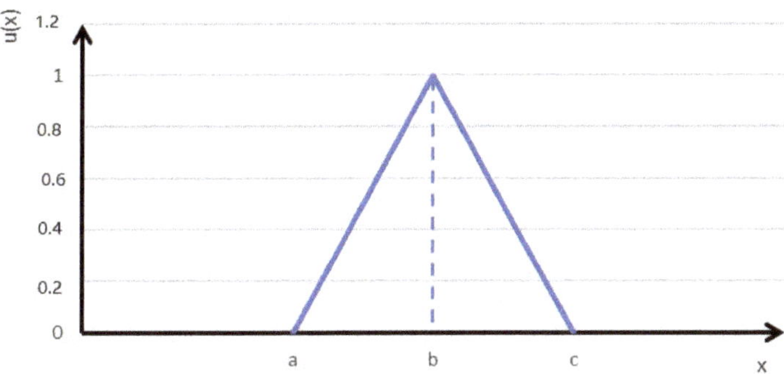

Figure 3. Graph of the triangular membership function described by the Formula (1).

It was assumed that two input variables ($\times1$ and $\times2$) and a single output variable (y) are related, respectively: {small, medium, large}, {short, medium, long} and {bad, medium, good}. What can be presented in the form of language rules:

R1W IF X1 is small and X2 is short, THEN Y is bad; also

R2W IF X1 is small and X2 is medium, then Y is bad, too

R3W IF X1 is medium and X2 is short, THEN Y is medium; also

R4W IF X1 is large and X2 is medium, THEN Y is medium; also

R5W IF X1 is large and X2 is long and Y is good

The rules can be presented in the decision table (Table 1) whereas, an example of a fuzzy partition is shown in Figure 4.

Table 1. Sample decision table.

	x1		
x2	small	medium	large
short	bad	medium	
medium	bad		medium
long			good

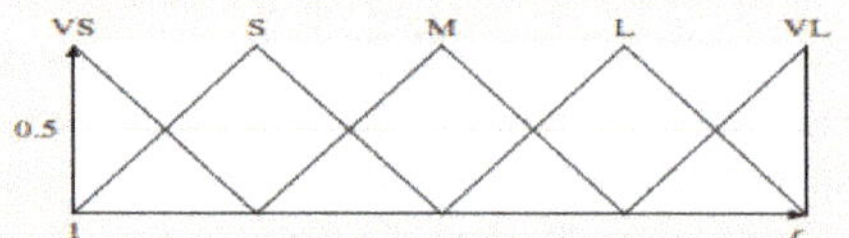

Figure 4. Example of a fuzzy partition where vs. -, S-, M-, L-, VL-.

We perform calculations for the values of V_1, V_i + 1 by reading the values from the graphs in Figures 3 and 4, according to the inference rule "min" specific rules were activated on the basis of which we set conclusions for the selected component that we evaluate. The next stage was the aggregation of conclusions, we should activate the selected rules for the selected component. In Mamdani's inference, which we use, there was a maximum operation as an operator of the aggregation of inference results obtained based on individual rules. For low average technology (range <0; 60>), the conclusion assumes a min value (0.67; medium technology)—lower value 0.67 or the value of the function, medium low technology. Fuzzy logic means that in the process of fuzzification, each rule was given a certain fuzzy value and must then be converted back to the real value, for this purpose we have defuzzification. In the work for defuzzification, a center of gravity method was proposed, which serves to sharpen the resulting fuzzy set and consists in determining the value of y *, which was the center of gravity of the area under the curve μ_{wyn} (y).

The Mamdani processing structure of fuzzy set inference methods consist of the following five elements:

- Input scaling, which transforms parameter values, enter variables from its domain to the one in which the input fuzzy partitions were defined;
- A fuzzy interface that converts explicit input into fuzzy values that serve as input to the fuzzy inference process;
- An inference engine that extracts data from blurred input data into several resulting fuzzy sets according to the information stored in the knowledge base;
- Defuzzification interface that converts fuzzy sets received from the inference process into a clear value;
- Output scaling that converts the defragmented value from the output domain of the fuzzy areas to the output variables, creating a global result of the fuzzy set inference method.

The reference model of the project was of the type: multiple inputs—multiple outputs MIMO. To compile results according to the above MATLAB software was used for the model.

The proposed DFA model of conduct based on fuzzy logic and the use of multiple entries—fuzzy rules (Figure 5) and multiple outputs enables efficient operation also in small-lot production conditions when there was no data from design verification by building many versions of prototypes and testing subsequent assumptions and design effects.

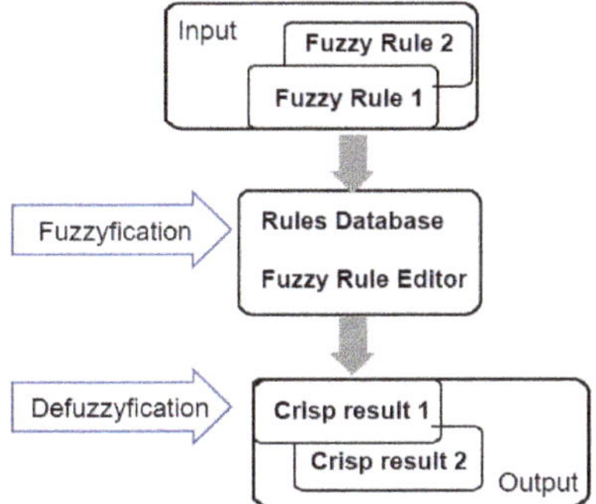

Figure 5. Reference model of the project is of the type: multiple entries—multiple outputs.

4. Implementation

4.1. Input Assumptions

Based on the analysis of the above methods of assessing the product's producibility, an improved proprietary approach was proposed in the process to shape the product's productiveness. The illustration of the presented proposals is presented on the example of a single-stage gear in Figure 6. General purpose gearboxes are designed in the form of a series of types from the point of view of market demand, production costs and delivery time to the customer. The gearbox is shown in Figure 6 was designed in a traditional way (welded body, many bolted joints, etc.).

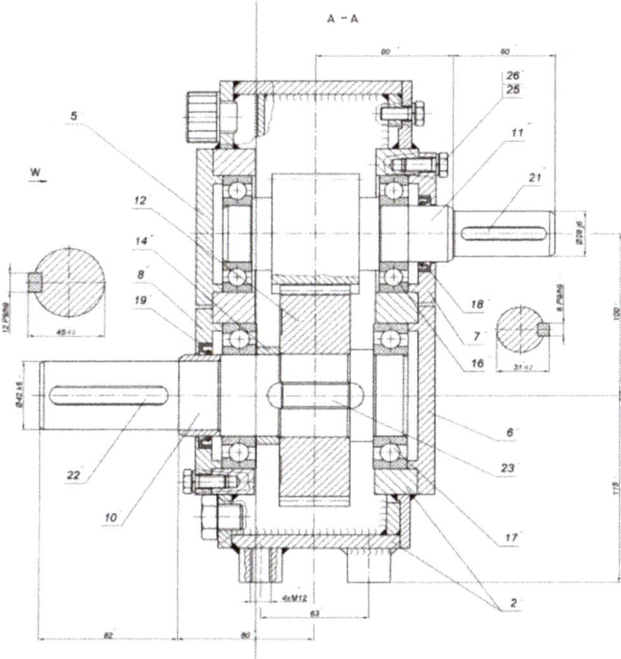

Figure 6. Diagram of the analyzed gearbox. 2—body; 5, 6, 7, 8—bearing caps; 10—shaft; 11—pinion; 12—tooth gear; 14—spacing rings; 17; 16—bearings; 18, 19; —seals; 21, 22, 23—keys; 25, 26—washers; screws.

A manufacturability analysis of the design was carried out for the adopted criteria presented in Figure 7. To illustrate the progress of the procedure in the method, the method of assessing the technological efficiency of the structure is more widely presented, on the example of the assembly of two elements—the gear housing and cover.

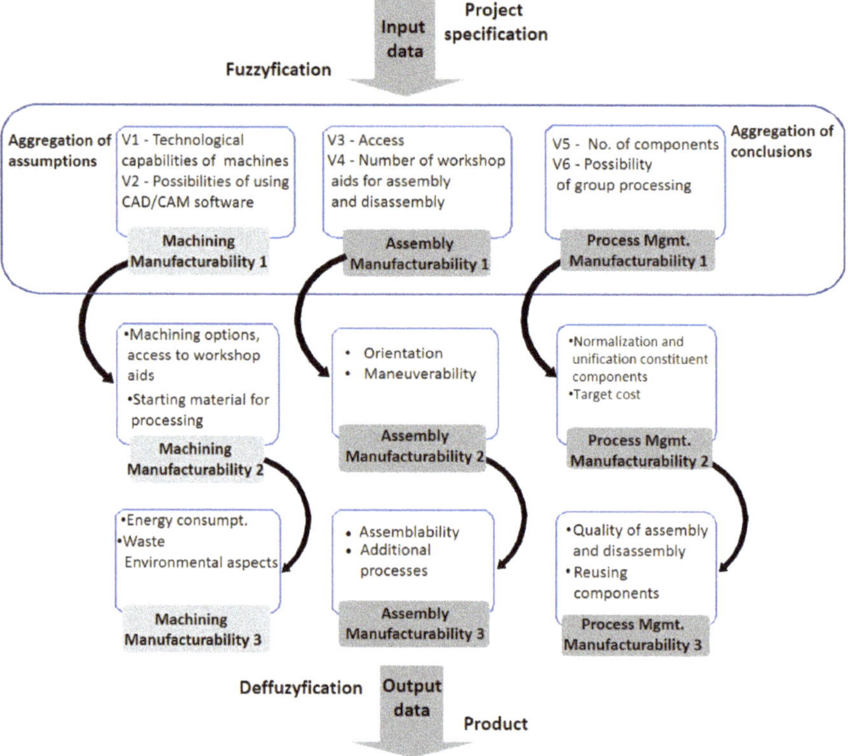

Figure 7. Model of the new method design for manufacturability based on three successive stages and substages of fuzzy inference (without feedback).

4.2. Assembly Manufacturability Fuzzy Assessment

The assessment was carried out in three substages—substage 1 (access, number of workshop aids), substage 2 (orientation, maneuverability), substage 3 (assemblability, processes).

4.2.1. Assembly Manufacturability Assessment—Substage 1

It was assumed that the assembly technology of the considered elements depends on two factors, which were: accessibility, number of workshop aids. The experts determined the ocean for the parameter "Access" = 20, "number of workshop aids" = 55. The functions of belonging linguistic variables for the given factors are given in Tables 2 and 3, the bases of rules for them are presented in Tables 4 and 5.

Table 2. Membership functions in tabular form of linguistic variables for "access".

Description—Access	Rank
Very difficult to access area, special care/tools/techniques required to remove the part without damaging it	0
limited surface/eyesight, extreme care required to take pictures without damage	30
The area has limited access, but some can be removed without damage	60
The area is easy to assemble, plenty of room for hands/tools	100

Table 3. Membership functions in tabular form of linguistic variables for "number of workshop aids".

Description—Number of Workshop Aids	Rank
Unnecessary	0
Easy to grasp	0
Orientation tools in 1 axis	30
Orientation tools in 2 axes	30
Orientation tools in both axis	60
Medium difficult tools	60
Heavy nesting or tangling	60
Requires a tool to grasp	60
Requires 2 operators	100
Requires special equipment	100

Table 4. Rules database for "access".

	Access			
	Very Difficult	Restricted	Medium Restricted	Easy
0	1	0	0	0
30	0	1	0	0
60	0	0	1	0
100	0	0	0	1

Table 5. Rules database for "number of workshop aids".

	Number of Workshop Aids			
	Easy	Require Orientation	Heavy/Equipment	Two Persons
0	1	0	0	0
30	0	1	0	0
60	0	0	1	0
100	0	0	0	1

The "Access" factor is described by formulas:

$$\mu_{VERY\ HARD}(x) = \begin{cases} \frac{30-x}{30-0} \ dla\ 0 < x < 30 \\ x = 0 \ dla\ 30 \leq x \leq 100 \end{cases} \tag{3}$$

$$\mu_{RESTRICTED}(x) = \begin{cases} \frac{x}{30-0} \ dla\ 0 < x < 30 \\ \frac{60-x}{60-30} \ dla\ 30 < x < 60 \\ x = 0 \ dla\ 60 \leq x \leq 100 \end{cases} \tag{4}$$

$$\mu_{MEDIUM\ RESTRICTED}(x) = \begin{cases} x = 0 \ dla\ x \leq 30 \\ \frac{x-30}{60-30} \ dla\ 30 < x < 60 \\ \frac{100-x}{100-60} \ dla\ 60 < x < 100 \end{cases} \tag{5}$$

$$\mu_{EASY}(x) = \left\{ \begin{array}{l} x = 0 \;\; dla \;\; x \leq 60 \\ \frac{x-60}{100-60} \; dla \; 60 < x < 100 \end{array} \right. \tag{6}$$

The fuzzy rules for assembly technology are presented in Table 6.

Table 6. Fuzzy rules table for assembly technology—substep 1.

1	If	Access easy	And	number of workshop aids two person/equipment	Then	Manufacturability medium law
2	If	Access easy	And	number of workshop aids heavy or equipment	Then	Manufacturability medium
3	If	Access easy	And	number of workshop aids require orientation	Then	Manufacturability high
4	If	Access easy	And	number of workshop aids easy	Then	Manufacturability high
5	If	Access medium restricted	And	number of workshop aids two person/equipment	Then	Manufacturability low
6	If	Access medium restricted	And	number of workshop aids heavy or equipment	Then	Manufacturability medium law
7	If	Access medium restricted	And	number of workshop aids require orientation	Then	Manufacturability medium
8	If	Access medium restricted	And	number of workshop aids easy	Then	Manufacturability medium
9	If	Access restricted	And	number of workshop aids two person/equipment	Then	Manufacturability low
10	If	Access restricted	And	number of workshop aids heavy or equipment	Then	Manufacturability medium law
11	If	Access restricted	And	number of workshop aids require orientation	Then	Manufacturability medium
12	If	Access restricted	And	number of workshop aids easy	Then	Manufacturability medium
13	If	Access very difficult	And	number of workshop aids two person/equipment	Then	Manufacturability low
14	If	Access very difficult	And	number of workshop aids heavy or equipment	Then	Manufacturability low
15	If	Access very difficult	And	number of workshop aids require orientation	Then	Manufacturability medium law
16	If	Access very difficult	And	number of workshop aids easy	To	Manufacturability medium

In order to make the method compared with traditional methods transparent, the evaluations and results were scaled. The best theoretical value for the design feasibility of the structure maybe 100. After scaling, this rating may have a maximum value of 1.00. The assessments of the efficiency according to the new method will be equal to x/100 where x was the given assessment of the structure's efficiency. For the body, for the values "access" = 20 and "number of workshop aids" = 55 based on Figure 8, according to the above-mentioned inference rule "min", the following rules were active:

- Rule 14 Access "very difficult" and number of workshop aids "heavy or equipment" in the degree of min (0.33, 0.17) = 0.17 (low technology);
- Rule 15 Access "very difficult" and number of workshop aids "require orientation" in the degree of min (0.33, 0.833) = 0.33 (medium low technology);
- Rule 10 "limited" access and number of workshop aids "heavy or equipment" in the degree of min (0.67, 0.17) = 0.17 (medium low technology);

- Rule 11 'limited' access and number of workshop aids 'require orientation' in the degree of min (0.67, 0.833) = 0.67 (medium technology).

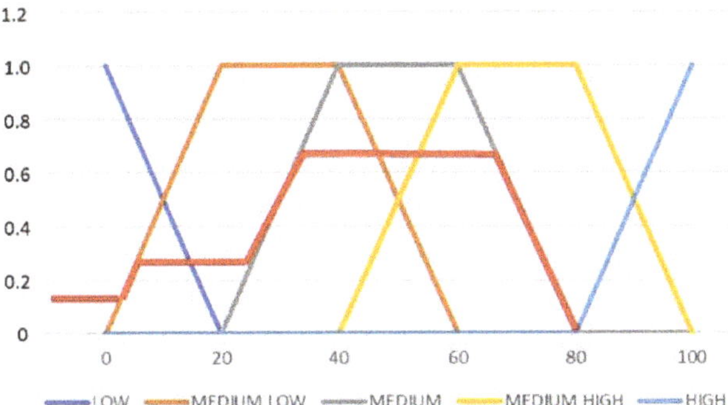

Figure 8. Aggregation of rules for assembly technology Substep 1.

The practical approach of aggregation of rules was that for example for Rule 11 for values 20 and 55 calculated value 0.67 defines surface area under function "medium on Figure 9. After taking into account rules 10, 11, 14 and 15, in Mamdani's inference there was a maximum operation as an operator of the aggregation of inference results obtained on the basis of individual rules, therefore rules 10 and 15 which have the same "medium low" rating, we choose MAX so we activate rule 15. Hence, activated were rules 11,14,15. Complete aggregated values for assembly technology in substep 1 are given in Figure 8.

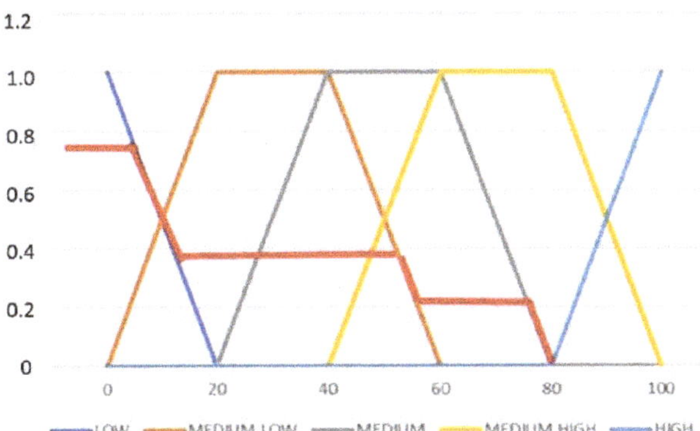

Figure 9. Aggregation of rules for assembly technology 2.

The next step was defuzzification (sharpening) of the parameter value to provide the predicted factor value. The basis of this step was the resulting membership function represented in a fuzzy form, while the inference should end with providing a specific numerical value, hence the need to sharpen. Various methods can be used to carry out this process: center of gravity, average maximum, first maximum, last maximum. The center of gravity method was selected:

$$\begin{cases} y = \frac{x}{20} \\ y = 0.17 \end{cases} ; \ 0.17 = \frac{x}{20}; \ x = 20 \cdot 0.17 = 3.4 \tag{7}$$

$$\begin{cases} y = \frac{x}{20} \\ y = 0.33 \end{cases} ; \ 0.33 = \frac{x}{20}; \ x = 20 \cdot 0.33 = 6.6 \tag{8}$$

$$\begin{cases} y = \frac{80-x}{80-60} \\ y = 0.67 \end{cases} ; \; 0.67 = \frac{80-x}{20}; \; x = 80 - 13.4 = 66.6 \tag{9}$$

$$\begin{cases} y = \frac{x-20}{40-20} \\ y = 0.33 \end{cases} ; \; 0.33 = \frac{x-20}{20}; \; x = 20 \cdot 0.33 + 20 = 26.6 \tag{10}$$

$$\begin{cases} y = \frac{x-20}{40-20} \\ y = 0.67 \end{cases} ; \; 0.67 = \frac{x-20}{20}; \; x = 20 \cdot 0.67 + 20 = 33.4 \tag{11}$$

Defuzzied center of gravity value:

$$r = \frac{r_1}{r_2} = \frac{\int_0^{80} y \cdot \mu_{B'}(y)dy}{\int_0^{80} \mu_{B'}(y)dy} \tag{12}$$

$$\begin{aligned} r = \pm \int_{3.4}^{6.6} y \cdot \frac{y}{20}\,dy \\ + \int_{26.6}^{33.4} y \cdot \frac{y-20}{20}\,dy + \int_{6.6}^{26.6} y \cdot 0.33\,dy \\ + \int_{33.4}^{66.6} y \cdot 0.67\,dy + \int_{66.6}^{80} y \cdot \frac{80-y}{20}\,dy \int_0^{80} \mu_{B'}(y)dy \end{aligned} \tag{13}$$

where:

$$r_1 = \left[\frac{y^2}{12}\right]_0^{3.4} + \left[\frac{y^3}{60}\right]_{3.4}^{6.6} + \left[\frac{y^2}{6}\right]_{6.6}^{26.6} + \left[\frac{1}{20} \cdot \left(\frac{y^2}{3} - 10y^2\right)\right]_{26.6}^{33.4} + \left[\frac{y^2}{3}\right]_{33.4}^{66.6} + \left[\frac{1}{20} \cdot \left(40y^2 - \frac{y^2}{3}\right)\right]_{66.6}^{80} \tag{14}$$

$$r_1 = 0.96 + 4.14 + 110.67 + 103.31 + 1106.67 + 319.02 = 1644.76$$

$$r_2 = \int_0^{80} \mu_{B'}(y)dy = P_1 + P_2 + P_3 \tag{15}$$

$$P_1 = (6.6 - 0) \cdot 0.17 = 1.1; \; P_2 = (20 - 0) \cdot 0.33 = 6.6; \; P_3 = \frac{[60+33.2] \cdot 0.67}{2} = 31.22$$

$$r_2 = \int_0^{80} \mu_{B'}(y)dy = 1.1 + 6.6 + 31.22 = 38.9 \tag{16}$$

$$r = \frac{1644.76}{38.9} = 42.2 \tag{17}$$

The assessment of technology for the 1st stage assumes for the adopted access assessment-20 and the number of workshop aids-55. The value of ~42.20 was determined.

4.2.2. Assembly Manufacturability Assessment—Substage 2

The component's technology is determined, assuming that it depends on two factors, which were: orientation, maneuverability. The functions of belonging linguistic variables for the given factors are given in Tables 7 and 8, the bases of rules for them are presented in Tables 9 and 10. The expert group made the following assessment: orientation—10, maneuverability—35.

Table 7. Membership functions in tabular form of linguistic variables for orientation.

Orientation	Rank
Not require orientation	100
Requires orientation in the assembly axis	60
Requires orientation orthogonal to the assembly axis	30
Requires orientation in the assembly axis and perpendicular to the assembly axis	0

Table 8. Membership functions in tabular form of linguistic variables for maneuverability.

Maneuverability	Rank
Easy to grasp (one hand)	0
Easy to grasp (BH)	0
Orientation to change (OH)	30
Orientation to change (BH)	30
Slippery	60
Flexible or mall	60
Heavy nesting or tangling	60
Requires a tool to handle	60
Requires two operators	100
Requires equipment to operate	100

Table 9. Rule base for orientation.

Orientation				
	Both Axis	**Perpendicular to Axis**	**In Axis**	**No Orientation**
0	1	0	0	0
30	0	1	0	0
60	0	0	1	0
100	0	0	0	1

Table 10. Rule base for maneuverability.

Maneuverability				
	Easy	**Require orientation**	**Heavy/Equipment**	**Two Person/Equipment**
0	1	0	0	0
30	0	1	0	0
60	0	0	1	0
100	0	0	0	1

Aggregation of rules for assembly technology 2 is shown in Figure 9.

The technological assessment for the 2nd stage assumes for the adopted assessment of orientation—10 and maneuverability—35. The value equal to −31.0 was determined.

4.2.3. Assembly Manufacturability Assessment—Substage 3

The technology of the 3rd component was determined, assuming that it depends on two factors, which were: assembly, processes. The functions of belonging linguistic variables for the given factors are given in Tables 11 and 12. The expert group made the following assessment: assemblability = 20, joining processes = 35.

Table 11. Membership functions in tabular form of linguistic variables for assemblability.

Assemblability	Rank
Difficult access and blind assembly	0
Special equipment	30
Requires two hands	60
No difficulty	100

Table 12. Membership functions in tabular form of linguistic variables for processes.

Joining Process	Rank
Place part	100
Snap fit	100
Light interference	60
Pressed	60
Manual screwing	60
Screwing with tooling	30
Automatic screwing	30
Riveting	30
Clinching	30
Soldering	0
Welding	0

Aggregation of rules for assembly technology 3 is shown in Figure 10.

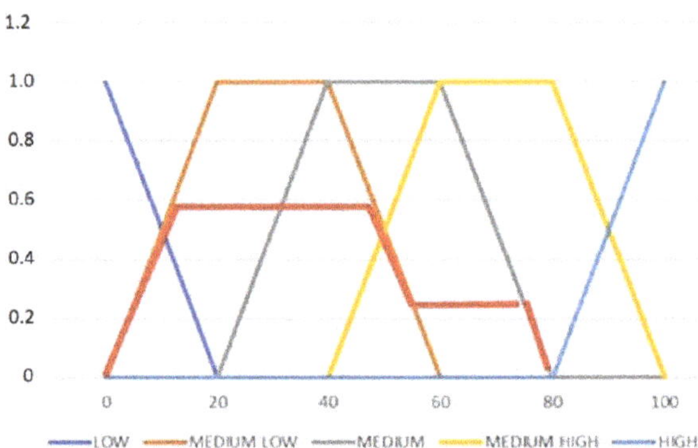

Figure 10. Aggregation of rules for assembly technology 3.

The technological assessment for the 3rd stage was assumed for the accepted assessment of assemblability—70 and joining processes—10. The value equal to −36.0 was determined.

4.3. Fuzzy Assessment of Design for Machinability—for Example

To decrease the number of components of a product may increase its complexity and increase its manufacturing costs. The final product can be easy to assemble and expensive to process its components.

The condition for the correct determination of the cost-related factors involved in the production process of a given element was information about the characteristics that this element has from the point of view of construction, production and organization of production. The main task that must be performed was to determine the value of the costs of implementing individual operations. The cost of product processing and organization of production includes material costs, costs of cooperation and processing of a given operation. Classification of elements should include its type, e.g., shaft, sleeve, specify dimensions, the accuracy of workmanship, etc. Based on technological similarity, the costs of individual operations can be determined in accordance with the data in the database of costs of operation of technologically closest components [36,39,42].

The assessment was based on a multi-level classification of elements, assemblies made in the enterprise, etc. (Figure 11). The new element was assigned to a given shape representative based on the designer's decision—Figures 12 and 13. Based on the shape and design parameters from the manufacturing processes database, the process of the element with the same shape code and parameters most like the parameters of the new element was searched. Having the process of manufacturing the nearest element at your disposal and data on the value of cost factors in connection with the time and then cost calculation system, you can specify the production costs of the designed element [34,36,42].

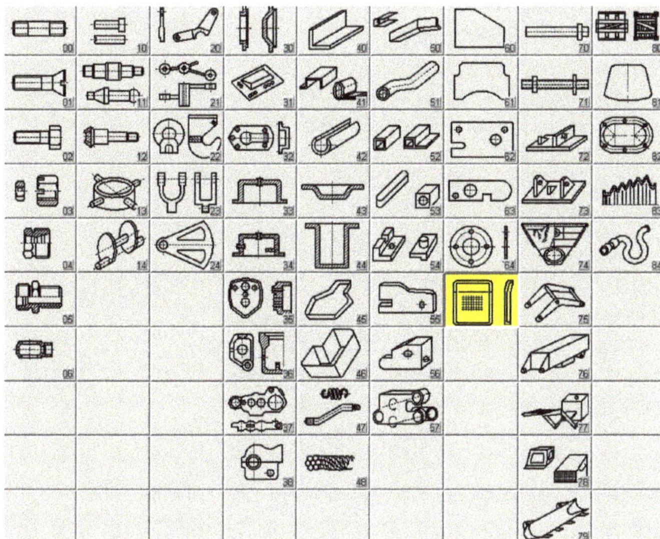

Figure 11. Example of the first level of the production item classifier–restrictions on unnecessary diversity.

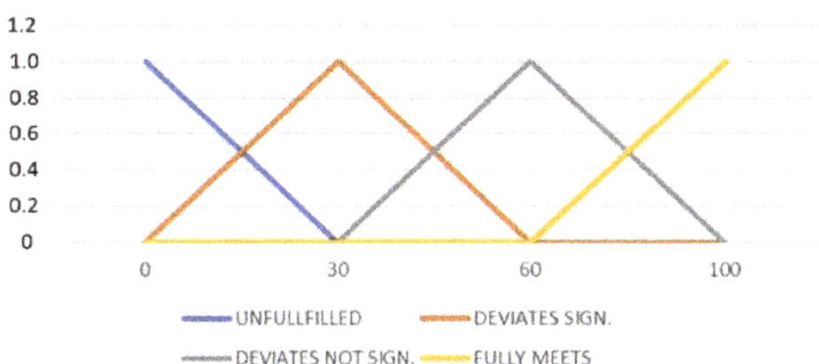

Figure 12. Graphical representation membership functions for linguistic variables for technological capabilities.

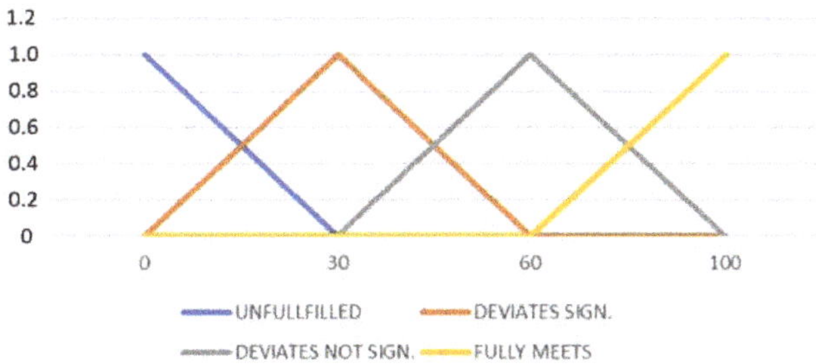

Figure 13. Graphical representation membership functions for linguistic variables for Software Capability.

Result of aggregation of rules for design for machining manufacturability 1 calculated as center of gravity of the surface under the curve presented in Figure 14—design for machining manufacturability 1 assessment for Substep 1 was 29.9.

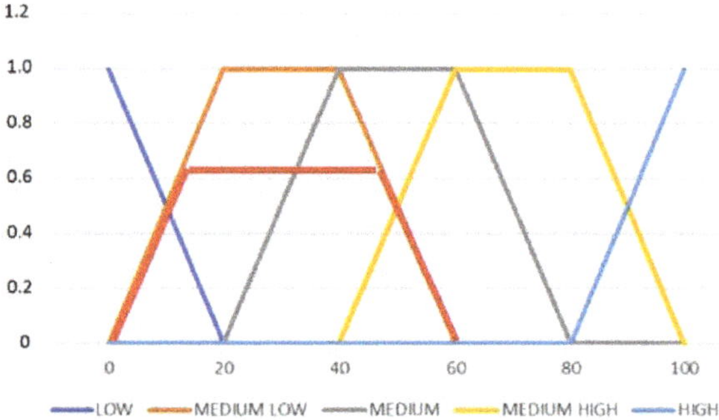

Figure 14. Aggregation of rules for machining technology 1.

The design for machining processing 2 assessment was determined in a similar procedure: tool machining capability V3 = 10, compliance requirements V4 = 35. Results of aggregation of rules for design for machining manufacturability 2 calculated as center of gravity of the surface under the curve are presented in Figure 15—the design for machining technology—machining processing assessment for substep 2 was 30.6.

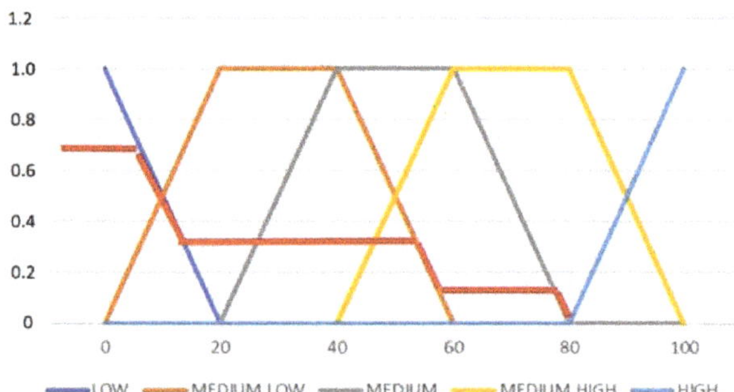

Figure 15. Aggregation of rules for machining technology 2.

The design for machining processing 3 assessment was determined in the same procedure: Energy consumption V5 = 70, waste, environmental aspects V6 = 10. Aggregation of rules for design for machining manufacturability 3 calculated as center of gravity of the surface under the curve presented in Figure 16—design for machining technology—machining processing assessment for substep 3 was 36.

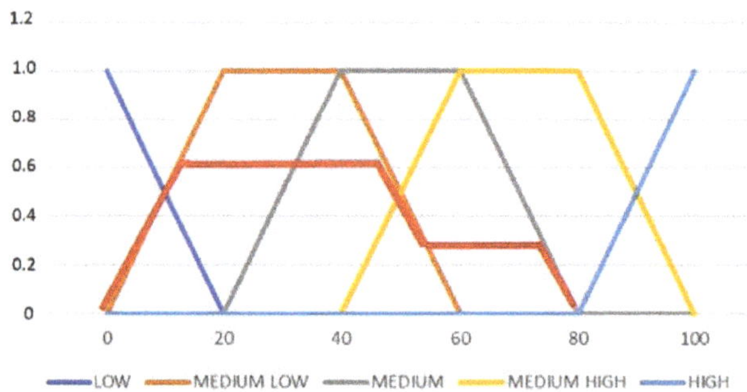

Figure 16. Aggregation of rules for design for machining technology 3.

4.4. Fuzzy Assessment of the Design for a Manufacturing Organization

In the next step called assessment of design for manufacturing organization 1 we perform calculations for the values of V_1 and V_2 by reading the values from the relevant graphs as Figures 14 and 15, according to the inference rule "min" specific rules were activated on the basis of which we set conclusions for the selected component that we evaluate. It depends on two factors, which were: number of components and the possibility of group processing, experts have determined the rating as Number of components $V_1 = 20$, the possibility of group processing $V_2 = 20$. The result of aggregation of rules for design for manufacturing organization 1 calculated as center of gravity of the surface under the curve design for manufacturing organization 1 was 40 (Figure 17).

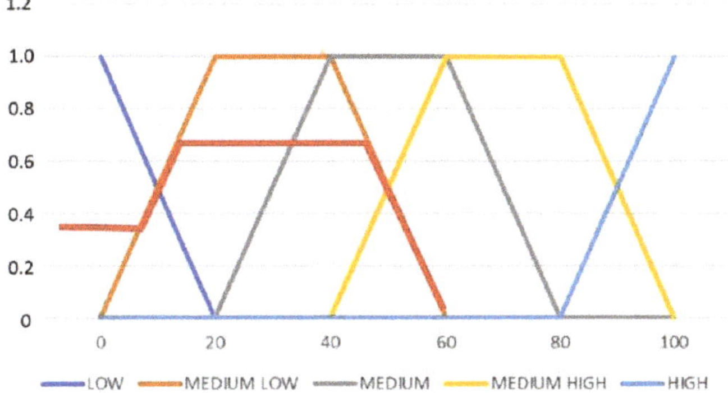

Figure 17. Aggregation of rules for design for manufacturing organization 1.

The design for manufacturing organization 2 assessment was determined in similar procedure. It depends on two factors, which were: component normalization $V_3 = 20$, target cost $V_4 = 55$.

Aggregation of rules for design for manufacturing organization 2 calculated as center of gravity of the surface under curve presented in Figure 18—design for manufacturing organization 2 was 31. design for manufacturing organization 3 assessment was determined in a similar procedure. It depends on two factors: quality of assembly $V_5 = 70$, reuse components $V_6 = 10$.

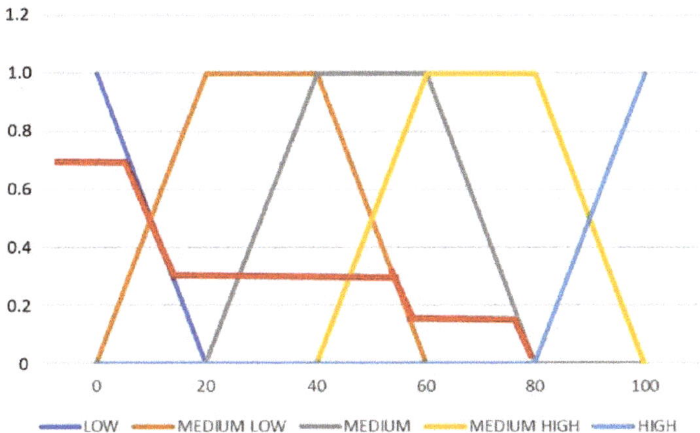

Figure 18. Aggregation of rules for design for manufacturing organization 2.

Aggregation of rules for design for manufacturing organization 3 calculated as center of gravity of the surface under curve presented in Figure 19—design for manufacturing organization 3 was 36.

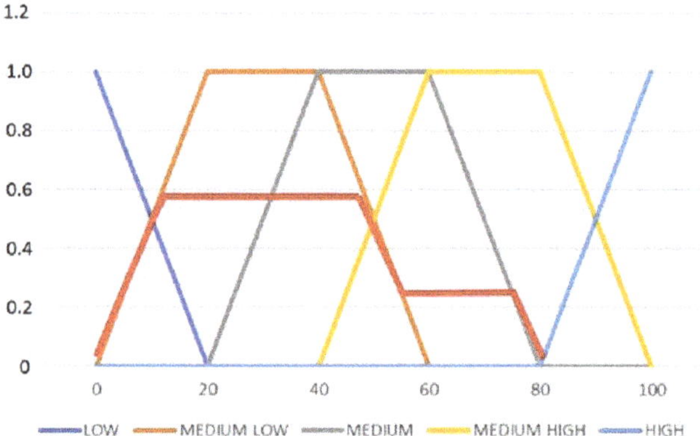

Figure 19. Aggregation of rules for assessment of design for manufacturing organization 3.

4.5. A Fuzzy Assessment of Design for Technology

A complete fuzzy analysis of the technology was carried out in the same stages as for the sample "body" component presented in previous chapters using MATLAB software. Each component of the analyzed transmission was assessed by a group of experts according to their best knowledge in the field of technology, organizational and cost options. Expert assessments were entered in Table 13 and were used in subsequent stages as input to fuzzy analyses. The stages of the analysis were identical to those presented in Figure 3. Calculations of the fuzzy technology assessment method were made using the fuzzy logic toolbox package, which is an addition to the MATLAB program. In the MATLAB FIS editor window, the number of entries and exits is defined and given a name. In this case, two inputs were specified in each step, for example, technological capabilities V1, software capability V2 and one output—The design for machining manufacturability 1. The logical method I (And), logical OR (Or), type of implication, type of aggregation (Aggregation), sharpening method (Defuzzification) were also defined. The analysis selected a uniform representation of the membership function. It can be obtained by using a membership function with a uniform shape and parametric definition of the function. In the case of the assessment of technology, triangular and trapezoidal functions were used.

It should be added that the parametric description of the triangular membership function is the most economical, it only requires three parameters, which are important in practical applications of the method in small lot production industry. After determining the membership sets, one should proceed to the next step of fuzzy analysis of manufacturability. It is creating a set of linguistic rules representing the relationships between system variables. The rules are the heart of the entire regulator. In MATLAB, the next part of the FIS editor is used—Rule Editor. Entered rules can be edited in several different ways. Rule Editor uses the words if, and, or, then, which are the closest to natural language. The final step is to read sharp results using "Rule viewer" or graphical explanation using "Surface viewer".

"Surface viewer" and its graphs—surface charts are additional tools useful in the assessment of construction. With the help of such charts, we can quickly obtain the result of the Technology component without the need for complex calculations. In the chart below "Assembly technology 1" if, for example, the "Access" and the "number of workshop aids" rating would change from (0.55, 0.2)—point A to (0.75, 0.8)—point B, then from the surface chart in Figure 20 we can read the value of Assembly technology 1 at the level of 0.6. In the process of selecting variants to improve the technology, this is a very useful tool that allows you to quickly assess how potential product changes can affect the result of the technology.

Table 13. Set of component assessments made by experts for established criteria.

Group	Machining Manufacturability 1		Machining Manufacturability 2		Machining Manufacturability 3		Assembly Manufacturability 1		Assembly Manufacturability 2		Assembly Manufacturability 3		Process Mgmt. Manufacturability 1		Process Mgmt. Manufacturability 2		Process Mgmt. Manufacturability 3	
	Technological Capabilities	Software Capability	Tool Machining Capability	Compliance Requirements	Energy Consumption	Waste, Environmental Aspects	Access	Number of Workshop Aids	Orientation	Manoeuvrability	Assemblability	Processes	Number of Components	Possibility of Group Processing	Normalization, Unification of Components	Target Cost	Quality of Assembly and Disassembly	Reusing Components
Gear	60	50	40	70	40	20	60	70	80	70	60	70	50	60	90	80	60	60
Main housing	20	55	10	35	70	10	20	55	10	35	70	10	20	20	10	35	70	10
Bearing	80	70	80	70	30	20	60	90	80	90	70	70	50	60	90	90	60	90
Bearing	80	70	80	70	30	20	60	90	80	90	70	70	50	60	90	90	60	90
Bearing	80	70	80	70	30	20	60	90	80	90	70	70	50	60	90	90	60	90
Bearing	80	70	80	70	30	20	60	90	80	90	70	70	50	60	90	90	60	90
Bearing	80	70	80	70	30	20	60	90	80	90	70	70	50	60	90	90	60	90
Vent	70	60	70	60	20	15	45	60	45	60	45	50	50	60	40	60	60	20
Oil sight	70	60	70	60	20	15	45	60	45	60	45	50	50	60	40	60	60	20
Washer	95	95	95	95	10	10	60	20	45	20	45	70	80	80	90	90	80	90
Washer	95	95	95	95	10	10	60	20	45	20	45	70	80	80	90	90	80	90
Washer	95	95	95	95	10	10	60	20	45	20	45	70	80	80	90	90	80	90
Washer	95	95	95	95	10	10	60	20	45	20	45	70	80	80	90	90	80	90
Washer	95	95	95	95	10	10	60	20	45	20	45	70	80	80	90	90	80	90
Washer	95	95	95	95	10	10	60	20	45	20	45	70	80	80	90	90	80	90
Washer	95	95	95	95	10	10	60	20	45	20	45	70	80	80	90	90	80	90
Washer	95	95	95	95	10	10	60	20	45	20	45	70	80	80	90	90	80	90
Washer	95	95	95	95	10	10	60	20	45	20	45	70	80	80	90	90	80	90
Washer	95	95	95	95	10	10	60	20	45	20	45	70	80	80	90	90	80	90
Cover	40	50	40	35	80	10	65	60	30	60	60	20	40	40	50	50	80	10
Cover	40	50	40	35	80	10	65	60	30	60	60	20	40	40	50	50	80	10
Cover	40	50	40	35	80	10	65	60	30	60	60	20	40	40	50	50	80	10
Cover	40	50	40	35	80	10	65	60	30	60	60	20	40	40	50	50	80	10
Cover	40	50	40	35	80	10	65	60	30	60	60	20	40	40	50	50	80	10
Add. Process	95	95	95	95	90	10	90	10	10	10	10	10	10	10	20	30	50	90
Add. Process	95	95	95	95	90	10	90	10	10	10	10	10	10	10	20	30	50	90
Add. Process	95	95	95	95	90	10	90	10	10	10	10	10	10	10	20	30	50	90
Add. Process	95	95	95	95	90	10	90	10	10	10	10	10	10	10	20	30	50	90
Add. Process	95	95	95	95	90	10	90	10	10	10	10	10	10	10	20	30	50	90
Add. Process	95	95	95	95	90	10	90	10	10	10	10	10	10	10	20	30	50	90
Screw	80	70	80	70	10	30	50	70	35	70	55	60	60	40	70	60	70	90
Screw	80	70	80	70	10	30	50	70	35	70	55	60	60	40	70	60	70	90
Screw	80	70	80	70	10	30	50	70	35	70	55	60	60	40	70	60	70	90
Screw	80	70	80	70	10	30	50	70	35	70	55	60	60	40	70	60	70	90
Screw	80	70	80	70	10	30	50	70	35	70	55	60	60	40	70	60	70	90
Screw	80	70	80	70	10	30	50	70	35	70	55	60	60	40	70	60	70	90
Screw	80	70	80	70	10	30	50	70	35	70	55	60	60	40	70	60	70	90
Screw	80	70	80	70	10	30	50	70	35	70	55	60	60	40	70	60	70	90
Screw	80	70	80	70	10	30	50	70	35	70	55	60	60	40	70	60	70	90
Screw	80	70	80	70	10	30	50	70	35	70	55	60	60	40	70	60	70	90
Nameplate	70	60	70	60	20	15	45	60	45	60	45	50	50	60	40	60	60	20
Shaft	50	35	40	70	40	20	60	50	80	50	40	70	35	50	40	60	60	60
Shaft	50	35	40	70	40	20	60	50	80	50	40	70	35	50	40	60	60	60
Shaft	50	35	40	70	40	20	60	50	80	50	40	70	35	50	40	60	60	60
Shaft	50	35	40	70	40	20	60	50	80	50	40	70	35	50	40	60	60	60
Shaft	50	35	40	70	40	20	60	50	80	50	40	70	35	50	40	60	60	60
Shaft	50	35	40	70	40	20	60	50	80	50	40	70	35	50	40	60	60	60
Groove	70	60	70	60	20	15	45	60	45	60	45	50	50	60	40	60	60	20
Groove	70	60	70	60	20	15	45	60	45	60	45	50	50	60	40	60	60	20

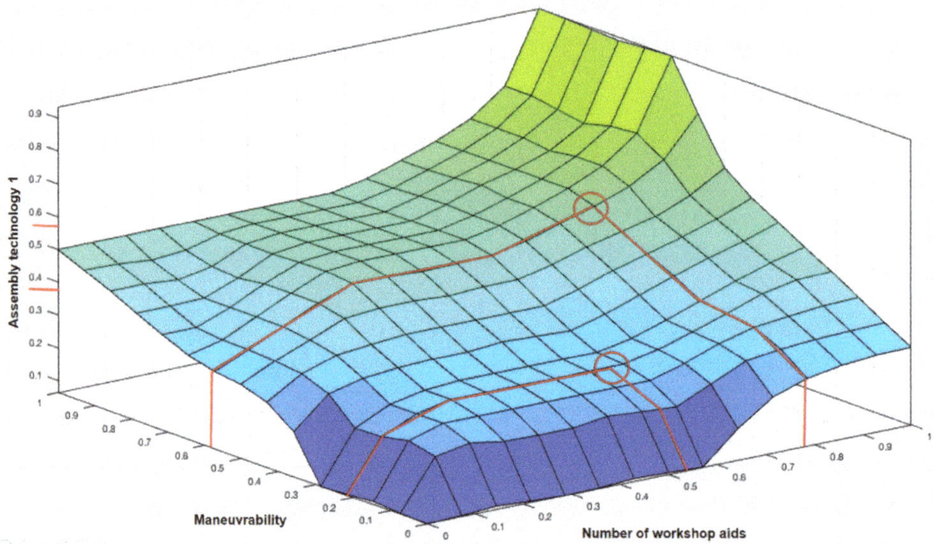

Figure 20. Surface viewer—The surface of the dependence of the variable "installation technology 1" on the input variables for "gears".

For each of the components, calculations were made using the above scheme MATLAB R2017b—the "fuzzy logic designer" module. The development of an approximate representation of knowledge and fuzzy inference methods enables the construction of models for assessing technology to support decision making in conditions of uncertainty and lack of complete information about the problems being solved. Premises and conclusions in these systems were developed using fuzzy logic elements. The knowledge contained in the system should come mainly from a field expert and the effectiveness and efficiency of the system operation depend mainly on the ability to model this knowledge by the system designer. The elements of fuzzy logic presented in the article were used in solving tasks in the field of technological preparation of production. An important problem was the correct definition of fuzzy sets by determining for them the course of belonging functions. [source-fuzzy logic toolbox—user's guide]

Table 14 presents a summary of the results of individual components obtained in the fuzzy transmission analysis. An acceptability criterion of 0.55 was adopted for each component (modeled on the recommendations of the Lucas method and other DFA methods as well as the opinions of experts from industrial practice), elements of lower value should be redesigned. The following transmission components need to be redesigned as a result of the above assessment: body, cover, breather, oil level indicator, covers, additional processes, nameplate, shaft assembly and inlets. Elements rated as not requiring redesign were gears, bearings, washers, screws.

Expert assessments mentioned above are shown in Table 13—those values were used as input to fuzzy analyze conducted in MATLAB—"fuzzy logic designer" module. The results are shown in Table 14.

Table 14. A set of fuzzy method results for individual components for specified criteria.

	Machining Manufacturability 1	Machining Manufacturability 2	Machining Manufacturability 3	Assembly Manufacturability 1	Assembly Manufacturability 2	Assembly Manufacturability 3	Process Mgmt. Manufacturability 1	Process Mgmt. Manufacturability 2	Process Mgmt. Manufacturability 3	
Gear	0.56	0.55	0.36	0.5	0.71	0.55	0.64	0.76	0.7	0.59
Main housing	0.3	0.3	0.36	0.42	0.31	0.36	0.4	0.31	0.36	0.35
Bearing	0.71	0.61	0.3	0.5	0.76	0.56	0.64	0.56	0.7	0.59
Bearing	0.71	0.61	0.3	0.5	0.76	0.56	0.64	0.56	0.7	0.59
Bearing	0.71	0.61	0.3	0.5	0.76	0.56	0.64	0.56	0.7	0.59
Bearing	0.71	0.61	0.3	0.5	0.76	0.56	0.64	0.56	0.7	0.59
Bearing	0.71	0.61	0.3	0.5	0.76	0.56	0.64	0.56	0.7	0.59
Vent	0.7	0.55	0.29	0.5	0.6	0.42	0.64	0.56	0.44	0.52
Oil sight	0.7	0.55	0.29	0.5	0.6	0.42	0.64	0.56	0.44	0.52
Washer	0.83	0.76	0.24	0.29	0.39	0.56	0.73	0.78	0.73	0.59
Washer	0.83	0.76	0.24	0.29	0.39	0.56	0.73	0.78	0.73	0.59
Washer	0.83	0.76	0.24	0.29	0.39	0.56	0.73	0.78	0.73	0.59
Washer	0.83	0.76	0.24	0.29	0.39	0.56	0.73	0.78	0.73	0.59
Washer	0.83	0.76	0.24	0.29	0.39	0.56	0.73	0.78	0.73	0.59
Washer	0.83	0.76	0.24	0.29	0.39	0.56	0.73	0.78	0.73	0.59
Washer	0.83	0.76	0.24	0.29	0.39	0.56	0.73	0.78	0.73	0.59
Washer	0.83	0.76	0.24	0.29	0.39	0.56	0.73	0.78	0.73	0.59
Cover	0.44	0.36	0.38	0.53	0.5	0.3	0.44	0.56	0.38	0.43
Cover	0.44	0.36	0.38	0.53	0.5	0.3	0.44	0.56	0.38	0.43
Cover	0.44	0.36	0.38	0.53	0.5	0.3	0.44	0.56	0.38	0.43
Cover	0.44	0.36	0.38	0.53	0.5	0.3	0.44	0.56	0.38	0.43
Cover	0.44	0.36	0.38	0.53	0.5	0.3	0.44	0.56	0.38	0.43
Add. Process	0.83	0.76	0.37	0.36	0.24	0.24	0.24	0.29	0.64	0.44
Add. Process	0.83	0.76	0.37	0.36	0.24	0.24	0.24	0.29	0.64	0.44
Add. Process	0.83	0.76	0.37	0.36	0.24	0.24	0.24	0.29	0.64	0.44
Add. Process	0.83	0.76	0.37	0.36	0.24	0.24	0.24	0.29	0.64	0.44
Add. Process	0.83	0.76	0.37	0.36	0.24	0.24	0.24	0.29	0.64	0.44
Add. Process	0.83	0.76	0.37	0.36	0.24	0.24	0.24	0.29	0.64	0.44
Screw	0.71	0.61	0.25	0.5	0.54	0.5	0.56	0.71	0.71	0.57
Screw	0.71	0.61	0.25	0.5	0.54	0.5	0.56	0.71	0.71	0.57
Screw	0.71	0.61	0.25	0.5	0.54	0.5	0.56	0.71	0.71	0.57
Screw	0.71	0.61	0.25	0.5	0.54	0.5	0.56	0.71	0.71	0.57
Screw	0.71	0.61	0.25	0.5	0.54	0.5	0.56	0.71	0.71	0.57
Screw	0.71	0.61	0.25	0.5	0.54	0.5	0.56	0.71	0.71	0.57
Screw	0.71	0.61	0.25	0.5	0.54	0.5	0.56	0.71	0.71	0.57
Screw	0.71	0.61	0.25	0.5	0.54	0.5	0.56	0.71	0.71	0.57
Screw	0.71	0.61	0.25	0.5	0.54	0.5	0.56	0.71	0.71	0.57
Nameplate	0.7	0.55	0.29	0.5	0.6	0.42	0.64	0.56	0.44	0.52
Shaft	0.47	0.55	0.36	0.44	0.62	0.55	0.47	0.56	0.7	0.52
Shaft	0.47	0.55	0.36	0.44	0.62	0.55	0.47	0.56	0.7	0.52
Shaft	0.47	0.55	0.36	0.44	0.62	0.55	0.47	0.56	0.7	0.52
Shaft	0.47	0.55	0.36	0.44	0.62	0.55	0.47	0.56	0.7	0.52
Shaft	0.47	0.55	0.36	0.44	0.62	0.55	0.47	0.56	0.7	0.52
Shaft	0.47	0.55	0.36	0.44	0.62	0.55	0.47	0.56	0.7	0.52
Groove	0.7	0.55	0.29	0.5	0.6	0.42	0.64	0.56	0.44	0.52
Groove	0.7	0.55	0.29	0.5	0.6	0.42	0.64	0.56	0.44	0.52

5. Results and Discussion

In the study, the indicators of the assessment of the manufacturability of the structure were determined for the sample product presented in Figure 6. As a result of the analysis after the proposed changes, the new form of the gear structure change is illustrated in Figure 21.

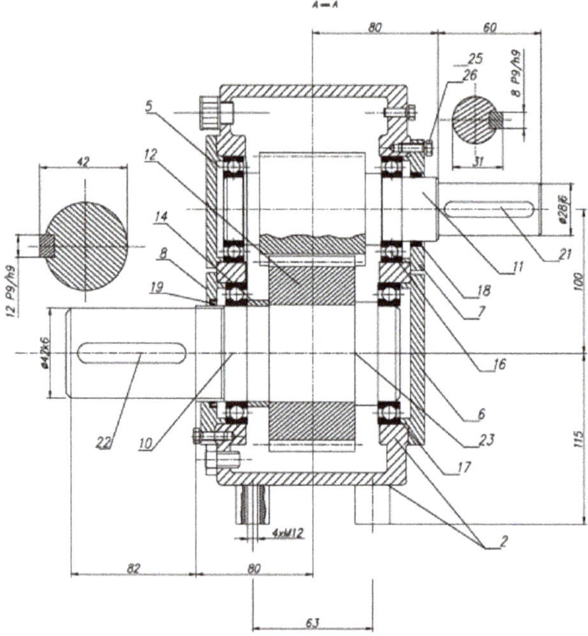

Figure 21. Construction form of the gearbox after the changes were made.

The study cites a comparison of new and currently used in mass production methods of construction technology in Table 15 and Figure 22 which presents the values of the indicators according to the traditional methods and the newly proposed method.

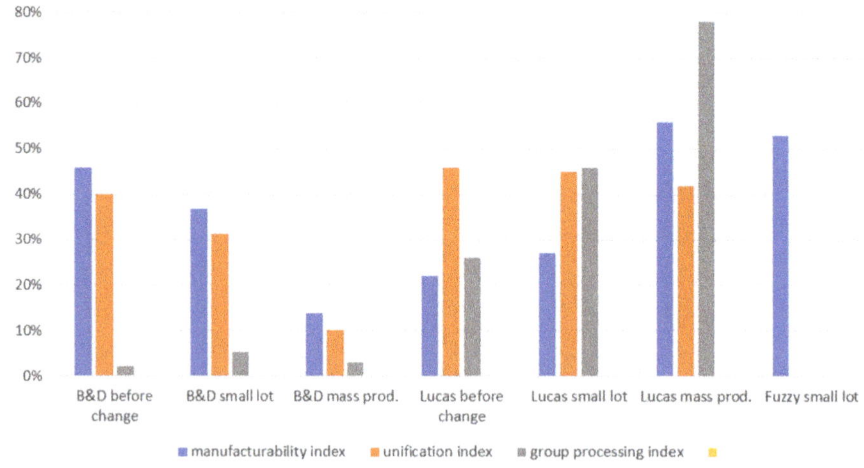

Figure 22. Comparison of methods for gear groups.

Table 15. Comparison of methods design for manufacturability.

Lucas																		Manufacturability index	0.3
Pressing to bearing 16	Pinion shaft	A	1		1	0	0.1	0.2	1.3	Pick up and hold down	1	0	0	0	0	0	0	1	1
										Pressing to bearing	2	0	0	0	0.7	0.7	0	3.4	4.4
Pick up	Main shaft	A	1		1	0	0.1	0.2	1.3	Pick up and place	1	0	0	0	0	0	0	1	5.4
Assembly	Wedge 23	B		1	1	0	0	0.2	1.2	Assembly on shaft subassy	1	0	0	0	0.7	0	0	1.7	7.1
Assembly on shaft subassy	Gear 5	B		1	1	0.4	0.1	0.2	1.7	Assembly on shaft subassy	2	0.1	0	0	0.7	0.7	0	3.5	10.6
Assembly on shaft subassy	Spacer sleeve 14	B		1	1	0	0	0	1	Assembly on shaft subassy	1	0.1	0	0	0	0	0	1.1	11.7
Preheating gear to 180 deg.	Preheating	B		1	0	0	0	0	0	Prehating	0	0	0	0	0	0	7.5	7.5	19.2
		6	2						6.5										59.4
Boothroyd																		Manufacturability index	**0.24**
Gear	pressing for bearing 16	1	37	194	360	0	88	6.35	41	7.5	13.85					Y	N	Y	1
Drive gear	Pick up and place	1	28	216	360	360	30	1.95			1.95					Y	N	Y	1
Groove	Assembly	1	16	40	180	360	20	1.8	01	2.5	4.3					N	N	N	0
Gear no. 5	Assembly on shaft subassy	1	41	136	180	360	88	6.35	31	5	11.35					N	N	N	0
Sleeve no. 14	Assembly on shaft subassy	1	4	60	180	0	00	1.13	01	2.5	3.63					N	N	N	0
Preheating	Preheating gear to 180 deg.	1							99	12	12					N	N	N	0
Shaft subassembly	Pressing to bearing	1	136	216	360	360	30	1.95	51	9	10.95					Y	Y	Y	1
				7							58.03								3
Fuzzy																		Manufacturability index	**0.53**
Shaft to bearing	Pressing to bearing	0.47		0.55		0.36		0.44		0.62		0.55		0.47		0.56		0.7	0.52
Shaft	Assembly	0.47		0.55		0.36		0.44		0.62		0.55		0.47		0.56		0.7	0.52
Groove	Assembly	0.7		0.55		0.29		0.5		0.6		0.42		0.64		0.56		0.44	0.52
Gear	Assembly on shaft subassy	0.56		0.55		0.36		0.5		0.71		0.55		0.64		0.76		0.7	0.59
Sleeeve	Assembly	0.71		0.61		0.3		0.5		0.76		0.56		0.64		0.56		0.7	0.59
Preheating	Preheating gear to 180 deg.	0.83		0.76		0.37		0.36		0.24		0.24		0.24		0.29		0.64	0.44
Shaft subassembly	Assembly	0.47		0.55		0.36		0.44		0.62		0.55		0.47		0.56		0.7	0.52

The comparison should read that the lower the score, the more you need to redesign/reduce the product design. From the comparison of the assessment, it can be concluded that the Boothroyd-Dewhurst method is the most stringent and focused on reducing/simplifying the details of the project components. At the same time, in the case of production which is not qualified for high-volume production, the result of such an assessment may be a product with a small number of components, but in a very complicated form and therefore will have a high cost of processing, quality and others in the field of production organization. However, the new fuzzy method, because it takes into account the treatment and its limitations and the index of production organization in which it directs the result towards small-lot production gives the least result which improves the technological efficiency of this sub-assembly for the assumptions of the model of fuzzy small-lot production.

The result of a single method does not give a picture of effectiveness nor a new fuzzy method. For this purpose, a comparison of a selected transmission fragment of the new method and existing methods was carried out. The selected fragment in the form of a drive shaft assembly and was compared in Table 3 from the overall assessment of the transmission. Comparisons show that for small lot production assessment B&D with 0.3 results, Lucas with 0.24 results, versus fuzzy defined for a small lot with 0.53 results, was a less restrictive approach of Fuzzy method. This was very important as small lot production usually has much less capital available.

The comparison should be read as follows, the lower the score, the more you need to redesign/reduce the product design. From the comparison of the assessment, it can be concluded that the Boothroyd & Dewhurst method was the most stringent and focused on reduce/simplify the components of the project. At the same time, in the case of production not qualified for high-volume production, the result of such assessment may be a product with a small number of components, but a very complicated form and therefore a high cost of processing and quality and other in the field of production organization. The Lucas method in a more balanced way assesses the above project, but the difference from Boothroyd-Dewhurst is not large, which means that it will also work best in mass production.

6. Conclusions and Comments

In this study, we have focused on the assessment of gearbox using a new developed fuzzy method and compared it to the most known Boothroyd and Lucas DFA methods. The purpose was to assess gearbox development for small lot production, propose design changes and evaluate its design. This was very important as small lot production usually has much less capital available. The method developed is open and other or additional criteria may be considered according to the production conditions of the company concerned.

The fuzzy method was more tuned to this volume level of process and it was an advantage of this method in comparison to other methods that were more suitable for only mass production. The flexibility of this method was one of the aims of creating it.

In standard technology analysis, according to B&D and Lucas DFA, this was associated with a reduction in the number of components that have no significant effect on the product functions which results in an improvement in terms of assembly time and costs. In the traditional arrangement, of the above mentioned the methods, they were oriented towards mass production.

The proposed proprietary method based on the analysis of the obtained values of the parameters of the assessment of the efficiency of the entire process enables:

- Considering—in addition to assembly—many other various factors, for example, availability of spare parts, production seriality, production conditions in the form of equipment types, available assembly techniques, level of automation, the scope of external cooperation orders
- The method can be used for smaller series of manufactured products;
- Assessment of technology in the form of given indicators and coefficients should be carried out by experts with extensive production experience;

Appl. Sci. **2020**, *10*, 3935

- Arousing designers' creativity when designing new products, rationalizing work at the stage of improving and expanding the range of implemented production.

The presented method is universal. The use of fuzzy logic allows expressing incomplete and uncertain information in natural language, in a simple way for humans based on expert knowledge and empirical data. The method considers the analysis of the production process in a holistic way.

Author Contributions: Conceptualization, J.M., T.S. and A.M.; data curation, T.S.; formal analysis, J.M., T.S. and A.M.; funding acquisition, J.M.; methodology, J.M.; resources, J.M., T.S. and A.M.; software, T.S.; validation, A.M.; visualization, A.M.; writing—original draft, J.M., T.S. and A.M.; writing—review and editing, J.M., T.S. and A.M. All authors have read and agreed to the published version of the manuscript.

Funding: This research has received no external funding.

Conflicts of Interest: The authors declare no conflict of interest.

References

1. Zorowski, C.F. PDM—A product assemblability merit analysis tool. In Proceedings of the Design Engineering Technical Conference, Columbus, OH, USA, 5–8 October 1986; pp. 5–8.
2. Barnes, C.J. *A Methodology for the Concurrent Design of Products and Their Assembly Sequence*; Cranfield University: Cranfield, UK, 1999; p. 27.
3. Gupta, P.; Trusko, B.E. *Global Innovation Science Handbook*; McGraw Hill Professional: New York, NY, USA, 2014; ISBN 0071834745.
4. Akao, Y. *Quality Function Deployment (QFD). Integrating Customer Requirements into Product Design*; Productivity Press: Cambridge, UK, 1990; Volume 369.
5. Evans, J.R.; Lindsay, W. *The Management and Control of Quality*; South-Western College: Chula Vista, CA, USA, 1999.
6. Kehoe, D. *Acceptance Sampling. //The Fundamentals of Quality Management*; Springer: Dordrecht, The Netherlands, 1996; pp. 226–227.
7. Palady, P.; Olyai, I. The status quo's failure in problem-solving. *Qual. Prog.* **2002**, *35*, 34–39.
8. Fayaz, M.; Ullah, I.; Kim, D.H. Underground risk index assessment and prediction using a simplified hierarchical fuzzy logic model and Kalman filter. *Processes* **2018**, *6*, 103. [CrossRef]
9. Miles, B.L. Design for Assembly—A key element within Design for Manufacture. *Proc. Inst. Mech. Eng.* **1989**, *1*, 29–38, ISSN 0954-4070. [CrossRef]
10. Yang, K.I.; El-Haik, B.S. *Design for Six Sigma*; Mcgraw-Hill: New York, NY, USA, 2003; pp. 184–186.
11. Andreasen, M.M.; Kähler, S. *Lund Design for Assembly*; Ifs: Lund, Sweden, 1988.
12. Boothroyd, G. *Product Design for Manufacture and Assembly. Computer-Aided Design*; Elsevier: Amsterdam, The Netherlands, 1994; Volume 26, pp. 505–520.
13. *Lock Project Management*, 9th ed.; Gower Publishing Ltd.: Hampshire, UK, 2009.
14. Sarwar, B.; Bajwa, I.S.; Ramzan, S.; Ramzan, B.; Kausar, M. Design and application of fuzzy logic based fire monitoring and warning systems for smart buildings. *Symmetry* **2018**, *10*, 615. [CrossRef]
15. Swift, K.G.; Booker, J.D. *Process Selection, from Design to Manufacture*; Butterworth-Heinemann: Oxford, UK, 2003; ISBN 978–0–7506–5437-1.
16. More, N.K.; Buktar, R.B.; Ali, S.M.; Samant, S. Design for Manufacture and Assembly (DFMA) Analysis of Burring Tool Assembly. *Int. Res. J. Eng. Technol.* **2015**, *2*, 2395-56.
17. Chu, W.S.; Kim, M.S.; Jang, K.H.; Song, J.H.; Rodrigue, H.; Chun, D.M.; Min, S. From design for manufacturing (DFM) to manufacturing for design (MFD) via hybrid manufacturing and smart factory: A review and perspective of paradigm shift. *Int. J. Precis. Eng. Manuf. Green Technol.* **2016**, *3*, 209–222. [CrossRef]
18. Bouissiere, F.; Cuiller, C.; Dereux, P.E.; Malchair, C.; Favi, C.; Formentini, G. Conceptual Design for Assembly in the aerospace industry: A method to assess manufacturing and assembly aspects of product architectures. In Proceedings of the Design Society: International Conference on Engineering Design, Delft, The Netherlands, 5–8 August 2019; Cambridge University Press: Cambridge, UK, 2019; Volume 1, pp. 2961–2970.

19. Matuszek, J.; Seneta, T. Algorithmization of new product process implementation in terms of mass production. *Mechanik* **2016**, *7*, 755–757. [CrossRef]

20. Matuszek, J. *Process Engineering*; Publisher Technical University of Lodz Branch Bielsko-Biala: Bielsko-Biala, Poland, 2000.

21. Zhang, H.; Han, X.; Li, R.; Qin, S.; Ding, G.; Yan, K. A new conceptual design method to support rapid and effective mapping from product design specification to concept design. *Int. J. Adv. Manuf. Technol.* **2016**, *87*, 2375–2389. [CrossRef]

22. El Wakil, S.D. *Processes and Design for Manufacturing*; CRC Press: Boca Raton, FL, USA, 2019; ISBN 978-1-138-58108-1.

23. Favi, C.; Germani, M.; Mandolini, M. Development of complex products and production strategies using a multi-objective conceptual design approach. *Int. J. Adv. Manuf. Technol.* **2017**, *95*, 1281–1291. [CrossRef]

24. Konnikov, E.A.; Konnikova, O.A.; Rodionov, D.G. Impact of 3D-Printing Technologies on the Transformation of Industrial Production in the Arctic Zone. *Resources* **2019**, *8*, 20. [CrossRef]

25. Suhariyanto, T.T.; Wahab, D.A.; Rahman, M.N.A. Product design evaluation using life cycle assessment and design for Assembly: A case study of a water leakage alarm. *Sustainability* **2018**, *10*, 2821. [CrossRef]

26. Chung, P.-H.; Ma, D.-M.; Shiau, J.-K. Design, Manufacturing, and Flight Testing of an Experimental Flying Wing UAV. *Appl. Sci.* **2019**, *9*, 3043. [CrossRef]

27. Francia, D.; Ponti, S.; Frizziero, L.; Liverani, A. Virtual Mechanical Product Disassembly Sequences Based on Disassembly Order Graphs and Time Measurement Units. *Appl. Sci.* **2019**, *9*, 3638. [CrossRef]

28. Gielisch, C.; Fritz, K.-P.; Noack, A.; Zimmermann, A. A Product Development Approach in the Field of Micro-Assembly with Emphasis on Conceptual Design. *Appl. Sci.* **2019**, *9*, 1920. [CrossRef]

29. Han, C.; Sin, I.; Kwon, H.; Park, S. The Role of the Process and Design Variables in Improving the Performance of Heat Exchanger Tube Expansion. *Appl. Sci.* **2018**, *8*, 756. [CrossRef]

30. Kim, Y.-J.; Heo, J.-Y.; Hong, K.-H.; Hoseok, I.; Lim, B.-Y.; Lee, C.-S. Computer-Aided Design and Manufacturing Technology for Identification of Optimal Nuss Procedure and Fabrication of Patient-Specific Nuss Bar for Minimally Invasive Surgery of Pectus Excavatum. *Appl. Sci.* **2019**, *9*, 42. [CrossRef]

31. Ding, K.; Avrutin, V.; Izyumskaya, N.; Özgür, Ü.; Morkoç, H. Micro-LEDs, a Manufacturability Perspective. *Appl. Sci.* **2019**, *9*, 1206. [CrossRef]

32. Rivera, C.A.; Poza, J.; Ugalde, G.; Almandoz, G. A Requirement Engineering Framework for Electric Motors Development. *Appl. Sci.* **2018**, *8*, 2391. [CrossRef]

33. Boothroyd, G.; Dewhurst, P. *Design for Assembly: A Designers Handbook*; University of Massachusetts: Amherst, MA, USA, 1983.

34. Herrmann, J.W.; Cooper, J.; Gupta, S.K.; Hayes, C.C.; Ishii, K.; Kazmer, D.; Wood, W.H. New directions in design for manufacturing. In Proceedings of the ASME 2004 International Design Engineering Technical Conferences and Computers and Information in Engineering Conference, American Society of Mechanical Engineers Digital Collection, Salt Lake City, UT, USA, 28 September–2 October 2004; pp. 853–861. [CrossRef]

35. Relich, M.; Świc, A.; Gola, A.A. Knowledge-Based Approach to Product Concept Screening. In Proceedings of the Distributed Computing and Artificial Intelligence, 12th International Conference, Advances in Intelligent Systems and Computing, Salamanca, Spain, 3–5 June 2015; Omatu, S., Malluhi, Q.M., González, S.R., Bocewicz, G., Bucciarelli, E., Giulioni, G., Iqba, F., Eds.; Springer: Cham, Switzerland, 2015; Volume 373.

36. Thekinen, J.; Panchal, J.H. Resource allocation in cloud-based design and manufacturing: A mechanism design approach. *J. Manuf. Syst.* **2017**, *43*, 327–338. [CrossRef]

37. Gebisa, A.W.; Lemu, H.G. Design for manufacturing to design for Additive Manufacturing: Analysis of implications for design optimality and product sustainability. *Procedia Manuf.* **2017**, *13*, 724–731. [CrossRef]

38. Mehta, M. *Holistic Consideration of Best Practices in Product Design, Quality, and Manufacturing Process Improvement through Design for Value*; American Society for Engineering Education: Atlanta, GA, USA, 2013.

39. Shetty, D.; Ali, A. A new design tool for DFA/DFD based on rating factors. *Assem. Autom.* **2015**, *35*, 348–357. [CrossRef]

40. Whitney, D.E. *Mechanical Assemblies—Their Design Manufacture and Role in Product Development*; Oxford University Press: Oxford, UK, 2004.

41. Kacprzyk, J.; Pedrycz, W. (Eds.) *Springer Handbook of Computational Intelligence*; Springer: Dordrecht, The Netherlands; Heidelberg, Germany; London, UK; New York, NY, USA, 2015.
42. Corsini, L.; Moultrie, J. An exploratory study into the impact of new digital design and manufacturing tools on the design process. In DS 87-2 Proceedings of the 21st International Conference on Engineering Design (ICED 17) Vol 2: Design Processes, Design Organizationand Management, Vancouver, BC, Canada, 21–25 August 2017; pp. 21–30, ISBN 978-1-904670-90-2.

Article

Productivity Improvement through Reengineering and Simulation: A Case Study in a Footwear-Industry

Rubén Calderón-Andrade [1], Eva Selene Hernández-Gress [2,*] and Marco Antonio Montufar Benítez [1]

[1] Engineering Academic Area, Universidad Autónoma del Estado de Hidalgo, Ciudad del Conocimiento; Carretera Pachuca Tulancingo km 4.5, Mineral de la Reforma, 42184 Hidalgo, Mexico; ca281824@uaeh.edu.mx (R.C.-A.); montufar@uaeh.edu.mx (M.A.M.B.)

[2] School of Engineering and Science, Tecnologico de Monterrey, Boulevard Felipe Ángeles No. 2003, Pachuca de Soto, 42080 Hidalgo, Mexico

* Correspondence: evahgress@tec.mx

Received: 15 July 2020; Accepted: 10 August 2020; Published: 12 August 2020

Featured Application: The present case study reports a methodology using Reengineering and Simulation in a real manufacturing production process. Future applications can be adapted to other manufacturing industries by integrating the most important principles and steps in their own context.

Abstract: Process reengineering is a very useful tool, specifically in industrial engineering where technological advances, information systems, customer requirements, and more have led to the need for radical change in some or all areas of an organization. The objective of this work is to show the usefulness of applying reengineering in the case of the footwear industry to make a proposal to change the problem area and the production decoration line as well as compare it with the current process using models of simulation performed in the Arena™ software. The proposal consisted of merging two production lines and comparing the current design with the proposal as well as comparing different parameters such as the use of resources and the production rate. The results indicated that the production rate increases by approximately 29% with the new design, using the same resources. In addition, using the OptQuest tool of the Arena™ software, it was found that with the new process, the production rate could be increased by up to 41% compared to the current process.

Keywords: reengineering; simulation; productivity

---

1. Introduction

Industries need to be more competitive in order to survive in the increasingly dynamic and global environment. For this reason, they require different methodological approaches to make their operations more efficient. Some problems that block such operations' efficiency and disable the ability to respond to the environment, could be administrative, financial, or a production process, among others. In this article, we will focus on improving such a production process.

A process is any activity that occurs within the company [1] and, in this case, if there are problems in the processes, it is difficult to meet customer expectations in quality, delivery time, among others, which are qualities that are currently necessary. In addition, a process is a collection of activities that has one or more inputs and generates an output that adds value to the customer [2], which is also a process that could be seen as chains of activities and decisions [3]. According to Laguna and Marklund [4], the essence of process design is related to doing things in the "correct" way. Correct in this context refers to the process being efficient and effective. Efficiency refers to the fact that the customer's requirements must be met at the correct time.

For Davenport [2], business must be viewed in terms of redesigning the process from the beginning to the end using resources that are available in the company. Using resources that the company has is important for industries in emergent economies that do not have money to invest, and desire major improvements in quality, flexibility, service levels, or productivity [5]. In our context, a business problem is not adequately employing the resources that the organization has: machinery, equipment, facilities, etc., to achieve a response to the client in terms of time, quality, and service.

Being productive is related to different factors. Some of them are considered within this research work, since they were used precisely for the solution of the case study. Among them, there are adequate flows of the product, arrangements of the plant, and correct allocation of human-machine resources to the different activities that the process requires. Productivity can be defined as the number of outputs or finished products per unit of time [4,6]. Productivity can also be calculated by dividing item prices by their costs [7] and can be compared with past values from the same company for analysis. In this work, we will use the amount of production per unit of time, but we will use it to carry out analysis of different scenarios.

The main objective of this article is describing a case study carried out in an industrial shoe company in the state of Hidalgo, Mexico. In this company, there were productivity problems caused by the poor flow of the product. To deal with it, the reengineering phases were applied and, through these, a proposal was presented, which was analyzed through simulation. The study company did not want to be identified in this article so it will be called "the company" from now on, but the result of this intervention was implemented. This article presents the applied methodology and the results found this work could help other companies that have no idea how to identify and solve problems within their organizations.

The rest of the article is organized in the following sections. First is the literature review, which cites some works related to reengineering and simulation. Second, in Section 3, the materials and methods were described. The methodology is presented in Section 4 for Reengineering and Section 5 describes the simulation. In Section 6, the most important results are analyzed. Lastly, in Section 7, the conclusions of the work are given.

2. Literature Review

Before focusing on reengineering and simulation, it is necessary to mention other disciplines. Reengineering is not the only discipline that is concerned with improving the operational performance of organizations. Total Quality Management (TQM) preceded and inspired reengineering. The focus is on continuously improving and sustaining the quality of products while reengineering focus on the improvement of processes. Operations management is a field relevant to production and manufacturing. This area includes probability theory, queuing theory, mathematical modeling, Markov chains, and simulation techniques for improving the efficiency from this perspective. Operations management could be applied to existing processes and reengineering generates a new one. Another technique is lean manufacturing that pursue waste elimination for eliminating activities that did not add value to the customer such as using value stream mapping. Both reengineering and lean manufacturing have the customer orientation as a principle. Another technique is Six Sigma that focuses on minimizing errors and defects when measuring process output. Many Six Sigma techniques are applied in reengineering. All these techniques not only have similarities with reengineering, but can be combined with it [3,4].

The task arrangement and the excessive control of them are not the appropriate approach to achieve better production times within a productive organization [8]. There are now methodologies where the client plays the most important role and the service provided can be the difference between a company and its competition. It is for this reason that a small improvement fails to position companies, as the offer is extremely huge, and customers are increasingly selective when making purchases. Reengineering achieves dramatic enhancements through a complete redesign of core business processes, combined with rapid implementation, which makes it a strategy when significant changes need to be achieved. At this point, the definition of best fit is to "start again" [1]. According to

Manganelli and Klein [9], reengineering is the answer when it is required to optimize workflows and productivity in an organization, as in the case presented in this study. Therefore, there are three types of improvement in the processes of organizations, which include a small improvement in one of the activities of the process, a redesign of the process, that involves changes in all operations, and reengineering, which is a radical redesign that destroys assumptions, builds again, and achieves dramatic changes in some performance measures [1,8].

For Hammer and Champy [1], reengineering is the fundamental revision and radical redesign of processes to achieve spectacular improvements in contemporary and critical measures of performance such as costs, quality, service, and speed. In this sense, it is always necessary to have measures of performance that provide clarity to know if the desired objective was accurate. Since the 1990s, when reengineering was first used, it has been successfully applied to different case studies. Ford, Mutual Benefit Life [8], Taco Bell [1], and others stand out. Hammer and Champy [1] considered a formal idea of what reengineering should be including the assumptions, what was expected in terms of achievement, and documented case studies. However, a formal methodology of how to achieve this was not detailed because reengineering must adapt to the particular situation of each company. As the years have passed, the concept has matured by applying it to different cases with significant results. Some of them are presented below.

Generally, Reengineering is combined with other methodologies because it is useful to analyze the root of the problems and present proposals for solutions based on its principles, but, to test if these proposals are adequate and implement them, it is necessary to use other tools. Nguyen [10] combines reengineering and lean manufacturing to improve an electronic assembly line by achieving a 40% decrease in the number of workers and 30% savings in production plant space. Other articles use basic tools but achieve significant changes. For example, with the application of reengineering to an air cargo handling company [11], flowcharts were used to analyze the number of activities, delays, etc. before and after applying reengineering, which achieves improvements in service. Reengineering and a balanced scorecard are used in the work of Turhan [12] where reengineering is applied to the supply chain. It is also used with a balanced scorecard in References [13,14] to make the diagnosis and present proposals to a bamboo panel construction company, which helps achieve improvements in the design of products, processes, and design of administrative activities.

Regardless of the number of companies involved in reengineering, the rate of failure in reengineering projects is more than 50 [1]. Some frequently mentioned problems related to reengineering include the inability to accurately predict the outcome of a radical change, and the inability to recognize the dynamic nature of the processes. Additionally, some publications argue that one major problem that contributes to the "failure" of reengineering projects is the lack of tools for evaluating the effects of designed solutions before implementation [15,16]. One way to do this is through simulation, since you have access to modify different input variables, while analyzing different parameters such as productivity, use of resources, along with others, and deciding on the best strategy. Simulation performance indicators are necessary to carry out the analysis [17]. The simulation also provides a graphical way of understanding the process flow, which is easily followed by the end user. It is possible to stop and modify and run again in such a way that each operation can be analyzed in a simple way [18].

Modeling through simulation is one of the most widely used techniques and can be considered a representation of a real system and in which it is experimented with the purpose of having a better understanding of its behavior and evaluating the impact of alternative strategies [19]. In today's dynamic environment, simulation helps us understand complex processes and can be used to make decisions within organizations [20].

There are many proven cases in the literature where reengineering and simulation have been used. Some of them related to production processes, Chen [21] applied reengineering to a construction company, where the traditional production is changed to a flow shop and the proposal is tested through simulation in Reference [22]. Reengineering and simulation are also combined to decrease bottleneck operations in a company that manufactures ceramic products by obtaining a 50% reduction in the time cycle and a reduction in delivery time from 3 to 1.5 days. Irani, Hulpic, and Giaglis [23] present the analysis in a production company, where reengineering was implemented using a simulation and innovation method to support decision-making. In a similar way, they have used both methodologies to service processes. See the works Sung [24] and Qiang et al. [25] in hospitals in which the first for improved surgical care and the second for hospital registration. Xiaoming and Xueqing [26] developed a framework for reengineering construction processes and corresponding methodologies that integrate lean principles and computer simulation techniques. Previous studies agree that simulation allows testing and analysis of different scenarios to understand their impact and assess feedback before moving forward with implementation plans.

Although other methodologies were analyzed when the intervention was being carried out in the organization, it was decided to use reengineering since it only requires simple activities such as training personnel. Another advantage is that human and economic resources that the company already has can be used, which makes it feasible. Moreover, the principles include destroying assumptions and starting again instead of not using technology without having analyzed if there are any problems in the process In addition to the numerous success stories in implementing it, they made it the ideal strategy to analyze the problem and issue a proposal. The simulation served to analyze this proposal with different scenarios before its implementation, as it is detailed in the next section. As a conclusion, there is no unique methodology for applying reengineering and sometimes it fails because it is not possible to test the effectiveness of the proposals made through reengineering, which is why we also address simulation. At the time the article was written, no references were found that used reengineering applied to an industrial footwear company.

3. Materials and Methods

As mentioned in the previous section, reengineering does not require too many resources since the human resources available to the organization can be used. Only the authors of this article were independent from it and the proposals issued were not small gradual changes. In addition to providing training, it also includes taking time of operations and attending meetings. These meetings were held in a small room provided by the company. The authors also carried out field work, design of the methodology, analysis of results, and more. To do this, they used the Arena™ 14.0 software to perform the simulation and Minitab™ 18.0 to perform the statistical analysis. The issue and the methodology used are detailed below.

3.1. Problem Description

The study described below was implemented in an industrial shoe company in the state of Hidalgo, Mexico. In the moment that we did the intervention in the company, they had a Make to Stock Model for doing business. In the industrialized society of mass production and marketing, this forecast needs standardization and efficient business management such as cost reduction and a fast response to costumers. Moreover, the company was trying to expand its market. For this reason, it was necessary to increment its productivity.

In this company, there were two decorating lines. Specifically, Line 1 presented productivity problems at the time of the study. It was necessary to work extra hours and still did not achieve the established production goal. Through observation, it was perceived that it was because the activities did not follow the natural flow of the process. However, a methodology was followed to analyze the problem. Reengineering served to conceptualize the problem and present the proposal, and the simulation helped validate it.

The proposal of this work is practical. It was carried out for the improvement of the processes and allowed a complete analysis of the current process with its respective deficiencies through reengineering, which were considered to make a change. Once the process improvement was selected, a proposal was made. This approach was tested through the simulation. Until the start of this research, any work focused on reengineering was found in an industrial footwear factory supported by simulation. This fusion of methodologies can be used in companies in another sector, especially small and medium-sized companies, with their respective adaptations.

3.2. Methodology

As already mentioned, the company's problem was low productivity on line 1 of decorating. Through process reengineering, a proposal was made to improve certain production indicators such as: quantity of products obtained at the end of a work shift, use of resources, and the inventory in the process of each activity. Data was collected and statistical analysis was used to establish patterns of behavior. In this sense, production times were taken in the activities of line 1 of decorating, which served to carry out the simulation and, thus, compare the current design of Line 1 against the proposal. The following sections briefly describe the methodology executed with reengineering and simulation in each of the stages.

4. Reengineering

After reviewing the literature and, since there is not a sole-step series for the reengineering application [1,2,5], a methodology was adapted to serve the specific case of the company, which is shown in Figure 1. This methodology, with some variations, had already been designed and used by one of the authors in a global intervention at a company that manufactures bamboo panels [14,15].

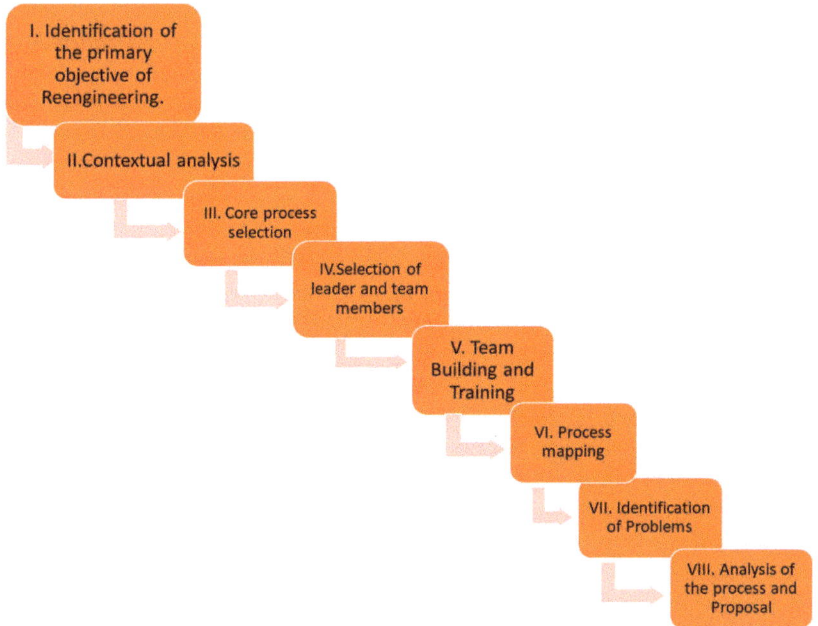

Figure 1. Reengineering methodology elaborated by the author.

Each of these stages is described below.

4.1. Identification of the Primary Objective of Reengineering

After observation, analysis and discussion with senior management, and carrying out a focus group, it was determined that the objective of the reengineering would be to increase the productivity of the company's Line 1 of Decorating since it never reached the production goal. It is necessary to mention that there was also another decorating line with the same machines and operations. Line 2 had achieved the production goals. Therefore, the problem of Line 1 was uncertain.

4.2. Contextual Analysis

This is the starting point to make a critical assessment of all those external things that affect the design or implementation of the project. In the case study, it was divided into 2 phases. First, the analysis of the shoe industry in Mexico and, second, the analysis of the company's main competitors in the state of Hidalgo.

4.2.1. Shoe Industry in Mexico

The manufacture of Mexican footwear is an important commercial activity in our country, which generates a highly competitive supply chain. According to the Ministry of Economy in Mexico [27], the following information is available.

- Four entities of the Republic concentrate 94% of the value of footwear production: Guanajuato 70%, Jalisco 15%, the State of Mexico 5%, and Mexico City, Federal District 3%.
- The footwear industry is the main link in the leather-footwear chain and is made up of nearly 7,400 manufacturing establishments (equivalent to 68.4% of the total production chain).
- About 41,500 shoe stores exist throughout the national territory.
- In 2014, 25.6 million pairs of shoes were exported with a value of $571.7 million.
- In August 2014, a framework was established to promote actions that stimulate the productivity and competitiveness of the industry as well as prevent and combat the underestimation of imported goods.
- These figures tell us that the footwear industry is growing, especially in terms of exports, so the company has opportunities in this regard. The reason for it must make its processes efficient in pursuance of competing internationally.

4.2.2. Company Competitors

At the time of the study, there were four companies dedicated to the manufacture of industrial footwear in the state of Hidalgo: KARTEK, TEMO, Tempac, and the "X" company. Competitor research is conducted because a reengineering tool is referencing (Benchmarking) [1,2], which means searching for companies that are doing something optimally and finding out how they do it to emulate them. The analysis was performed by considering four factors: antiquity, personnel, strategic planning, and products. We realized that the four companies offer a wide variety of industrial footwear models offering quality. Two of the companies have the continuous improvement in their products and services as their mission. In the third aspect, they define their mission based on technology and the last one considers the client's requirements. Only TEMO has the vision of being an internationally recognized company. KARTEK raises its vision as being the leading shoe producer nationwide. The study company sets its goals with a higher percentage of customer satisfaction and delivery of its products on time, while KARTEK is based on customer satisfaction and being a profitable company. The Official Mexican Standard NOM-113-STPS (1994) governed all four and only two are certified (ISO 9001-2008). An advantage of the study company is that it is certified, but in order to achieve its objective of delivering products on time, workers must stay out of their work shift with Line 1 of Decorating being the problem area. With contextual analysis, the information accomplished was that competitors use similar technologies and have similar certifications. It was also found that the industry in Mexico is growing and it is necessary to organize the resources to have a better response to the customers.

4.3. Core Process Selection

According to Hammer and Champy [1], when the objective is to improve the process, a detailed analysis is not necessary. Rather, a general analysis is required that detects the critical aspects that demonstrate the causes of the deficiencies of the current process. The company operates with a process or functional organization, since the areas are grouped according to the equipment or machinery by similar functions, such as the Strobel department, the stitching department, the assembly department, the PU injection, the decoration area, and more, as shown in Figure 2.

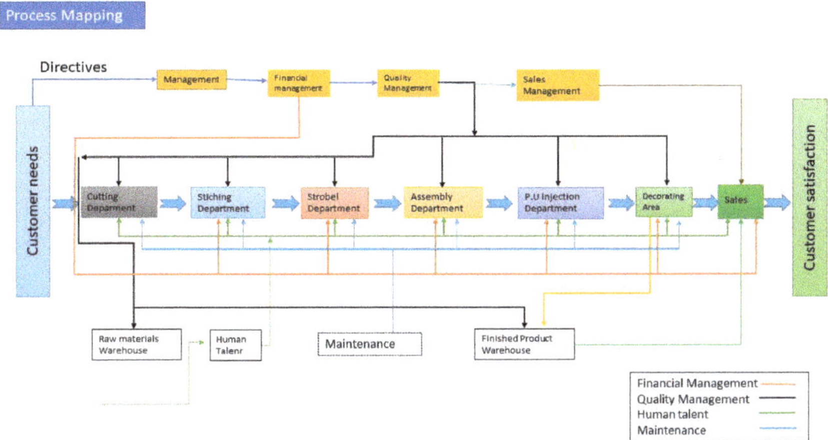

Figure 2. Process mapping elaborated by the author. The colors represent the flow of materials and information by the department.

Regarding the process, before the product reaches the decorating area, it is processed in the Desma machine where a mixture is injected, for the elaboration of the sole of the industrial shoe. The company has 2 Desma. Each of them produces 1600 pairs of shoes in an 8-h workday. These shoes are stored in the P.U. Injection Machines department, where operators transfer them to the decorating area. This is the activity that generates the longest production time due to the lack of instruments. Since shipments of raw material to the decorating area are in large quantities, it is necessary to make them through batches by generating a large volume of warehouse usage, which causes delays in the process. The main operations are Flaming, Cleaning, Painting, and Applying Gloss Polish. Such a process is detailed below.

4.4. Selection of Leader and Team Members

The process for choosing the reengineering leader consisted of conducting different interviews with the general management and those involved in the decorating process, to assess how well they know the process, and investigate the scope they have in decision-making. According to Hammer and Champy [1], the characteristics of the team members are observed in Figure 3.

In consensus with the director of the company and after conducting the interviews, the team was defined. The information is presented in Table 1.

4.5. Team Building and Training

At this stage, it is necessary to discard the old patterns of thinking about how a process should be, and to generate new forms of operation and new possibilities that fundamentally modify the current process. The owner of the process and the reengineering team were trained in "doing reengineering." It was mentioned that reengineering does not intend to automate the current processes to make them faster. Rather it seeks to create a new agile process that is capable of satisfying the needs of the internal-external client. In addition, other main seminars that are shown in Figure 4 were developed based on Hammer and Champy's [1] techniques to conceive ideas. In these seminars, the importance of "search and destroy assumptions" and "search opportunities for creative application of technology."

Subsequently, techniques such as nominal groups, brainstorming, and focus groups were used to define the strategies that would lead to the improvement proposal.

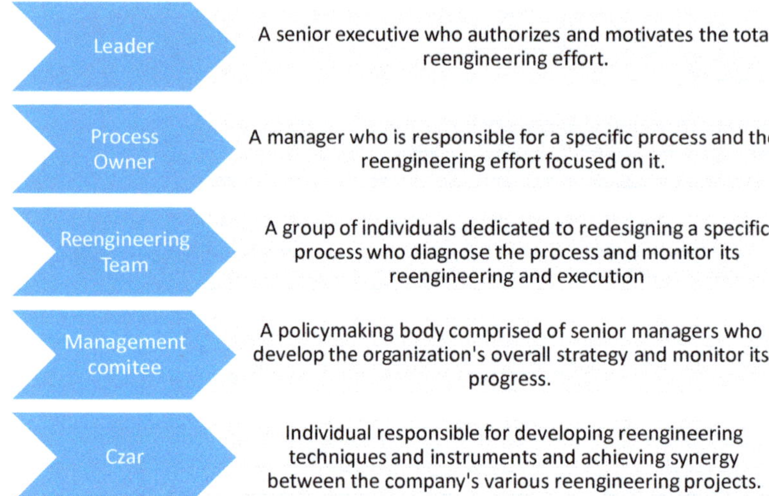

Figure 3. Characteristics of the roles in reengineering elaborated by the authors, according to Hammer and Champy [1].

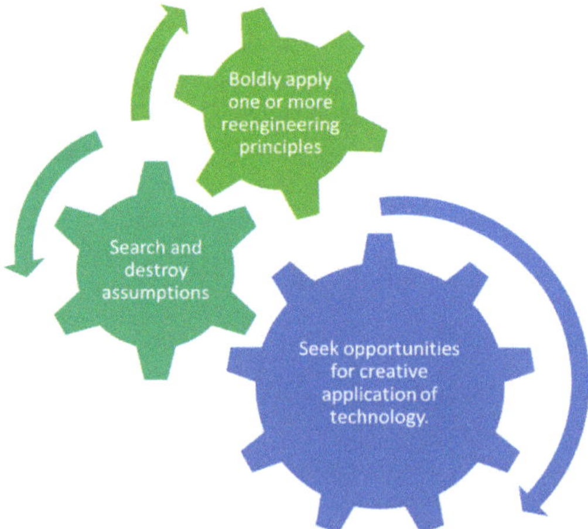

Figure 4. Seminars elaborated by the author, taking Hammer and Champy [1] as a reference.

Table 1. Selection of the members of the reengineering team.

Role	Current Position in the Company	Profile
Leader	Quality manager	He wants to reinvent the company and improve the decoration area. He has authority and it is him who implants in the staff a vision of the type of organization that is desired. It is involved in the creation of the team. It ensures the acceptance or rejection of proposals, which are not radical.
Process owner	Production manager	It ensures that the reengineering proposals are implemented by obtaining the necessary resources and the cooperation of the company members. He is a motivating and inspirational adviser to the team.
Reengineering Team	External to the process: 1. Chief of Engineering 2. Head of Human Resources 3. Warehouse manager of raw materials and finished product Internal to the process: 4. Coordinator of the decoration area 5. Quality Inspector	They supervise planning and brainstorming. Internal members have credibility with their peers, which makes it easier for ideas to be accepted.
Management committee	1. Head of Human Resources. 2. Head of Purchasing 3. Chief accounting officer 4. Chief Executive	They were responsible for advising and solving the conflicts that arise between the process owner and his team during reengineering, in addition to allocating the available resources.
Czar	He does not belong to the company to have objective opinions.	Trains the process owner and the reengineering team as well as coordinates all the activities.

Moreover, the employees of the decorating area were trained in what was reengineering to better accept the changes that were proposed. Some significant ideas that were obtained after applying the different techniques were the following. It is necessary to highlight that they did it without considering restrictions in space, costs, and human resources.

1. That the worker can choose Desma machine according to its availability. Currently, the worker uses the machine according to the belonging line (Desma machine 1 for line 1, and the Desma machine 2 for line 2, even though the other is near and empty). Hence, that causes the process not to follow an adequate flow and the transport time is increased.
2. Relocate the Desma from Line 1 following the flow of the process because it is currently far from the other operations on Line 1.
3. Merge both lines of decorating.
4. Implement an automated transport system through a conveyor belt to streamline the transport of the product in process and material.

In order to select the best idea, the process map detailed below was elaborated on.

4.6. Process Mapping

Along with the proposal, the process map of the decoration area was made. The process is briefly described below, and a flow diagram is presented in Figure 5. The shoe is taken to the first activity of the decorating area, where the operator performs the flaming, which consists of removing the excess seams that the shoes have through the flaming machine. The next step is carrying the shoe to the work table where 3 to 4 operators are in charge of cleaning, which consists of removing burrs through a razor by taking care that the shoe is not scratched and cleaning the exterior dust. The third activity is placing paint on the soles. Then another operator applies the polish gloss. Then an operator places a paper inside the shoe for interior protection, and the shoelace is placed on the shoe and, thus, the shoe is finished. If the shoe is clean, it is packed and labeled according to the characteristics of the shoe. Otherwise, a cleaning operator is asked to clean it again and when it is ready it is packed. Lastly, an operator stows the boxes, until obtaining five pairs of shoes and placing a safety clasp through the strapping machine to transport to the "Finished Product Warehouse" area. In addition, the team takes times of the operations of both decorating lines in order to analyze the process.

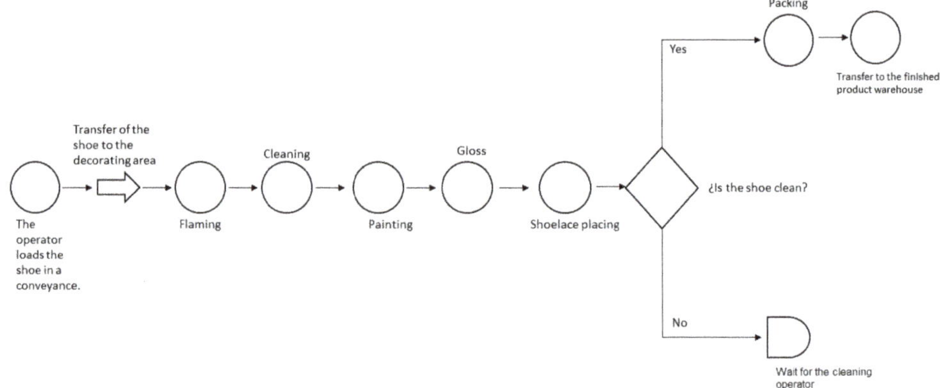

Figure 5. Process flow diagram elaborated by the author.

4.7. Identification of Problems

Based on the previous stages, the reengineering team defined that the main problem was that the decorating area has two lines, where the same activities are carried out during the shoe manufacturing process. One of the two lines has a U type flow line (Line 2) and, the other line lacks of design (Line 1), including the latter being the one that has many conflicts during the process due to reduced spaces between one activity. The has a bad sequence in the flow of the material and inadequate order of operations and equipment. For this reason, it is one of the areas with the greatest conflicts within the company and in which the proposal detailed in the next section will focus.

4.8. Analysis of the Process and Proposal

In this phase, the proposal was generated, considering the results of the previous phases. Of the ideas that were generated in the training phase of the team, it was selected to merge both lines of scenery due to their economic viability. The others were discarded because they would generate problems among the workers or because of the economic cost to the company, which, at that time, it could not assume.

In the proposal, a change is made in the distribution of machinery and equipment. The old plant layout is shown in Figure 6. The area and its activities: flaming, cleaning, painting, gloss, fitting, packing and stowage is presented. This area is divided in two line 1 and line 2, according to what was observed in the company. In these lines, there is no logical sequence in the workflow and in the route of the materials, which causes delays in production. The Desma 1 and Desma 2 machines are distributed in different areas with one at 25 meters and the other at 15 meters, respectively, which is due to their technical conditions that cannot be moved. In the third activity, painting on the soles of the shoe, line 1, also has flow problems. It is necessary to pass the shoes under a table to get to the next activity where paint is applied to the sole of the shoe. Later, another operator performs the application of polish gloss, where both lines share the machinery, which causes accumulation of shoes.

After analyzing the current process, an improvement to the production process is proposed in which a change is made in the distribution of machinery and equipment. The proposed plant layout is shown in Figure 7, where the work area and a new U-type material flow path can be seen, which occupies the same dimensions, allowing for greater mobility and eliminating unnecessary work such as passing the shoes under a table to get to the next activity. Additionally, a merger of both lines of work is proposed, becoming a single line of work within the decorating area with a new distribution and with more space between activities. The flow will be as follows. At the beginning, two operators go to the injection department and take 100 pairs of shoes from the warehouse, which are transported in a box to the decorating area. The shoe is taken to the first activity in the area, flaming, in this activity. Unlike the previous process, you can have 100 instead of 50 pairs of shoes, allowing the shoe to be closer to the operation and the three operators to take it and process it faster, which sends it to the next activity: cleaning. In this activity, eleven people clean the shoe and, unlike the previous configuration, the shoe is stored in an orderly manner by preventing the worker from being surrounded by shoes.

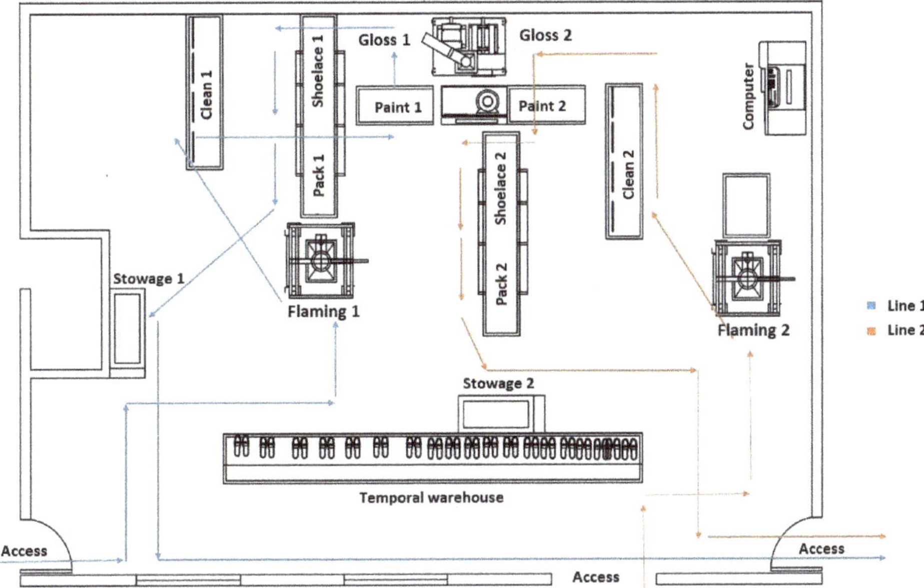

Figure 6. Flow of the current process elaborated by the author based on what was observed in the company.

As shown in Figure 6, where we observe the poor design of the work area, a bad flow of material and a total disorder of the area. Once the shoe is cleaned, it is taken to the painting activity in which three operators participate in painting the sole, so that they can be processed in the application of the gloss by two operators. Instantly, two operators place a paper inside the shoe for interior protection, and the shoelace is placed in it, while obtaining the finished shoe. Subsequently, it is packaged and labeled, according to the characteristics of the shoe. Lastly, three operators stow the boxes, until obtaining five pairs of shoes and placing a safety clasp, to transport them to the "Finished Product Warehouse" area.

The proposal was validated with a Simulation in the Software Arena™ and is described in the next section. Once the proposal was validated through simulation, a meeting was held. There, the leader spoke of the need to redesign the decorating area. It is necessary to mention that the designation of the activities, necessary resources, and the time to achieve it were decided by the company and did not allow that the activities and results achieved were published. Therefore, only the proposal validation is detailed in the following section. Although, the proposal was implemented and brought good results, even superior to those generated through simulation.

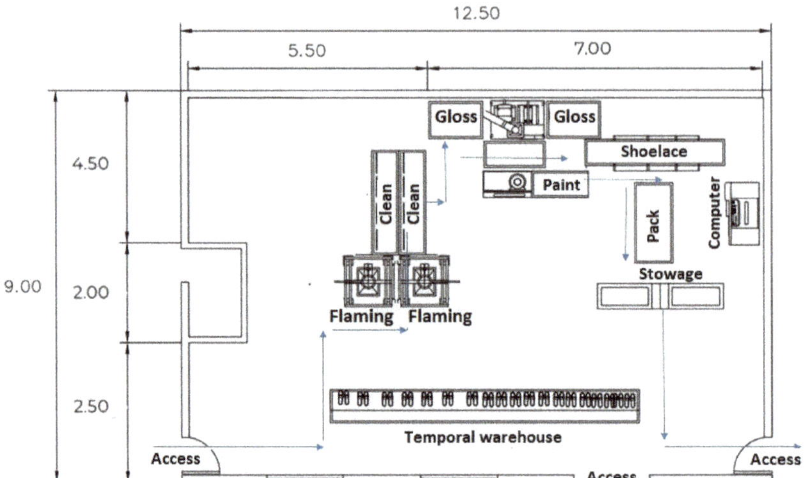

Figure 7. Flow of the proposal process, merger of both lines, elaborated by the author.

5. Simulation

In order to compare the current process with the proposal defined in the previous section, a discrete event simulation is used. In this case, it is a dynamic simulation model since it represents a system as it evolves over time and is stochastic since it has random variables. These state variables change only at discrete points in time at which certain parameters are scored. An event is defined as a situation that causes the state of the system to change instantaneously such as the arrival of an entity (for shoes) to an operation on the production line [4,28]. The simulation is executed with the methodology suggested by Kelton [28].

5.1. System Definition

At this stage, the current process of the decoration area is replicated. People, machinery, location, and working conditions are considered in the system, so that the model is as similar as possible to the real system. The objective is to compare, once the proposal has been made, whether it improves with respect to the current process.

5.2. Model Formulation

Current: To the simulation conceptual model, different criteria were analyzed through reengineering. The layout and the process diagram detailed in the previous stage and the human resources operating the process were considered (see Table 2). These workers operate in 8-h shifts by contemplating a 30-min break for eating. There are two flaming machines, one gloss machine, and two strapping machines.

Proposal: In the simulation model, the possibility of locating all the equipment and machinery with a U-type flow path was studied, and the company's two decorating lines are merged. The same machinery and human resources are used per operation as in the current system.

Table 2. Human resources by operation, elaborated by the author.

Human Resources Per Activity			
Activity	Line 1	Line 2	Total
Flaming	2	2	4
Cleaning	4	4	8
Painting	1	1	2
Gloss	1	1	2
Shoelace placing	1	1	2
Packing	2	2	4
Transport	1	1	2

5.3. Data Collection

Times were taken, specifically 73 records in each operation during February to April 2019, considering the hours when there was low and high production. Neither the first nor the last hour of the shift were used. The data was checked for detecting and discarding atypical data. These data were entered in the input analyzer, which is a tool of the Arena™ software. This allows determining which specific distribution fits the data, and which uses the chi square χ^2 and the Kolmogorov-Smirnov test as goodness-of-fit. The distributions that accept the null hypothesis in were used for obtaining the different distributions of each activity, which will be used to simulate the process. Figure 8 shows the analysis of the flaming operation, including the times taken, the analysis of fit test, and the expression that was used to model the production process in Arena™.

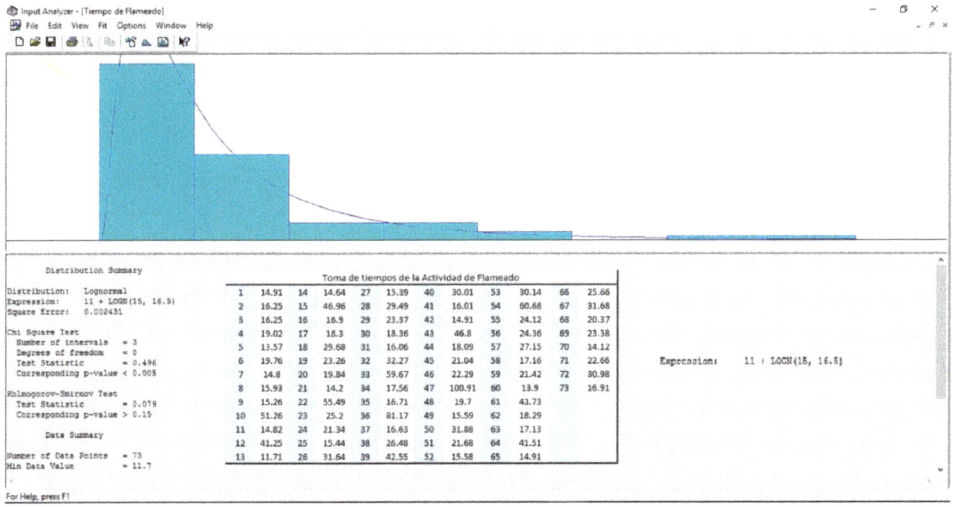

Distribution Summary

Distribution: Lognormal
Expression: 11 + LOGN(15, 16.5)
Square Error: 0.002431

Chi Square Test
  Number of intervals   = 3
  Degrees of freedom   = 0
  Test Statistic   = 0.496
  Corresponding p-value   < 0.005

Kolmogorov-Smirnov Test
  Test Statistic   = 0.079
  Corresponding p-value   > 0.15

Data Summary

Number of Data Points   = 73
Min Data Value   = 11.7

Expression: 11 + LOGN(15, 16.5)

Toma de tiempos de la Actividad de Flameado

#	Time	#	Time	#	Time	#	Time	#	Time	#	Time
1	14.91	14	14.64	27	15.39	40	30.01	53	30.14	66	25.66
2	16.25	15	46.96	28	29.49	41	16.01	54	60.68	67	31.68
3	16.25	16	16.9	29	23.37	42	14.91	55	24.12	68	20.37
4	19.02	17	18.3	30	18.36	43	46.8	56	24.36	69	23.38
5	13.57	18	29.68	31	16.06	44	18.09	57	27.15	70	14.12
6	19.76	19	23.26	32	32.27	45	21.04	58	17.16	71	22.66
7	14.8	20	19.84	33	59.67	46	22.29	59	21.42	72	30.98
8	15.93	21	14.2	34	17.56	47	100.91	60	13.9	73	16.91
9	15.26	22	55.49	35	16.71	48	19.7	61	43.73		
10	51.26	23	25.2	36	81.17	49	15.59	62	18.29		
11	14.82	24	21.34	37	16.63	50	31.88	63	17.13		
12	41.25	25	15.44	38	26.48	51	21.68	64	41.51		
13	11.71	26	31.64	39	42.55	52	15.58	65	14.91		

For Help, press F1

Figure 8. Probability distribution found for the flaming activity, elaboration in the Arena™ software.

Proposal: For doing the simulation model, the same distributions of the operations that were calculated with the times collected were considered.

5.4. Model Implementation in the Computer

The model was developed in Arena ™ software. In Figure 9, the simulation model of the current process is visualized, which consists of 42 modules in total. See also Figure A1.

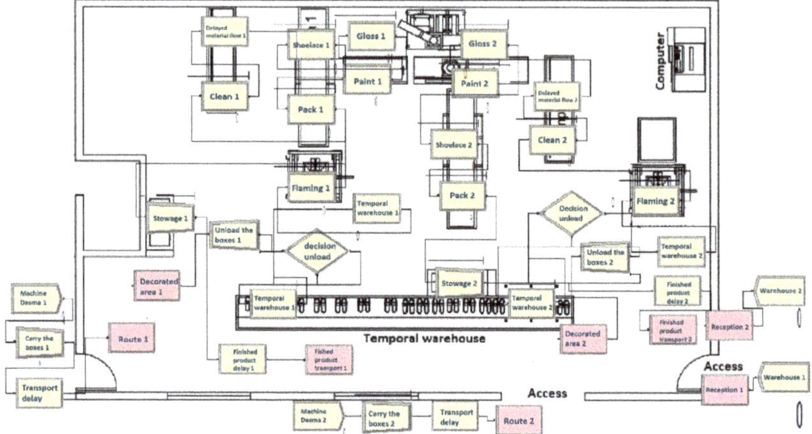

Figure 9. Arena ™ Model, current system, elaborated by the author.

- 2 *Create* modules that represent the Desma machines,
- 4 *Batch* modules that allow the shoes to be put together,
- 6 *Delay* modules where they represent the delays,
- 4 *Station* modules where the arrival of raw material during the process,
- 4 *Route* modules that allow identifying the routes,
- 2 *Hold* modules, which allow certain restrictions,
- 2 *Separate* modules used to unload the transport to a specific place,
- 2 *Decide* modules that allow making decisions during the process,
- 2 *Signal* modules that allow signaling the *Hold* module,
- 12 *Process* modules that represent the six main activities of the process,
- 2 Dispose modules that are used to represent the Finished Product Warehouse.

Proposal: The two lines are merged, and the U-shaped arrangement is used (see Figure 10). Figure A2 shows the running model.

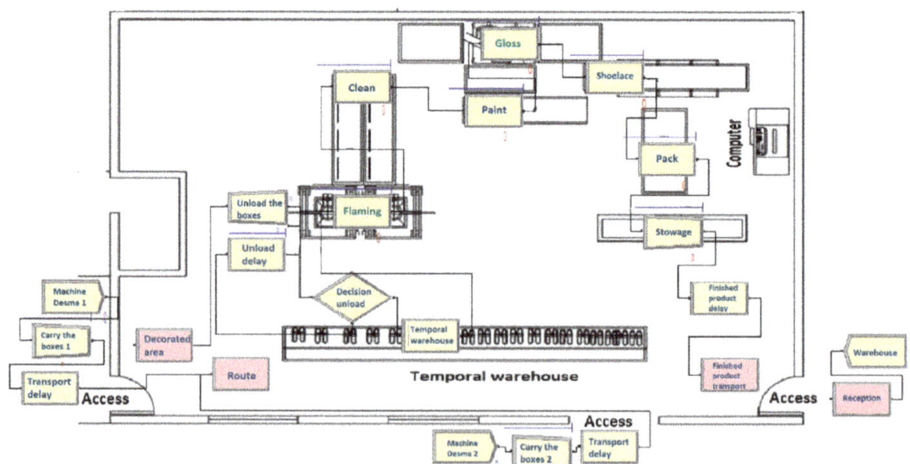

Figure 10. Arena ™ model of the proposal elaborated by the author.

5.5. Verification

Current: In this part, before running the model, aspects such as time, which should be in seconds for all processes, and distances in minutes were reviewed. In addition, the connectors between the consecutive activities followed the flow chart of Figure 9, which is the computational model. Likewise, there were no errors when running the model. The same is revised in the proposal, but refers to Figure 10.

5.6. Validation

To validate the simulation, factory floor-data information was used, where a minimum of 1700 and a maximum of 1750 pairs per day were observed. Therefore, before carrying out the experimentation, several tests were run, and details were corrected that lead us to reproduce what happens in the current process. An example of this is that the warehouse between Desma machine and the Flaming process cannot exceed 50 pairs of shoes and that was corrected in the simulation model. The average production per day in each activity is shown in Table 3.

Table 3. Average final production per day, current process, Arena ™ Software report.

Average Production Per Day-Current Process			
Activity	Line 1	Line 2	Total
Flaming	975	991	1966.00
Cleaning	834	863	1697.00
Painting	816	850	1666.00
Gloss	816	840	1656.00
Shoelace placing	816	848	1662.00
Packing	814	847	1661.00

Because validation is a comparison of the simulation model report against actual data, and the company does not provide us with data after the implementation, this step is not performed in the proposal.

5.7. Experimentation

The model was run for about 30 replications. Each one starts and ends according to the same rules and uses the same sets of parameters. In the real context, each replication represents a shift. It was decided to use 30 replications because the greater the number of replications, the more reliable the inference is regarding what is happening. There is convincing evidence that a sample size of n = 30 is sufficient to overcome the bias of the population distribution and provides approximately a normal sampling distribution of the random variables [28], which ensures that the sample can infer what is happening in the population. In this case, the simulation can infer what happens in the real process. In the proposal, 30 replications were made as well.

5.8. Analysis of Results

The indicators to compare the current process with the proposal are the percentage of use of each activity and the units produced at the end of each process, which includes, in this specific case, the pairs of shoes that are processed in the decorating area. As we observe in Table 4, the utilization percentage are registered, line 1 is of a lower percentage compared to line 2, and, when merging both lines into one, we can see that the utilization percentage of each activity increases considerably, which outpaces the current process. See Figures A3 and A4 for more information on the simulation in Arena™.

Table 4. Comparison of process indicators, average utilization per day: Current-Proposal.

	Average Utilization Per Day		
	Current Process		Proposal
Activity	Line 1	Line 2	Same Resources
Flaming	94%	95%	92%
Cleaning	94%	95%	94%
Painting	63%	65%	55%
Gloss	46%	48%	61%
Shoelace placing	56%	58%	73%
Packing	26%	27%	35%

The average production per day is presented in Table 5, where each activity appears with its respective production. The proposal has the same human resources than the current line 1 and line 2. See Figures A5 and A6 for more information on the simulation in Arena ™.

Table 5. Comparison of process indicators, average production per day: Current-Proposal.

	Average Production Per Day			
	Current Process			Proposal
Activity	Line 1	Line 2	Total	Same Human Resources
Flaming	975.00	991.00	1966.00	2856.00
Cleaning	834.00	863.00	1697.00	2154.00
Painting	816.00	850.00	1666.00	2144.00
Gloss	816.00	840.00	1656.00	2142.00
Shoelace placing	814.00	848.00	1662.00	2139.00
Packing	814.00	847.00	1661.00	2138.00

Additionally, the inventory in the process is shown in Table 6 and it is useful to validate the simulation model. In the flaming and cleaning activities, whose utilization is higher, there are waiting lines, while, in activities where utilization is less, they do not exist.

Table 6. Inventory in the process by operation: Current-Proposal.

	Inventory in Process Per Operation			
	Current Process			Proposal
Activity	Line 1	Line 2	Total	Same Human Resources
Flaming	17	17	34	46
Cleaning	284	277	561	326
Painting	0	0	0	0
Gloss	0	0	0	0
Shoelace placing	0	0	0	0
Packing	0	0	0	1

When performing an analysis of human resources per day, it was observed that human resources are underutilized. Due to this, it was decided to use OptQuest to maximize production using human resources as a constraint (see Table 7). Once the tool has been used to optimize, there is an increase to 41% of the average final production per day, which is equivalent to 2350 pairs of shoes (Table 8). This preserves the same number of workers. See Figure A7, Figure A8, and Figure A9 for more information on the simulation in Arena™.

Table 7. Comparison of human resource use per day Current-Proposal-Optimization.

	Human Resources				
	Current Process			Proposal	
Activity	Line 1	Line 2	Total [1]	Same [2] Human Resources	Optimization[3]
Flaming	2	2	4	4	3
Cleaning	5	5	10	10	11
Painting	1	1	2	2	3
Gloss	1	1	2	2	2
Shoelace placing	1	1	2	2	2
Packing	2	2	4	4	3

[1] The total is the sum of human resources per activity of line 1 and 2 in the current process. [2] In the proposal, the same resources were used. 3. The optimization is the OpQuest suggestion to maximize production.

Table 8. Comparison of average production per day Current-Proposal-Optimization.

	Average Production Per Day				
	Current Process			Proposal	
Activity	Line 1	Line 2	Total [1]	Same Human Resources [2]	Optimization [3]
Flaming	975	991	1966	2856	2861
Cleaning	834	863	1697	2154	2369
Painting	816	850	1666	2144	2358
Gloss	816	840	1656	2142	2355
Shoelace placing	816	848	1662	2139	2352
Packing	814	847	1661	2138	2350

[1.] The total is the sum of average production per activity of line 1 and 2 in the current process. [2.] The average production with the same resources in the proposal. [3.] The optimization is the average production with the resource's suggestion of OpQuest.

In order to identify data affected by errors, the Grubbs test was applied for the data of the current process in Table 9 and for the data of the proposal in Table 10. Thirty simulations with one run were performed and the average production was recorded per work shift for both the current system with two lines (see Table 9) and the proposal with a single line (Table 10).

First, a Kolmogorov-Smirnov test was applied to evaluate if the data can be reasonably approximated by a normal distribution before applying the Grubbs test. It was concluded that both the data of the current process and those of the proposal have a normal distribution because the p value is greater than the value of $\alpha = 0.05$. Then, the Grubbs test was made (see Figure 11).

Table 9. Average production/day of the current process.

Current Process—Average Production Per Day											
Run	Line 1	Line 2	Total	Run	Line 1	Line 2	Total	Run	Line 1	Line 2	Total
1	794	846	1640	11	837	856	1693	21	808	834	1642
2	807	842	1649	12	836	839	1675	22	812	851	1663
3	781	852	1633	13	810	837	1647	23	831	835	1666
4	818	852	1670	14	825	858	1683	24	821	856	1677
5	817	837	1654	15	810	860	1670	25	822	858	1680
6	792	846	1638	16	815	867	1682	26	821	832	1653
7	804	841	1645	17	815	834	1649	27	834	853	1687
8	794	844	1638	18	819	834	1653	28	824	874	1698
9	837	841	1678	19	773	831	1604	29	808	831	1639
10	807	828	1635	20	828	876	1704	30	815	861	1676

Table 10. Average production/day of the proposed process.

Proposal—Average Production Per Day					
Run	Pairs of Shoes	Run	Pairs of Shoes	Run	Pairs of Shoes
1	2162	11	2155	21	2133
2	2159	12	2149	22	2148
3	2140	13	2108	23	2130
4	2140	14	2147	24	2125
5	2140	15	2164	25	2145
6	2089	16	2125	26	2127
7	2139	17	2129	27	2179
8	2146	18	2137	28	2175
9	2146	19	2123	29	2149
10	2130	20	2155	30	2120

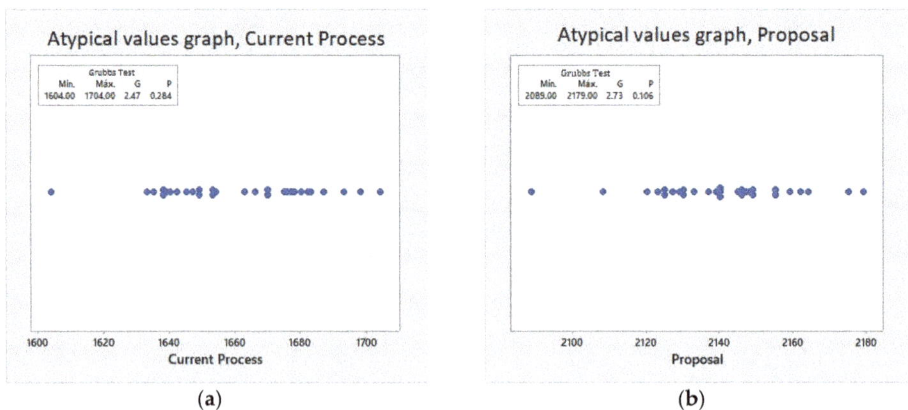

Figure 11. Grubbs test (**a**) Atypical values, current process. (**b**) Atypical values, proposal.

Grubbs's test is defined for the hypothesis:

$$\mathbf{H_0} : \textit{There are no outliers in the data set}$$
$$\mathbf{H_1} : \textit{There is at least one outlier in the data set} \tag{1}$$

It was concluded that both the data of the current process and the proposal do not have outliers because the p value is greater than the value of $\alpha = 0.05$. Additionally, to formally compare if the proposed method (Table 10) is better in terms of average production per work shift than the current one (Table 9), the following hypothesis test is proposed. The average production/day of the proposed process is higher than the average production/day of the current process, where $\mu_1 =$ proposed process and $\mu_2 =$ current process.

$$\mathbf{H_0} : \ \mu_1 = \mu_2$$
$$\mathbf{H_1} : \ \mu_1 > \mu_2 \tag{2}$$

To perform this test, it is necessary to know if the population variances are the same or different even when they are unknown [29]. Therefore, the following hypothesis test is performed in Minitab™ and presented in Table 11.

Table 11. Equality of variances test.

Null hypothesis		$\mathbf{H_0}$: $\sigma_1{}^2/\sigma_2{}^2 = 1$		
Alternative hypothesis		$\mathbf{H_1}$: $\sigma_1{}^2/\sigma_2{}^2 \neq 1$		
Significance level		$\alpha = 0.05$		
Method	Test Statistical	DF1	DF2	p-Value
F	0.67	29	29	0.290

Because the value of $p = 0.290$ is greater than the significance level $\alpha = 0.05$, there is statistical evidence to accept the null hypothesis, that is, the population variances, although unknown, are equal. This information is used to verify if the means are equal or if the mean of the proposal is greater than that of the current process defined in Equation (2). These calculations are also done with Minitab™ and presented in Table 12.

Table 12. Hypothesis Testing for Comparing Medians of two Populations.

Null hypothesis		$\mathbf{H_0}$: $\mu_1 - \mu_2 = 0$
Alternative hypothesis		$\mathbf{H_1}$: $\mu_1 - \mu_2 > 0$
Significance level		$\alpha = 0.05$
Test T	DF	Valor p
F	58	0.000

Since the value $p = 0$ is less than the significance level $\alpha = 0.05$, there is statistical evidence to reject the null hypothesis, that is, the mean of the average production per day is higher than the mean of the current process. Lastly, to know how superior it is, the confidence interval for the difference of the means is presented, $469 < \mu\_1 - \mu\_2 < 491$. This interval determines the number of pairs of shoes that the proposed process can increase when compared to the current process with 95% reliability.

5.9. Implementation

In order to perform the implementation, the company was recommended to continue with the reengineering team, which is, in this case, the quality manager who served as the reengineering leader. The owner of the process was the production manager. The Reengineering team was formed by the Engineering, Human Resources, Warehouse of Raw Materials and Finished Product chiefs.

They were the outside members. Additionally, the inside members were the coordinator of the decoration area and the quality inspector. The steering committee was made up of the chiefs of Human Resources, Purchasing, Accounting, and Engineering. Lastly, the Reengineering Czar had two main functions: to train and support the process owner and the reengineering team, and to coordinate all the reengineering activities that were launched.

To start with the implementation, all the reengineering collaborators met, and the leader spoke to them about the need to redesign the decorating area. The problem was exposed and the proposal that merges both lines was explained, giving rise to a new process flow and a new plant distribution. The amount of resources calculated in Table 8, in the optimization column, were used. The activities, who was accountable for them, approximate time and resources were defined among all, which gave rise to the Gantt Chart of the implementation. The company only granted permission to discuss the implementation in a general way, but implemented it in 35 days. Its historical results show a higher productivity increase than that reported in the proposal analysis with the simulation.

6. Discussion

As already mentioned, the indicators to study were the percentage of use of each activity and units produced at the end of each process, which, in this case, are the pairs of shoes that are processed in the decorating area. As we observe in Table 4, where the results of the utilization percentage are registered, line 1 is of a lower percentage compared to line 2, and, when merging both lines into one, we can see that the utilization percentage of each activity increases considerably, outpacing the current process. It is necessary to point out that the painting activity is underutilized. This is because, in the previous activity of cleaning, high inventories were generated in the process. Both the original and the proposed models were simulated in Arena™, using the same probability distributions and the original personnel to make the comparison.

In Table 8, it is observed that Line 1 ends the first activity with a production of 975 pairs of shoes during one shift. The following is the cleaning of the shoe with a total of 834, and so on. It is observed that the production is decreasing in each activity, which causes high inventories in the process. At the end of production, 814 pairs of shoes are obtained from Line 1 and 847 pairs of shoes are obtained in Line 2, which is a total of 1661 pairs of shoes obtained per day from the current process. Regarding the proposal, we can see that, at the end of the first activity, the current process is greater with a total of 2856 pairs of shoes, and, at the end of the process, in packaging, the proposal delivers more production, 2138 against 1661 pairs, which increases 29% of the current production. There are few inventories in the process in the proposal since it is observed that the pairs produced at the end of each activity are approximately constant. However, since this model uses the same probability distributions and human resources, even the painting activity is underused. For solving this, another OptQuest tool was used, which is an optimization module designed to facilitate its integration in applications that require the optimization of highly complex systems. In our case, we decided to maximize resources and number of staff that work in the decorating area. These are 24 people distributed in the different activities, as can be seen in Table 7. Once the tool has been used to optimize, there is an increase to 41% of the final average production per day, which is equivalent to 2350 pairs of shoes (see Table 8). Regarding the validation of the data, we have factory floor-data before the implementation but not after. This is a limitation of our work because validation is making a comparison between the results reported in the arena and real data. Therefore, the proposal could not be validated in only the current process.

7. Conclusions

In this research work, two methodologies were used to solve problems within industries. First, reengineering to conceptualize the problem and generate solution proposals. Second, simulation to compare the proposal conceived with the current process. Both methodologies were used to deal with the current problems presented by the company, and to issue a solution proposal. Despite the fact that, in this work, there is no theoretical contribution, there are practical contributions not only to help the improvement of the processes, but also to do a complete analysis of the current process with its respective deficiencies, which were analyzed. Once the process where the improvement would be made was selected, a proposal was made that was tested through the simulation. Until the time of this research, there were no articles focused on reengineering in an industrial footwear plant supported by simulation. Moreover, this fusion of methodologies can be used in companies in another sector with their respective adaptations.

After performing the simulation experiments, the results indicate that the production rate increases by approximately 29% with the new configuration, and up to 41% when the human resources are optimized through OptQuest. The company implemented the configuration of the proposal, or more precisely the optimization, which brings significant improvements such as making better use of the factory space, eliminating unnecessary work, allowing better mobility, better use of resources, and more. Some of the limitations of this work was dealing with people by convincing them to get them out of preconceived ideas. The other was that we have factory floor-data after the implementation. In future work, we would like to implement the methodology but only when using other process analysis techniques.

Author Contributions: Conceptualization, E.S.H.-G., M.A.M.B., and R.C.-A. Methodology, E.S.H.-G. Software, R.C.-A. and M.A.M.B. Validation, E.S.H.-G. and R.C.-A. Formal analysis, E.S.H.-G. and M.A.M.B. Investigation, R.C.-A. and E.S.H.-G. Resources, E.S.H.-G. Data curation, R.C.-A. Writing—original draft preparation, E.S.H.-G. and R.C.-A. Writing—review and editing, E.S.H.-G. All authors have read and agreed to the published version of the manuscript for the term explanation.

Funding: This research did not receive external funding.

Appendix A. Simulation in Process

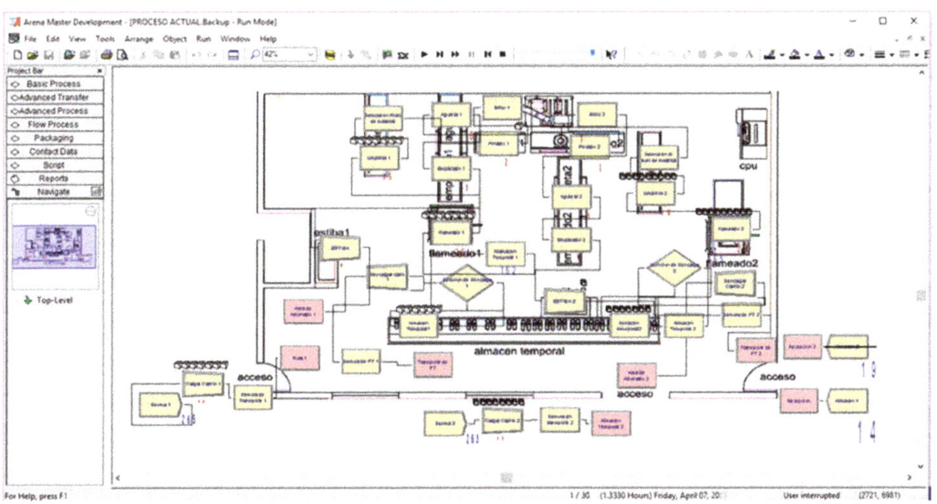

Figure A1. Simulation in process. Current configuration.

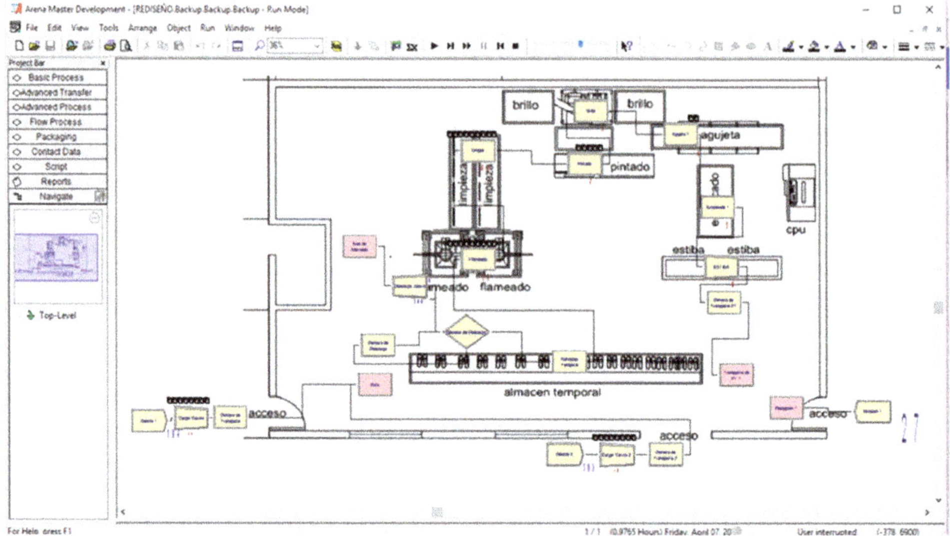

Figure A2. Simulation in process. Proposal.

Appendix B. Utilization Per Activity

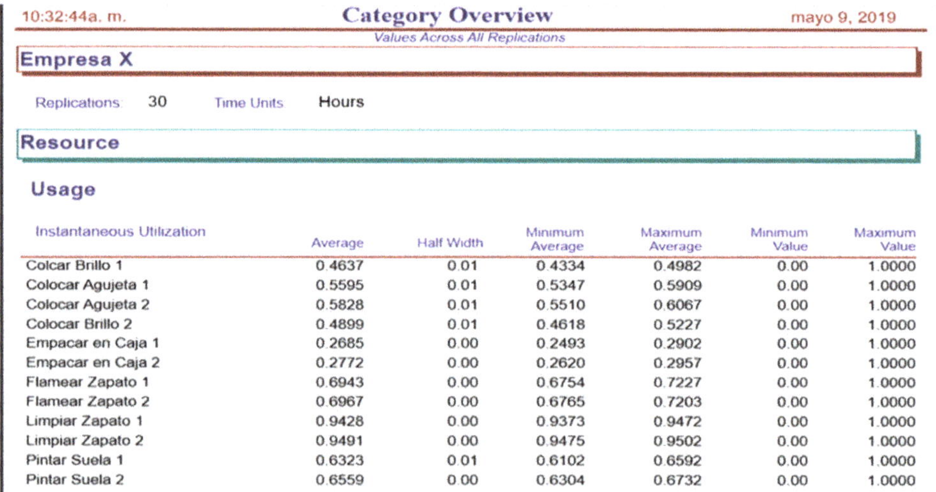

Figure A3. Utilization per activity. Current process.

Replications: 30   Time Units: Hours

Resource

Usage

Instantaneous Utilization	Average	Half Width	Minimum Average	Maximum Average	Minimum Value	Maximum Value
Aplicar Brillo	0.6098	0.00	0.5914	0.6389	0.00	1.0000
Colocar Agujeta	0.7298	0.00	0.7135	0.7467	0.00	1.0000
Empacar en Caja	0.3469	0.00	0.3294	0.3675	0.00	1.0000
Flamear Zapato	0.9219	0.00	0.9082	0.9403	0.00	1.0000
Limpiar Zapato	0.9440	0.00	0.9399	0.9500	0.00	1.0000
Pintar Suela	0.5510	0.00	0.5370	0.5647	0.00	1.0000

Figure A4. Utilization per activity. Proposal

Appendix C. Final Production Per Shift

Usage

Total Number Seized	Average	Half Width	Minimum Average	Maximum Average
Colcar Brillo 1	819.73	5.06	792.00	850.00
Colocar Agujeta 1	818.90	5.07	791.00	850.00
Colocar Agujeta 2	851.63	4.39	830.00	877.00
Colocar Brillo 2	852.40	4.32	831.00	877.00
Empacar en Caja 1	818.30	5.08	790.00	849.00
Empacar en Caja 2	850.93	4.34	830.00	876.00
Flamear Zapato 1	1433.43	2.30	1428.00	1450.00
Flamear Zapato 2	1444.17	1.60	1435.00	1450.00
Limpiar Zapato 1	838.70	5.05	814.00	870.00
Limpiar Zapato 2	865.60	4.28	846.00	889.00
Pintar Suela 1	820.50	5.05	793.00	851.00
Pintar Suela 2	853.17	4.31	832.00	878.00

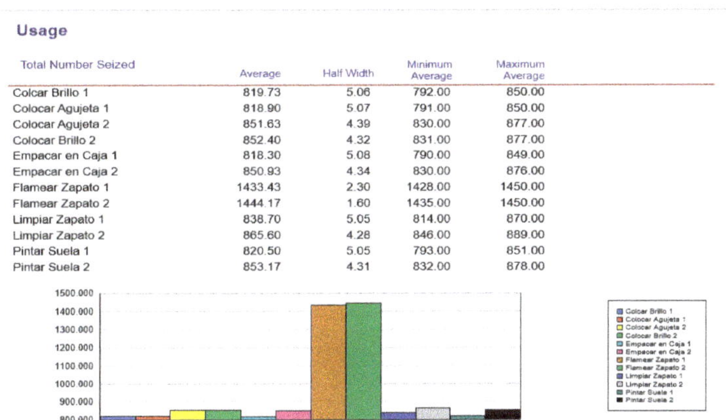

Figure A5. Final production of each activity during the process with a 7.5 h shift, Current.

Total Number Seized	Average	Half Width	Minimum Average	Maximum Average
Aplicar Brillo	2141.37	6.50	2117.00	2197.00
Colocar Agujeta	2138.93	6.57	2112.00	2193.00
Empacar en Caja	2137.13	6.59	2110.00	2191.00
Flamear Zapato	2855.30	8.62	2753.00	2880.00
Limpiar Zapato	2153.60	6.58	2128.00	2210.00
Pintar Suela	2143.40	6.53	2118.00	2200.00

Figure A6. Final production of each activity during the process with a 7.5 h shift, Proposal.

Appendix D. OptQuest Optimization

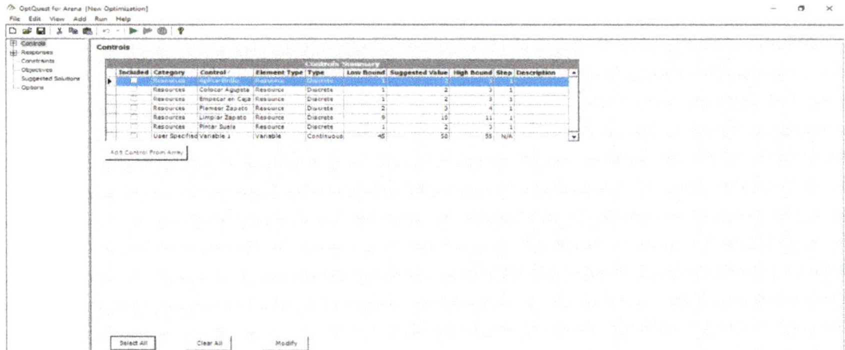

Figure A7. Selection of resources, restrictions, and objective function.

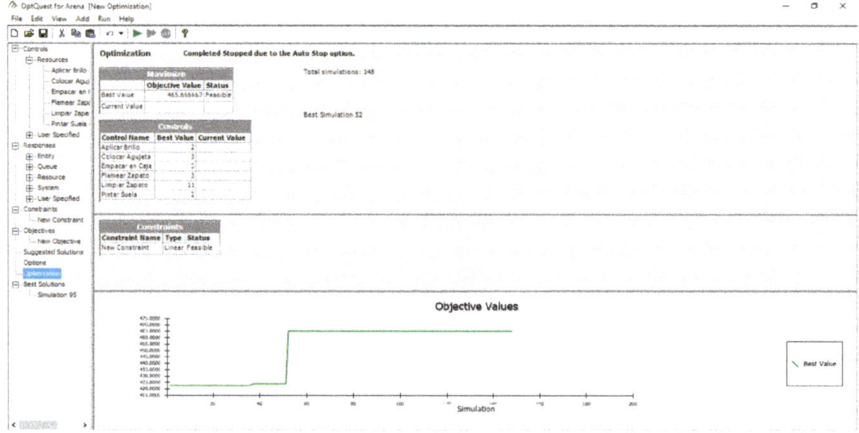

Figure A8. Optimization, considering the target value, current value and best value.

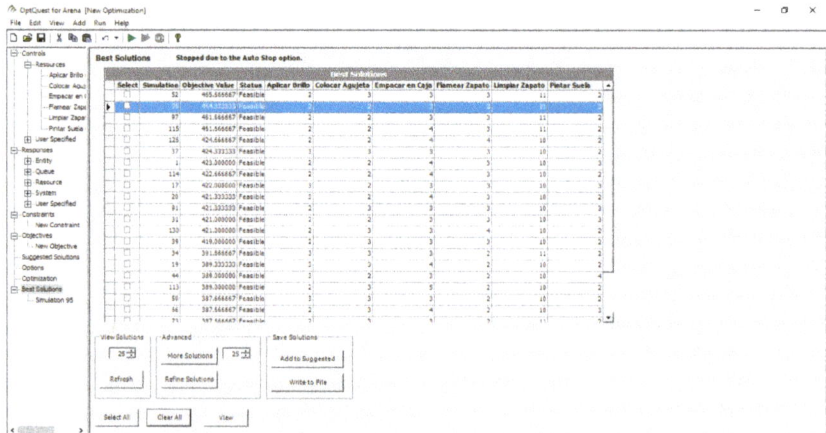

Figure A9. Comparison of results. In simulation 52, the best value is obtained with 466 pairs of shoes.

References

1. Hammer, M.; Champy, J. *Reengineering the Corporation: A Manifesto for Business Revolution*, 1st ed.; Harper Collins Publishers: New York, NY, USA, 1993.
2. Davenport, T. *Process Innovation: Reengineering Work Through Information Technology*, 1st ed.; Harvard Business School Press: Boston, MA, USA, 1993.
3. Dumas, M.; La Rosa, M.; Mendling, J.; Reijers, H. *Fundamentals of Business Process Management*, 1st ed.; Springer-Verlag: Heidelberg/Berlin, Germany, 2013.
4. Laguna, M.; Marklund, J. *Business Process Modeling*, 2nd ed.; CRC Press: New York, NY, USA, 2013.
5. Elzinga, J.; Gulledge, T.; Lee, C. *Business Process Engineering: Advancing the State of the Art*, 1st ed.; Kluwer Academic Publishers: Norwell, MA, USA, 1999.
6. Niebel, B.; Freivalds, A. *Niebel's Methods, Standards, & Work Design*, 13th ed.; McGraw-Hill Education: Philadelphia, PA, USA, 2013.
7. Chen, T.; Wang, Y.C. Evaluating sustainable advantages in productivity with a systematic procedure. *Int. J. Adv. Manuf. Technol.* **2013**, *87*. [CrossRef]
8. Hammer, M. Reengineering Work: Don't Automate, Obliterate. *Harv. Bus. Rev.* **1993**, 1–16.
9. Manganelli, R.; Klein, M. A framework for reengineering. *MIT Sloan Manag. Rev.* **1994**, *83*, 10–35.
10. Nguyen, M.N.; Do, N.H. Reengineering assembly line with lean techniques. In *Procedia CIRP, Proceedings of 13th Global Conference on Sustainable Manufacturing Decoupling Growth from Resource Use, Ho Chi Minh City, Vietnam, 16–18 September 2015*; Elsevier: Amsterdam, The Netherlands, 2016; Volume 40, pp. 590–595.
11. Khan, M.R.R. Business process reengineering of an air cargo handling process. *Int. J. Prod. Econ.* **2000**, *63*, 99–108. [CrossRef]
12. Turhan, Z.D.; Vayvay, O.; Birgun, S. Supply chain reengineering in a point company using axiomatic design. *Int. J. Adv. Manuf. Technol.* **2011**, *57*, 421–435. [CrossRef]
13. Gress, E.S.H.; Reyes, A.O.O.; Gonzalez, J.G.; Arment, J.R.C. Proposal of an embedded methodology that uses organizational diagnosis and reengineering: Case of bamboo panel company. *Adv. Sci. Technol. Eng. Syst.* **2017**, *2*, 1626–1633. [CrossRef]
14. Gress, E.S.H.; Reyes, A.O.O.; Alarcon, J.A.Z.; Oliva, M.S.T.; Munoz, B.M.; Reyna, S.B.R. Systemic Diagnosis and Strategy-Based Performance Indicators Bamboo Panel Company Case. In Proceedings of the 6th International Conference on Industrial Technology and Management (ICITM), Cambridge, UK, 7–10 March 2017; pp. 123–133.
15. Irani, Z.; Hlupic, V.; Baldwin, L.P.; Love, P.E.D. Re-engineering manufacturing processes through simulation modelling. *Logist. Inf. Manag.* **2000**, *13*, 7–13. [CrossRef]
16. Tumay, K. Business Process Simulation. In Proceedings of the WSC' 95 Winter Simulation Conference, Arlington, TX, USA, 3–6 December 1995; Alexopoulos, A., Kang, K., Lilegdon, W.R., Goldsman, D., Eds.; pp. 55–60.
17. Gejo, G.J.; Gallego, G.S.; García, M. Development of a pull production control method for ETO companies and simulation for the metallurgical industry. *Appl. Sci.* **2019**, *10*, 274. [CrossRef]
18. Iannino, V.; Mocci, C.; Vannocci, M.; Colla, V.; Caputo, A.; Ferraris, F. An event drive agent-based simulation model for industrial processes. *Appl. Sci.* **2020**, *10*, 4343. [CrossRef]
19. Ferreira, L.P.; Gomez, E.A.; Lourido, G.C.P.; Quintas, J.D.; Tiahjono, B. Analysis and optimization of a network of closed-loop automobile assembly line using simulation. *Int. J. Adv. Manuf. Technol.* **2011**, *59*, 351–366. [CrossRef]
20. Grznar, P.; Gregor, M.; Krajcovic, M.; Mozol, S.; Schickerle, M.; Vavrik, V.; Bielik, T. Modeling and simulation of processes in a factory of the future. *Appl. Sci.* **2020**, *10*, 4503. [CrossRef]
21. Chen, J.H.; Yang, L.R.; Tai, H.W. Process reengineering and improvement for building precast production. *Autom. Constr.* **2016**, *68*, 249–258. [CrossRef]
22. Balan, S. Using simulation for process reengineering in a refractory ceramics manufacturing—A case study. *Int. J. Adv. Manuf. Technol.* **2017**, *93*, 1761–1770. [CrossRef]
23. Irani, Z.; Hulpic, V.; Giaglis, G. Editorial: Business process reengineering: A modeling perspective. *Int. J. Flex. Manuf. Syst.* **2001**, *13*, 99–104. [CrossRef]
24. Kumar, A.; Shim, S. Using computer simulation for surgical care process reengineering in hospitals. *INFOR Inf. Syst. Oper. Res.* **2005**, *43*, 303–319. [CrossRef]

25. Su, Q.; Yao, X.; Su, P.; Shi, J.; Zhu, Y.; Xue, L. Hospital registration process reengineering. *J. Healthc. Eng.* **2010**, *1*, 67–82. [CrossRef]
26. Mao, X.; Zhang, X. Construction process reengineering by integrating lean principles and computer simulation techniques. *J. Constr. Eng. Manag.* **2008**, *134*, 371–381. [CrossRef]
27. La Industria del Calzado en México. Secretaría de Economía, Gobierno de México. Available online: https://www.gob.mx/se/articulos/la-industria-del-calzado-en-mexico (accessed on 1 July 2020).
28. Kelton, D.; Sadowski, R.; Sturrock, D. *Simulation with Arena*, 4th ed.; Mc Graw Hill: New York, NY, USA, 2008.
29. Devore, J. *Probability and Statistics for Engineering and the Sciences*, 8th ed.; Cengage Learning: Long Beach, CA, USA, 2010.

Article

Application of Logistic Regression for Production Machinery Efficiency Evaluation

Anna Borucka * and Małgorzata Grzelak

Faculty of Security, Logistics and Management, Military University of Technology, 00-908 Warszawa, Poland; malgorzata.grzelak@wat.edu.pl
* Correspondence: anna.borucka@wat.edu.pl; Tel.: +48-261-837-060

Received: 11 October 2019; Accepted: 5 November 2019; Published: 8 November 2019

Abstract: Production companies operate in a complex economic, technological, social and political environment. There are a number of factors contributing to a satisfactory market position, the most important one being a properly defined and implemented strategy. It needs, however, to be continuously monitored and, if necessary, modified. One of the elements subject to such evaluation is the efficiency of the production processes, which has become the genesis of this article. In response to the methods presented in the literature, a proposal using the logistic regression method for this purpose is presented. The dichotomous form of the dependent variable makes it possible to make such an evaluation in an unambiguous manner and to determine the significance and influence of selected factors on the result thereof.

Keywords: efficiency; production processes; machinery; production maintenance; logistic regression

1. Introduction

Market success of a manufacturing company is shaped primarily by the demand for the manufactured products and the rate of return on capital employed [1,2]. Poor machine efficiency and frequent downtimes can lead to a reduction in production levels, resulting in lost market opportunities, increased operating costs and reduced profits [3]. It is therefore necessary to apply appropriate methods and tools to support management and to organize maintenance services in an adequate manner to ensure that the production system operates at the assumed levels of productivity and efficiency [4].

Effectiveness is an important element in the analysis of the production process [5–7], often considered in scientific publications. It is assessed on the basis of various measures. In practice, numerous mathematical models and tools are used to support the assessment of the performance of machinery. The most frequently used measures for analyzing the efficiency of technical facilities are those resulting from three general models of operation assessment, i.e., the reliability model, the operational efficiency OEE (overall equipment effectiveness) model and the organizational and technical KPI (key performance indicators) model [8]. In addition, methods and tools for its evaluation can be classified in five main areas, i.e., operational, market, financial, technical or dynamic [9,10]. Particularly important from the point of view of machinery efficiency diagnostics is operational efficiency, and the research in this area focuses primarily on the search for opportunities to reduce the consumption of production resources. These include analysis of labor productivity growth, cost reduction, minimization of losses and shortening of production cycles. Studies available in the literature indicate the application of a number of methods and tools in this area, such as methods of productivity and profitability indicators, analysis of efficiency and degree of work stations' utilization, cost calculation of activities, study of spatial efficiency of production organization and economic evaluation of the production structure [11].

Maintaining the company's machinery stock at an appropriate level requires continuous monitoring and evaluation of the adopted effectiveness indicators. A number of companies have

MES (manufacturing execution system) systems in place, which enable ongoing control of the above parameters. There are also companies (including those examined by the authors) that do not have such software and therefore proper evaluation of efficiency parameters is difficult. Such analyses are supported by mathematical tools and methods, which also include modeling with the use of logistic regression, as presented in this article. The subject of this research was a plastics manufacturing company, while the main objective was to evaluate the effectiveness of the production process based on selected factors that may significantly affect the level of machinery efficiency. The analysis was carried out on the basis of information on the performance of the company's production system recorded from 1 September 2015 to 31 August 2017.

Monitoring the effectiveness of utilization of the available machinery allows production reserves or waste in the processes underway to be identified [12–14]. The basis for successful assessment is an appropriate selection of measures and indicators. The analysis of literature made it possible to distinguish those which were of the greatest importance both in theoretical and industrial-practical aspects. Three general models should be distinguished:

- operational efficiency model OEE (overall equipment effectiveness),
- reliability model,
- organizational and technical model—KPI [15].

Within the operational efficiency model, a frequently employed parameter (which was monitored in the examined entity as well) is the overall equipment effectiveness (OEE) indicator, which is widely described in the literature [16–19]. The available studies most often present the theoretical aspects of its calculation and indicate the categories of losses that may occur during the process of machinery and equipment use in relation to ideal conditions [19–21]. Analyses are also available to demonstrate the practical implementation of this parameter in manufacturing companies [12,22,23].

The OEE index is a product of three components [23,24], i.e., readiness and efficiency of machinery and quality of the manufactured products. It is therefore a general, comprehensive assessment, most often presented in percentage form. According to Seichi Nakajime from the Japan Institute of Plant Maintenance [25,26], OEE should remain at 85.41%, but it should be stressed that each enterprise operates in a specific environment; thus, this indicator will be different for each entity, depending on its size, profile and industry, and will not take on the same value in two different operating units [9,27]. Therefore, in practice the above indicator has evolved into different forms of application depending on the sector in which a given entity operates, adjusting to the needs of the environment. The following indicators should be mentioned: OFE (overall factory effectiveness), OPE (overall plant effectiveness), OTE (overall throughput effectiveness), PEE (production equipment effectiveness), OAE (overall asset effectiveness) or TEEP (total equipment effectiveness performance) [21].

The reliability model allows measures in statistical terms to be determined, on the basis of a time analysis of the performance of technical facilities. In practice, these refer to the technical condition of machines, as well as to the activities of maintenance staff. These are MTBF (mean time between failures), MTTR (mean time to repair) or MTTF (mean time to failure) [15].

The organizational and technical KPI model includes a set of measures enabling a comprehensive assessment of the efficiency and effectiveness of the implemented processes. It includes 72 indicators classified in three areas: economic (e.g., total relative cost of maintenance), technical (availability of facilities for preventive works) and organizational (number of maintenance staff) [28].

In relation to the analyzed company, indicators associated with the operational effectiveness model, related to efficiency, will be preferable in the context of machinery stock management; therefore, they have become the subject of this analysis. Following the literature in this field [13,29], it was assumed that efficiency in production processes is the quotient of the actual efficiency to the nominal efficiency, as specified in the following ratio (1):

$$W_Q = \frac{Q_r}{Q_O},$$

(1)

where:

W_Q—machinery operational efficiency indicator,

Q_r—actual (achieved) efficiency (pcs./h),

Q_O—theoretical efficiency, as defined in the technical documentation (pcs./h).

Thus calculated, it indicates the degree of efficient use of the production line for each operation and, as such, indicates areas for improvement. Efficiency is most often presented in percentage form. This does not always allow for its quick and unambiguous assessment. In forecasting studies, it is assumed to be a quantitative variable, which limits the availability of some modeling methods. Therefore, with regard to the analyzed company, according to the authors, a better approach would be to analyze the efficiency from the individual point of view of each company by determining its satisfactory level and reacting only if it is not achieved.

Numerous studies using logistic regression models with regard to machine maintenance are available in the literature. The main objective of the proposed tools is to assess the technical condition of technical objects along with reliability parameters [30,31], predict upcoming failures [32,33] and estimate the service life of machinery [34]. For example, Yan and Lee assessed the performance of an elevator door system in real time and identified the types of possible failures [30]. Kozłowski et al. developed a model classifying the condition of a cutting tool blade and predicting its durability [31]. Lee et al. studied the reliability of a cutting tool using a combination of logistic regression and acoustic emission methods [32]. Caesarendra combined methods of logistic regression and relevance vector machine to evaluate performance degradation and to predict failure times based on simulation and experimental data [33], whereas Chen et al., on the basis of vibration characteristics of cutting tools, developed a universal model enabling the analysis of reliability and performance for machine tools [34].

This article proposes a model of logistic regression to be used for analysis and evaluation of the level of efficiency of executed processes. The research covered the process of manufacturing garbage bags in a company operating several production plants located in Poland and Ukraine. It was carried out in three main stages, in line with the CBM (condition-based maintenance) strategy. The basis for the research were the work and inspection cards of roll making machines provided by the company, which came from one of the plants and covered the period from 1 September 2015 to 31 August 2017. These provided information in two main categories. Event data indicated what events occurred during the operation of the machine (i.e., the need to repair, replacement of worn parts or breakdowns). On the other hand, the condition monitoring data provided information about the current technical condition of the facilities and the need for preventive measures (e.g., adjustment of Teflon blades). The processing of the above information and interpretation thereof made it possible to identify factors shaping the efficiency of the machinery stock. Then, on their basis, a model for the evaluation of machinery efficiency was built. It was assumed that its satisfactory level was 90%. This value is based on the daily production cycle, which also includes the breaks required by the Labor Code, daily service and the preparation of the machine for operation. Finally, the manner in which the model can support decision making in the area of improvement of the production processes was indicated [35].

Due to the specificity of manufactured products, i.e., serial products with standard parameters, the company operates in the MTS (make to stock) production system. The plant works on a three-shift basis, with each shift lasting 8 h. The process of model parameter estimation and the results obtained are presented in subsequent sections of this article.

2. Logistic Regression Model

Logistic regression is a model that allows the influence of several variables $X_1, X_2, \ldots, X_k$ on the dichotomous variable Y in the mathematical form to be presented. The logistic regression model is based on a logistical function that takes the following form:

$$f(x) = \frac{e^x}{1 + e^x} = \frac{1}{1 + e^{-x}}, \tag{2}$$

where e is the Euler number, and x is the value of the explanatory variable X.

The use of logistic regression is supported by the fact that it is not required to meet many assumptions that are formulated in relation to linear regression and general linear models. These include, first of all, the linearity of the relationship between a dependent and an independent variable, as well as the normality and homoscedasticity of the distribution of independent variables. In addition, observations must be reported using metric measurement systems.

The logistic regression model can be written in several ways. Assuming that Y stands for a dichotomous variable with values 1, for the occurrence of the event we are interested in (success), and 0, for the opposite case (failure), the logistic regression model is described by Equation (3):

$$P(Y = 1|x_1, x_2, \ldots, x_k) = \frac{e^{\beta_0 + \sum_{i=1}^{k} \beta_i \cdot x_i}}{1 + e^{\beta_0 + \sum_{i=1}^{k} \beta_i \cdot x_i}}, \tag{3}$$

where β_i $i = 0, \ldots, k$ are logistic regression factors, while $x_1, x_2, \ldots, x_k$ are independent variables, which can be measurable or qualitative.

An equivalent form of the logistic regression equation can be written as the odds for the occurrence of the event (success) we are interested in:

$$\frac{P(Y = 1|X)}{1 - P(Y = 1|X)} = e^{\beta_0 + \sum_{i=1}^{k} \beta_i \cdot x_i}. \tag{4}$$

In a special case, for one independent variable the logistic regression equation takes the following form:

$$P(Y = 1|X) = \frac{e^{\beta_0 + \beta_1 \cdot x_1}}{1 + e^{\beta_0 + \beta_1 \cdot x_1}}. \tag{5}$$

If, in turn, both sides of the Equation (5) are logarithmized, the logit form of the logistic model will be obtained:

$$logit\, P(Y = 1|X) = ln\frac{P(Y = 1|X)}{1 - P(Y = 1|X)} = \beta_0 + \beta_1 \cdot x_1. \tag{6}$$

The condition necessary for logistic regression is a sufficiently large sample, the number of which should be $n > 10(k + 1)$, where k is the number of parameters.

Important concepts related to logistic regression are the odds and the odds ratio. The odds are defined as the probability of an event occurring $P(A)$ divided by the probability of an event not occurring, $1 - P(A)$:

$$(Odds)S(A) = \frac{P(A)}{P(non - A)} = \frac{P(A)}{1 - P(A)}. \tag{7}$$

The odds ratio, in turn, marked OR, is defined as the odds of one event occurring $S(A)$ divided by the odds of another event occurring $S(B)$:

$$OR_{AxB} = \frac{S(A)}{S(B)} = \frac{P(A)}{1 - P(A)} : \frac{P(B)}{1 - P(B)}. \tag{8}$$

3. Estimation of Markov Logistic Model Parameters

The first stage of the study was to define possible explanatory variables in order to determine which of them could be used in the model. The following explanatory variables were selected: shift, device, occurrence of failure (yes or no) and no production order (yes or no).

The shift predictor was analyzed first. First of all, the normality of distribution and homogeneity of the variance of the efficiency dependent variable during individual shifts was examined in order to determine the possible methods of statistical analysis. The distributions in all groups turned out to be inconsistent with the normal distribution, which is confirmed by the graphs in Figure 1 and the calculated chi-square test statistic values, presented in Table 1.

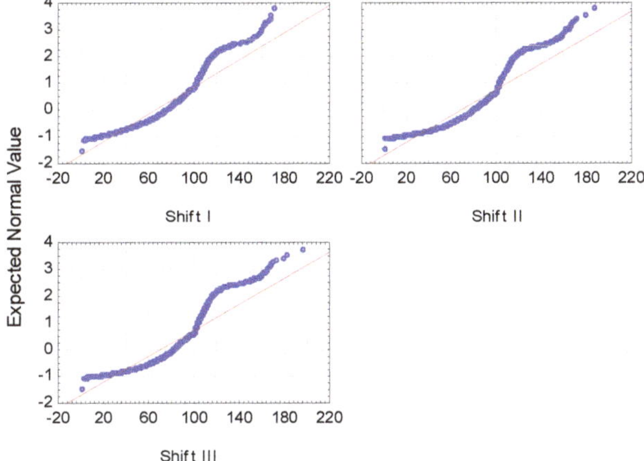

Figure 1. Graphs of normality of distribution of the efficiency variable grouped by shifts.

Table 1. Chi-square test statistic values of the efficiency variable grouped by shifts.

	Chi-Square	Degrees of Freedom	*p*-Value
Shift = 1st	4355.45	df = 8	p = 0.00
Shift = 2nd	5932.71	df = 8	p = 0.00
Shift = 3rd	6474.96	df = 8	p = 0.00

Next, the homogeneity of variance in individual groups was checked; the Levene and Brown-Forsythe test was used for this purpose. The obtained results are presented in Table 2.

Table 2. Results of the Levene and Brown-Forsythe tests of the efficiency variable grouped by shifts.

Average—1st	Average—2nd	Average—3rd	Levene F(1,df)	Levene *p*	Brn-Fors F(1,df)	Brn-Fors *p*
66.54	69.84		1.54	0.21	0.01	0.9
66.54		70.06	1.07	0.3	0.35	0.55
	69.84	70.06	0.039	0.84	0.47	0.49

Although the homogeneity of variance was confirmed in all groups, due to the lack of normality of distributions, the Mann–Whitney test was used to examine the significance of differences between individual averages, and the results thereof are presented in Table 3.

Table 3. Results of the Mann-Whitney test for the difference between the average efficiency of individual shifts.

1st Rank—Sum	2nd Rank—Sum	3rd Rank—Sum	U	*p*	Z	*p*
67,690,346	71,212,432		32,380,940	0.00	−7.54	0.00
67,299,784		70,787,487	31,990,378	0.00	−8.17	0.00
	67,922,665	67,864,295	33,771,685	0.57	−0.57	0.57

The analyses showed that there were no significant differences between the efficiency of the second and third shift, so a decision was made to combine them. However, the values obtained for the first shift differ significantly from those obtained for the other shifts, therefore this group was left without interference. These conclusions are confirmed by Figure 2 showing the differences described.

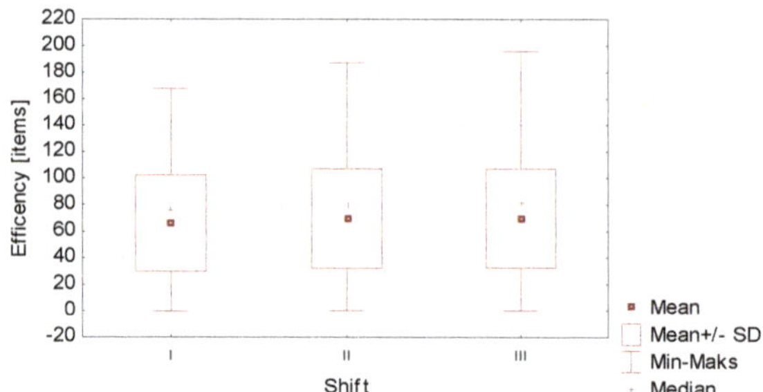

Figure 2. Frame diagram of the efficiency variable grouped by shifts.

The same test was performed for the device variable. The machines analyzed were of a single type and came from a single production batch, which suggests that their productivity would be similar. In order to confirm the equality of averages, the analysis of distribution normality and variance equality in individual groups (this time defined by the device variable) was carried out again in order to select a proper statistical distribution. The results of the normality test did not confirm the conformity. All the calculated chi-square test statistic values did not allow the zero hypothesis of the compatibility of the examined distribution with the normal one to be accepted. A definite deviation is confirmed by Figure 3.

The analysis of the equality of variance using the Levene and Brown-Forsythe tests showed that variances are not equal in some groups. Consequently, the Mann–Whitney test was used to check the difference between averages, the results of which are presented in Table 4.

Table 4. Results of the Mann-Whitney test for the difference between the average efficiency of individual shifts.

	H4	H5	H6	H12	H14	H21	H22	H23	H24	H25
H2	0.000	0.083	0.768	0.000	0.000	0.000	0.000	0.000	0.166	0.473
H4		0.000	0.000	0.007	0.012	0.000	0.000	0.000	0.000	0.000
H5			0.040	0.000	0.000	0.000	0.000	0.019	0.019	0.022
H6				0.000	0.000	0.000	0.000	0.000	0.069	0.716
H12					0.000	0.063	0.166	0.001	0.000	0.000
H14						0.000	0.000	0.000	0.000	0.000
H21							0.502	0.114	0.000	0.000
H22								0.107	0.000	0.000
H23									0.085	0.000
H24										0.056

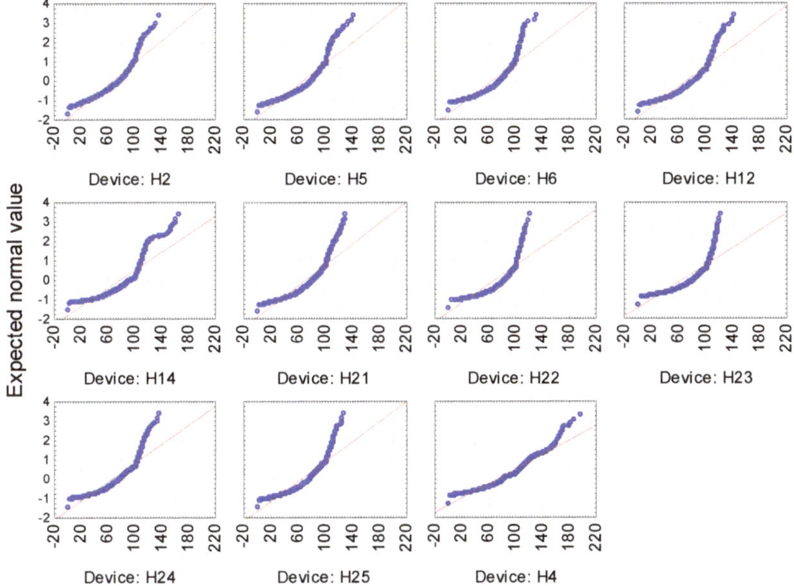

Figure 3. Graphs of normality of distribution of the efficiency variable grouped by the device variable.

Since the average efficiency varied for virtually every pair of devices, a decision was made not to combine them and to include each of them in the study. After defining the form of independent variables, the impact of each of them on the dependent variable, i.e., efficiency, was checked, but presented in a dichotomous form, as an assessment of whether the level achieved was satisfactory for the company. In line with the expectations of the Management Board, it was assumed that the assessment was positive if the productivity was equal to or above 90%. In other cases, the assessment would be negative. The chi-square test allowed for a statistical and substantive study of the relationship between variables. In all cases, the calculated test statistic did not allow the zero hypothesis on the lack of relationships between variables to be accepted. It was therefore rejected in favor of the alternative hypothesis of the existence of a relationship, the strength of which was measured using Yule's Φ (for binary tables) and Cramér's V coefficient (for tables more complex than 2×2). The obtained results are presented in Table 5.

The observed relationships between variables, although significant, are not strong. This is also confirmed by the graphs of interaction of individual dependent variables with the explained variable (Figure 4). Nevertheless, from the point of view of the analyzed company, the diagnosed bonds should not take place at all. A uniform and efficient operation of all devices is expected, so even minor deviations are undesirable and require further investigation.

The calculations carried out (Table 5) and the charts (Figure 4) confirm that the model variables were selected correctly. This allows the parameters of the logistic regression model to be estimated, the values of which are presented in Table 6.

Table 5. Results of the tests of significance and strength of the relationship between the predictors and the efficiency variable.

Statistics	Chi-Square	df	p
Shift			
Pearson's Chi^2	82.35	df = 1	$p = 0.00$
NW Chi^2	83.12	df = 1	$p = 0.00$
Yates Chi^2	82.10	df = 1	$p = 0.00$
Fi for 2 × 2 tables	0.06		
Contingency coefficient	0.06		
Device			
Pearson's Chi^2	538.57	df = 10	$p = 0.00$
NW Chi^2	532.84	df = 10	$p = 0.00$
Contingency coefficient	0.15		
Cramér's V	0.15		
Failure			
Pearson's Chi^2	786.17	df = 1	$p = 0.00$
NW Chi^2	1062.94	df = 1	$p = 0.00$
Yates Chi^2	784.68	df = 1	$p = 0.00$
Fi for 2 × 2 tables	−0.18		
Contingency coefficient	0.18		
Order			
Pearson's Chi^2	1756.69	df = 1	$p = 0.0000$
NW Chi^2	2576.74	df = 1	$p = 0.0000$
Yates Chi^2	1754.97	df = 1	$p = 0.0000$
Fi for 2 × 2 tables	−0.27		
Contingency coefficient	0.26		

Table 6. Parameters of the logistic regression model and their evaluation.

Effect	Modeled Probability Evaluation of Effectiveness—Positive Result Distribution: Binomial, Binding Function: LOGIT						
	Effect Level	Parameter	Estimated Standard Error	Wald's Test Statistics	p	95.00% CI	−95.00% CI
Absolute term		$\beta_0 = -7.549$	0.288	689.03	0.00	−8.112	−6.985
Shift	1st	$\beta_1 = -0.292$	0.031	90.59	0.00	−0.352	−0.232
Device	H2	$\beta_2 = -1.241$	0.067	343.29	0.00	−1.373	−1.110
Device	H5	$\beta_3 = -1.066$	0.067	256.69	0.00	−1.196	−0.936
Device	H6	$\beta_4 = -1.153$	0.068	291.56	0.00	−1.286	−1.021
Device	H12	$\beta_5 = -0.668$	0.065	104.14	0.00	−0.796	−0.539
Device	H21	$\beta_6 = -0.809$	0.065	153.40	0.00	−0.937	−0.681
Device	H22	$\beta_7 = -0.496$	0.067	55.25	0.00	−0.627	−0.365
Device	H23	$\beta_8 = -0.525$	0.067	60.84	0.00	−0.656	−0.393
Device	H24	$\beta_9 = -0.873$	0.067	168.66	0.00	−1.004	−0.741
Device	H25	$\beta_{10} = -1.167$	0.068	296.30	0.00	−1.300	−1.034
Device	H4	$\beta_{11} = -0.317$	0.072	19.323	0.00	−0.458	−0.176
Order	no	$\beta_{12} = 5.047$	0.251	403.059	0.00	4.554	5.539
Failure	no	$\beta_{13} = 3.08$	0.135	520.933	0.00	2.816	3.345

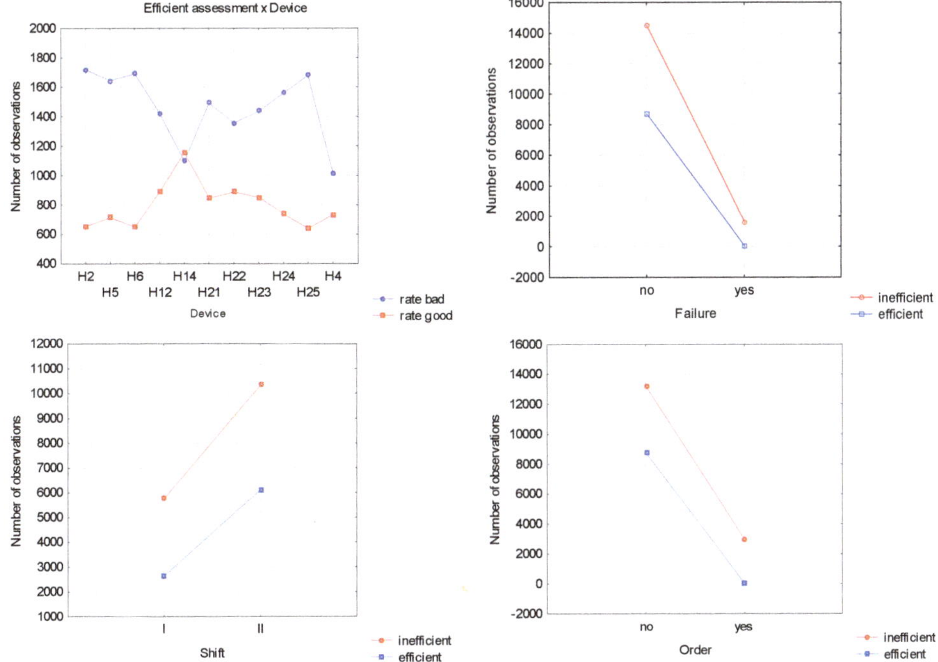

Figure 4. Interaction charts of dependent variable and predictors.

All calculated parameters turned out to be statistically significant, which is confirmed by the calculated Wald's statistic value and the associated probability value p, which for each line is lower than the assumed level of significance $\alpha = 0.05$. (Table 6). This means that all the distinguished factors significantly affect the evaluation of production efficiency. This allows the equation of the logistic regression model to be written in the following form:

$$P\left(efficiency = yes\middle|X\right) = \frac{e^a}{1 + e^a},\tag{9}$$

where

$$a = -7.549 - 0.292 \, shift \, I - 1.241 * H2 - 1.066 * H5 - 1.153 * H6 - 0.668 *$$
$$H12 - 0.809 * H21 - 0.496 * H22 - 0.525 * H23 - 0.873 * H24 - 1.167 *\tag{10}$$
$$H25 - 0.317 * H4 \; + 5.047 * no \, order \; + 3.08 * no \, failure.$$

The logistic regression curve is shown in Figure 5.

The logistic regression equation presented above can also take equivalent forms:

- logistic regression logit function:

$$logit \, P(efficient = 1\middle|X) = \ln\frac{P(efficient = 1|X)}{1 - P(efficient = 1|X)} = -7.549 - 0.292 \, 1st \, shift -$$
$$1.241 * H2 - 1.006 * H5 - 1.153 * H6 - 0.668 * H12 - 0.809 * H21 - 0.496 *\tag{11}$$
$$H22 - 0.525 * H23 - 0.873 * H24 - 1.167 * H25 - 0.317 * H4 \; + 5.047 *$$
$$no \, order \; + 3.08 * no \, failure,$$

- in the form of the odds:

$$P\frac{P(efficient = 1|X)}{1 - P(efficient = 1|X)} = e^a,\tag{12}$$

where

$$a = -7.549 - 0.292 \; shift\,I - 1.241 * H2 - 1.066 * H5 - 1.153 * H6 - 0.668 *$$
$$H12 - 0.809 * H21 - 0.496 * H22 - 0.525 * H23 - 0.873 * H24 - 1.167 * \tag{13}$$
$$H25 - 0.317 * H4 + 5.047 * no\;order + 3.08 * no\;failure.$$

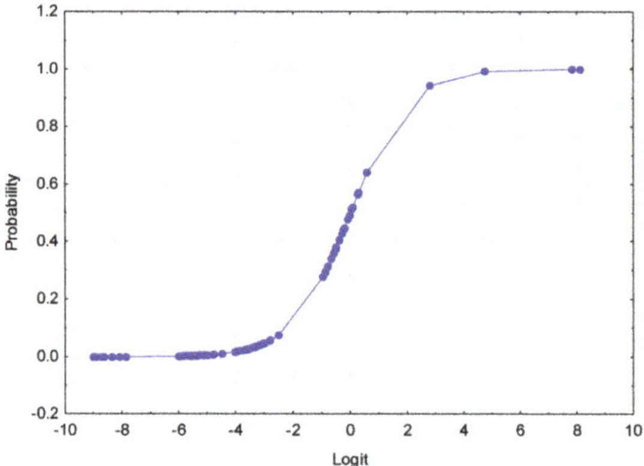

Figure 5. Logistic regression curve.

An important element of the evaluation of the studied process is the calculation of the odds of an event occurrence (in this case a satisfactory result of efficiency). The sign at the estimated parameter of the logistic regression model indicates whether the analyzed odds are greater (plus) or smaller (minus) in relation to the reference level. The scale of this change is indicated by the unit odds ratio shown for each parameter in Table 7.

Table 7. Odds ratios for individual predictors.

Effect	Effect Level	Odds Ratio	95.00% Cl	−95.00% Cl	p
Shift	1st	0.747	0.704	0.793	0.00
Device	H2	0.289	0.253	0.330	0.00
Device	H5	0.344	0.302	0.392	0.00
Device	H6	0.316	0.276	0.360	0.00
Device	H12	0.513	0.451	0.583	0.00
Device	H21	0.445	0.392	0.506	0.00
Device	H22	0.609	0.534	0.694	0.00
Device	H23	0.592	0.519	0.675	0.00
Device	H24	0.418	0.366	0.477	0.00
Device	H25	0.311	0.273	0.356	0.00
Device	H4	0.729	0.633	0.839	0.00
No order—yes no	no	155.489	95.002	254.487	0.00
Failure—yes no	no	21.7679	16.7083	28.3595	0.00

The odds ratio for the 1st shift is 0.75, which means that compared to the 2nd shift, the odds of achieving satisfactory efficiency is 0.75 times lower. In other words, the odds for the proper efficiency is 1.34 times higher in the case of the 2nd shift. The H14 machine was used as a reference when analyzing the impact on the efficiency of individual devices. Its efficiency is the highest in the test sample, so all the model coefficients obtained are negative, which means less odds of achieving a positive result. The odds ratios are given in column 3 of Table 6. The worst result was obtained for the H2 machine with an odds ratio of 0.289, which means an almost 3.5-fold increase in the odds of achieving satisfactory efficiency when replacing H2 with H14.

The results for the other machines, showing how much the efficiency of each machine should be increased in order to obtain the efficiency evaluation as in the case of the H12 machine, are shown in Table 8.

Table 8. Odds ratios for individual predictors.

Machine Number	H2	H5	H6	H12	H21
OR	3.460	2.904	3.169	1.950	2.246
Machine Number	H22	H23	H24	H25	H4
OR	1.642	1.690	2.393	3.212	1.372

The last two parameters in Table 7 refer to the absence of an order or failure, as their occurrence has a negative impact on the efficiency. Where there is no downtime, the odds of achieving the expected efficiency are 155 times greater than otherwise. Similarly, the occurrence of a failure has a similar effect, but its absence is not as spectacular. There is a 21-fold increase in the odds if the failure does not occur.

The presented model can also be used for predictive purposes, allowing for forecasting the probability of achieving the predicted success (here, the assumed efficiency). It is therefore important to assess the quality of the prediction. For this purpose, it is helpful to determine the so-called cut-off point π_0. This parameter allows the observed dichotomous values of a dependent variable to be compared with the continuous probability values calculated on the basis of the model. This value falls within the range $(0, 1)$ and is defined as follows [36] when:

$$\hat{\pi}(x) = \hat{P}(Y = 1|x) > \pi_0, \tag{14}$$

it is assumed that an event has occurred $(\hat{y} = 1)$. In the opposite situation, when

$$\hat{\pi}(x) \leq \pi_0, \tag{15}$$

it is assumed that an event has not occurred $(\hat{y} = 0)$.

Prediction ideally occurs when sensitivity and specificity are equal to 1, which means no false positive or negative results. In real life research, the point corresponding to a case where a model best discriminates occurrences is called the optimal cut-off point. It is determined using the Youden's index (J), which takes the following form:

$$J = sensitivity + specificity - 1. \tag{16}$$

The optimum cut-off point corresponds to the case where the J value reaches its maximum. For the case under consideration, the proposed cut-off point is shown in Table 9.

Table 9. Cut-off point of the logistic regression model concerned.

Number of Observations	Cut-off Point	True Positive	True Negative	False Positive	False Negative	SE	1-SP	Youden's Index
5	0.51	2864	13,903	2210	5905	0.33	0.14	0.19

For the proposed cut-off point, the sensitivity is 0.32, and the specificity is 0.86. There are 16,767 well classified cases (2864 true positive and 13,903 true negative) and 8115 badly classified cases (2210 false positive and 5905 false negative cases).

Based on the above table it is possible to assess the effectiveness of model prediction in relation to successes and failures, using tools among which one can distinguish such statistics as accuracy, sensitivity or specificity, ROC (receiver operating characteristic) curve or values of rank correlations.

The simplest measure is accuracy, calculated according to the following formula:

$$Accuracy = ACC = \frac{TP + TN}{TP + TN + FP + FN}, \tag{17}$$

where:

TP—number of true positive results,

TN—number of true negative results,

FP—number of false positive results,

FN—number of false negative results.

For the model concerned,

$$ACC = \frac{2864 + 13903}{2864 + 5905 + 2210 + 13903} = 0.674 = 67.4\%. \tag{18}$$

However, the sensitivity *SE* and specificity *SP* are most often considered in such analyses but treated as pairs, which, after being marked on the plane and after connecting the points with segments, form the so-called ROC curve. For the analyzed model this curve is presented in Figure 6.

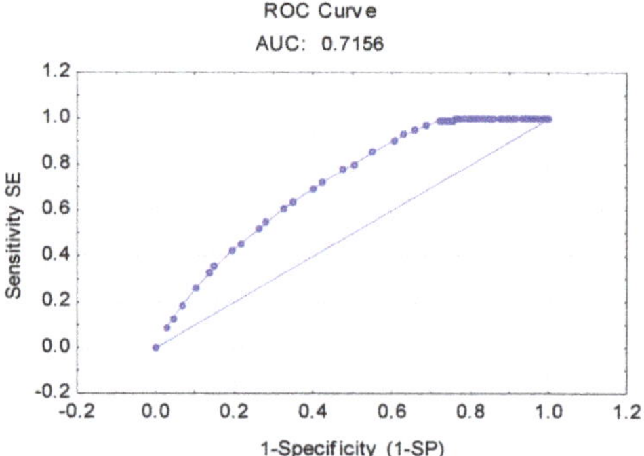

Figure 6. ROC curve for the model concerned.

The most important parameter for assessing the ROC curve is AUC—area under the ROC curve. It takes values from 0 to 1. The interpretation of the result was based on the Kleinbaum and Klein classification (Table 10), according to which discrimination is sufficient [37].

Table 10. Cut-off point of the logistic regression model concerned.

AUC Value	Score
0.9 < AUC < 1.0	Excellent discrimination
0.8 < AUC < 0.9	Good discrimination
0.7 < AUC < 0.8	Sufficient discrimination
0.6 < AUC < 0.7	Weak discrimination
0.5 < AUC < 0.6	Insufficient discrimination

The model can therefore be considered satisfactory, although it is recommended rather for qualitative analysis of processes and modification of production strategy on its basis.

Production in the company in question is carried out in a continuous three-shift system. Regardless of the production plan, the plant is fully manned and all machines are in operation at all times. The acceptable level of efficiency assumed by the management should be 90%, which allows all scheduled and expected downtime to be taken into account. The study indicates that in most cases this level is not reached. In almost 25,000 out of over 156,000 observations, this level was not reached.

It turned out that the efficiency of the first shift was lower compared to the second and third ones, which suggests the need to diagnose the causes of circumstances or even to restructure the shift system. The use of individual machines also has a negative impact on efficiency. It turns out that many of them represent much lower efficiency than the one taken as a reference point, the efficiency of which was the highest (but also less than 100%). Of course, a lack of orders also leads to a significant reduction in productivity. Such a result, next to failures that occur, indicates the need for a detailed analysis of the machinery. It might be advisable to exclude several machines from the production process so that production is closely correlated with demand. Unused machines could constitute a reserve in case of failure and thus increase the level of readiness of the machinery.

4. Conclusions

The reliability of machinery and equipment is an essential part of the proper functioning of a business. Modern technologies support the maintenance of an adequate level of readiness and suitability of the technical infrastructure. They not only facilitate production control and implementation of modern operating strategies, but also ensure continuous monitoring of processes and detection of any disturbances or failures. The activities carried out in this area boil down to balancing the maintenance of full operational efficiency and continuity of production and ensuring an acceptable level of costs of these activities.

This is a difficult task, particularly in the case of older generation machinery stock, which is deprived of support from computerized production management systems—as the one presented in this article. In any case, however, the aim is to ensure that machines function perfectly without failures and that products are manufactured without defects. On the other hand, it is also important to ensure the efficiency of the equipment use and balanced workload, which proved to be a problem in the analyzed company. This was the reason for undertaking research in this field and the basis for mathematical analysis of the effectiveness of the manufacturing process.

The lack of IT systems for controlling and monitoring the production process results in the inability to archive data on an ongoing basis, which makes it much more difficult to control processes, and sometimes it is conducive to abandonment thereof. Poor quality of recorded documentation imposes significant limitations on the use of mathematical tools as well; therefore, the authors wanted to present simultaneously that even in such a situation it is possible to create mathematical models that would improve the efficiency of the machinery stock. The available data proved to be sufficient to achieve the research objective, which was to formulate a model providing an unambiguous answer as to whether the efficiency of the equipment used in production is acceptable from the point of view of the assumptions made by the company (in this case 90%).

This was made possible through the use of logistic regression, which, above all, does not require the meeting of assumptions made by other mathematical models, e.g., linear regression and general linear models. The advantage of this method is also the form of the dependent variable. The predictor here is a dichotomous variable and its values can be interpreted as the probability of an event occurring. The organization of production in the analyzed company has remained unchanged for many years. All machines work three shifts every day. Regardless of the orders placed and the market demand, full staff is employed. Machines are taken out of the process only in cases of random incidents. Lack of modification of the adopted procedures and control of the implemented processes results—as demonstrated in the research—in the process not being effective and the use of machinery not being optimal.

The logistic regression model made it possible to identify the causes influencing machinery efficiency. It turned out that the load is not identical during every shift, the productivity is much lower during the first shift in comparison to the other shifts. Restructuring the shift system and limiting the production process to only two shifts or modifying the working time could increase the productivity of machinery, optimize the use of human resources and reduce the costs of the production process. The load on the individual machines also appeared to be disproportionate. Increased use of one piece of equipment may result in an increase in the frequency of breakdowns, increase the costs of repairs and reduce the total life of the equipment, so it would be advisable to evenly distribute production across all equipment. The reduction in productivity was also caused by a decline in the number of production orders. The lack of production control with regard to orders received and the maintenance of a continuous three-shift readiness exposes the company to costs and favors the aforementioned disproportionate workload.

The studied company has no IT systems in place that enable comprehensive monitoring of production processes. Therefore, the unquestionable advantage of the proposed model is the provision of additional information allowing decisions to be made on the production and use of machinery. They can also encourage the implementation of modern solutions and the abandonment of traditional, outdated methods of recording and archiving data.

In companies that use specialist MES systems based on real-time information on manufacturing execution at subsequent workstations, the proposed model can improve monitoring of productivity drops below the adopted level and activate preventive actions. Additionally, the implementation of data obtained from the IT system may allow to the model to be extended with additional parameters, which is important from the point of view of individual companies.

The aim of the article was to investigate the possibility of developing a model for the analysis and evaluation of the level of efficiency of ongoing production processes, as well as to indicate the method of logistic regression as a tool supporting decision-making in this respect. The model developed for the analyzed company indicated the need for a strict correlation between the demand for a product and the production process. Adopted strategies require verification and modification.

The proposed model may also serve as a basis for setting directions for improvement of the production process, by maximizing the use of the machinery stock and reducing the idle time of both employees and equipment. Re-application of the logistic regression model constructed on the basis of the observation of the process after introduction of changes allows the effectiveness and efficiency of implemented solutions to be evaluated.

Author Contributions: Conceptualization, A.B. and M.G.; Formal analysis, A.B.; Methodology, A.B.; Resources, M.G.; Writing—original draft, A.B. and M.G.; Writing—review & editing, A.B.

Funding: This research received no external funding.

Conflicts of Interest: The authors declare no conflicts of interest.

References

1. Świć, A.; Gola, A. Economic Analysis of Casing Parts Production in a Flexible Manufacturing System. *Actual Probl. Econ.* **2013**, *141*, 526–533.
2. Gola, A. Reliability analysis of reconfigurable manufacturing system structures using computer simulation methods. *Eksploat. I Niezawodn. Maint. Reliab.* **2019**, *21*, 90–102. [CrossRef]
3. Jasiulewicz-Kaczmarek, M.; Saniuk, A. How to make maintenance processes more efficient using lean tools? *Adv. Intell. Syst. Comput.* **2017**, *605*, 9–20.
4. Antosz, K.; Stadnicka, D. Evaluation measures of machine operation effectiveness in large enterprises: Study results. *Eksploat. I Niezawodn. Maint. Reliab.* **2015**, *17*, 107–117. [CrossRef]
5. Chien, C.-F.; Dou, R.; Fu, W. Strategic capacity planning for smart production: Decision modeling under demand uncertainty. *Appl. Soft Comput.* **2018**, 68900–68909. [CrossRef]
6. Iscioglu, F.; Kocak, A. Dynamic reliability analysis of a multi-state manufacturing system. *Eksploat. I Niezawodn. Maint. Reliab.* **2019**, *21*, 451–459. [CrossRef]
7. Liu, T.; Chen, C.; Kuo, Y.; Kuo, M.; Hsueh, P. Developing a monitoring system of production effectiveness and traceability for semi-automatic production process. In Proceedings of the International Conference on Fuzzy Theory and Its Applications (iFUZZY), Pingtung, Taiwan, 12–15 November 2017; pp. 1–5.
8. Loska, A. Przegląd modeli ocen eksploatacyjnych systemów technicznych. *Komput. Zintegr. Zarz.* **2011**, *2*, 37–46.
9. Kang, N.; Zhao, C.; Li, J.; Horst, J.A. A Hierarchical structure of key performance indicators for operation management and continuous improvement in production systems. *Int. J. Prod. Res.* **2016**, *54*, 6333–6350. [CrossRef]

10. Koliński, A.; Śliwczyński, B.; Golińska-Dawson, P. The assessment of the economic efficiency of production process-simulation approach. In Proceedings of the 24th International Conference on Production Research (ICPR 2017), Poznań, Poland, 30 July–3 August 2017; pp. 283–289.

11. Chen, W.; Wang, Z.; Chan, F.T.S. Robust production capacity planning under uncertain wafer lots transfer probabilities for semiconductor automated material handling systems. *Eur. J. Oper. Res.* **2017**, *261*, 929–940. [CrossRef]

12. Gola, A.; Kosicka, E.; Daniewski, K.; Mazurkiewicz, D. Analiza błędów przy ocenie wskaźnika OEE na przykładzie linii rozlewu butelkowego. *Innow. W Zarz. I Inż. Prod.* **2016**, *2*, 654–662.

13. Aleš, Z.; Pavlů, J.; Legát, V.; Mošna, F.; Jurča, V. Methodology of overall equipment effectiveness calculation in the context of Industry 4.0 environment. *Eksploat. I Niezawodn. Maint. Reliab.* **2019**, *21*, 411–418. [CrossRef]

14. Wilczarska, J. Efektywność i bezpieczeństwo użytkowania maszyn. *Inż. I Apar. Chem. Chem. Eng. Equip.* **2012**, *51*, 41–43.

15. Szwedzka, K.; Jasiulewicz-Kaczmarek, M. Determining maintenance services using production performance indicators. *Res. Logist. Prod.* **2016**, *6*, 361–374. [CrossRef]

16. Jasiulewicz-Kaczmarek, M.; Piechowski, M. Practical Aspects of OEE in Automotive Company–Case Study. In Proceedings of the 3rd International Conference on Management Science and Management Innovation (MSMI 2016), Guilin, China, 13–14 August 2016; pp. 213–218.

17. Oechsner, R.; Pfeffer, M.; Pfitzner, L.; Binder, H.; Muller, E.; Vonderstrass, T. From overall equipment efficiency (OEE) to overall Fab effectiveness (OFE). *Mat. Sci. Semicond. Process.* **2002**, *5*, 333–339. [CrossRef]

18. Wang, T.J.; Pan, H.C. Improving the OEE and UPH data quality by Automated Data Collection for the semiconductor assembly industry. *Expert Syst. Appl.* **2011**, *38*, 5764–5773. [CrossRef]

19. Ylipää, T.; Skoogh, A.; Bokrantz, J.; Gopalakrishnan, M. Identification of maintenance improvement potential using OEE assessment. *Int. J. Prod. Perform. Manag.* **2017**, *66*, 126–143. [CrossRef]

20. Jonsson, P.; Lesshammer, M. Evaluation and improvement of manufacturing performance measurement systems–the role of OEE. *Int. J. Oper. Prod. Manag.* **1999**, *19*, 55–78. [CrossRef]

21. Muchiri, P.; Pintelon, L. Performance measurement using overall equipment effectiveness (OEE): Literature review and practical application discussion. *Int. J. Prod. Res.* **2008**, *46*, 3517–3535. [CrossRef]

22. Antosz, K.; Pacana, A. Wskaźnikowa ocena efektywności funkcjonowania maszyn na przykładzie wybranego przedsiębiorstwa–studium przypadku. *Innow. W Zarz. I Inż. Prod.* **2016**, *2*, 634–6461.

23. Panagiotis, H.T. Evaluation of overall equipment effectiveness in the beverage industry: A case study. *Int. J. Prod. Res.* **2013**, *51*, 515–523.

24. Gola, A. Economic Aspects of Manufacturing Systems Design. *Actual Probl. Econ.* **2014**, *6*, 205–2012.

25. Kubik, S.; Kornicki, K. *OEE dla operatorów. Całkowita Efektywność Wyposażenia*; ProdPress.com: Wrocław, Poland, 2009.

26. Panagiotis, H.T. Evaluation of maintenance management through the overall equipment effectiveness of a yogurt production line in a medium-sized Italian company. *Int. J. Prod. Qual. Manag.* **2015**, *16*, 123–135.

27. Luo, H.; Wang, K.; Kong, X.T.R.; Lu, S.; Qu, T. Synchronized production and logistics via ubiquitous computing technology. *Robot. Comput. Integr. Manuf.* **2017**, *45*, 99–115. [CrossRef]

28. Muchiri, P.; Pinteleon, L.; Martin, H.; De Meyer, A.M. Empirical analysis of maintenance performance measurement in Belgian industries. *Int. J. Prod. Res.* **2010**, *20*, 5905–5924. [CrossRef]

29. Kalliski, M.; Krahé, D.; Beisheim, B.; Krämer, S.; Engell, S. Resource Efficiency Indicators for Real-Time Monitoring and Optimization of Integrated Chemical Production Plants. In *Computer Aided Chemical Engineering*; Gernaey, K.V., Huusom, J.K., Gani, R., Eds.; Elsevier: Amsterdam, The Netherlands, 2015; Volume 37, pp. 1949–1954.

30. Yan, J.; Lee, J. Degradation assessment and fault modes classification using logistics regression. *J. Manuf. Sci. Eng.* **2005**, *127*, 912–914. [CrossRef]

31. Kozłowski, E.; Mazurkiewicz, D.; Żabiński, T.; Pruncal, S.; Sęp, J. Assessment model of cutting tool condition for real-time supervision system. *Eksploat. I Niezawodn. Maint. Reliab.* **2019**, *21*, 679–685. [CrossRef]

32. Yan, J.; Ko, M.; Lee, J. A prognostic algorithm for machine performance assessment and its application. *Prod. Plan. Control* **2004**, *15*, 796–801. [CrossRef]

33. Caesarendra, W.; Widodo, A.; Yang, B.S. Application of relevance vector machine and logistic regression for machine degradation assessment. *Mech. Syst. Signal Process.* **2010**, *24*, 1161–1171. [CrossRef]

34. Chen, B.; Chen, X.; Li, B. Reliability estimation for cutting tool based on logistics regression model using vibration signals. *Mech. Syst. Signal Process.* **2011**, *25*, 2516–2537. [CrossRef]
35. Jardine, A.K.S.; Lin, D.; Banjevic, D. A review on machinery diagnostics and prognostics implementing conditio-based maintenance. *Mech. Syst. Signal Process.* **2006**, *20*, 1483–1510. [CrossRef]
36. Stanisz, A. *Modele Regresji Logistycznej. Zastosowanie W Medycynie, Naukach Przyrodniczych I Społecznych*; Statsoft Polska: Kraków, Poland, 2016.
37. Kleinbaum, D.G.; Klein, M. *Logistic Regression A Self-Learning Text*; Springer: Berlin, Germany, 2010.

Article

Evaluation of Machinery Readiness Using Semi-Markov Processes

Andrzej Świderski [1], Anna Borucka [2,*], Małgorzata Grzelak [2] and Leszek Gil [3]

[1] Motor Transport Institute, 03-301 Warsaw, Poland; andrzej.swiderski@its.waw.pl
[2] Logistics and Management, Faculty of Security, Military University of Technology, 00-908 Warsaw, Poland; malgorzata.grzelak@wat.edu.pl
[3] Department of Mechanics and Machine Building, University of Economics and Innovation, 20-209 Lublin, Poland; leszek.gil@wsei.lublin.pl
* Correspondence: anna.borucka@wat.edu.pl

Received: 24 January 2020; Accepted: 20 February 2020; Published: 24 February 2020

Abstract: This article uses Markov and semi-Markov models as some of the most popular tools to estimate readiness and reliability. They allow to evaluate of both individual elements as well as entire systems—including production systems—as multi-state structures. To be able to distinguish states with varying degrees of technical readiness in complicated and complex objects (systems) allows to determine their individual impact on the tasks performed, as well as on the total reliability. The application of the Markov process requires, for the process dwell times in the individual states, to be random variables of exponential distribution and the fulfilling Markov's property of the independence of these states. Omitting these assumptions may lead to erroneous results, which was the authors' intention to show. The article presents a comparison of the results of the examination of the process of non-parametric distribution with an analysis in which its exponential form was (groundlessly) assumed. Significantly different results were obtained. The aim was to draw attention to the inconsistencies obtained and to the importance of a preliminary assessment of the data collected for examination. The diagnostics of the machine readiness operating in the studied production company was additionally performed. This allowed to evaluate its operational potential, especially in the context of solving process optimization problems.

Keywords: semi-Markov model; Markov model; empirical data distribution; readiness; production machines

---

1. Introduction

1.1. Background Introduction to the Study

Ensuring desired availability of all machine tools in a production line is an important issue [1,2]. It stands for their ability to obtain and maintain the functional state necessary to produce the required performance [3–5]. The technical readiness of machines is an important element of the company diagnostics and should be estimated, as its evaluation helps shape the capacity of a production line. High machine tools reliability translates into no unnecessary downtime and, consequently, greater process efficiency. Machine tools must be technically sound, adequately controlled and supplied with necessary materials, energy and information [6]. The availability of a machine tool is determined using a probability theory-based reliability model. In probability theory, the state of an object is defined as the result of one and only one event in a sequence of trials of finite or computable set of elementary events excluding each other in pairs [7]. This makes it possible to use the tools of probability calculus and mathematical statistics to analyze technical systems. When machine tools are in operation they stochastically transit from one state to another. As a result, transition probabilities are associated with all machine tools in a production line. Therefore, Markov chain and its derivatives are often used to set a model of reliability. Some of the relevant articles where Markov chain-based reliability models are used to study the availability of machines tools in a production line are described below.

The use of Markov processes and their generalization—semi-Markov processes—are popular. Their use is dictated by the multi-states condition of the technical objects and the assumption that the assessment of individual functional states the object is in, is a better measure than the readiness of the object as a whole. However, the use of these models is subject to restrictions. First of all, it is necessary to fulfil the Markov property which states that the probability of a future state is independent of the past states, and depends only on the present state. Identifying a model without meeting this assumption may lead to false conclusions, which is suggested by many authors [8,9]. They point out that ignoring the Markov property examination will results in incorrect analysis results, e.g., Shi et al. [10], Zhang et al. [8], or Kozłowski et al. [11]. Therefore, it is necessary to examine the randomness of sequences of subsequent operational states, as it is done by Yang et al. [12] or Komorowski and Raffa [13].

In addition, Markov's models require meeting the assumption that the unconditional process dwell times in the individual states and the conditional durations of an individual state, are random variables of exponential distribution, provided that the next one is one of the remaining states [14,15]. Many authors point out that proper matching of distributions affects the reliability of results [13,16]. They use Markov's model for exponential distributions [17,18], and the semi-Markov model for the remaining ones, e.g., Weibull [19] or Gamma [20]. Adoption of only the assumption on the form of distribution without a statistical survey of the collected sample may lead to wrong conclusions.

1.2. The Aim of the Study

The need to check Markov's property is discussed in more detail in the literature [10,11], while less attention is paid to the distribution of variables studied. Therefore, this publication compares the results of process examination according to the semi-Markov model for variables of non-parametric distribution with the analysis according to the Markov model, in which their exponential form was (falsely) assumed. The differences in the results obtained clearly indicate that it is necessary to carry out a preliminary test before choosing the right model. Failure to meet the assumptions leads to an inaccurate analysis of the process.

The aim of the article was also to evaluate the readiness of a production machine, which is an important element of the analyzed production process. The results obtained made it possible to determine the probabilities of transitions between the individual states distinguished in the production process, as well as to define limit probabilities and the technical readiness coefficient. This allows to assess the compliance of the functioning of the analyzed process with the schedule adopted in the company, or to evaluate the results of production abilities. The proposed models can also be used to simulate the production process, e.g., at the design phase.

The article consists of five sections. The first one presents an analysis of the literature on the application of Markov models for studying the technical readiness of machine tools in a production line. Section 2 presents a mathematical formulation of the research problem. In Section 3, a description of the studied company was given and data analysis was carried out in terms of studying the Markov property and the form of distribution of variables. Section 4 presents a case study containing the estimation of Markov and semi-Markov models parameters, as well as a numerical example and accurate calculations according to the developed model. The article ends with conclusions describing the goals achieved and indicating the added value of the study.

2. Mathematical Modeling

Definition 1. *Let us consider a random process with a finite state space $S = \{1,..., s\}$, $s < \infty$. Let $(\Omega, \mathcal{F}, P)$ be a probabilistic space and $\{X(t) : t \in T\}$ a stochastic process defined for $(\Omega, \mathcal{F}, P)$, taking values from the finite* *or calculable set S. Process $\{X(t) : t \in T\}$ is called a Markov process if for each i, j, i_0, $i_1,...,i_{n-1} \in S$ and for each t_0, t_1, ..., $t_n, t_{n+1} \in T$ meeting the requirement $t_0 < t_1 < t_n < t_{n+1}$ the dependency given below is met:*

$$P(X(t_{n+1}) = j|X(t_n) = i, X(t_{n-1}) = i_{n-1}, \ldots, X(t_0) = i_0) = P(X(t_{n+1}) = j|X(t_n) = i), \qquad (1)$$

Assuming that $t_n = u$, $t_{n+1} = \tau$, then the conditional probability:

$$P(X(\tau) = j|X(u) = i) = p_{ij}(u, s), \qquad (2)$$

for $i, j \in S$, where $p_{ij}(u,s)$ denotes the probability of transition from state i at time u, to state j at time s.

Assuming that t_0, t_1, ..., t_{n-1} denote time (instants) from the past, t_n denotes the present instant, and t_{n+1} the time in the future, the equation says that the future does not depend on the past when the present is known, thus the probability of the future state is independent of the past states, but only of the present state. This property is called Markov's property, and the stochastic process that satisfies it, a memoryless process. If instants of time are discrete, $T = N_0 = \{0, 1, 2, \ldots\}$, then we are dealing with Markov's chain, and when the process is realized in a continuous time $T = R_+ = [0, \infty)$, it is a continuous-time Markov process.

For the stochastic process $\{X(t) : t > 0\}$ taking values from the finite or countable set S with fixed and right-hand continuous phase trajectories in some sections, and for $\tau_0 = 0$ which marks the start of the process and τ_1, τ_2, ... which denotes successive times of change of states, the random variable:

$$T_i = \tau_{n+1} - \tau_n|X(\tau_n) = i, i \in S, \qquad (3)$$

denotes the waiting time in the state i when a successor state is unknown. From the Chapman–Kolmogorov equation, it follows [21,22] that the process dwelling times in the individual states constitute random variables with exponential distributions and with the parameter $\lambda_i > 0$:

$$G_i(t) = P(T_i \leq t) = P(\tau_{n+1} - \tau_n \leq t|X(\tau_n) = i) = 1 - e^{-\lambda_i t}, t \geq 0, i \in S, \qquad (4)$$

where G_i is a cumulative probability distribution of a random variable T_i [23] when a successor state is unknown.

The generalization of Markov processes are semi-Markov processes, for which dwelling times in the individual states can have arbitrary distributions, concentrated in the set $[0, \infty]$. Based on [24,25] it was assumed in this article to define the semi-Markov process with a finite set of states starting from Markov renewal process.

In the probabilistic space $(\Omega, \mathcal{F}, P)$ random variables are defined for each $n \in N$:

$$\xi_n : \Omega \to S, \qquad (5)$$

$$\vartheta_n : \Omega \to R_+ \qquad (6)$$

A two-dimensional sequence of random variables $\{(\xi_n, \vartheta_n) : n \in N\}$ is referred to as the Markov renewal process if for each $n \in N, i, j \in S, t \in R_+$:

$$P\{\xi_{n+1} = j, \vartheta_{n+1} < t/\xi_n = i, \xi_{n-1}, \ldots \xi_0, \vartheta_n, \ldots \vartheta_0\} = P\{\xi_{n+1} = j, \vartheta_{n+1} < t/\xi_n = i\}, \qquad (7)$$

and

$$P\{\xi_0 = i, \vartheta_0 = 0\} = P\{\xi_0 = i\}, \qquad (8)$$

This definition shows that the Markov renewal process is a specific case of the two-dimensional Markov process. Transition probabilities of this process depend solely on the discrete value of the coordinate. The Markov renewal process $\{(\xi_n, \vartheta_n) : n \in N\}$ is called homogeneous if the probabilities:

$$P\{\xi_{n+1} = j, \vartheta_{n+1} < t / \xi_n = i\} = Q_{ij}(t), \tag{9}$$

Do not depend on n.

From the above definition, it follows that for each pair $(i, j) \in SxS$ function $Q_{ij}(t)$ is [24,25]:

- non-decreasing,
- right-hand continuous,
- $Q_{ij}(0) = 0$,
- $Q_{ij}(t) \leq 1$,
- $\sum_{j \in S} \lim_{t \to \infty} Q_{ij}(t) = 1$.

Functional matrix:

$$Q(t) = \left[Q_{ij}(t) \right], \ i, j \in S, \tag{10}$$

is called the renewal kernel of the semi-Markov process and together with the initial distribution:

$$p_i = P\{\xi_n = 1\}, \ i \in S, \tag{11}$$

characterizes the homogeneous Markov renewal process.

Semi-Markov process is defined based on the homogeneous Markov renewal process $\{(\xi_n, \vartheta_n) : n \in N\}$. Let:

$$\tau_0 = \vartheta_0 = 0, \tag{12}$$

$$\tau_n = \vartheta_1 + \ldots + \vartheta_n, \tag{13}$$

$$\tau_\infty = \sup\{\tau_n : n \in N_0\}. \tag{14}$$

The stochastic process $\{X(t) : t \in R_+\}$, which assumes a constant value in the range $(\tau_{n+1}), n \in N$:

$$X(t) = \xi_n, \tag{15}$$

is called the semi-Markov process.

Markov and semi-Markov models are particularly often used to assess the readiness and reliability of technical facilities or their individual components [26–28]. Various systems, including production ones [29,30], are analyzed both in terms of maintaining operability [31], production organization [32] as well as shaping of the demand [33]. This article analyzes the production system from the point of view of machine readiness to perform production tasks.

3. Data Handling

3.1. Description of the Company Studied

The subject of the research is a company manufacturing plastic garbage bags. It is a three shift serial production, with 8-h shifts. The roller welding machines, which weld and perforate finished rolls of polyethylene film, constitute a critical element of the whole process. Among all the machines in the production line, the efficiency of the roller welding machines is the lowest, they have the highest failure rate and their downtimes lead to substantial increase in costs, making them a bottleneck in the process. This is why they became the subject of the study. The analysis was carried out on the example of a selected model, marked with the H2 symbol.

The analysis of the activities carried out when operating the roller welding machines allowed to distinguish the states of the machine. They are presented in the Table 1.

Table 1. Operating states under consideration.

State	
S_1	Operation
S_2	Failure
S_3	Downtime due to lack of orders
S_4	Maintenance activities (cleaning, reorientation, knife adjustment, inspection, roll change, consultation, raw material preparation)
S_5	Scheduled employee breaks
S_6	Downtime due to lack of raw materials

Among the selected states, those directly related to the production process should be distinguished: S_1—manufacturing process, S_4—necessary maintenance activities to keep the machine in good working order and to prepare it for the manufacturing process. Planned employee breaks, resulting from the Labor Code (S_5), also constitute a necessary element of the manufacturing process. Other states should be identified as undesirable. These include the stoppage in the manufacturing process, due to lack of orders, S_3, and lack of raw materials, S_6.

The relationships between the individual states are shown in Figure 1.

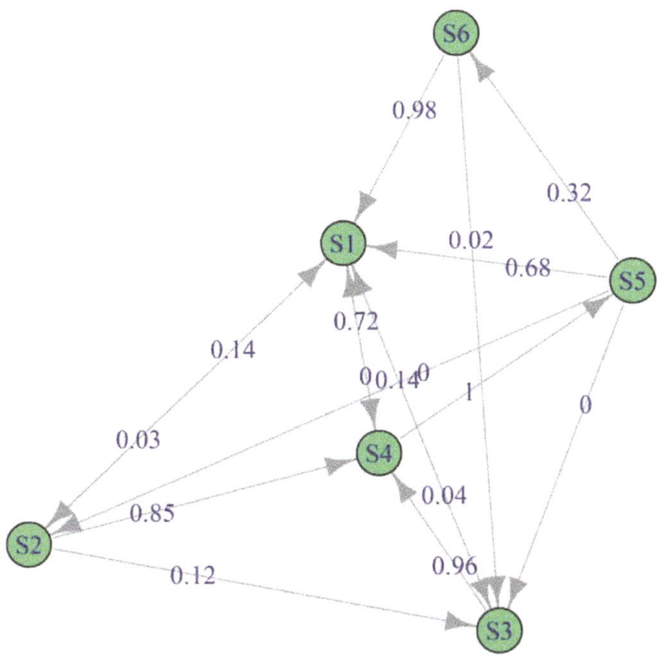

Figure 1. Diagram of the inter-state transitions of the process studied.

3.2. Studying Markov's Property

In the first stage, the lack of memory characteristic of the process was assessed. The goodness of fit test χ^2 was used for the study, defining at the level of significance $\alpha = 0.05$ the zero hypothesis assuming that the chain studied meets the Markov property, and the alternative hypothesis that the Markov property is not met [11]. The test statistic $\chi^2 = 228.7$, while related to it p-value = 0.264, which means that there are no grounds to reject the zero hypothesis on the chain meeting the Markov property.

3.3. Fit of Distributions

The next step was to assess the form of distributions of the individual states. Considerations in this respect were presented using the example of state S_6, and for the others the same was done. The goodness of fit to the selected theoretical distributions that were considered most likely was verified based on a Cullen and Frey graph, presented for state S_6 in Figure 2.

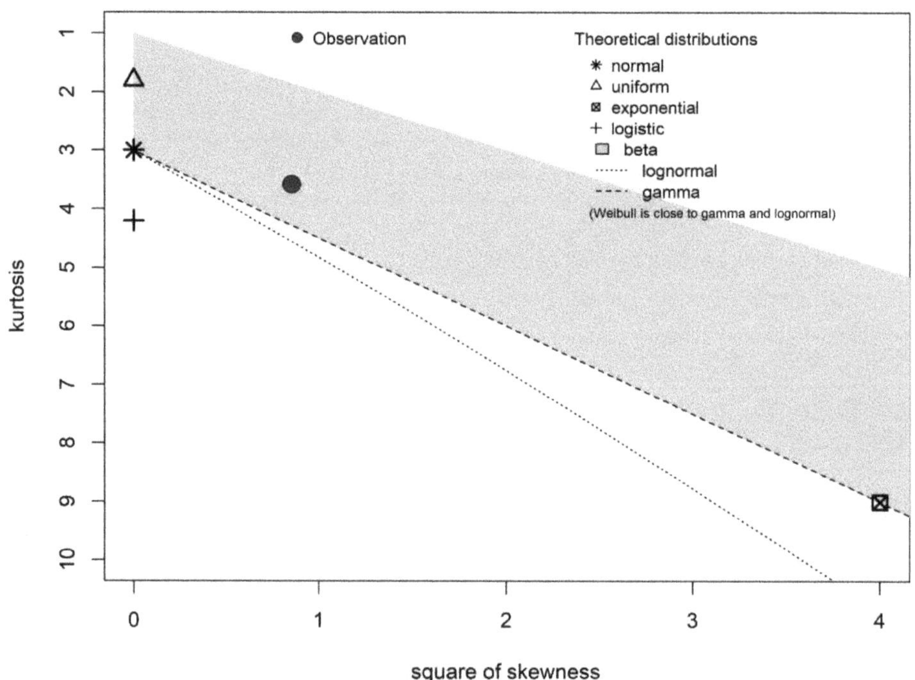

Figure 2. Cullen and Frey graph for the state S6.

For further analysis, the Weibull and Beta distributions were selected, for which the estimated parameters are presented in Table 2, while the goodness of fit of empirical data to the individual distributions is presented in Figure 3.

Table 2. Estimated model parameters according to Weibull and Beta distributions.

Weibull Distribution	Distribution Parameters		Kolmogorov–Smirnov Test Statistic	*p*-Value	Akaike Criterion
	Scale	Shape			
Parameters	186.57	1.66	D = 0.07	0.385	1872.38
Std. Error	9.53	0.11			
Beta distribution	shape 1	shape 2			
Parameters	1.93	9.76	D = 0.06	0.56	−297.29
Std. Error	0.21	1.13			

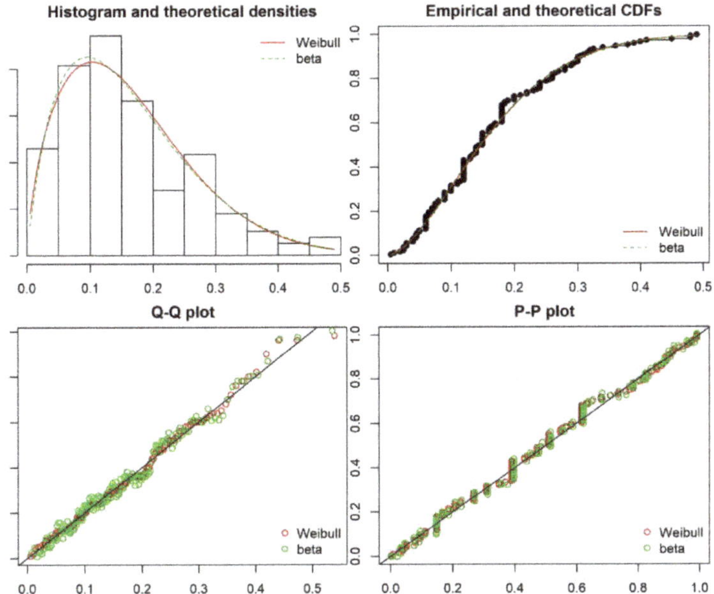

Figure 3. The assessment of goodness of fit of the empirical data of S_6 state to the Weibull and Beta distributions.

For each of them the *AIC* (Akaike information criterion) was calculated according to the formula (16) and based on that, the one with the better fit was selected.

$$AIC = -2\ln L + 2k,\tag{16}$$

where k—number of parameters in the model, L—credibility function.

The same calculations were made for the other states. The proposed distributions are presented in Table 3.

Table 3. Goodness of fit results for states S_1–S_6.

State	Distribution	Test Statistic KS	p-Value
S_1	non-parametric		
S_2	Beta	D = 0.07	0.793
S_3	non-parametric		
S_4	Gamma	D = 0.03	0.726
S_5	non-parametric		
S_6	Beta	D = 0.07	0.385

Not all the distributions could be fitted to the parametric ones. Moreover, none of the distributions belong to the family of exponential distributions, which is a condition for using the Markov process [25]. The form of distributions makes the parameters estimation possible based on the semi-Markov model only. In order to compare whether meeting the condition of the distribution form affects the obtained results, the subsequent part of the article compares the limit values of the probabilities of the object's dwelling time according to two different models.

4. Case Study

4.1. Estimation of the Semi-Markov Model Parameters

First of all, based on the actual relationship between the states defined in Figure 1, the transition probability matrix was calculated. If n_i denotes the number of instants of the system waiting in state s_i, while n_{ij} denotes the number of state transitions from state s_i to state s_j, then the transition estimator from state s_i to state s_j shall be determined from the formula:

$$\hat{p}_{ij} = \frac{n_{ij}}{n_i}. \tag{17}$$

The distribution of probability of changes of the distinguished operating states (in one step), assuming that each graph arch of the exploitation process representation (Equations (2) and (10)) corresponds to the value of probability p_{ij}, is presented in Table 4.

Table 4. The p_{ij} inter-states transition probability matrix.

p_{ij}	S_1	S_2	S_3	S_4	S_5	S_6
S_1	0	0.141	0.143	0.716	0	0
S_2	0.029	0	0.117	0.854	0	0
S_3	0.043	0	0	0.957	0	0
S_4	0.003	0	0	0	0.997	0
S_5	0.676	0.004	0.001	0	0	0.319
S_6	0.982	0	0.018	0	0	0

For the process studied, some limits exist:

$$\lim_{n \to \infty} p_{ij}(n) = \pi_j \; i, \; j = 1, 2, \ldots, 6 \; i \neq j, \tag{18}$$

where $p_{ij}(n)$—probability of transition from state S_i to state S_j in n steps.

Definition 2. *A probability distribution* [25]:

$$\pi = \left[\pi_j : \; j \in S \right], \tag{19}$$

Satisfying a system of linear equations:

$$\sum_{i \in S} \pi_i \cdot p_{ij} = \pi_j, \quad j \in S, \tag{20}$$

and

$$\sum_{i \in S} \pi_i = 1, \tag{21}$$

is said to be a stationary probability distribution of the Markov chain witch transition matrix $P = \left[p_{ij} : i, \; j \in S \right]$. In matrix form, the Equation (20) takes the following form:

$$\Pi^T P = \Pi^T \leftrightarrow \left(P^T - I \right) \cdot \Pi = 0, \tag{22}$$

Stationary probabilities π_j were calculated in accordance with Equation (22). For the process studied, for the 6-state model, the determination of the stationary probabilities π_j required solving the following matrix equation:

$$
\begin{bmatrix} \pi_1 \\ \pi_2 \\ \pi_3 \\ \pi_4 \\ \pi_5 \\ \pi_6 \end{bmatrix}^T
\begin{bmatrix}
0 & p_{12} & p_{13} & p_{14} & 0 & 0 \\
p_{21} & 0 & p_{23} & p_{24} & 0 & 0 \\
p_{31} & 0 & 0 & p_{34} & 0 & 0 \\
p_{41} & 0 & 0 & 0 & p_{45} & 0 \\
p_{51} & p_{52} & p_{53} & 0 & 0 & p_{56} \\
p_{61} & 0 & p_{63} & 0 & 0 & 0
\end{bmatrix}
=
\begin{bmatrix} \pi_1 \\ \pi_2 \\ \pi_3 \\ \pi_4 \\ \pi_5 \\ \pi_6 \end{bmatrix}^T ,
\tag{23}
$$

with the normalization condition:

$$
\pi_1 + \pi_2 + \pi_3 + \pi_4 + \pi_5 + \pi_6 = 1,
\tag{24}
$$

which is equivalent to the following system of equations:

$$
\begin{cases}
\pi_2 \cdot p_{21} + \pi_3 \cdot p_{31} + \pi_4 \cdot p_{41} + \pi_5 \cdot p_{51} + \pi_6 \cdot p_{61} = \pi_1 \\
\pi_1 \cdot p_{12} + \pi_5 \cdot p_{52} = \pi_2 \\
\pi_1 \cdot p_{13} + \pi_2 \cdot p_{23} + \pi_5 \cdot p_{53} + \pi_6 \cdot p_{63} = \pi_3 \\
\pi_1 \cdot p_{14} + \pi_2 \cdot p_{24} + \pi_3 \cdot p_{34} = \pi_4 \\
\pi_4 \cdot p_{45} = \pi_5 \\
\pi_5 \cdot p_{56} = \pi_6 \\
\pi_1 + \pi_2 + \pi_3 + \pi_4 + \pi_5 + \pi_6 = 1
\end{cases}
\tag{25}
$$

After substituting the figures we get:

$$
\begin{cases}
\pi_2 \cdot 0.029 + \pi_3 \cdot 0.043 + \pi_4 \cdot 0.003 + \pi_5 \cdot 0.676 + \pi_6 \cdot 0.982 = \pi_1 \\
\pi_1 \cdot 0.141 + \pi_5 \cdot 0.004 = \pi_2 \\
\pi_1 \cdot 0.143 + \pi_2 \cdot 0.117 + \pi_5 \cdot 0.001 + \pi_6 \cdot 0.018 = \pi_3 \\
\pi_1 \cdot 0.716 + \pi_2 \cdot 0.854 + \pi_3 \cdot 0.957 = \pi_4 \\
\pi_4 \cdot 0.997 = \pi_5 \\
\pi_5 \cdot 0.319 = \pi_6 \\
\pi_1 + \pi_2 + \pi_3 + \pi_4 + \pi_5 + \pi_6 = 1
\end{cases}
\tag{26}
$$

The solution are the stationary probabilities presented in Table 5.

Table 5. Stationary probabilities of the Markov chain.

	S_1	S_2	S_3	S_4	S_5	S_6
p_{ij}	0.276	0.04	0.046	0.276	0.275	0.088

The analysis of the stationary distribution showed (Table 5) that the limit highest transitions probabilities concern states (S_1, S_4, S_5) related to standard activities resulting from the production process technology (all over 27%). Indications of undesirable conditions such as failure (S_2) or downtime (S_3, S_6) range from 4% to 8%, which is a good result.

The calculated limit probabilities relate to the frequency of observations in the sample and do not take into account the duration of individual states, therefore, the limit distribution of the semi-Markov process represents more significant diagnostics. It can be determined using the stationary distribution of the Markov chain and the expected duration of the process states [24,25]. Then the limit probabilities of semi-Markov process are expressed by the formula:

$$P_j = \lim_{t \to \infty} P(t) = \frac{\pi_j E(T_j)}{\sum_{j \in S} \pi_j E(T_j)}. \tag{27}$$

The solution requires to calculate the following forms from the sample of average conditional durations of the process states:

$$T = [\overline{T}_{ij}], \quad i, j = 1, 2, \ldots, 6. \tag{28}$$

that are presented in Table 6.

Table 6. Average conditional durations of the semi-Markov process states.

$\overline{T}_{ij}$ [minutes]	S_1	S_2	S_3	S_4	S_5	S6
S_1		795.48	637.63	1082.14		
S_2	4800		256.67	277.27		
S_3	4020			508.6		
S_4	21.5				248.62	
S_5	42.33	40	15			40.36
S_6	156.3		226.25			

Based on the transitions probabilities matrix $P = [p_{ij}]$ (Table 5) and the matrix of average conditional durations of the states of the process $T = [T_{ij}]$ of random variables $\overline{T}_{ij}$ (Table 6), dependencies describing average unconditional durations of the process states were determined $\overline{T}_j$ according to the formula:

$$\overline{T}_j = \sum_{i=1}^{6} p_{ij} \overline{T}_{ij}. \tag{29}$$

For this purpose, the following equation system was solved:

$$\begin{cases} \overline{T}_1 = p_{12} \cdot T_{12} + p_{13} \cdot T_{13} + p_{14} \cdot T_{14} \\ \overline{T}_2 = p_{21} \cdot T_{21} + p_{23} \cdot T_{23} + p_{24} \cdot T_{24} \\ \overline{T}_3 = p_{31} \cdot T_{31} + p_{34} \cdot T_{34} \\ \overline{T}_4 = p_{41} \cdot T_{41} + p_{45} \cdot T_{45} \\ \overline{T}_5 = p_{51} \cdot T_{51} + p_{52} \cdot T_{52} + p_{53} \cdot T_{53} + p_{56} \cdot T_{56} \\ \overline{T}_6 = p_{61} \cdot T_{61} + p_{63} \cdot T_{63} \end{cases} \tag{30}$$

The obtained expected values of unconditional dwelling $\overline{T}_i$ times of the process X(t) in the individual operational states are presented in Table 7.

Table 7. Unconditional times $\overline{T}_i$ [minutes] for the 6-state model.

	$\overline{T}_1$	$\overline{T}_2$	$\overline{T}_3$	$\overline{T}_4$	$\overline{T}_5$	$\overline{T}_6$
$\overline{T}_i$ [minutes]	978.16	406.02	659.59	247.96	41.67	157.56

The calculated random variables T_i have finite, positive expected values. This allows to calculate, based on theorem (27), the limit probabilities P_j which are presented in Table 8.

Table 8. Values of limit probabilities P_j of the 6-state model.

P_1	0.6579	P_1 [%]	65.79
P_2	0.0396	P_2 [%]	3.96
P_3	0.0740	P_3 [%]	7.4
P_4	0.1667	P_4 [%]	16.67
P_5	0.0279	P_5 [%]	2.79
P_6	0.0337	P_6 [%]	3.37

Thus determined probabilities P_j are limit probabilities determining that the system will remain, for a longer period ($t \rightarrow \infty$), in the given operational state. This prognosis is more satisfactory than for the frequency of the states occurring. The highest values are achieved by state S_1, i.e., operation (over 65%) and less than 17% by state S_4, which stems from the necessity to perform maintenance activities. The remaining limit values are satisfactorily small, which shows the correct operation of the machines.

The technical readiness factor was also determined in the form of the sum of appropriate probabilities of reliability states [34]. For the system under analysis, S_1, S_4, S_5 were considered as fitness states, while the states S_2 S_3 and S_6 as unfitness states. Then, the readiness of the 6-element semi-Markov model can be calculated as the sum of limit probabilities of the respective states:

$$K = \sum_j P_j = P_1 + P_4 + P_5, \tag{31}$$

This gives $K = 0.85$, which means that the machine is in the readiness state for over 85% of the time, which is a very good result.

4.2. Calculations According to the Markov Model

Markov processes concern exponential distributions, the most popular ones in reliability theory [11,35]. They are described by two parameters, which fully define them. The first of these is the already calculated probability matrix of interstate transitions p_{ij} (Table 4).

The second, important parameter is the function describing the transitions of objects between states, called the process transition intensity $\lambda_{ij}(t)$, which characterizes the rate of changes in the probability of transition $p_{ij}(t)$ [36].

$$\lambda_{ij}(t) = \lim_{\Delta t \to 0} \frac{1}{\Delta t} p_{ij}(t, \, t + \Delta t) \text{ for } i, \, j = 0, 1, 2, \ldots i \neq j, \tag{32}$$

For homogeneous Markov processes, the transition intensity is constant and equal to the inverse of the expected duration t_{ij} of the state S_i before S_j [37]:

$$\lambda_{ij}(t) = \frac{1}{\mathrm{E}(t_{ij})}, \tag{33}$$

where:

$\lambda_{ij}(t)$—intensity of transitions from the state i to state j,
$\mathrm{E}(t_{ij})$—expected duration value t_{ij}.

The intensities $\lambda_{ii} \leq 0$ for $i = j$ are defined as a complement to the sum of transition intensity from state S_i for $i \neq j$ to 0:

$$\lambda_{ii} + \Sigma_j \, \lambda_{ii} = 0 \tag{34}$$

thus:

$$\lambda_{ii} = -\Sigma_j \, \lambda_{ii}. \tag{35}$$

The modules $|\lambda_{ii}| = -\lambda_{ii}$ are called the exit intensities from the state S_i.

Calculated according to the above formulas (33)–(35), the element λ_{ij} of the matrix Λ of transition intensity is shown in Table 9.

Table 9. The transition intensity matrix of the process studied.

λ_{ij} [1/*minutes*]	S_1	S_2	S_3	S_4	S_5	S_6
S1	−0.0037	0.0013	0.0016	0.0009	0	0
S2	0.0002	−0.0077	0.0039	0.0036	0	0
S3	0.0002	0	−0.0022	0.0020	0	0
S4	0.0465	0	0	−0.0505	0.0040	0
S5	0.0236	0.0250	0.0667	0	−0.1401	0.0248
S6	0.0064	0	0.0044	0	0	−0.0108

Then, using the relationship (36), ergodic probabilities p_j were calculated for the Markov model in continuous time.

$$\prod{}^{T} * \Lambda = 0, \tag{36}$$

where:

- $\prod^{T} = \left[p_j\right]^{T} = [p_1; ; p_{ns}]$—transposed vector of limit probabilities p_j,
- $|\Lambda|$—transition intensity matrix:
- $\sum_j p_j = 1$—the normalization condition.

This way, for the process studied, we obtain the following matrix Equation (37):

$$
\begin{bmatrix} p_1 \\ p_2 \\ p_3 \\ p_4 \\ p_5 \\ p_6 \end{bmatrix}^{T}
\cdot
\begin{bmatrix}
-\lambda_{11} & \lambda_{12} & \lambda_{13} & \lambda_{14} & 0 & 0 \\
\lambda_{21} & -\lambda_{22} & \lambda_{23} & \lambda_{24} & 0 & 0 \\
\lambda_{31} & 0 & -\lambda_{33} & \lambda_{34} & 0 & 0 \\
\lambda_{41} & 0 & 0 & -\lambda_{44} & p_{45} & 0 \\
\lambda_{51} & \lambda_{52} & \lambda_{53} & 0 & -\lambda_{55} & \lambda_{56} \\
\lambda_{61} & 0 & \lambda_{63} & 0 & 0 & -\lambda_{66}
\end{bmatrix}
=
\begin{bmatrix} 0 \\ 0 \\ 0 \\ 0 \\ 0 \\ 0 \end{bmatrix},
\tag{37}
$$

Taking into account the normalization condition: $\sum_{j=1}^{6} p_j = 1$, we get the limit probabilities p_j of the system's dwelling time in the states S_1–S_6, which are shown in Table 10.

Table 10. Limit probabilities p_j in the continuous physical time for the Markov process.

	S_1	S_2	S_3	S_4	S_5	S_6
p_j	0.4228	0.0742	0.4695	0.0306	0.0009	0.0020
$p_{j\%}$	42.28	7.42	46.95	3.06	0.09	0.20

The results obtained deviate from the values determined for the semi-Markov process, disturbingly revealing that the system studied tends primarily to remain in the downtime state (S_3). The state in which the production takes place (S_1) comes only second and takes the value lower by over 35% in relation to calculations made according to the semi-Markov process. A comparison of the other results is presented in Table 11. The highest difference concerns state S_3, and is over 534%.

Table 11. Comparison of results for the Markov and semi-Markov models.

	S_1	S_2	S_3	S_4	S_5	S_6
p_j semi-Markov model	0.6589	0.0396	0.074	0.1667	0.0279	0.0337
p_j Markov model	0.4228	0.0742	0.4695	0.0306	0.0009	0.0020
difference in %	−35.74	87.41	534.50	−81.65	−96.87	−94.05

The technical readiness coefficient was also calculated based on the (31), which for the Markov process amounts to 45% ($K = 0.45$)—almost half the size of what was determined according to the semi-Markov process.

5. Conclusions

The study achieved two important goals. The first of them was a presentation of the method of evaluating the readiness of a selected element of a machine tools in the production system. The analysis according to the Markov chain allowed to determine the probabilities of interstate transitions, which reflect the frequency of the occurrence of individual states. The highest values were achieved for relations S_6–S_4, S_3–S_4, S_4–S_5. They suggest a high incidence of unsuitability states—S_3 and S_6—and the need to determine their causes and reduce their occurrence.

The limit values of transition probability were also calculated. The analysis of stationary distribution showed that the greatest indications concern states related to activities resulting from production process technology (S_1, S_4, S_5), which is a good result.

However, a complete evaluation is only ensured by an analysis according to the semi-Markov process, taking into account the average dwell times of an object in the individual operating states. The calculated probability limits, examining the behavior of the object for $t \to \infty$, were the highest for state S_1—operation (over 65%) and state S_4—service (almost 17%). The remaining limit values were found to be satisfactorily low, which means that the operation of the machine should be considered as proper. The calculated technical readiness rate of 85% should also be viewed as positive.

Such an analysis not only provides information on the assessment of the current and expected functioning of the machine, but also reveals areas where modifications can be made in order to increase the level of availability and, as a result, ensure more efficient execution of production orders.

Another goal was to compare the results according to the assumptions made, concerning the forms of distribution of the examined variables. In the literature this analysis is often omitted and it is assumed that the examined variable has an exponential distribution. This allows to use Markov's processes, whose parameter estimation is simpler and is described in more detail in publications. Such an assumption—as the study has shown—may lead to different results and effectively to form an incorrect assessment of the process/system studied. The intention of the authors was to indicate that omitting an important stage of statistical analysis of the collected data and assuming a priori the form of distributions does not guarantee the correctness of the obtained analyses.

In the presented study, the differences in the values of the calculated limit probabilities are large, reaching even over 530%. The overall evaluation of system readiness indicates a value lower by 46% in the case of the Markov process analysis.

However, the problem is not only the value of the calculated probabilities, but also the main aim of the system. According to the semi-Markov process, the system tends primarily to occupy state S_1 (operation) which is a satisfactory result, emphasizing the proper implementation of tasks. The results according to the Markov process show that the system tends to occupy mainly state S_3—downtime, which indicates mismanagement and system inactivity.

The goals set by the authors have been achieved, but it should be stressed out that the results obtained concern only one selected machine. As part of further research, it is worth considering a comprehensive analysis of the entire production system using the method indicated in the article. It will provide complete information on its readiness, determine the level of impact of individual elements (machines), and identify areas for improvement.

Author Contributions: Conceptualization, A.B. and L.G.; Formal analysis, A.B. and A.Ś.; Methodology, A.B.; Resources, M.G.; Writing—original draft, A.B. and A.Ś.; Writing—review & editing, A.B. and M.G. All authors have read and agreed to the published version of the manuscript.

Funding: This research received no external funding.

Conflicts of Interest: The authors declare no conflict of interest.

References

1. Kozłowski, E.; Mazurkiewicz, D.; Żabiński, T.; Prucnal, S.; Sęp, J. Assessment model of cutting tool condition for real-time supervision system. *Eksploat. Niezawodn. Maint. Reliab.* **2019**, *21*, 679–685. [CrossRef]
2. Kosicka, E.; Kozłowski, E.; Mazurkiewicz, D. The use of stationary tests for analysis of monitored residual processes. *Eksploat. Niezawodn. Maint. Reliab.* **2015**, *17*, 604–609. [CrossRef]
3. Jasiulewicz-Kaczmarek, M.; Żywica, P. The concept of maintenance sustainability performance assessment by integrating balanced scorecard with non-additive fuzzy integral. *Eksploat. Niezawodn. Maint. Reliab.* **2018**, *20*, 650–661. [CrossRef]
4. Antosz, K. Maintenance—Identification and analysis of the competency gap. *Eksploat. Niezawodn. Maint. Reliab.* **2018**, *20*, 484–494. [CrossRef]
5. Kishawy, H.A.; Hegab, E.; Umer, U.; Mohany, A. Application of acoustic emissions in machining processes: Analysis and critical review. *Int. J. Adv. Manuf. Technol.* **2018**, *98*, 1391–1407. [CrossRef]
6. Rymarczyk, T.; Kozłowski, E.; Kłosowski, G.; Niderla, K. Logistic Regression for Machine Learning in Process Tomography. *Sensors* **2019**, *19*, 3400. [CrossRef]
7. Fisz, M. *Rachunek Prawdopodobieństwa i Statystyka Matematyczna*; PWN: Warszawa, Poland, 1967.
8. Zhang, Y.; Zhang, Q.; Yu, R. Markov property of Markov chains and its test. In Proceedings of the International Conference on Machine Learning and Cybernetics, Qingdao, China, 11–14 July 2010; pp. 1864–1867. [CrossRef]
9. Chen, B.; Hong, Y. Testing for the Markov property in time series. *Econ. Theory* **2012**, *28*, 130–178. [CrossRef]
10. Shi, S.; Lin, N.; Zhang, Y.; Cheng, J.; Huang, C.; Liu, L.; Lu, B. Research on Markov property analysis of driving cycles and its application. *Transp. Res. Part D Transp. Environ.* **2016**, *47*, 171–181. [CrossRef]
11. Kozłowski, E.; Borucka, A.; Świderski, A. Application of the logistic regression for determining transition probability matrix of operating states in the transport systems. *Eksploat. Niezawodn. Maint. Reliab.* **2020**, *22*, 192–200. [CrossRef]
12. Yang, H.; Nair, V.N.; Chen, J.; Sudjianto, A. Assessing Markov property in multistatetransition models with applications to credit risk modeling. *Appl. Stoch. Models Business Ind.* **2019**, *35*, 552–570. [CrossRef]
13. Komorowski, M.; Raffa, J. Markov Models and Cost Effectiveness Analysis: Applications in Medical Research. In *Secondary Analysis of Electronic Health Records*; Springer: Cham, Switzerland, 2016; pp. 351–367.
14. Gercbach, I.B.; Kordonski, C.B. *Modele Niezawodnościowe Obiektów Technicznych*; WNT: Warszawa, Poland, 1968.
15. Gichman, I.I.; Skorochod, A.W. *Wstęp Do Teorii Procesów Stochastycznych*; PWN: Warszawa, Poland, 1968.
16. Li, Y.; Dong, Y.; Zhang, H.; Zhao, H.; Shi, H.; Zhao, X. Spectrum Usage Prediction Based on High-order Markov Model for Cognitive Radio Networks. In Proceedings of the 10th IEEE International Conference on Computer and Information Technology, Bradford, UK, 29 June–1 July 2010; pp. 2784–2788. [CrossRef]
17. Perman, M.; Senegacnik, A.; Tuma, M. Semi-Markov models with an application to power-plant reliability analysis. *IEEE Transac. Reliab.* **1997**, *46*, 526–532. [CrossRef]
18. Lana, X.; Burgueño, A. Daily dry–wet behaviour in Catalonia (NE Spain) from the viewpoint of Markov chains. *Int. J. Climatol.* **1998**, *18*, 793–815. [CrossRef]
19. Pang, W.; Forster, J.J.; Troutt, M.D. Estimation of Wind Speed Distribution Using Markov Chain Monte Carlo Techniques. *J. Appl. Meteorol.* **2010**, *40*, 1476–1484. [CrossRef]
20. Love, C.E.; Zhang, Z.G.; Zitron, M.A.; Guo, R. A discrete semi-Markov decision model to determine the optimal repair/replacement policy under general repairs. *Eur. J. Oper. Res.* **2010**, *125*, 398–409. [CrossRef]
21. Doob, J.L. *Stochastic Processes*; John Wiley & Sons/Chapman & Hall: New York, NY, USA, 1953.
22. Iosifescu, M. *Finite Markov Processes and Their Applications*; John Wiley & Sons: Chichester, UK, 1980.
23. Howard, R.A. *Dynamic Probabilistic System, Volume II: Semi-Markow and Decision Processes*; Wiley: New York, NY, USA, 1971.
24. Jaźwiński, J.; Grabski, F. *Niektóre Problemy Modelowania Systemów Transportowych*; Biblioteka Problemów Eksploatacji: Warszawa, Poland, 2003.
25. Grabski, F. *Semi-Markow Processes: Applications in System Reliability and Maintenance*; Elsevier: Gdynia, Poland, 2014.

26. Geng, J.; Xu, S.; Niu, J.; Wei, K. Research on technical condition evaluation of equipments based on matter element theory and hidden Markov model. In Proceedings of the 4th Annual International Workshop on Materials Science and Engineering (IWMSE2018), Xi'an, China, 18–20 May 2018; Volume 381, p. 012134. [CrossRef]
27. Yuriy, E.; Obzherin, M.; Siodorov, S. Application of hidden Markov models for analyzing the dynamics of technical systems. *AIP Conf. Proc.* **2019**, *2188*, 050019. [CrossRef]
28. Iscioglu, F.; Kocak, A. Dynamic reliability analysis of a multi-state manufacturing system. *Eksploat. Niezawodn. Maint. Reliab.* **2019**, *21*, 451–459. [CrossRef]
29. Liu, C.; Duan, H.; Chen, P.; Duan, L. Improve Production Efficiency and Predict Machine Tool Status using Markov Chain and Hidden Markov Model. In Proceedings of the 8th International Conference of Computer Science and Information Technology (CIST), Amman, Jordan, 26–28 May 2018; pp. 276–281. [CrossRef]
30. Gola, A. Reliability analysis of reconfigurable manufacturing system structures using computer simulation methods. *Eksploat. Niezawodn. Maint. Reliab.* **2019**, *21*, 90–102. [CrossRef]
31. Wang, C.H.; Sheu, S.H. Determining the optimal production-maintenance policy with inspection errors: Using a Markov chain. *Comput. Oper. Res.* **2003**, *30*, 1–17. [CrossRef]
32. Dung, K.; Lei, J.; Chan, F.; Hui, J.; Zhang, F.; Wang, Y. Hidden Markov model-based autonomous manufacturing task orchestration in smart shop floors. *Robot. Comput. Integr. Manuf.* **2020**, *61*, 101845. [CrossRef]
33. Xiang, L.; Guan, J.; Wu, S. Measuring the impact of final demand on global production system based on Markov process. *Phys. A Stat. Mech. Appl.* **2018**, *502*, 148–163. [CrossRef]
34. Żurek, J.; Tomaszewska, J. Analiza system eksploatacji z punktu widzenia gotowości. *Prace Nauk. Politech. Warsz.* **2016**, *114*, 471–477.
35. Pavlov, N.; Golev, A.; Rahney, A.; Kyurkchiev, N. A Note on the Generalized Inverted Exponential Software Reliability Model. *Int. J. Adv. Res. Comput. Commun. Eng.* **2018**, *7*, 484–487.
36. Borucka, A.; Niewczas, A.; Hasilova, K. Forecasting the readiness of special vehicles using the semi-Markov model. *Eksploat. Niezawodn. Maint. Reliab.* **2019**, *21*, 662–669. [CrossRef]
37. Filipowicz, B. *Modele Stochastyczne w Badaniach opeRacyjnych, aNaliza i Synteza Systemów Obsługi i Sieci Kolejkowych*; Wydawnictwa Naukowo-Techniczne: Warszawa, Poland, 1996.

Article

The Use of Artificial Intelligence Methods to Assess the Effectiveness of Lean Maintenance Concept Implementation in Manufacturing Enterprises

Katarzyna Antosz [1], Lukasz Pasko [2] and Arkadiusz Gola [3,*]

[1] Department of Manufacturing Processes and Production Engineering, Faculty of Mechanical Engineering and Aeronautics, Rzeszow University of Technology, Al. Powstańców Warszawy 8, 35-959 Rzeszów, Poland; kcktmiop@prz.edu.pl

[2] Department of Computer Science, Faculty of Mechanical Engineering and Aeronautics, Rzeszow University of Technology, Al. Powstańców Warszawy 8, 35-959 Rzeszów, Poland; lpasko@prz.edu.pl

[3] Department of Production Computerisation and Robotisation, Faculty of Mechanical Engineering, Lublin University of Technology, ul. Nadbystrzycka 36, 20-618 Lublin, Poland

* Correspondence: a.gola@pollub.pl

Received: 19 October 2020; Accepted: 6 November 2020; Published: 8 November 2020

Abstract: The increase in the performance and effectiveness of maintenance processes is a continuous aim of production enterprises. The elimination of unexpected failures, which generate excessive costs and production losses, is emphasized. The elements that influence the efficiency of maintenance are not only the choice of an appropriate conservation strategy but also the use of appropriate methods and tools to support the decision-making process in this area. The research problem, which was considered in the paper, is an insufficient means of assessing the degree of the implementation of lean maintenance. This problem results in not only the possibility of achieving high efficiency of the exploited machines, but, foremost, it influences a decision process and the formulation of maintenance policy of an enterprise. The purpose of this paper is to present the possibility of using intelligent systems to support decision-making processes in the implementation of the lean maintenance concept, which allows the increase in the operational efficiency of the company's technical infrastructure. In particular, artificial intelligence methods were used to search for relationships between specific activities carried out under the implementation of lean maintenance and the results obtained. Decision trees and rough set theory were used for the analysis. The decision trees were made for the average value of the overall equipment effectiveness (OEE) indicator. The rough set theory was used to assess the degree of utilization of the lean maintenance strategy. Decision rules were generated based on the proposed algorithms, using RSES software, and their correctness was assessed.

Keywords: decision-making process; lean maintenance; effectiveness; decision trees; rough set theory

---

1. Introduction

Obtaining appropriate reliability and quality of products requires proper methods of enterprise management, production, and the means necessary for its implementation [1]. These methods allow for the coordination and integration of all company functions [2]. One of the elements affecting the high quality of a product is the condition of technological machines maintained in enterprises [3–5]. Its suitability and technical condition largely determine the quality and competitiveness of a product.

Maintenance management is a critical issue amongst management activities of manufacturing organizations [6–8]. Therefore, in recent years, intensive efforts have been made to propose and improve maintenance strategies that aim to extend the useful life of every piece of existent equipment, increase its availability, and guarantee higher levels of reliability [9–11].

In recent decades, maintenance was regarded as a necessary evil in managing an organization, because it was limited to the appropriate functions that are usually performed in emergency situations, such as a machine failure. However, this practice is no longer acceptable, because the role of maintenance has been recognized as a strategic element of generating revenues for the organization [1].

The maintenance process in enterprises was not always carried out in a way that ensures the minimization of outlays while maximizing the achieved effects in service and maintenance processes. In practice, obtaining the maximum benefits from the operation of a technological machine system requires an optimal solution to a number of tasks [12,13]. For a larger number of machines, it creates the need for appropriate system modelling, simulation research, and optimization of partial and complex tasks based on the adopted optimization criteria [14]. Accordingly, the company must establish a maintenance system that enables these activities to be carried out in an optimal manner from the point of view of resource pro-vision and its effectiveness [15]. These activities are usually carried out in accordance with a specific exploitation strategy that has been developed with the development of production systems. Most of the enterprises with foreign capital managed to organize effective maintenance [16]. However, small and medium enterprises are still looking for the right method for their reorganization as well as for the right way of supervising technological machines and equipment that would allow for improving effectiveness as well as for using it in a manufacturing process. That is why, some organizations have started implementing lean methods and tools in the area of maintenance defined as lean maintenance [17]. Lean maintenance is a proactive maintenance strategy whose main goal is to support reliability in the most effective, cost-effective way possible, which means keeping costs to a minimum while ensuring high efficiency and productivity. This philosophy is mainly based on the concept of total productive maintenance (TPM), the idea of which is to involve all employees at every level of the organization in maintenance and management tasks [18].

This paper presents the possibilities of using intelligent systems to support decision-making processes in the implementation of the lean maintenance concept. The aim of the article is to indicate the methods and tools of lean maintenance, which have the greatest impact on increasing the effectiveness of the enterprise. The obtained research results indicate which lean maintenance methods and tools should be implemented in the company in the first place to increase the efficiency, quality, and availability of its production processes. This problem is particularly important from the point of view of small- and medium-sized enterprises, which do not always have adequate human or financial resources to implement modern concepts in a wide range. The methodology used in the work is based on the use of artificial intelligence methods (decision trees, rough set theory), which allows for the identification of factors influencing the effectiveness of lean maintenance implementation by enterprises. The second chapter of the article contains the background. Section 3 describes the problem formulation and methodology. Section 4 presents the results of research on the use of the overall equipment effectiveness (OEE) indicator in enterprises and the concept of using artificial intelligence (AI) methods to assess the effectiveness of the implementation of the lean maintenance concept. The work is summarized with conclusions and a proposal for further work.

2. Literature Review

2.1. Lean Manufacturing as the Foundation of Lean Maintenance

Global industry in the 21st century motivates companies to seek and implement a more competitive production system. Many of them implement or plan to implement lean manufacturing. Lean manufacturing philosophy is mainly used in industry to increase efficiency and productivity. It was developed in the 1990s and is mainly based on the Toyota Production System (TPS) [19].

The basis of this concept is the elimination of unnecessary losses that have a significant impact on productivity and profit. These losses can be divided into three main types: Muda, Mura, and Muri. Muda identifies seven types of waste, which include transportation, supplies, redundant movement, waiting, overproduction, over processing, and defects. Mura means unevenness, non-uniformity, and irregularity and is the reason for the existence of any of the seven wastes. Finally, Muri means overburden, beyond one's power, excessiveness, impossible, or unreasonableness and can result from Mura and, in some cases, can be caused by excessive removal of Muda from the process [20–22].

Production using the lean manufacturing philosophy should consist of reducing the amount of losses related to people, inventory, time to market, and production space, so as to obtain a highly reactive demand for customer needs, while producing high-quality products in the most efficient and cost-effective manner [23]. Lean manufacturing can be a cost reduction mechanism, and if properly implemented, it will make it a world-class organization [24], and importantly, lean manufacturing can be used in all industries [19,25].

Many organizations have embarked on the practice of using "lean tools" primarily to eliminate wasted production. It is widely recognized that organizations that have applied lean manufacturing methods have significant cost and quality advantages over those that continue to use traditional manufacturing [23]. It turns out that organizations pay more and more attention to maintenance, which is why some organizations have started to practice lean maintenance in addition to lean manufacturing.

Lean maintenance is a concept that implements activities aimed at increasing the effectiveness of technical infrastructure. These activities are related to the elimination of losses in maintenance, such us [18,19]:

- Unproductive works—execution of works that do not increase the reliability of the technical infrastructure;
- Delays in the implementation of works—waiting for the availability of technical infrastructure in order to carry out preventive actions;
- Unnecessary motion—unnecessary trips to stores with spare parts and searching for the required tools to get the job done;
- Poor inventory management—lack of having an appropriate number of needed spare parts in a specific time;
- Reworking—repeating tasks due to poor quality of performance;
- Insufficient use of resources—inadequate use of available resources and skills of maintenance teams;
- Machine misuse—malfunction or intentional operating strategies leading to maintenance work that does not need to be performed;
- Ineffective data management—collecting data that are useless, and not those that are important.

Duran et al. [26] show the connection between the sources of waste in lean manufacturing and lean maintenance (Table 1).

Table 1. Relationships between sources of waste in lean manufacturing and lean maintenance.

Waste in Lean Manufacturing	Waste in Lean Maintenance
material transport	transport of spare parts and tools
manufacture of non-required goods	implementation of non-required maintenance activities (over maintenance)
waiting between operations or in the course of an operation	the waiting between maintenance actions or procedures
stocks of materials	excessive stocks of spare parts
over processing	excessive or too frequent maintenance activities
non-conforming products	rework

The elimination of waste is most often carried out by implementing lean tools such as 5S, standardized work, Kaizen, Poka-Yoke, and value stream mapping (VSM) [27,28]. They are most commonly applied to make production processes more effective and to reduce lead-time or cost of production, but they can also be applied for maintenance operations. As the most common examples of application lean tools in the maintenance area, the implementation of the standardized work for maintenance operators, the Andon system to initiate corrective maintenance, or using VSM to identify and eliminate waste in maintenance operations can be distinguished [29]. However, the fundamental elements of lean philosophy and total productive maintenance (TPM) must be implemented before application of such specific tools [30].

2.2. Decision Support Systems in Maintenance Management

The problem of supporting decision processes in maintenance management has been the topic of many studies for over a dozen years now. Bashiri, Badri, and Hejazi [31] and Zhaoyang, Jianfeng, Zongzhi, Jianhu, and Weifeng [32] highlighted the role of risk-based maintenance in the maintenance management process. Cruz and Rincon point out that the maintenance process is at risk due to equipment failure [33]. Rinaldi, Portillo, Khalid, Henriques, Thies, Gato, and Johanning emphasized the importance of quantitative reliability, availability, and maintainability at early design stages [34]. Additionally, Wang, Furst, Cohen, Keil, Ridgway, and Stiefel predicted the risk of equipment failure using the Monte Carlo method [35], while in later works, they proposed approaches to monitoring disruptions and risk using ontologies and multi-agent systems [36].

Taghipour, Banjevic, and Jardine have proposed a method to identify and prioritize critical devices to mitigate functional failures [37]. Li, Parikh, He, Qian et al. incorporated machine learning techniques into the process of predicting failures, taking into account historical and real-time data analysis [38].

Jamshidi, Rahimi, Aitkadi, and Ruiz used fuzzy failure modes and analysis of effects to prioritize the operation of machinery, equipment, and classification [39], while Carnero and Gomez suggested the use of a multi-criteria model to increase the efficiency of the maintenance process [40].

Zeineb, Malek, Ahmand Ikram, and Faouzi, taking into account the total cost of ownership, used the Analytic Hierarchy Process (AHP) method to establish an appropriate-optimal maintenance program [41]. Moreover, fuzzy analytic hierarchy process for performing diagnostic and prescription tasks was discussed by Duran, Capaldo, and Duran Acevedo [27].

Lin, Yuan, and Tovilla use a continuous Markov chain model in a stochastic decision model that combines the effectiveness of maintenance activities and natural changes in state [42]. Jasiulewicz-Kaczmarek and Żywica use the non-additive fuzzy integral and balanced scorecard in the maintenance process [43]. In [44], the authors proposed a scorecard model that allows for monitoring the maintenance process in an enterprise. Finally, the importance of modern IT technologies in the maintenance decision-making process was also emphasized by Kosicka, Gola, and Pawlak [45].

Although the literature on the subject presents many solutions supporting decision-making in maintenance management, intelligent systems that are dedicated to supporting the implementation of the lean maintenance concept are not presented. For the time being, some limited results were presented by Antosz, Pasko, and Gola during the 13th IFAC Workshop on Intelligent Manufacturing Systems (Oshawa, ON, Canada) [46]. This explains why the research problem that was considered in the paper is an insufficient means of assessing the degree of the implementation of lean maintenance. This problem results in not only the possibility of achieving high efficiency of the exploited machines, but, foremost, it influences a decision process and the formulation of maintenance policy of an enterprise. In the context of the work conducted, the methodology of assessing lean maintenance was presented.

3. The Work Methodology

The research was carried out in two stages. The first stage of the study considered collecting the information on the systems of technical infrastructure management, in particular, the methods and tools of lean maintenance as well as the ability to identify the factors affecting the efficiency of their application. This stage in detail includes:

- Identification of the degree of the use of specific methods and tools of the lean maintenance concept in manufacturing enterprises;
- Identification of the results obtained by the production enterprises that are implementing the lean maintenance concept;

- Identification of the factors affecting the results obtained after the implementation of the lean maintenance concept in enterprises;
- Indication of the relationship between the activities undertaken as part of the implementation of the lean maintenance concept and the results achieved.

The detailed work methodology is presented on Figure 1.

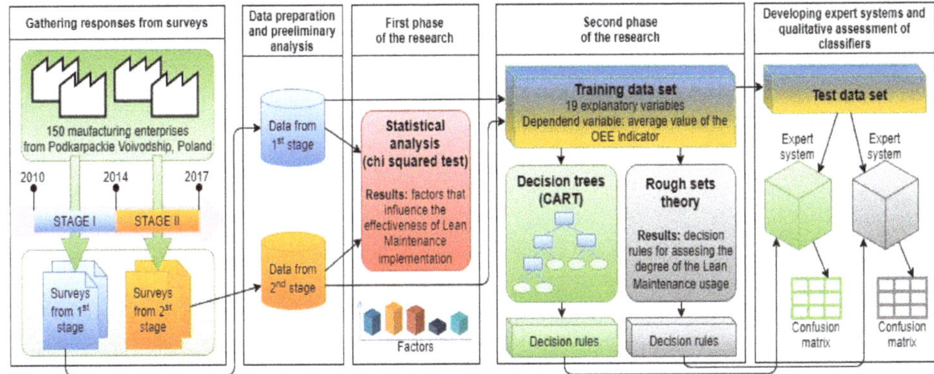

Figure 1. The detailed work methodology of the research.

The studies were carried out in 150 manufacturing enterprises in Podkarpackie Voivodship (Poland). Qualitative as well as quantitative research methods were used to analyze the results obtained. Additionally, a statistical analysis with the chi square test of the obtained results allowed us to identify the factors that influence the lean maintenance implementation effectiveness. In the second phase of the study, the concept of using artificial intelligence (AI) methods was proposed in order to assess the effectiveness of the lean maintenance concept implementation.

Artificial intelligence methods were used to search for the relationship between specific activities carried out under the implementation of lean maintenance and the results obtained. Decision trees and the rough set theory were used for the analysis. Decision trees were made for the variable of the average value of the OEE indicator. Decision trees enabled the generation of decision rules that can be the basis for determining the directions and effects of implementing lean maintenance in manufacturing enterprises. The rough set theory was used in order to assess the degree of the lean maintenance concept usage.

4. The Assessment of Lean Maintenance Effectiveness Concept in Enterprises

4.1. Use of the OEE Indicator in Enterprises—Study Results

The aim of the first stage of the research was to collect information on the use of lean maintenance methods and tools in enterprises, such as total productive maintenance (TPM), single minute exchange of die (SMED), 5S, and OEE indicator. The research, which was carried out in two stages, stage I—in 2010–2014 and stage II in 2014–2017, covered 150 production companies in the Podkarpackie Voivodship. The following criteria were taken into account when classifying the surveyed enterprises: size of the organization, production, type, industry, type of ownership, its capital, the company's condition, and the type of machines owned. Among the analyzed enterprises, there were those that carried out several types of production or operated in several industries. Among the analyzed enterprises, the biggest group were large enterprises (stage I: 46%, stage II: 52%). The next group were medium-sized enterprises (stage I: 27%, stage II: 32%). Among the surveyed enterprises, those from the metal processing industry dominated (stage I: 22.77%, stage II: 22.41%), followed by the aviation industry (stage I: 23.76%, stage II: 24.14%) and the automotive industry (stage I: 18.81%, stage II:

20.69%) (Figure 2). The majority were private enterprises (stage I: 94%, stage II: 98%) with Polish capital (stage I: 44%, stage II: 2%) or majority foreign capital (stage I: 44%, stage II: 25%) (Figure 3).

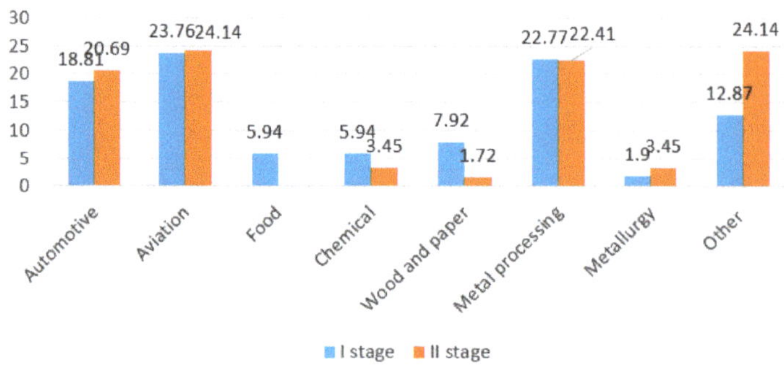

Figure 2. Structure of the surveyed research enterprises—type of industry.

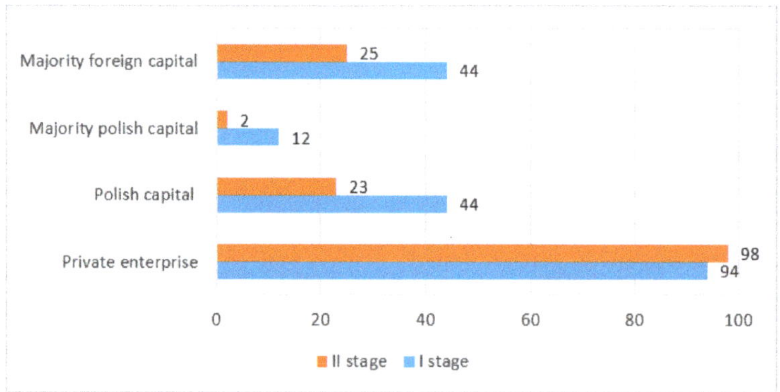

Figure 3. Structure of the surveyed research enterprises—property and capital.

Unit production dominated among the surveyed enterprises (stage I: 28.44%, stage II: 28%) and low- and medium-batch production, respectively, (stage I: 24.77%, stage II: 24%) and (stage I: 21.10%, stage II: 28%) (Figure 4).

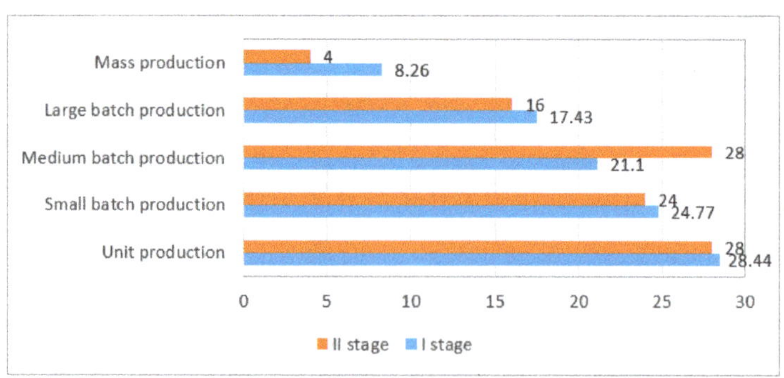

Figure 4. Structure of the surveyed research enterprises—type of production.

A survey method was used for the research. The designed questionnaires allowed us to obtain the data in the same form from all the respondents who did them independently. The survey included the representatives of medium and top management as well as the workers directly responsible for the supervision process of technological machines and devices and the chosen machine operators. The survey was realized in the form of conjunctive closed questions, which included a list of the prepared, provided-in-advance answers presented to a respondent with a multiple response item in which more than one option might be chosen. Additionally, other answers could be given if they were not among the provided options.

Within the conducted survey, the identification of the measures used for the effectiveness assessment of the implemented LM methods and tools as well as the benefits from their use noticed by enterprises were thoroughly analyzed. Collecting the information on the used types of measures for the effectiveness assessment of machine operation was the area of the conducted studies. The OEE indicator is one of the measures recommended in the literature. While assessing the effectiveness of the possessed machines and the implementation of the TPM method, this parameter is crucial. However, as the studies show, it is not always used [47,48]. The OEE indicator was one of the main study areas. The aim of the studies was to investigate if the OEE indicator was calculated in enterprises and how its value changed after the TPM method implementation.

Figure 5 shows that most of the analyzed enterprises still do not apply the OEE indicator (stage II: 60.38%, stage I: 73.96%). Only a few percent of the enterprises use this indicator for all machines (stage II: 7.55%, stage I: 5.21%).

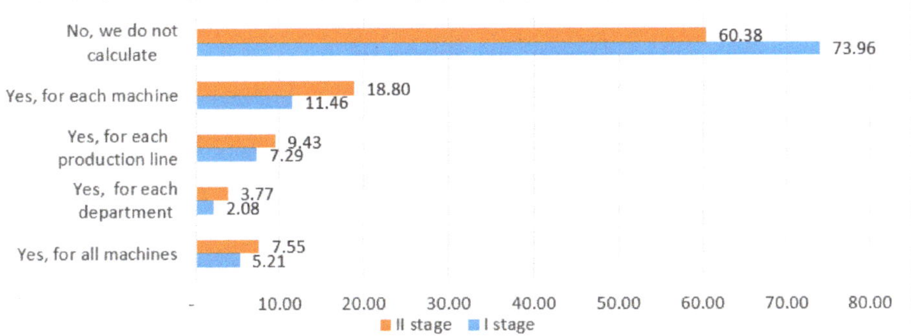

Figure 5. Is the overall equipment effectiveness (OEE) indicator calculated?—study results.

Figure 6 shows the size of the enterprises where this indicator in not used. The second stage of the studies indicates that the indicator is the most often not used in the medium size enterprises (stage I: 7.55%, stage II: 5.21%). However, its use increased significantly in micro companies (stage I: 10.26%, stage II: 5.88%).

In addition, the fact this indicator is not most often used in the enterprises with unit production (stage I: 29.49%, stage II: 29.41%) as well as with medium-batch production (stage I: 23.08%, stage II: 29.41%) was identified. Its use increased significantly in mass production (stage I: 8.97%, stage II: 2.94%). Furthermore, the indicator is most often not used in the enterprises of aviation, metal processing, and automotive industries. However, all the enterprises of the food industry that took part in the second stage of the studies declared its application. A crucial issue during the conducted studies was to obtain the information on the rate of calculating the OEE indicator. The rate of obtaining such information is essential, because the OEE indicator values inform us on an ongoing basis about productivity of the possessed machines. If the information is collected too seldom, a prompt reaction will not be possible in cases when the use of machines decreases.

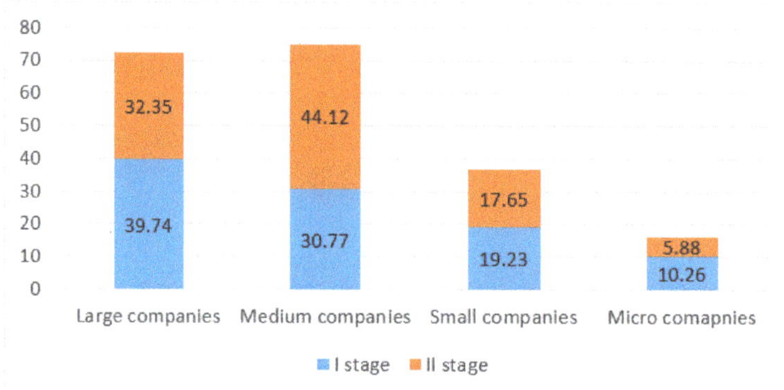

Figure 6. The percentage of enterprises that do not use the OEE indicator—the enterprise size.

The conducted studies show that in many enterprises the OEE indicator is calculated once a month (stage I: 26.09%, stage II: 25.00%). However, more and more often, the OEE indicator is calculated once per shift—the increase of over 17%, less often once a day—the decrease of over 16%. The obtained results show that large enterprises calculate the OEE value once per shift most often, medium enterprises once a month, small and micro enterprises once a week. The rate of calculating the indicator does not differ significantly in case of conventional and numerical machines (most often once per shift). Comparing the study results from the stages I and II, the rate of calculating changed from once a month to once a week for the machines described as "other".

Another element of the realized studies was collecting the information, which considered the value of the OEE indicator. Its value is important, because it allows us to conduct initially a general analysis of the effectiveness of the possessed machines. The value over 85% is considered as the world level value of this indicator [49]. Analyzing the obtained results, it was stated that the number of companies that declared an average value of the OEE indicator at the level of 50–70% (stage I: 18.18%, stage II: 33.33%) and at the level of 30–50% (stage I: 9.09%, stage II: 25.00%) increased significantly. In stage II of the studies, none of the analyzed companies declared the OEE values below 30%. The highest OEE indicator values of over 70% are obtained in large enterprises for numerical machines, in the aviation and automotive industry with major foreign capital. The lowest OEE indicator values, below 30%, are obtained in small enterprises for the machines described as "other", in the metal processing industry with Polish capital.

4.2. Identification of Factors Influencing the Effectiveness of Lean Maintenance Implementation

The aim of this stage was to identify the factors, which have an influence on the OEE value in analyzed enterprises. For provided analyses, the statistical chi-squared test was used. The following hypotheses were proposed zero hypotheses (H0), which means that there is not a significant difference in the solutions used in particular enterprises and alternative hypotheses H1, as there is a difference in the solutions used in particular enterprises.

These hypotheses can be written as

$$H0 = p_1 = p_2 = p_3 = \ldots \ldots = p_n \tag{1}$$

and

$$(H1) = p_1 \neq p_2 \neq p_3 \neq \ldots \ldots \neq p_n \tag{2}$$

The obtained *p*-value decided about accepting or rejecting H0, and therefore about accepting the alternative Hypothesis H1. If:

1. *p*-value < 0.05—H0 is rejected; thus, the alternative Hypothesis H1 is accepted.
2. *p*-value ≥ 0.05—H0 is accepted.

Table 2 shows the posed research hypotheses for the values of the OEE indicator as well as the obtained *p*-values.

Table 2. Research hypotheses—the effects obtained for the implementation of the lean maintenance methods and tools.

Hypothesis Number	Hypothesis	*p*-Value OEE
Hypothesis 1	There is no difference in the values of measures obtained in the enterprises of different sizes	0.588
Hypothesis 2	There is no difference in the values of measures obtained in the enterprises of different production types (SB—small-batch, UP—unit production, MB—medium-batch, LB—large-batch, MP—mass production)	0.316
Hypothesis 3	There is no difference in the values of measures obtained in the enterprises of different industries	**0.041**
Hypothesis 4	There is no difference in the values of measures obtained in the enterprises of different ownership	**0.048**
Hypothesis 5	There is no difference in the values of measures obtained in the enterprises in different condition	0.967
Hypothesis 6	There is no difference in the values of measures obtained in the enterprises of different capital	0.235
Hypothesis 7	There is no difference in the values of measures obtained in the enterprises with different types of machines owned	0.339
Hypothesis 8	There is no difference in the values of measures obtained in the enterprises that are implementing *5S* method	0.422
Hypothesis 9	There is no difference in the values of measures obtained in the enterprises that are implementing the *SMED* method	0.535
Hypothesis 10	There is no difference in the values of measures obtained in the enterprises that use Kanban for spare parts	0.348
Hypothesis 11	There is no difference in the values of measures obtained in the enterprises with different ways of supervision	0.854
Hypothesis 12	There is no difference in the values of measures obtained in the enterprises with different types of supervision	0.305
Hypothesis 13	There is no difference in the values of measures obtained in the enterprises with different mean time to repair	**0.025**
Hypothesis 14	There is no difference in the values of measures obtained in the enterprises with a different number of actions that prevent unplanned downtimes	0.707

For the analyzed Hypotheses 3 and 4, there is a statistically validated difference in the value of the OEE indicator (*p*-value OEE = 0.041 and OEE = 0.048—Hypothesis H0 rejected, Hypothesis H1 accepted). It means that, in the studied enterprises, the value of OEE depends on the type of ownership and the enterprise industry. In case of the concerned Hypothesis 13, there is also a statistically validated difference in the value of the OEE indicator (*p*-value OEE = 0.025, *p*-value LA = 0.005—Hypothesis H0 rejected, Hypothesis H1 accepted). It means that, in the studied enterprises, the value of OEE depends on the mean time to repair. The detailed results of the obtained studies were presented in the work [50].

On this basis, the following conclusions were drawn: the factors that influence the use of lean maintenance methods and tools are for example industry and the capital owned. The presented analyses allowed us to highlight the actual activities undertaken in the management of technical infrastructure and existing problems, and, thus, the possibility of identifying factors that increase the efficiency of lean maintenance. It should be noted that the studies often showed that single factors do not have a significant impact on the studied areas, although their interaction with other factors may have a substantial impact on the analyzed area. However, the problem is that analyzing a process with many variables is very difficult. Therefore, in the second stage of the study, the concept of using artificial intelligence (AI) methods in order to assess the effectiveness of the lean maintenance concept implementation was proposed.

5. Intelligent System to Support the Decision-Making Process in Lean Maintenance Management

5.1. Selection of Decision Variables in the Evaluation Process

To assess the effectiveness of lean maintenance tools, the factors influencing the dependent variable were identified. In the research on the assessment of the effectiveness of lean maintenance tools, one dependent variable and 19 explanatory variables (predictors) were determined. In the conducted research, one dependent variable was assumed: the average value of the OEE indicator.

Due to the large variety (combination) of response options, three additional indicators were introduced in the surveyed enterprises: maintenance strategy indicator (MSI), number of preventive activities (NPA), and number of TPM activities (NTPMA) indicator. The NPA number is the number of actions to prevent unplanned downtime, calculated as the total value of actions carried out simultaneously by the enterprise. During the survey (data collection) process, the company could choose several activities from the following:

1. Implementation of autonomous service (by the operator).
2. Implementation of preventive maintenance.
3. Forecasting activities based on the condition of machines (e.g., vibration analysis).
4. Additional operator training.
5. Additional training of maintenance service employees.
6. Equipping the maintenance services with specialized instruments (e.g., for measuring vibrations, for measuring the noise level).
7. Exchange of machines for new ones.
8. Modernization of machines.
9. Increasing the number of employees of maintenance services.
10. Outsourcing some maintenance activities to external companies.

Depending on how many activities are carried out by the enterprise at the same time, the indicator may range from 1 to 10. In addition, during the survey (data collection) process, the company could choose several activities implemented as part of the implementation of the TPM method, recommended in the literature on the subject, from the following:

1. Training of selected employees.
2. Training of all employees.
3. Implementation in a selected pilot area (position, line, etc.).
4. TPM workshops in the selected pilot area (stand, line, etc.).
5. Assessment of machines in terms of meeting health and safety requirements.
6. Assessment of the technical condition of machines.
7. Identification of non-conformities on machines.
8. Development of the inspection schedule.
9. Development of a renovation schedule.
10. Development of the scope of preventive service (for maintenance services).
11. Development of the scope of autonomous service (for the operator).

$$NTPMA = \frac{\sum_{i=1}^{11} x_i}{\max number\ of\ activities} * 100\% \tag{3}$$

Depending on the value obtained, the indicator had four levels: low, medium, high and very high (Table 3).

Table 3. The levels of number of TPM activities (NTPMA) indicator.

The Value of NTPMA Indicator	0–25%	26–50%	51–75%	More than 75%
Level	Low	Medium	High	Very high

The last index developed is the MSI index. With this indicator, it is possible to determine what technical infrastructure management strategy is applied by the enterprise. During the study (data collection), the company could choose several activities defining the realized activities implemented under the corrective maintenance (CM), preventive maintenance (PM), and condition-based maintenance (CBM) strategies. In order to define the index for possible variants of answers, numerical values ranging from 1 to 7 were introduced (Table 4). The lowest value was given to the action implemented in accordance with the CM strategy as the least effective strategy. However, the highest efficiency (value 7) was adopted for the operation: continuous monitoring of the condition of all machines (e.g., noise, vibrations, temperature) (CBM).

Table 4. Maintenance strategy—realized activities.

Maintenance Strategy—Realized Activities	Value
Only failures are removed on a regular basis (CM)	1
Inspections during the warranty period (PM)	2
Scheduled inspections by maintenance services (PM)	3
Scheduled inspections and repairs carried out by maintenance services (PM)	4
Machine condition assessment by the operator before starting work—autonomous maintenance (PM)	5
Continuous condition monitoring of selected machines (e.g., noise, vibration, temperature) (CBM)	6
Continuous condition monitoring of all machines (e.g., noise, vibration, temperature) (CBM)	7

The value of MSI indicator is calculated as the sum of the value of activities by the number of implemented activities (4).

$$MSI = \frac{\sum_{i=1}^{n} x_i}{n} \tag{4}$$

The MSI indicator may take values from 1 to 7. Value 1 means mainly the CM strategy, value 3.5—PM strategy, value 7—CBM strategy. When the value of the ratio is <3.5, it means the implementation of a mixed strategy, mainly CM–PM; when >3.5, it means the implementation of mainly a mixed strategy PM–CBM. At the same time, when closer to the value of 3.5, PM is the prevailing strategy. In order to ensure the adequacy of the adopted indicator, the variants of the strategy implemented by the examined enterprises were analyzed. For individual values of the indicator, implemented strategy variants (sequence of implemented actions) were assigned. The distribution of variants of the implemented strategies (distribution close to the normal distribution) allows us to confirm the validity of the adopted indicator (Table 5).

Table 5. The values of maintenance strategy indicator (MSI) indicator.

The Value of MSI Indicator	1	1.5	2	2.5	3	3.5	4	4.5	5	5.5	6	6.5	7
Range	1.00–1.24	1.25–1.74	1.75–2.24	2.25–2.74	2.75–3.24	3.25–3.74	3.75–4.24	4.25–4.74	4.75–5.24	5.25–5.74	5.75–6.24	6.25–6.74	More than 6.75
Maintenance strategy (realized actions)	1	12	13	125	3	2345	23456	36	456	457	6		7
				14	235	245	457	1367	2567		57		
				23	12345	236	4	3456	3457				
					15	235	345	346					
					135		245	2457					
					24		35						
	CM		CM—PM		PM				PM—CBM				CBM

The Statistica Data Miner system was used to conduct the analyzes. This system enables the preparation of data in the form of a training and test set, intuitive guidance through the model building and fitting procedure, and a clear visualization of test results.

5.2. Decision Trees in the Assessment of the Effectiveness of Lean Maintenance

Due to qualitative nature of the dependent variables, classification decision trees with the use of the classification and regression trees (CART) algorithm were used.

Not all surveyed enterprises used the same solutions, methods, and tools, therefore the main criterion for selecting this method was its insensitivity to the occurrence of atypical observations, which are believed to come from a different population, and the possibility of its effective use in datasets characterized by numerous shortcomings in independent variables. Additionally, the following advantages of CART classification trees determined the choice of the method:

- Taking into account non-monotonic dependencies through successive divisions with respect to the same variable;
- Simple interpretation of results in comparison with other methods;
- Suitability for tasks, where the a priori knowledge of which variables are related and how they are uncertain and intuitive;
- Non-parametric and non-linear;
- Estimating and ranking the importance of individual predictors (input variables) in the process of shaping the value of the dependent variable;
- Very useful for classification issues.

The CART tree for a dependent variable—a mean value of the OEE indicator—was designed for 24 enterprises out of the studied group of enterprises, which had analyzed this indicator and implemented the TPM method. The following explanatory variables (predictors) were assumed: enterprise size, production type, industry, ownership type, capital, company condition, machine type, 5S implementation, 5S activities, SMED implementation, way of supervision, maintenance strategy, actions undertaken to prevent unplanned downtimes (number of prevent actions—NPA), machine classification, spare parts classification, actions within TPM implementation (NTPMA indicator), and mean time to repair. The following were assumed while creating the tree: equal costs of the incorrect classification, Gini coefficient, stop rule, and a minimum size criterion in the divided node $n \geq 2$, which will allow for a detailed analysis of a tree structure and for 10-fold cross validation as a quality measure. A tree consisting of 12 divided nodes and 13 end nodes was chosen for the analysis. In order to assess the quality of the chosen tree, its validation for a new dataset was conducted. Thirteen decision rules may be defined for the created tree, which has 13 end nodes. The chosen decision rules, for which the highest values of OEE were reached (over 85% and for the range 70 to 85%) with the use of additional lean maintenance methods and tools, were presented below. Decision rules established on the basis of the decision tree are:

1. If an enterprise represents metal processing, aviation, or paper and wood industry, does not run partial supervision by outsourcing, and possesses an MSI indicator at the level of ≠4,5 and realizes machine classification, then it reaches an average value of the OEE indicator of over 85% (node 12).

2. If an enterprise represents an industry other than metal processing, aviation, or paper and wood. possesses an MSI indicator at the level of 3.5–5.6 and NPA number > 3, and the mean time to repair is below 1 h, then it reaches a mean value of the OEE indicator within the range from 70 to 85%.

3. If an enterprise represents an industry other than metal processing, aviation, or paper and wood and possesses an MSI indicator at the level other than that of 3.5–5.6 and an NTPMA indicator at any level other than high, then it reaches a mean value of the OEE indicator within the range from 70 to 85%.

4. If an enterprise represents an industry other than metal processing, aviation, or paper and wood, possesses an MSI indicator at any level other than that of 3.5–5.6a and an NTPMA indicator at a high level, and realizes supervision on its own with the service through outsourcing, then it reaches a mean value of the OEE indicator within the range from 70 to 85%.

5. If an enterprise represents an industry other than metal processing, aviation, or paper and wood, possesses an MSI indicator at any level other than that of 3.5–5.6 and an NTPMA indicator at a level other than high, and realizes supervision in a way other than on its own with the service through outsourcing, then it reaches a mean value of the OEE indicator of over 85%.

In order to evaluate the generated decision-making rules, research was again carried out in 20 randomly selected enterprises. Then, an expert system was designed and made (using PC-Shell—an expert system shell from the Aitech Sphinx software), taking into account the generated decision rules. Then, the general classification ability of the generated decision rules was tested using qualitative measures. Two blocks—aspects and rules—were used to develop the knowledge base in the system. The aspect block was used to declare the decision attributes and their values. On the other hand, the explanatory variables placed in the decision tree nodes are the decision attributes. The results of system inference were represented by the result attribute (target attribute). Finally, the value of the received attribute "OEE value" is presented in a separate window. The quality analysis consisted of developing binary matrices of classifiers' errors determined for the classes that most commonly appear in the conducted studies. In the developed binary matrices (confusion matrices) (Table 6), the class analyzed at a particular moment was assumed as positive, while the remaining classes were treated as negative.

Table 6. Confusion matrix.

Real Classes	Predicted Classes	
	Positive	**Negative**
Positive	TP (True positive)	FN (False negative)
Negative	FP (False positive)	TN (True negative)

Tables 7 and 8 present confusion matrices for the classifier—the value of OEE for the two most-emerging classes: 30–50% and 70–85%.

Table 7. Confusion matrix for the classifier value of the OEE indicator—30–50% class.

Real Classes	Predicted Classes	
	Positive	**Negative**
Positive	7	0
Negative	0	13

Table 8. Confusion matrix for the classifier value of the OEE indicator—70–85% class.

Real Classes	Predicted Classes	
	Positive	**Negative**
Positive	5	1
Negative	2	12

Based on the confusion matrix, numerical indicators presented in Table 9 can be designated. In detail, these indicators have been presented and discussed, among others in the works [51–53].

Table 9. Indicators used to test the quality of classifiers [54].

Indicator	Designation	Formula
Acc	Accuracy	$Acc = \frac{TP+TN}{TP+TN+FP+FN}$
Err	Overall error rate	$Err = \frac{FP+FN}{TP+TN+FP+FN}$
TPR	True positives rate	$TPR = \frac{TP}{TP+FN}$
TNR	True negatives rate	$TNR = \frac{TN}{TN+FP}$
PPV	Positive predictive value	$PPV = \frac{TP}{TP+FP}$
NPV	Negative predictive value	$NPV = \frac{TN}{TN+FN}$
FPR	False positive rate	$FPR = \frac{FP}{FP+TN} = 1 - TNR$
FDR	False discovery rate	$FDR = \frac{FP}{FP+TP}$
FNR	False negatives rate	$FNR = \frac{FN}{TP+FN} = 1 - TPR$
MCC	Matthew's correlation coefficient	$MCC = \frac{TP \times TN - FP \times FN}{\sqrt{(TP+FN)(TP+FP)(FN+TN)(FP+TN)}}$
F1	F1-score	$F1 = \frac{2 \times PPV \times TPR}{PPV+TPR}$
J	Youden's J statistic	$J = TPR + TNR - 1$

On the basis of the developed binary matrices, for each of them, the values of the twelve indicators showing the classifiers' quality were calculated. Table 10 presents the results for the highlighted classifier classes.

Table 10. Indicators used to test the quality of classifiers.

Indicators			Acc	TPR	TNR	PPV	NPV	MCC	F1	J	Err	FPR	FDR	FNR
Classifier: an average OEE value	Marked class	30–50%	1.00	1.00	1.00	1.00	1.00	1.00	1.00	1.00	0.00	0.00	0.00	0.00
		70–85%	0.85	0.83	0.86	0.71	0.92	0.66	0.77	0.69	0.15	0.67	0.29	0.17

The obtained indicator values the assessment of a classification measure, e.g., of an error (Err) at the level of 0.00 and 0.15, proved high usefulness of the developed classifiers, and thereby, their possibility to be applied by manufacturing enterprises for the effective assessment of the lean maintenance methods and tools implementation.

5.3. The Theory of Rough Sets to Support the Lean Maintenance Assessment

The rough set theory is one of the fastest growing branches of data exploration. It allows for a formal approach to all phenomena related to knowledge processing, therefore it is used as a methodology in the process of knowledge discovery from data. In particular, it can be used to test the imprecision and uncertainty in the data analysis process. It enables finding the relationship between explanatory variables (conditional attributes) and explained variables (decision attributes), which facilitates supporting decision-making based on data. It is also used to reduce dimensionality, consisting of removing from the dataset those explanatory variables that do not significantly affect the explained variables. Knowledge derived from data based on the rough set theory is recorded in the form of decision rules [55]. Details on the formal description of the rough set theory can be found, among others, at work [56]. Often, the purpose of the decision-making system based on rough sets is to search for hidden, and therefore, implicit rules that have not worked well during the selection made by an expert (or experts) [55–57]. Approximate sets are used to process the so-called unclear data with the use of intuitively understood inference rules. They can be used to search for hidden dependencies in input data, including decision support in the scope of cases that can be described with discrete attributes.

In this paper, the rough set theory was used to assess the degree of lean maintenance use. The same set of input data was used for the assessment as in the decision trees. Due to the presence of the so-called incomplete data, the use of rough sets improved the accuracy of the solution. Various types of algorithms were used to interfere with the rules.

The use of the rough set theory, and thus incomplete data, increased the number of analyzed enterprises from 24 included in decision trees to 34. An additional 10 analyzed enterprises were characterized by a set of variables, for which at least one variable did not have a specific value (no answer). By using decision tables in the rough set theory, it is possible to include more data when generating rules. This allows for the identification of new dependencies between the variables. To make the assessment, the rules were validated. In order to generate decision rules on the basis of the rough set theory, Rough Set Exploration System (RSES) software was used. The software was developed at the Institute of Mathematics of Warsaw University.

RSES software allows one to generate decision rules with the means of four algorithms: exhaustive algorithm, genetic algorithm, covering algorithm, and learning from examples module version 2 (LEM2). They were described in the works [57,58]. Furthermore, the software contains a number of other options, which, e.g., assign reductions for a given computer system. A reduct is a set of R attributes, where $R \subset A$, which allows to differentiate pairs of objects in a computer system, and at the same time, no other R proper subset possesses this property. Reductions are calculated with an exhaustive or genetic algorithm. On the basis of the assigned reductions, it is possible to create decision rules as well.

For each of decision classes, RSES software calculates three indicators, which indicate classification quality:

- Accuracy—the ratio of properly classified objects of a given class to all objects belonging to this class;
- Coverage—the ratio of the objects classified to a given class with decision rules to all objects belonging to this class;
- True positive rate—the ratio of the properly classified objects of a given class to all objects that were classified to this class.

Accuracy and coverage are also calculated jointly for all decision classes (for the whole set of rules).

Decision rules for the described variable "an average OEE value" were generated by means of all four algorithms available in RSES. The scheme of the conducted study is shown in Figure 7. The OEE symbol designates a decision table which contains 34 studied objects (enterprises). Each object is described by 17 explanatory variables: an enterprise size, production type, industry, ownership type, capital, actions undertaken to prevent unplanned downtimes (NPA number—MSI indicator), machine category, spare parts category, actions in the TPM implementation (NTPMA indicator), and mean time to repair. The described variable "an average OEE value" played in the study the role of a decision attribute. The remaining symbols in the scheme are described in Table 11.

While formulating the decision rules, the parameters of genetic and covering algorithms were chosen in such a way that the accuracy and coverage of the created set were equal to 1. Table 12 includes the information on a number of rules in each of the four sets of rules. For each of the rules, a rule match is calculated. It is equal to the number of objects from the learning set and matching the forerunner of the rule.

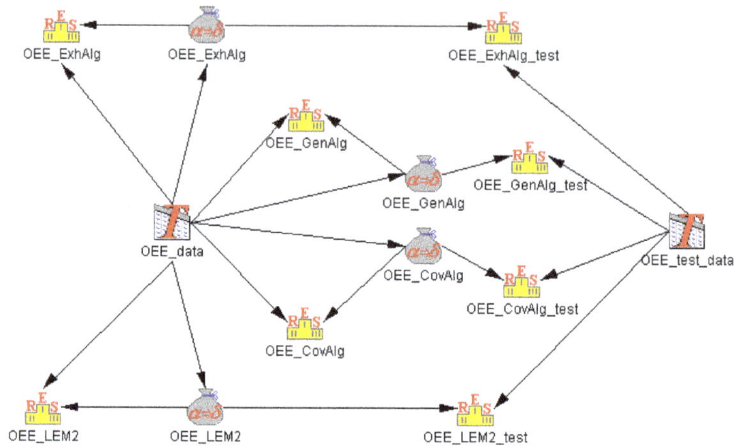

Figure 7. The scheme for the explained variable "an average OEE value".

Table 11. Description from the study scheme (Figure 3).

Symbol	Symbol Name	Description
	OEE_ExhAlg	A set of decision rules generated with an exhaustive algorithm.
	OEE_GenAlg	A set of decision rules generated with an exhaustive algorithm working in an exact mode with a population size equal to 10.
	OEE_CovAlg	A set of decision rules generated with a covering algorithm with the coverage parameter equal to 0.0001.
	OEE_LEM2	A set of decision rules generated with LEM2 algorithm with the coverage parameter equal to 1.
		A confusion matrix that includes the results of the classification prepared with a set of rules from which the arrow on the scheme/diagram leads. The classification is realized on the objects from the OEE decision table.

Table 12. The number of rules in each of the created sets for the described variable "an average OEE value".

Name of a Rule Set	Number of Rules
OEE_ExhALg	2606
OEE_GenAlg	325
OEE_CovAlg	55
OEE_LEM2	21

Each of the four rule sets was used for the classification of the data from the OEE decision table. The classification was accomplished by a standard voting method. The manner of such voting is as follows: each of the generated rules determines the value of a variable described for the considered object (an enterprise). The calculated match value of particular rules is treated during the voting as an importance—the higher the match of a given rule, the more important is its vote. That is why it is more influential on a final voting result than the vote of the rule with a lower match. Eventually, the object is assigned such a value of an explanatory variable that won the weighted voting.

The result of the classification was presented in the form of a confusion matrix. The confusion matrix that includes the results of the classification accomplished by the rule set generated by an exhaustive algorithm was presented in Figure 8. Matrix rows correspond to the real decision classes (the values of the variable described). However, matrix columns are the results of the classification that was accomplished by the generated rules. All 34 objects that are in the decision board were classified properly. It is reflected in the values located only on the main diagonal of the confusion matrix. The last three columns in the described figure show the information on the number of objects belonging to a given decision class (no. of obj.), accuracy, and coverage. The last row of the table is the true positive rate calculated for each class individually. The bottom part of the window presents the number of all studied objects and the accuracy and coverage calculated for all decision classes altogether.

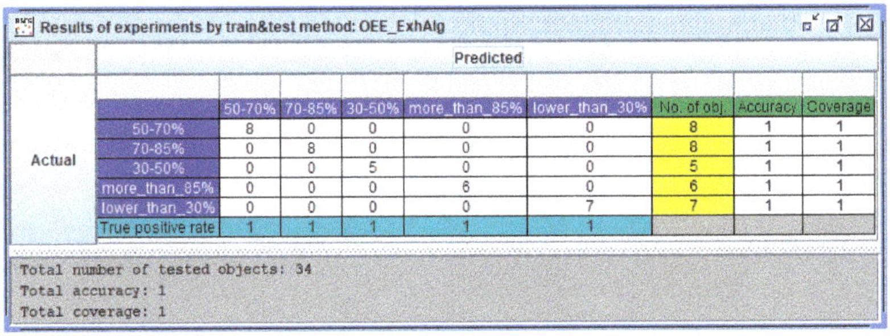

Results of experiments by train&test method: OEE_ExhAlg

Actual	Predicted							
		50-70%	70-85%	30-50%	more_than_85%	lower_than_30%	No. of obj.	Accuracy	Coverage
	50-70%	8	0	0	0	0	8	1	1
	70-85%	0	8	0	0	0	8	1	1
	30-50%	0	0	5	0	0	5	1	1
	more_than_85%	0	0	0	6	0	6	1	1
	lower_than_30%	0	0	0	0	7	7	1	1
	True positive rate	1	1	1	1	1			

Total number of tested objects: 34
Total accuracy: 1
Total coverage: 1

Figure 8. Confusion matrix for the rules generated by an exhaustive algorithm.

Confusion matrices were also created for the classification results based on the three remaining rule sets. Each of these matrices included the same results (accuracy = 1), such as the matrix presented in Figure 4, which indicates that there are no classification errors also for other classifications.

The assessment of the developed decision rules (decision trees and rough set theory) was carried out in the following stages: generation of a decision table and confusion matrix; development of an expert system based on the generated decision rules; use of the obtained study results to test the overall classification capacity; use of the obtained study results to test the overall classification capacity of decision-making rules using the developed expert system; and qualitative assessment of the results obtained using classification quality measures.

The results of the surveys from 20 companies of the Podkarpackie Voivodship were reused to validate the decision-making rules. Among the companies analyzed, the largest group included large companies (85%) from the aviation industry (50%). The majority of them were private companies (95%), with a majority of foreign capital (85%). Large-batch production (45%) dominated among the companies surveyed. On the basis of the results obtained, a decision table was created. The created decision table was introduced into the RSES system. It allowed us to create a confusion matrix for the explanatory variable "an average OEE value". Maximum coverage calculated for all decision classes in total that equals 1 and accuracy of 0.30 for the explanatory variable "an average OEE value" were achieved using a set of rules generated by a genetic algorithm (Figure 9).

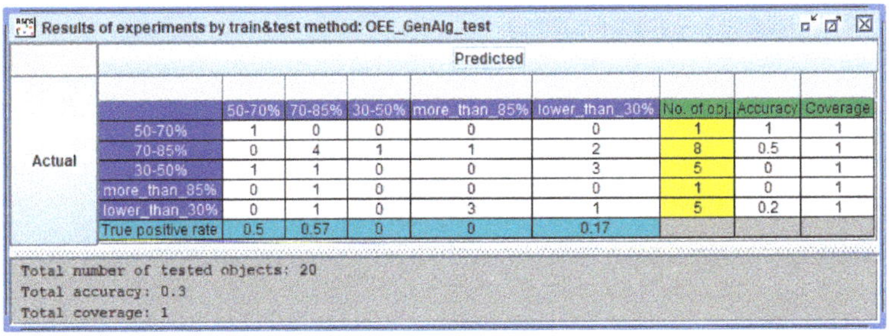

Results of experiments by train&test method: OEE_GenAlg_test

Actual	Predicted							
		50-70%	70-85%	30-50%	more_than_85%	lower_than_30%	No. of obj.	Accuracy	Coverage
	50-70%	1	0	0	0	0	1	1	1
	70-85%	0	4	1	1	2	8	0.5	1
	30-50%	1	1	0	0	3	5	0	1
	more_than_85%	0	1	0	0	0	1	0	1
	lower_than_30%	0	1	0	3	1	5	0.2	1
	True positive rate	0.5	0.57	0	0	0.17			

Total number of tested objects: 20
Total accuracy: 0.3
Total coverage: 1

Figure 9. Confusion matrix for the rules generated by an genetic algorithm.

In order to carry out the next stage of validation, an expert system was developed. The decision-making rules generated by all algorithms were implemented in the knowledge base that was created for the system needs. The knowledge base for the expert system was developed with PC-Shell software, which is part of the Aitech SPHINX integrated artificial intelligence suite and Aitech HybRex software.

In order to test the overall quality of the classifier for all algorithms, confusion matrices (Tables 13–16) were used. These matrices were developed by comparing the results obtained by the studied companies with the result generated by the developed expert system. The classification was carried out again according to the following voting method:

- The conformity of the actual class value and predicted class value was considered as a valid result;
- The cases when the value of the class predicted by an expert system was one class lower than the actual class value was allowed as a valid result, e.g., the actual class is a range of 30–50% and the predicted value is 10–30%. In practice, this means that an enterprise can expect projected variables to be at a minimum level of 10–30%. However, in fact, it may be higher.

Table 13. Confusion matrix for the rules generated by an exhaustive algorithm.

		Predicted					No. of obj.	Accuracy	Coverage
		Lower than 30%	30–50%	50–70%	70–85%	More than 85%			
Actual	Lower than 30%	0	0	3	2	0	5	0	1
	30–50%	0	3	0	1	1	5	0.6	1
	50–70%	0	0	0	1	0	1	0	1
	70–85%	0	0	0	7	1	8	0.875	1
	More than 85%	0	0	0	0	1	1	1	1
	True positive rate	0	1	0	0.64	0.33			

Table 14. Confusion matrix for the rules generated by a genetic algorithm.

		Predicted					No. of obj.	Accuracy	Coverage
		Lower than 30%	30–50%	50–70%	70–85%	More than 85%			
Actual	Lower than 30%	1	0	0	3	1	5	0.2	1
	30–50%	0	0	1	2	1	5	0	0.8
	50–70%	0	0	1	0	0	1	1	1
	70–85%	0	0	0	5	3	8	0.625	1
	More than 85%	0	0	0	0	1	1	1	1
	True positive rate	1	0	0.5	0.5	0.17			

Table 15. Confusion matrix for the rules generated by a covering algorithm.

		Predicted					No. of obj.	Accuracy	Coverage
		Lower than 30%	30–50%	50–70%	70–85%	More than 85%			
Actual	Lower than 30%	1	0	2	1	1	5	0.2	1
	30–50%	0	1	1	1	2	5	0.2	1
	50–70%	0	0	1	0	0	1	1	1
	70–85%	3	1	0	3	1	8	0.75	1
	More than 85%	0	0	1	0	0	1	0	1
	True positive rate	0.25	0.5	0.2	0.6	0			

Table 16. Confusion matrix for LEM2—generated rules.

		Predicted					No. of obj.	Accuracy	Coverage
		Lower than 30%	30–50%	50–70%	70–85%	More than 85%			
Actual	Lower than 30%	1	0	0	0	1	5	0.5	0.4
	30–50%	0	1	0	0	2	5	0.333	0.6
	50–70%	1	0	0	0	0	1	0	1
	70–85%	2	0	0	3	0	8	0.6	0.625
	More than 85%	0	0	0	0	0	1	0	0
	True positive rate	0.25	1	0	1	0			

When analyzing particular confusion matrices, it should be noted that the best results for the most common classes (30–50% and 70–85%) of the accuracy value were obtained for the rules generated by an exhaustive algorithm. The accuracy value was 0.6 and 0.875, respectively. In order to accurately assess the quality of the classifiers based on binary matrices, the values of the twelve indicators were calculated for each of the matrices according to the Table 9.

6. Analysis of the Obtained Results

In order to assess the obtained results, the values of the indicators used to test the quality of classifiers were compared. In the Table 17, a comparison of the values obtained for the models acquired with decision trees (DT) and the rough set theory (RST) was presented. Indicators from Acc to J should acquire the highest possible values up to 1, while the remaining ones should acquire the lowest possible value to 0. The green color indicates those indicator values that obtained more favorable values for particular classes. Yellow, on the other hand, indicates that the values obtained for particular indicators were identical.

Table 17. Comparison of the obtained indicator values for testing the quality of classifiers for the models obtained with decision trees (DT) and the rough set theory (RST).

	Classifier: An Average OEE Value									
	Marked Class									
Indicators	30–50%					70–85%				
	DT	RST				DT	RST			
		LEM2	Exh.Alg.	Gen.Alg.	Cov.Alg.		LEM2	Exh.Alg.	Gen.Alg.	Cov.Alg.
Acc	1.00	0.90	0.90	0.79	0.75	0.85	0.90	0.75	0.58	0.65
TPR	1.00	1.00	0.60	0.00	0.20	0.83	1.00	0.88	0.62	0.37
TNR	1.00	0.88	1.00	1.00	0.94	0.86	0.86	0.67	0.55	0.83
PPV	1.00	0.60	1.00	0.00	0.50	0.71	0.75	0.64	0.50	0.60
NPV	1.00	1.00	0.88	0.79	0.78	0.92	1.00	0.89	0.67	0.67
MCC	1.00	0.73	0.73	0.00	0.20	0.66	0.80	0.53	0.17	0.24
F1	1.00	0.75	0.75	0.00	0.29	0.77	0.86	0.74	0.56	0.46
J	1.00	0.88	0.60	0.00	0.13	0.69	0.86	0.54	0.17	0.20
Err	0.00	0.10	0.10	0.21	0.25	0.15	0.10	0.25	0.42	0.35
FPR	0.00	1.00	0.00	0.00	0.20	0.67	1.00	0.80	0.62	0.29
FDR	0.00	0.40	0.00	0.00	0.50	0.29	0.25	0.36	0.50	0.40
FNR	0.00	0.00	0.40	1.00	0.80	0.17	0.00	0.12	0.38	0.62

Legend:

    - indicator values that obtained more favorable results,

    - indicator values that obtained the same results.

When analyzing the results presented in the table, it should be noted that the first indicator (accuracy—Acc) shows that the classifier developed with DT for the class 30–50% allocates objects to this class to which they actually belong (Acc = 1) with the most likelihood. By contrast, the lowest Acc value obtained for the RST classifier was a genetic algorithm for the class 70–85% (Acc = 0.58). The sensitivity (TPR) of the classifiers presents itself in a slightly different way. The ability to detect the objects from the highlighted class is the highest for LEM2 algorithm for both highlighted classes and for DT in the class of 30–50%. The lowest ability for a genetic algorithm in the class of 30–50% (TPR = 0.00). By analyzing specificity (TNR), you can see that, for the highlighted class of 30–50%, the results are the highest (TNR = 1.00) for DT, exhaustive, and genetic algorithms. Precision (PPV) is similarly the highest for the highlighted class of 30–50% for DT and exhaustive algorithms (PPV = 1), and the lowest, again, for the highlighted class of 30–50% for a genetic algorithm. Negative predictive value (NVP), or the probability of the membership of the object recognized by a classifier as non-highlighted to the actual non-highlighted class, is the highest in case of the LEM2 algorithm for both highlighted classes and DT in the class of 30–50% (NVP = 1). It is the lowest for the class of 70–85% for covering and genetic algorithms (NPV = 0.67).

The results of the compared values of the Matthew's correlation coefficient (MCC) measure (correlation coefficient between real classes and projected classes by the model) indicate that the best result was achieved in the class of 30–50% for DT (MCC = 1). The lowest result was for a genetic algorithm in the class of 70–85% (MCC = 0.17). By analyzing the results obtained for F1 (harmonic mean of precision and sensitivity of a model) and J (the sum of sensitivity and specificity reduced by 1) for all the models, it can be seen that the best classifier is DT in the class of 30–50%, while the worst is a genetic algorithm for the same class.

The remaining indicators should take the lowest possible values. A general classifier error (Err) is more favorable in case of using the DT classifier for the class of 30–50% (Err = 0.00), and the worst for the class of 70–85% is a genetic algorithm. The FPR indicator (probability of false alarms, i.e., the objects incorrectly assigned to a highlighted class, among all objects actually non-highlighted) and the FDR indicator (probability of false alarms among all the objects recognized by the classifier as highlighted) achieves the best values for the class of 30–50% (FPR = 0 and FDR = 0). However, the best FNR values (the probability of missing the highlighted objects, that is, their assignment by the classifier to the non-highlighted class) were also obtained for the class of 30–50% (FNR = 0).

7. Conclusions

The main research problem that was considered in the paper was an insufficient means of assessing the degree of the implementation of lean maintenance. To find the solution of identified problem, artificial intelligence methods such as decision trees and the rough set theory were used. When analyzing the results presented, it should be noted that the models generated with the rough set theory achieved much better results than in decision trees. The decision rules generated by DT showed better values for all indicators for the classifier for the class of 30–50%. However, better values for the class of 70–85% were achieved for RST, mainly for LEM2 algorithm. The number of rules generated by the LEM2 algorithm is the smallest compared to the other algorithms. This shows that a large number of rules is not needed to get good prediction results in the investigated problem.

The resulting indicators for testing the quality of classifiers confirmed the high usefulness of the generated decision rules, both those using decision trees and the rough set theory. The developed dependencies allow us to assess which results a given company can expect after the implementation of specific lean maintenance methods and tools, and which lean maintenance methods and tools should be used to achieve the intended goals. These dependencies may be the basis for determining the directions and effects of implementing lean maintenance in manufacturing companies. Additionally, an expert system in the form of a software application, developed on the basis of the generated dependencies (decision rules), allows for the selection of appropriate actions in order to obtain the best results after implementing lean maintenance.

The presented studies can be used by enterprises to build and organize maintenance processes, to select an appropriate action strategy, but above all, to improve already implemented activities in this area. Although the research was conducted in a limited area, it was based on common assumptions, principles, and objectives of implementing the lean maintenance concept in the enterprise. Therefore, the presented solutions are useful for practical use by all production companies for forecasting and assessing the effectiveness of implementing lean maintenance methods and tools, regardless of the region.

Moreover, the positive results obtained during the conduct of the described study lead to the conclusion that the activities in these areas should be continued. In particular, there ought to be studies considering the assessment of the effectiveness of using other methods and tools recommended in the literature within lean maintenance implementation; the possibility of extending functionality designed in a computer application; and the use of other methods of data exploration for generating decision rules and comparing their classification quality.

Author Contributions: K.A. gave the theoretical and substantive background for the provided research and conceived and designed the experiments; L.P. made an experimental verification of the proposed approach; A.G. provided technical guidance and gave critical review for this paper. All authors have read and agreed to the published version of the manuscript.

Funding: This research received no external funding.

Conflicts of Interest: The authors declare no conflict of interest.

References

1. Jasiulewicz-Kaczmarek, M.; Gola, A. Maintenance 4.0 technologies for sustainable manufacturing—An overview. *IFAC-PapersOnLine* **2019**, *52*, 91–96. [CrossRef]
2. Danilczuk, W.; Gola, A. Computer-aided material demand planning using ERP systems and business intelligence technology. *Appl. Comput. Sci.* **2020**, *16*, 42–55.
3. Gornicka, D.; Burduk, A. Improvement of production processes with the use of simulation models. *Intell. Syst. Comput.* **2018**, *657*, 265–274.
4. Kotowska, J.; Markowski, M.; Burduk, A. Optimization of the supply of components for mass production with the use of the ant colony algorithm. In Proceedings of the 1st International Conference on Intelligent Systems in Production Engineering and Maintenance (ISPEM), Wroclaw, Poland, 18 August 2017; Volume 637, pp. 347–357.
5. Valis, D.; Mazurkiewicz, D. Application of selected levy processes for degradation modelling of long range mine belt using real-time data. *Arch. Civ. Mech. Eng.* **2018**, *18*, 1430–1440. [CrossRef]
6. Sa Ribeiro, D.R.; Forcellini, F.A.; Pereira, M.; Xavier, F.A. An overview about the implementing of lean maintenance in manufacturing processes. *J. Lean Syst.* **2019**, *4*, 44–59.
7. Mouzani, I.A.; Bouami, D. The integration of lean manufacturing and lean maintenance to improve production efficiency. *Int. J. Mech. Prod. Eng. Res. Dev.* **2019**, *9*, 593–604.
8. Antosz, K. Maintenance–identification and analysis of the competency gap. *Eksploat. Niezawodn. Maint. Reliab.* **2018**, *20*, 484–494. [CrossRef]
9. Sobaszek, Ł.; Gola, A.; Świć, A. Time-based machine failure prediction in multi-machine manufacturing systems. *Eksploat. Niezawodn.* **2020**, *22*, 52–56. [CrossRef]
10. Prabowo, H.A.; Adesta, E.Y.T. A study of total productive maintenance (TPM) and lean manufacturing tools and their impact on manufacturing performance. *Int. J. Recent Technol. Eng.* **2019**, *7*, 39–43.
11. Loska, A. Exploitation assessment of selected technical objects using taxonomic methods. *Eksploat. Niezawodn. Maint. Reliab.* **2013**, *15*, 1–8.
12. Szwarc, E.; Bocewicz, G.; Banaszak, Z.; Wikarek, J. Competence allocation planning robust to unexpected staff absenteeism. *Eksploat. Niezawodn. Maint. Reliab.* **2019**, *21*, 440–450. [CrossRef]
13. Sobaszek, Ł.; Gola, A.; Kozłowski, E. Application of survival function in robust scheduling of production jobs. In Proceedings of the 2017 Federated Conference on Computer Science and Information Systems (FEDCSIS), Prague, Czech Republic, 3–6 September 2017; Ganzha, M., Maciaszek, M., Paprzycki, M., Eds.; IEEE: New York, NY, USA, 2017; pp. 575–578.
14. Gola, A.; Kłosowski, G. Development of computer-controlled material handling model by means of fuzzy logic and genetic algorithms. *Neurocomputing* **2019**, *338*, 381–392. [CrossRef]
15. Jasiulewicz-Kaczmarek, M.; Saniuk, A. How to make maintenance processes more efficient using lean tools? *Adv. Intell. Syst. Comput.* **2018**, *605*, 9–20.
16. Antosz, K.; Stadnicka, D. The results of the study concerning the identification of the activities realized in the management of the technical infrastructure in large enterprises. *Eksploat. Niezawodn. Maint. Reliab.* **2014**, *16*, 112–119.
17. Ramos, E.; Mesia, R.; Alva, C.; Miyashiro, R. Applying lean maintenance to optimize manufacturing processes in the supply chain: A Peruvian print company case. *Int. J. Supply Chain Manag.* **2020**, *9*, 264–281.
18. Clarke, G.; Mulryan, G.; Liggan, P. Lean maintenance. A risk-based approach, pharmaceutical engineering. *Off. Mag. ISPE* **2010**, *30*, 1–6.
19. Womack, J.; Jones, D.T.; Roos, D. *The Machine that Changed the World*; Rawson Associates: New York, NY, USA, 1990.

20. Ohno, T. *Toyota Production System—Beyond Large Scale Production*; Productivity Press: New York, NY, USA, 1988.

21. Womack, J.; Jones, D. *Lean Thinking: Banish Waste and Create Wealth in Your Corporation*; Simon & Schuster Inc.: London, UK, 2003.

22. Melton, T. The benefits of lean manufacturing. What lean thinking has to offer the process industries. *Chem. Eng. Res. Des.* **2005**, *83*, 662–673. [CrossRef]

23. Pavnaskar, S.J.; Gershenson, J.K.; Jambekar, A.B. Classification scheme for lean manufacturing tools. *Int. J. Prod. Res.* **2003**, *41*, 3075–3090. [CrossRef]

24. Papadopoulu, T.C.; Ozbayrak, M. Leanness: Experiences from the journey to date. *J. Manuf. Technol. Manag.* **2005**, *16*, 784–806. [CrossRef]

25. Billesbach, T.J. A study of the implementation of just in time in the United States. *Prod. Inventory Manag. J.* **1991**, *32*, 1–4.

26. Duran, O.; Capaldo, A.; Acevado, P.A.D. Lean maintenance applied to improve maintenance efficiency in thermoelectric power plants. *Energies* **2017**, *10*, 1653. [CrossRef]

27. Jasiulewicz-Kaczmarek, M. Sustainability: Orientation in maintenance management. Case study. In *EcoProduction and Logistics-Environmental Issues in Logistics and Manufacturing*; Golinska, P., Ed.; Springer: Berlin/Heidelberg, Germany, 2013; pp. 135–154.

28. Leksic, I.; Stefanic, N.; Veza, I. The impact of using different lean manufacturing tools on waste reduction. *Adv. Prod. Eng. Manag.* **2020**, *15*, 81–92. [CrossRef]

29. Levitt, J. *Lean Maintenance*; Industrial Press: New York, NY, USA, 2008.

30. Baluch, N.; Abdullah, C.S.; Mohtar, S. TPM and LEAN maintenance—A critical review. *Interdiscip. J. Contemp. Res. Bus.* **2012**, *4*, 850–857.

31. Bashiri, M.; Badri, H.; Hejazi, T.H. Selecting optimum maintenance strategy by fuzzy interactive linear assignment method. *Appl. Math. Model.* **2011**, *35*, 152–164. [CrossRef]

32. Zhaoyang, T.; Jianfeng, L.; Zongzhi, W.; Jianhu, Z.; Weifeng, H. An evaluation of maintenance strategy using risk based inspection. *Saf. Sci.* **2011**, *49*, 852–860.

33. Cruz, A.M.; Rincon, A.M.R. Medical device maintenance outsourcing: Have operation management research and management theories forgotten the medical engineering community? A mapping review. *Eur. J. Oper. Res.* **2012**, *221*, 186–197. [CrossRef]

34. Rinaldi, G.; Portillo, J.C.C.; Khalid, F.; Henriques, J.C.C.; Thies, P.R.; Gato, L.M.C.; Johanning, L. Multivariate analysis of the reliability, availability, and maintainability characterizations of a Spar-Buoy wave energy converter farm. *J. Ocena Eng. Mar. Energy* **2018**, *4*, 199–215. [CrossRef]

35. Wang, B.; Furst, E.; Cohen, T.; Keil, O.R.; Ridgway, M.; Stiefel, R. Medical equipment management strategies. *Biomed. Instrum. Technol.* **2006**, *40*, 233–237. [CrossRef]

36. Bayar, N.; Darmoul, S.; Hajri-Gabouj, S.; Pierreval, H. Using immune designed ontologies to monitor disruptions in manufacturing systems. *Comput. Ind.* **2016**, *81*, 67–81. [CrossRef]

37. Taghipour, S.; Banjevic, D.; Jardine, A. Prioritization of medical equipment for maintenance decisions. *J. Oper. Res. Soc.* **2011**, *62*, 1666–1687. [CrossRef]

38. Li, H.F.; Parikh, D.; He, Q.; Qian, B.Y.; Li, Z.G.; Fang, D.P.; Hampapur, A. Improving rail network velocity: A machine learning approach to predictive maintenance. *Transport. Res. C Emer.* **2014**, *45*, 17–26. [CrossRef]

39. Jamshidi, A.; Rahimi, S.A.; Aitkadi, D.; Ruiz, A. A comprehensive fuzzy risk-based maintenance frame-work for prioritization of medical devices. *Appl. Soft Comput.* **2015**, *32*, 322–334. [CrossRef]

40. Carnero, M.C.; Gomez, A. A multicriteria decision making approach applied to improving maintenance policies in healthcare organizations. *BMC Med. Inform. Decis. Mak.* **2016**, *16*, 47. [CrossRef]

41. Zeineb, B.H.; Malek, M.; Ahmad, A.H.; Ikram, K.; Faouzi, M. Quantitative techniques for medical equipment maintenance management. *Eur. J. Ind. Eng.* **2017**, *10*, 703–723.

42. Lin, P.; Yuan, X.-X.; Tovilla, E. Integrative modeling of performance deterioration and maintenance effectiveness for infrastructure assets with missing condition data. *Comput. Aided Civ. Infrastruct. Eng.* **2019**, *34*, 677–695. [CrossRef]

43. Jasiulewicz-Kaczmarek, M.; Żywica, P. The concept of maintenance sustainability performance assessment by integrating balanced scorecard with non-additive fuzzy integral. *Eksploat. Niezawodn.* **2018**, *20*, 650–661. [CrossRef]

44. Rodríguez-Padial, N.; Marín, M.; Domingo, R. An approach to evaluate tactical decision–making in industrial maintenance. *Procedia Manuf.* **2017**, *13*, 1051–1058. [CrossRef]

45. Kosicka, E.; Gola, A.; Pawlak, J. Application-based support of machine maintenance. *IFAC-PapersOnLine* **2019**, *52*, 131–135. [CrossRef]

46. Antosz, K.; Pasko, L.; Gola, A. The use of intelligent systems to support the decision-making process in lean maintenance management. *IFAC-PapersOnLine* **2019**, *52*, 148–153. [CrossRef]

47. Antosz, K.; Stadnicka, D. TPM in large enterprises: Study results. *Int. J. Ind. Manuf. Eng.* **2013**, *7*, 2101–2108.

48. Stadnicka, D.; Antosz, K. Overall equipment effectiveness: Analysis of different ways of calculations and improvements. In *Advances in Manufacturing. Lecture Notes in Mechanical Engineering*; Hamrol, A., Ciszak, O., Legutko, S., Jurczyk, M., Eds.; Springer: Cham, Germany, 2018.

49. Scodanibbio, C. World–Class TPM–How to Calculate Overall Equipment Efficiency. Carlo Scodanibbio 2008/2009. 2009. Available online: www.scodanibbio.com (accessed on 17 August 2020).

50. Antosz, K. *Metodyka Modelowania, Oceny i Doskonalenia Koncepcji Lean Maintenance*; Oficyna Wydawnicza Politechniki Rzeszowskiej: Rzeszów, Poland, 2019.

51. Costa, E.P.; Lorena, A.C.; Carvalho, A.C.P.L.F.; Freitas, A.A. A review of performance evaluation measures for hierarchical classifiers. In *Evaluation Methods for Machine Learning II: Papers from the AAAI-2007 Workshop*; AAAI Press: Palo Alto, CA, USA, 2007; pp. 182–196.

52. Fawcelt, T. An introduction to ROC analysis. *Pattern Recogn. Lett.* **2006**, *27*, 861–874. [CrossRef]

53. Sokolova, M.; Lapalme, G. A systematic analysis of performance measures for classification tasks. *Inf. Process. Manag.* **2009**, *45*, 427–437. [CrossRef]

54. Pasko, Ł.; Setlak, G. Badanie jakości predykcyjnej segmentacji rynku. *Zesz. Nauk. Politech. Śląskiej Ser. Inform.* **2016**, *37*, 83–97.

55. Pawlak, Z. *Rough Sets: Theoretical Aspects of Reasoning about Data*; Theory and Decision Library D; Springer: Dordrecht, The Netherlands, 1991.

56. Beaubouefa, T.; Petryb, F.E.; Arorac, G. Information-theoretic measures of uncertainty for rough sets and rough relational databases. *Inf. Sci.* **1998**, *109*, 185–195. [CrossRef]

57. Bazan, J.G.; Nguyen, H.S.; Nguyen, S.H.; Synak, P.; Wróblewski, J. Rough set algorithms in classification problem. In *Rough Set Methods and Applications: New Developments in Knowledge Discovery in Information Systems, Studies in Fuzziness and Soft Computing*; Polkowski, L., Tsumoto, S., Lin, T.Y., Eds.; Physica-Verlag HD: Heidelberg, Germany, 2000; pp. 49–88.

58. Grzymala-Busse, J.W. A new version of the rule induction system LERS. *Fundam. Inform.* **1997**, *31*, 27–39. [CrossRef]

Publisher's Note: MDPI stays neutral with regard to jurisdictional claims in published maps and institutional affiliations.

 *applied sciences*

Article

Improving a Manufacturing Process Using the 8Ds Method. A Case Study in a Manufacturing Company

Arturo Realyvásquez-Vargas [1], **Karina Cecilia Arredondo-Soto** [2], **Jorge Luis García-Alcaraz** [3,*] and **Emilio Jiménez Macías** [4]

[1] Department of Industrial Engineering, Tecnológico Nacional de México/Instituto Tecnológico de Tijuana, Calzada del Tecnológico S/N, Tijuana 22414, Baja California, Mexico; arturo.realyvazquez@tectijuana.edu.mx

[2] Chemical Sciences and Engineering Faculty, Universidad Autónoma de Baja California, Calzada Universidad #14418, Parque Industrial Internacional, Tijuana 22390, Baja California, Mexico; karina.arredondo@uabc.edu.mx

[3] Department of Industrial and Manufacturing Engineering, Autonomous University of Ciudad Juarez, Ave. del Charro 450 Norte. Col. Partido Romero, Ciudad Juárez 32310, Chihuahua, Mexico

[4] Department of Electrical Engineering, University of La Rioja, Edificio Departamental—C/San Jose de Calasanz 31, 26004 Logroño, La Rioja, Spain; emilio.jimenez@unirioja.es

* Correspondence: jorge.garcia@uacj.mx; Tel.: +52-656-6884843 (ext. 5433)

Received: 31 January 2020; Accepted: 30 March 2020; Published: 2 April 2020

Featured Application: The present case study reports the 8D technique applied to a real manufacturing production process. Future applications can be adapted to other manufacturing industries by integrating the most important variables in their own contexts.

Abstract: Customer satisfaction is a key element for survival and competitiveness in industrial companies. This paper describes a case study in a manufacturing company that deals with several customer complaints due to defective custom cable assemblies that are integrated in an engine. The goal of this research is to find a solution to this problem, as well as prevent its recurrence by implementing the eight disciplines (8Ds) method in order to: (1) develop a team, (2) describe the problem, (3) develop an interim containment action, (4) determine and verify root causes, (5) develop permanent corrective actions, (6) define and implement corrective actions, (7) prevent recurrences, and (8) recognize and congratulate teamwork as well as individual contributions. Therefore, a software tool is proposed to conduct a functional test on assembly lines. After the test, the problem was successfully reduced and detected, because from 67 engines that were identified with problems, 51 were redesigned before being sent to customers, consequently decreasing the number of defective products by 75%, whereas the remaining 16 engines were replaced by new engines. In conclusion, the research goal was accomplished, and the 8Ds method proved to be a helpful model with which to increase employees' motivation and involvement during the problem-solving process.

Keywords: 8 disciplines method; custom cable assemblies; defects; functional test; customer satisfaction

---

1. Introduction

In manufacturing industries, waste refers to the activities that consume resources but that do not directly add value to the product or service for the customer [1]. According to the literature review, there are seven categories of waste in manufacturing that negatively affect the quality of products, delivery times, and unit cost [2,3]. These wastes are overproduction, inventory, over-processing, motion, waiting, transport, and defects [4,5]. Regarding the defects, during the manufacturing processes, companies receive material or components from their suppliers. Then, those materials or components are changed to obtain a final product, which must be delivered to customers on time and without defects [6]. However, defects continue being present in the manufacturing industry nowadays. In fact, several authors mention that defects are the main cause of damages in final products or other components [7–10], which represent a critical situation for the industrial and manufacturing sector [11].

Moreover, customer satisfaction is a requirement that must be considered for any distributor business that is intending to remain globally competitive [12,13]. Nevertheless, if managers want to fulfill customer needs, an appropriate product design process must be included [14]. In this sense, one of main customer needs is a non-defective, quality product [15], since product defects lead to customer dissatisfaction, sales decreases, low financial profits, and greater unit costs [16,17]. In order to improve the effectiveness and efficiency of the production process, offer quality products, and avoid the latest problems, manufacturing companies rely on a wide range of methods and techniques for production improvement [18], including the six sigma management philosophy, DMAIC (i.e., define, measure, analyze, improve, and control) [19], process flow charting (PFC) [20], the Deming or PDCA cycle (i.e., plan, do, check, act) [21,22], and the eight disciplines (8Ds) method [23], among others.

Specifically, the 8Ds are focused on: (D1) develop a team, (D2) describe the problem, (D3) develop an interim containment action, (D4) determine and verify root causes, (D5) Choose/verify permanent corrective actions, (D6) implement and validate corrective actions, (D7) prevent recurrences, and (D8) recognize and congratulate teamwork as well as individual contributions, which is a powerful method because it helps with creating appropriate activities in order to identify the root causes of a problem, and provides permanent solutions to eliminate them. In addition, the 8Ds method is a special tool of ISO/TS 16949:2009 that has been broadly applied in automotive industry for service, including the issues concerning supplier qualification confirmation, process deviations, maintenance, customer complaints, and purchases.

The 8Ds method has been adopted widely in the manufacturing world [24]. For instance, several authors have applied it to solve problems of defects. Some of these authors are: Mitreva et al. [25], who applied it for solving a problem in a LED diode that does not perform its function in a circuit board. Likewise, Titu [26] implemented the 8Ds method to reduce complaints about a defective part; consequently, 60 days after corrective actions were implemented, no other product was identified with this type of defect, and customers decided to withdraw the complaint. Additionally, Kumar and Adaveesh [24] conducted a study in a spring and stamping manufacturing plant for solving a high rejection rate (i.e., 17.07%) of valve springs due to defects. In order to solve this problem, the 8Ds method was applied, and as a result the rejection rate decreased significantly in 6 months, by 4.91%.

Research Problem

A maquiladora is a factory that operates under preferential tariff programs established in Mexico that has headquarters in other countries and performs assembly operations with high hand labor required. Materials, assembly components, and production equipment used in maquiladoras are allowed to enter Mexico duty-free. Currently, in Mexico there are 5144 maquiladoras giving 2,678,633 direct jobs. However, Baja California state has 914 (17.76% from national) maquiladoras giving 333,392 direct jobs [27].

Those companies are using several techniques and methodologies for solving manufacturing problems in production lines. This paper reports a case study applied in a manufacturing company located in Tijuana, Mexico, dealing with the manufacturing of electric custom cables. Each cable is tested for quality through a series of computer-assisted programs for a complete inspection. This strategy allows the company to build and maintain long-term relationships with its customers, thereby helping the company reach its goals and be successful. However, the company has lately experienced problematic defects; as a result, customers are complaining due to 67 returned assemblies.

The problem concerns a stepper motor (see Figure 1), one of the main assembly components, which has a part number that will be called part number A. Customers provide the motors to the company, which introduces them into the production process; next, the motor cables are cut at a specific length, and the plate and terminals are riveted; then, the terminals are inserted into connector units in which a functional test is performed; finally, some defects that are found in this assembly process include cable inversion, incorrect cable length, and lack of an ID tag. In order to solve these problems, the 8Ds method is implemented to decrease the rate of defective products, and to increase customer satisfaction. Therefore, the objective of this paper is to prove the efficiency of the 8Ds method through a case study.

Figure 1. Stepper motor.

A case study is conducted because according to Easton [28], the critical realism approach (CRA) states that a single case study research method is enough to generalize theoretical and empirical findings, giving a new, rigorous, and coherent philosophical position that helps develop the theoretical and research process. Similarly, Tsang [29] states that CRA highlights the impacts of a case study on the theoretical process, empirical generalization, and theoretical evidence. Additionally, Tsang presents the fallibility of knowledge, which establishes that all developed theory requires being subjected to empirical evidence and evaluations; in that sense, case studies are appropriate research strategies to illustrate and analyze proposed theories. Therefore, only one case study is enough to generalize results [30]. Recently, several case studies in the manufacturing sector have been published in journals with a high impact factor. These case studies include the application of methodologies such as value-stream mapping [31,32], the plan-do-check-act (PDCA) cycle [22], lean six sigma [19], and standardized work [33], to mention few.

Specifically, this research implements the single case study approach, since the main contribution is that it allows generalizing the positive impact of the 8Ds methodology on defect reduction in the manufacturing processes with a single case study, which is supported by the CRA. Then, this paper contributes to illustrating how a single and easy technique can be applied for improving a production system in the maquiladora industry.

The rest of the paper is organized into five sections: Section 2 reports the literature review about the 8Ds method and its successful implementations from case studies; Section 3 addresses a description of materials and methods that are implemented in the present case study; Section 4 shows the findings obtained; and finally, Section 5 presents the conclusions and industrial implications regarding the 8Ds implementation.

2. Literature Review

The 8Ds is a teamwork-oriented problem-solving method that aims at identifying the root cause of a problem to solve it through a corrective-action-guided procedure [23]. From a business perspective, the 8Ds method seeks to find the main problems' root causes, identify their possible solutions, and assess their impacts on companies [34]. Originally, the 8Ds method was developed at Ford Motor Company; it was introduced in 1987 to a manual entitled "Team Oriented Problem Solving" (TOPS) [35]. Since then, the method has been applied mainly in automotive industries to solve product and service-related problems, such as defects, customer complaints, manufacturing process deviations, returned purchases, poor machinery maintenance, and supplier qualification issues, among others [34,35].

According to Chelsom et al. [36] and Vargas [37], the 8Ds method can be applied to any type of problem or activity in order to provide assistance to achieving effective communication among departments that share a common objective. However, the 8Ds method is popularly applied to solve quality problems; it is typically required when at least one of the following events are presented [38]:

- The company receives customer complaints.
- Safety or regulatory issues have been discovered.
- Internal rejects, waste, scrap, underperformance, or test failures occur at abnormal levels.
- Warranty concerns indicate greater-than-expected failure rates.

The literature review mentions several successful case studies wherein the 8Ds method was applied. For instance, Mitreva et al. [25] implemented the 8Ds method to solve the problem of an LED diode that did not perform its function in a circuit board; they reported a decrease of operational defects after its implementation, and an increase the efficiency of software packages in the application of statistical methods and techniques. In the same way, Bremmer [39] applied the 8Ds method and other techniques to analyze Scania's global supply chain; how the company could guarantee the quality of products was demonstrated. As a result, this author found the problem and its root causes.

Similarly, Pacheco-Pacheco [40] sought to optimize delivery times of alteration clothing (Alto de basta and Alto de camisa) products in a tailor shop by implementing the 8Ds method. It was found that production times decreased by 2.46% in two mix products. In both products, delivery delay times decreased by 33.33%. Finally, Zasadzień [41] employed the 8Ds method to reduce machine downtimes that were caused by bottlenecks. In summary, Table 1 presents the successful case studies wherein the 8Ds method was implemented.

Table 1. Recent case studies applying the 8Ds method.

Author	Implementation of 8Ds	Results
Mitreva et al. [25]	The study applies the 8Ds method to solve the problem of a LED diode that does not perform its function in the circuit board.	Employees' responsibility was improved towards carrying out business processes. Fewer operational defects were shown. Software packages efficiency increase in the application of statistical methods and techniques. Employees' participation increased. Employees' commitment towards quality improvement. Full managerial commitment. Ability to solve problems at all levels increased. Slightly, but significant improvements in the production processes and products. Business processes were optimized. Low organizational job levels were incorporated to the decision-making process.
Bremmer [39]	The research analyzes the Scania's global supply chain and determines how the corporation can guarantee the quality of products by applying 8Ds and other methodologies.	The current production process at Scania is working, but it is requiring some improvements, especially due to the expected growth of the North Bound Flow (NBF).

Table 1. *Cont.*

Author	Implementation of 8Ds	Results
Kumar and Singh [42]	The study explores the hospitality industry of Delhi and Rajasthan, in India. Specifically, the research addresses the issue of employee turnover in the housekeeping department by identifying both causes and solutions with the help of an 8Ds model for problem solving.	In the hospitality industry, the 8Ds method can be positively adopted to solve problems, especially in terms of employee turnover in the housekeeping department.
Zasadzień [43]	The research seeks to solve problems that are identified in the process of railway carriage renovation by implementing the 8Ds method.	The 8Ds method enabled to identify causes of problems in the railway repair process, as well as allowed the author to develop improvement actions, which considerably streamlined the analyzed process.
Mitreva et al. [44]	This work analyzes the quality assurance system of an automotive company to determine its efficiency. Specifically, the authors studied the company's business process management strategies (identification, documentation, and control), as well as verified whether the system's efficiency documentation had been properly developed or not.	The quality and a better productivity at the lowest costs in operation were defined.
Titu [26]	This study relies on the 8Ds method to solve the complaint about a defective part. The study takes place in SC COMPA S.A., a company based in Sibiu, Romania.	60 days after corrective actions were implemented, there were no other pieces identified with this type of defect. Thus, the customer decided to withdraw complaints.
Fuli et al. [45]	The research develops a quality improvement procedure for automotive companies based on quality management practices. The 8Ds method and the Six Sigma pilot programs were implemented.	The results indicated that the proposed procedure is effective among the studied in Chinese and South African automotive industries.
Nicolae et al. [46]	This work proposes a solution to decrease the response time for the 8Ds method by: (a) warning workstations and warehouses about the appearance of a customer complaint, as well as (b) using a software program for the computerized management of some documents that are needed for the 8Ds analysis.	There are some of the main results: a decreased in the communication time between the quality teamwork and the staff in the manufacturing process, since when a customer complaint is received, it is solved. A faster process of collecting information on manufacturing processes during the 8Ds analysis. A better quality of information that can lead to the resolution of non-compliance was obtained. Less 8Ds analysis time, especially in the first phase of the method. A brand-new customer interface that informs customers about the problem-solving steps that are being taken. The platform is more consistent with the common guidelines for reporting 8Ds analyses.
Kumar and Adaveesh [24]	The six-month study was conducted in a spring and stamping company. The research found a high rejection rate (i.e., 17.07%) of valve springs due to defects. Thus, the 8Ds method was implemented to reduce the rate in 4.91%.	The product rejection rate decreased significantly in 6 months: from 17.07%, in January 2014, to 4.91%, in July 2014.

Table 1. *Cont.*

Author	Implementation of 8Ds	Results
Roque and Berenice [47]	This work relies on the 8Ds method to design and implement new processes for manufacturing dental units with current technology for a company named Briggith. The goal was to ensure the company's subsistence in the current market.	The standardization of raw materials and variables that intervene in the process was possible. Design of new, lighter, and modern structures. Design of an overall electronic control method for the variables identified in the production process. Compliance with quality standards established in the project.
Škůrková [48]	The research focuses on reducing scrap costs in an industrial company. The author implemented a series of methodologies, including the 8Ds method.	Causes of scrap costs were identified, and corrective actions were taken to reduce such costs.
Wichawong and Chongstitvatana [49]	The research introduces a knowledge management system for failure analysis of hard disks that applies a case-based reasoning. The 8Ds method was implemented for problem solving to design a document template.	The document template was successfully designed. The system reported a high customer satisfaction rate, as well as searching effectiveness was acceptable. In summary, the system was successful.
Vargas [37]	This work implements the 8Ds method to solve the problem of sudden stoppages in a continuous vacuum batch cooker that is used in a Brazilian sugar and alcohol company.	An effective method combined with quality tools for detecting and solving the problem and eliminating its recurrence was implemented. The application of the 8Ds method increased the company's performance, as well as and contributed to the continuous improvement of its production process. However, the method could also be applied in other type of processes to increase the company's competitiveness in terms of quality and safety.
Zasadzień [41]	The study implements a quality engineering method to improve the company's maintenance processes in a Silesian production plant. Specifically, the research implements the 8Ds method to reduce machine downtimes caused by bottlenecks.	Machine downtimes caused by bottlenecks were significantly reduced.
Pachecho-Pacheco [40]	The research seeks to optimize delivery times of alteration clothing (Alto de basta and Alto de camisa) products in a tailor shop by implementing the 8Ds method.	Production times decreased by 2.46%, for Alto de basta products (i.e., from 13.30 min to 12.98 min), and by 21.16% for Alto de camisa products (i.e., from 8.49 min to 6.69 min). In both products, delivery delay times decreased by 33.33%: from 3 days to 2 days.

Although the 8Ds method is flexible—it can be adapted to different situations—and has several successful applications, it has some disadvantages, such as [50]:

- It can be time consuming and difficult to develop.
- Employees that are involved in its implementation should receive appropriate training about it.
- Constant communication among the participants and the application of a continuous improvement program are required.

3. Materials and Methods

In order to conduct the present case study, the following materials were used: Microsoft Excel® spreadsheets [51], AutoCAD® [52], Visual Basic® [53] software programs, a PDCA form, and a visual aid form. As for the methodology, the 8Ds method was applied and its steps are presented in Figure 2.

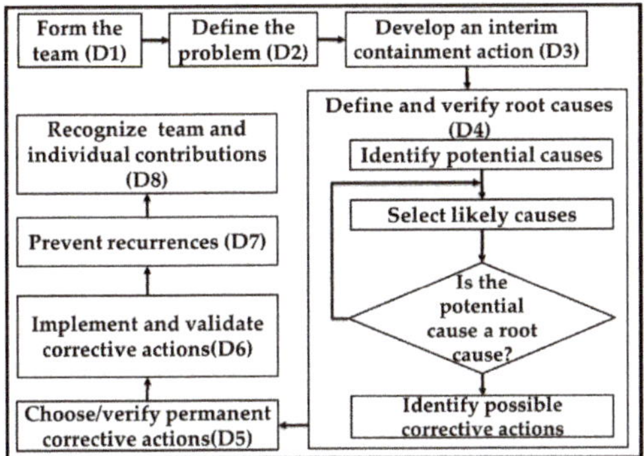

Figure 2. Steps for the 8Ds problem-solving process. Adapted from Joshuva and Pinto [35].

Some similar case studies to this research have used the Kano model as a tool to classify and prioritize customer needs based on how they affect customer satisfaction [54]. However, according to experts, the Kano model has several deficiencies, which discouraged its use in this case study. For instance, it is known that, to conveniently quantify the Kano model, customer satisfaction or dissatisfaction levels toward a product or service must be measured by using the customer satisfaction scale (see Table 2) of positive or negative comments with product or service attributes [55,56]. However, some experts claim that the satisfaction scale is asymmetric, since a positive answer is stronger than a negative answer, which reduces the impact of a negative assessment [54,55].

Table 2. Satisfaction scale of positive and negative comments.

		I Don't Like It	I Can Live with It	I Am Neutral	It Must Be This Way	I Like It Very Much
Product or service attribute	Without the attribute	1	0.5	0	−0.25	−0.5
	With the attribute	−0.5	−0.25	0	0.5	1

Another inconvenience with the Kano model is that it does not consider customer perceptions towards a product or service attributes. Particularly, it provides limited decision support for designers [57], and it is administered through a reduplicative survey, which is time-consuming. In addition, the classification obtained after analyzing the survey results is based merely on subjective assessments; therefore, it may be biased. Finally, it has been claimed that Kano's different classification schemes may influence resource allocation and product design strategy, not only customer satisfaction, and it inherently emphasizes customer and market perspectives, but does not consider the capacity of the company [54,57].

An alternative to the Kano model is the 8Ds method, which relies on facts rather than opinions [37,58]. Specifically, the 8Ds method adopts an objective approach, whereas the Kano model is based on a subjective approach. In this case study, the 8Ds method is applied to solve the identified problem.

3.1. Develop a Teamwork (D1)

Proper planning will always guarantee a better start; therefore, the following criteria should be applied before integrating 8Ds teamwork [38]:

- Collect information regarding symptoms, such as the ID number and description of the claimed part, failure date, customer and supplier numbers, and a short, descriptive analysis of the problem [39,59,60].

- Use a symptoms checklist to ask the correct questions.
- Identify the need for an emergency response action (ERA), which protects customers from further exposure to undesired symptoms.

Moreover, the 8Ds method involves organizing a cross functional teamwork that must have enough knowledge about the product/process to successfully deal with customer complaints or quality deviations in the problem-solving phase [23,35]. Additionally, the teamwork must be interdisciplinary—integrated by operators from several departments (i.e., manufacturing, engineering, and marketing) and different knowledge fields to create a solid task force [61], because the experience of the members is a key element to implementing any problem-solving method [62].

In addition, a teamwork leader is assigned, who ensures that all activities are being carried out and the 8Ds report is regularly updated. Additionally, there should be a champion; this is a person in a management position with enough authority to assist and lead the teamwork when it encounters difficulties or in case additional resources are required [59]. Similarly, any permanent solution may require subsequent teamwork involvement [36]. Based on these facts, manufacturing companies employ hundreds, or even thousands of people with different types of skill sets, ideas, and values, who must be useful for the company.

3.2. Describe the Problem (D2)

This step involves explaining the problem that affects quality or does not meet customer satisfaction [23]. The problem should be explained in detail, identifying in quantifiable terms the who, what, when, where, why, how, and how many problems are involved in the problem (i.e., 5W+2H) [35].

3.3. Develop an Interim Containment Action (D3)

Since 8Ds teamwork members have enough knowledge on the product/process, possible corrective actions must be undertaken in order to control the problem and avoid its expansion. Teamwork members should define and implement those intermediate actions that will protect the customer from the problem until permanent corrective actions are implemented. Additionally, interim containment actions should follow the ISO/TS 16949:2009 quality system and rely on the current approach to appropriately determine and verify the effectiveness of these actions. (ISO/TS 16949:2009 is a technical specification which defines the quality management system requirements for the design, development, production, relevant installation, and service of automotive-related products [23]). In addition, this step is aimed to preserve evidence and stop the outcome from being irremediably enlarged before the problem can be solved and the goal achieved. Some tasks must be monitored to ensure compliance with the requirements, such as documenting, control planning, scheduling, and assigning the specific needs according to the problem that is being solved [23].

3.4. Define and Verify Root Causes (D4)

This step refers to identification of all the applicable causes that could explain why the problem occurred, as well as the reasons why the problem was not perceived the first time it occurred. All causes shall be verified or proved, and not determined by assumptions. Experts recommend using the Ishikawa's five-whys diagrams to map causes against the identified effect [35]. The 5W2H method is used to make diagrams about customer requirements, review the problem-solving process, and analyze the problem [23].

3.5. Develop Permanent Corrective Actions (D5)

Depending on the different causes of the problem, several suitable strategies ought to be proposed. Therefore, either results must be reviewed and the required adjustments have to be made, or some permanent corrective actions must be taken [23]. Finally, a quantitative method ought to be performed through pre-production programs to confirm that the selected corrections will solve the problem [35].

3.6. Implement and Validate Corrective Actions (D6)

In this step, the best corrective actions are defined and implemented to ensure that the target is reached and the problem is solved. In addition, it is necessary to control or monitor any potential effects [23,34,35].

3.7. Prevent Recurrences (D7)

In this step, management systems, operation systems, practices, and procedures should be modified and controlled to prevent their recurrence or any other similar problems, avoiding customer complaints [35].

3.8. Recognize and Congratulate Teamwork as Well as Individual Contributions (D8)

Finally, in this step, the problem is solved; therefore, the knowledge and results are shared. Additionally, the collective efforts from team members are recognized, providing positive feedback and being formally recognized. Training and education records are established and the plan-do-check-act (PDCA) cycle is followed to attain higher customer satisfaction [23,34,35].

The 8Ds method has been successfully implemented in a wide range of case studies across multiple settings. Table 1 presents a recent literature review conducted on the practical applications of the 8Ds method.

3.9. Supplementary Tools in 8D Method

3.9.1. Ishikawa Diagram

The Ishikawa diagram is also known as a cause–effect diagram, fishbone diagram, or root cause analysis diagram, and was developed by Kaoru Ishikawa in the 1960s [63,64]. It helps to visualize a problem and categorize its root causes; it is considered as one of the seven basic quality management tools. The head of the diagram lists the problem to be studied, whereas the fish bones are arrows connected to the spine that list the causes that contribute to the problem. The arrows are interpreted as causal relationships.

According to Da Fonseca et al. [65], the diagram ramifications represent the possible sources of the problem that are related to some factors, such as materials work methods, workforce, measurements, machinery/equipment, and environment. The Ishikawa diagram offers multiple advantages, among which the following can be highlighted [66]. It:

- Classifies all causes that are related to a problem.
- Shortens a relatively large problem.
- Encourages the participation of all the teamwork members in the analysis and creation of project management dynamics.
- Increases the role of teamwork in the problem-solving process.
- Identifies the areas that require more in-depth research when some information is missing.
- Provides elements to develop an adequate solution to a problem.
- Offers a concise view of cause-and-effect relationships.

In this case study, an Ishikawa diagram is designed to find the root causes of the problem. For instance, it is supposed that there is an absenteeism problem in a manufacturing company; therefore, managers want to know the different causes of this problem that are related to the factors previously mentioned. Once the causes of the problem are identified, they are categorized by their factors, as shown in Figure 3.

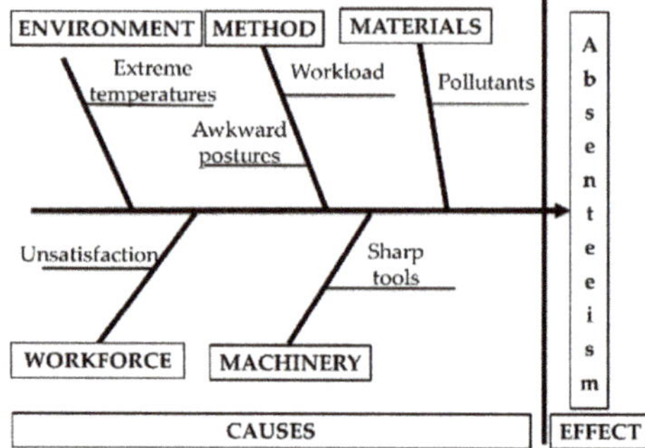

Figure 3. Ishikawa diagram to illustrate the causes of an absenteeism problem.

3.9.2. Pareto Chart

The Pareto chart is a special type of bar graph in which each bar represents a different category or part of a problem [67]. It was developed by the Italian scientist Wilfredo Pareto, who found that 80% of the wealth was held by 20% of the people in Italy [68]. The Pareto chart illustrates the frequency distribution of descriptive data that are classified into categories. The categories are placed on the horizontal axis, whereas the frequencies are placed on the vertical axis [67,68]. The categories are arranged in a descending order, from left to right, while a line represents the frequencies in cumulative percentage. The highest bars of the chart represent the categories that contribute the most to the problem.

Furthermore, Pareto charts help identify how certain factors influence on a problem along with other factors; in other words, Pareto charts help identify the best opportunities for improvement [69]. Experts recommend using Pareto charts for two particular purposes: to decompose a problem into categories or factors and to identify the key categories that contribute the most to a specific problem [67]. For instance, continuing with the example of absenteeism in a manufacturing company, the six causes shown in Figure 3 were ordered according to their frequencies, as shown in Figure 4. Based on this order, managers should try to eliminate the first three causes (extreme temperature, sharp tools, and workload), since they represent the 80.47% of all causes of absenteeism.

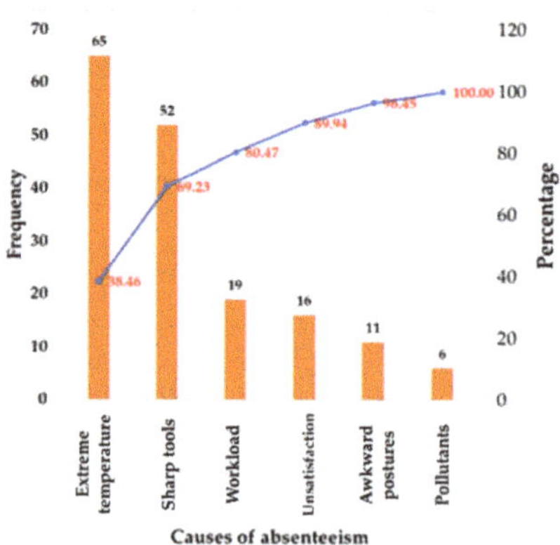

Figure 4. Example of a Pareto chart application for causes of absenteeism.

In the present case study, a Pareto chart is created for a better understanding of the key causes that contribute to the problem of non-working custom cable assemblies.

4. Results

The results obtained for each stage of the 8Ds methodology are shown as follows:

4.1. Develop the Teamwork (D1)

The teamwork included a maintenance engineer, a processes engineer, an intern engineer, a production line manager, and two quality inspectors. The principal teamwork goals were to determine an adequate manufacturing process for part number A and to define the root causes of the defects. In order to achieve these goals, a task was assigned to each teamwork member, as summarized in Table 3. Note that each PDCA cycle step comprised at least one discipline, since the 8Ds method follows the logic of this cycle [50,70]. Additionally, disciplines are assigned to different teamwork members; i.e., no more than one discipline was assigned to more than one member.

Table 3. Task assignment.

The 8Ds Methodology	PDCA Cycle	Teamwork Member
Develop the teamwork (D1)	Plan	Maintenance engineer
Describe the problem (D2)		
Develop an interim containment action (D3)	Do	
Define and verify root causes (D4)		Production line manager
Develop permanent corrective actions (D5)		Intern engineer
Implement and validate corrective actions (D6):	Check	Quality inspector 1
Prevent recurrences (D7):	Act	Quality inspector 2
Recognize and congratulate the teamwork as well as individual contributions (D8):		All involved employees

Once the tasks have been assigned to the teamwork members, they have to implement an efficient communication system to keep each other informed, and as a result, guarantee the involvement of all the members in the problem-solving process. Similarly, a PDCA form was designed on Microsoft Excel® for each teamwork member to report their corresponding tasks from the PDCA cycle.

4.2. Describe the Problem (D2)

As previously mentioned, 67 cable assemblies were returned to the company by customers, who complained about either the product's poor performance or regarding unacceptable features. The main problem was that the assembly did not work; however, that can be due to several types of defects. Table 4 lists the six different types of defects that were found in the cable assemblies.

Table 4. Defects found in the rejected cable assemblies.

Defect	Frequency	Percentage	Cumulative Percentage
Inverted cables	35	52%	52%
Disfigured motor	10	15%	67%
Noisy motor	9	13%	81%
Motor does not work	7	10%	91%
Lack of ID tag	4	6%	97%
Wrong cable length	2	3%	100%
Total	67	100%	

Specifically, the data in Table 4 were used to create a Pareto diagram, as shown in Figure 5. The diagram helped define which problems or defects had to be prioritized, according to their frequencies. In this sense, the most frequent defect was inverted cables, followed by a disfigured motor. Even though both wrong cable length and the lack of an ID tag were less frequent problems, they had to be solved from the root cause as well.

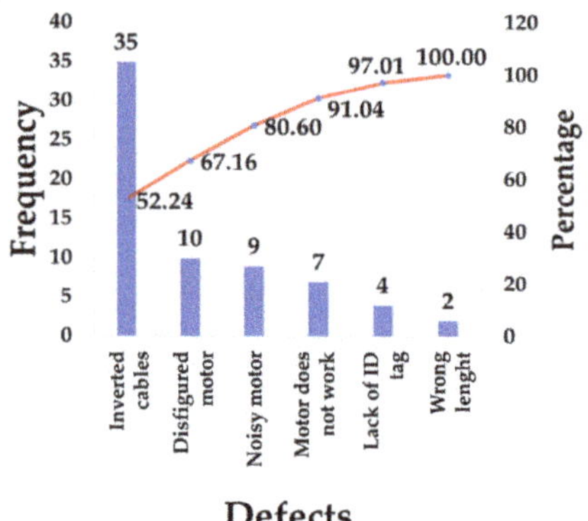

Figure 5. Pareto diagram of cable assembly defects.

4.3. Develop an Interim Containment Action (D3)

Both interim and rapid interventions were implemented to solve most of the six problems, including those concerning inverted cables, disfigured motors, lack of ID tag, and wrong cable length. A series of interim visual aids were developed to help employees assemble the components. Regarding inverted cables and disfigured motors, a document is created to report the conditions of both the stepper motor and the cables before and after being handled by the employee. Additionally, as Figure 6 presents, a provisional sign is created for helping employees to insert the assembly cables not only in the correct positions, but also in the right entry holes by using the colors of the cables as references. Similarly, the sign is intended to help employees guarantee that each cable's final end is the one that is required by customers.

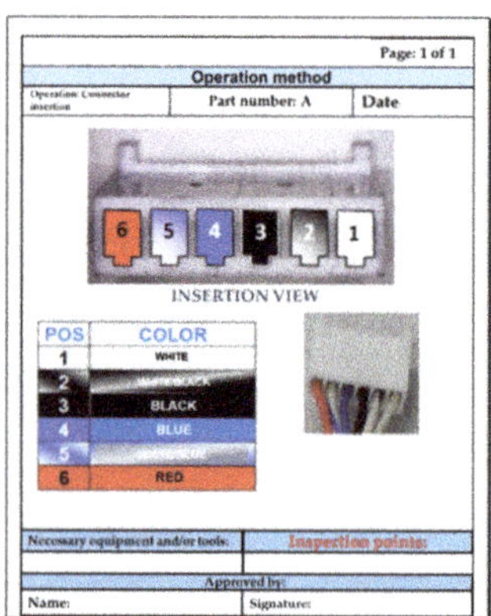

Figure 6. Provisional aid for cable insertion.

Finally, AutoCAD® was used to design a customizable 1:1 scale 2D template of a drawing provided by customers for the assemblies to verify that customers' demands would be accomplished, as shown in Figure 7. Perhaps the greatest advantage of this electronic template is that it can be stored in a database and updated for new specifications (i.e., new cable length) if required. The updates can be performed quickly and effectively without compromising the template function. After implementing this system of solutions (i.e., the spreadsheet, the sign, and the 2D template), it was noticed that the most insignificant errors were immediately fixed; consequently, four of the six problems were solved. In order to confirm this, a quality inspector assessed the assemblies and later confirmed that the problems had been successfully solved.

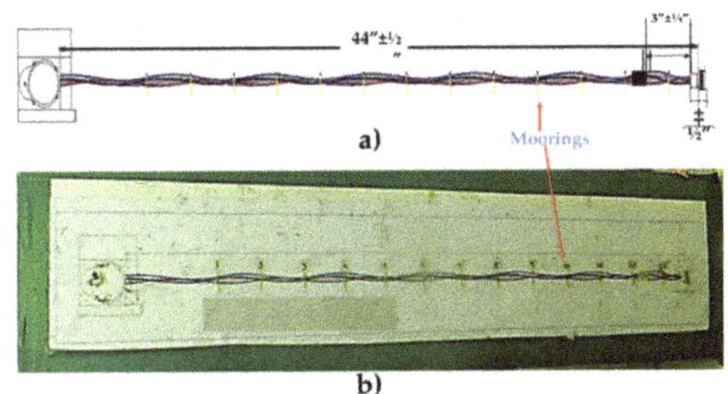

Figure 7. Template of the drawing provided by the customer: (**a**) picture of the template made in AutoCAD; (**b**) picture of the printed template.

4.4. Determine and Verify Root Causes (D4)

This discipline aims to find the root causes of problems. According to Škůrková (2017), cause–effect diagrams can be used to map causes with their corresponding effects or problems. The general problem in this case study is that the assembly does not work; hence, a fishbone diagram is developed—also known as Ishikawa diagram—as depicted in Figure 8 to identify the root cause. As can be observed, several causes were identified across five aspects: materials, methods, environment, workforce, and machinery.

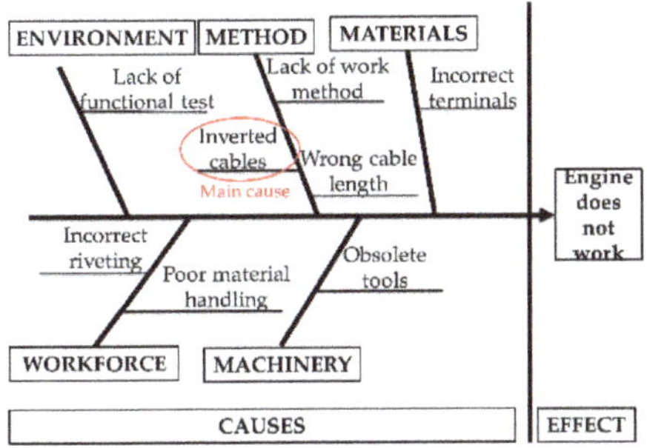

Figure 8. Cause-effect diagram to find out the root cause of the problem.

Regarding the environment, the reason why the returned assemblies were defective is because the company lacked a functional test to confirm that they worked. However, to perform this evaluation, the cables first had to be correctly assembled—and even then, it would have been impossible to know if the assemblies worked properly. As for the materials, it was found that the cable end terminals were incorrect, since the warehouse employees mistakenly provided the wrong types to the operators. Additionally, two more problem causes were associated with the work method. Usually, the motors supplied to the company come with already-integrated cables, and the employees only need to cut these cables as specified by the customers, and then rivet the excess. However, sometimes the cables are not always cut at the right length or inverted.

In terms of machinery, it was found that the tools of the company were obsolete and needed required to be replaced. Finally, regarding the workforce aspect, the diagram indicates that the assembly cables were not always riveted properly, yet correct riveting makes it possible for the motor to be connected to the cables, which in turn enables the functional test to be successfully performed. Similarly, it was found that the employees may poorly handle the motors, and in the case of the rejected assemblies, this could have an impact on their performance. Another possible cause of having defective assemblies is that the motors might have been damaged during their delivery.

Moreover, since most of the assemblies were returned because of inverted cables, this issue is considered as the main root cause of the problem (see Figure 3). In most of the assemblies, the black, white, and blue cables had been inverted. At first, this can be a problem related to the company work method; however, a functional test could have also solved the problem. In addition, with a functional test, the company could have prevented non-working motors and abnormal noise problems. During functional tests, motors usually display a "not working" message, in which case the position of the cables must be thoroughly reviewed. Finally, to prevent the problem from re-occurring, a program on Visual Basic® was developed to conduct motor functional tests (see Appendix A). The test uses binary values (0 and 1) that allow employees to confirm an assembly's functionality before it is delivered to the customer. Figure 9 introduces the truth table for the motor, with values 1 = true (ON) and 0 = false (OFF).

4 White	3 Black	2 Blue	1 Red
1	0	1	0
1	0	0	1
0	1	0	1
0	1	1	0

1= ON, 0 = OFF

Figure 9. Truth values for the motor.

The binary values are translated into decimal values to be used in the program; first a formula table is built, as depicted in Figure 10, wherein each row corresponds to one cable. Then, in each row, the first ten powers of 2 are displayed, i.e., $2^0 = 1, 2^1 = 2, ..., 2^9 = 512$, from right to left, and it is assigned one binary value from Figure 9 corresponding to a power of two, starting at $2^0 = 1$. Finally, each binary value in each row is multiplied by its corresponding power of 2, and the sum of the products is the resulting decimal value that is reported on the right side of each row. Once the four decimal values were obtained, they were used in the program commands to be executed, consequently beginning a new project according to customer specifications.

Row 1

512	256	128	64	32	16	8	4	2	1	= 10
						1	0	1	0	

Row 2

512	256	128	64	32	16	8	4	2	1	= 9
						1	0	0	1	

Row 3

512	256	128	64	32	16	8	4	2	1	= 5
						0	1	0	1	

Row 4

512	256	128	64	32	16	8	4	2	1	= 6
						0	1	1	0	

Figure 10. Translation of binary values to decimal values.

Once the Visual Basic® program was designed, the Parmon's parallel port monitor application was used to verify that the decimal values were correct when the program was executed, as shown in Figure 10. The Dec column contains the decimal values corresponding to the binary values from the binary column. In all the decimal values shown in Figure 11, the motor being tested was turned on. Once the motor finished its cycle, the program indicates that the motor is turning in the opposite direction regarding the position it had started in. The goal of this test is to confirm that the motor works properly without abnormal noise.

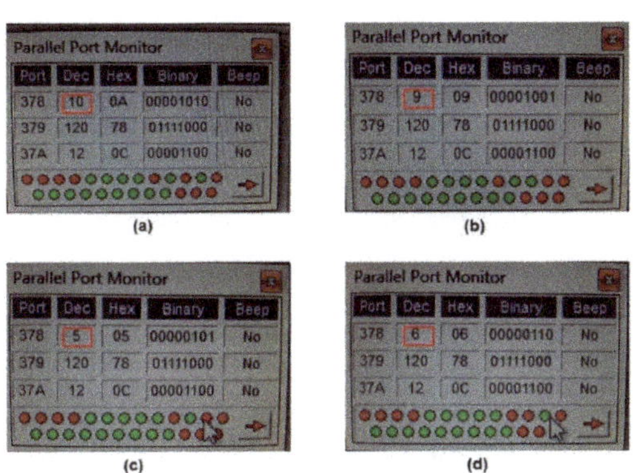

Figure 11. Parmon's parallel port monitor application and motor functioning: (**a**) decimal value 10 enabled, the motor starts turning; (**b**) decimal value 9 enabled, the motor continues its cycle; (**c**) decimal value 5 enabled, the motor continues its cycle; (**d**) decimal value 6 enabled, the motor finishes its cycle.

4.5. Develop Permanent Corrective Actions (D5)

In this discipline, the following corrective actions are implemented:

- Connector insertion method: This action was implemented because inverted cables were the root cause of the problem.
- Functional test: This action was implemented, since the lack of a functional test was one of the reasons why the assemblies had inverted cables. Functional tests can help solve the problems of noisy motors and dysfunctional motors.
- Template: This action was demanded by customers, because the template guarantees that the assembly component features match the customer specifications.

The three corrective actions significantly improved the production system, since they helped solve problems of inverted cables, noisy motors, dysfunctional motors, and wrong component features. Additionally, the brand-new insertion method was added in the datasheet of part number A, and it was stored in an electronic file to be updated when necessary. However, one important factor to consider is that, regardless of whether the motor was properly assembled or not, it was still likely to fail or generate abnormal noise.

4.6. Implement and Validate Corrective Actions (D6)

An operation method for the functional test was developed (see Appendix B). Specifically, each connector being tested only had to be connected to the box containing the driver. The process time established by the customer was 7.28 min, but it is managed to decrease in 4.61 min (i.e., 36.68% less time) after the process was documented and a functional test was conducted. In the end, the operation method helped employees avoid mistakes when assembling the cable. The corrective actions were validated by comparing the analysis results from the defective assemblies before and after implementing these actions. Actually, the defective products decreased by 76%, which validates the implemented corrective actions [24].

4.7. Prevent Recurrences (D7)

The manufacturing process of part number A comprises eight tasks: manual cable cutting, semi-automatic cable riveting, cable end terminal insertion, cable labeling, performing electrical and functional tests, conducting final inspection, packaging, and shipping. Once these tasks were identified, a series of checklists was designed to monitor their successful completion and ensure continuity in the manufacturing process. At the shipping stage, all this documentation was assigned a customer revision number, which would allow the resulting datasheet to be immediately updated as customer specifications change, thereby informing the production, quality, and cutting departments of such updates.

Finally, in this discipline, an executable version of the Visual Basic® program was developed. The program forbid employees from changing any of its settings, since it only allows them to open it and perform the test in a pre-configured mode to prevent misconfiguration problems.

4.8. Recognize Teamwork and Individual Contributions (D8)

In this stage, all the teamwork members were acknowledged for their individual and group performances. Although each member had his/her own ideas, and different suggestions were proposed during the problem-solving process, the teamwork remained united and worked towards a common goal.

5. Conclusions and Industrial Implications

The principal goal of this work was successfully accomplished. The 8Ds method implemented in the manufacturing company managed to decrease the number of assembly defects in part number A from 67 to 16, which represents a decrease of 76.12%. Figure 12 shows a comparison about the frequency of each defect before and after implementing the 8Ds method. Note that the frequency of all defects decreased. For example, the frequency of inverted cables, the most common defect, decreased from 35 to 2. Similarly, the frequency of motor disfigured decreased from 10 to 3, and the noisy motor decreased from 9 to 3, to just mention the higher frequency defects.

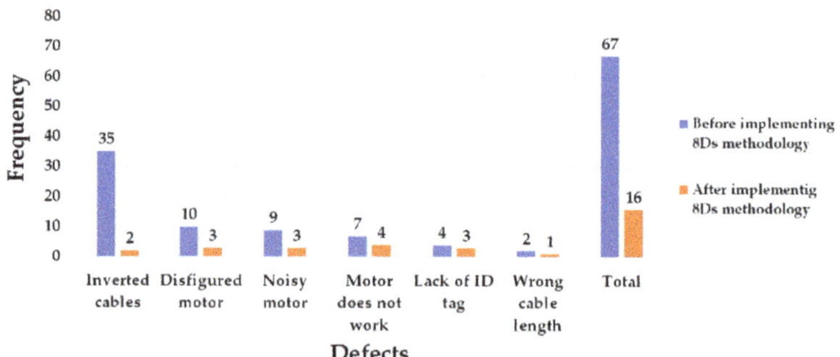

Figure 12. Comparison of the frequency of defects before and after applying the 8Ds method.

Simultaneously, the 8Ds method implementation allowed increasing customer satisfaction. In the 16 case studies reported in Section 3, the 8Ds method was applied to help corporations to comply with delivery times, reduce scrap and defect costs, implement new processes or develop new products, improve quality assurance systems, minimize supply chain and customer complaints, and improve services. However, solving these types of problems involves having a solid and effective communication system among the affected departments, which should also share a common goal.

In addition, by implementing the 8Ds method, the company managed to decrease production time, machine downtimes, scrap costs, operational defects, the rate of late deliveries, and customer complaints. Regarding the manufacturing system, the 8Ds method increased efficiency and productivity in the application of statistical methods and techniques at low operational costs. Table 5 shows a comparison of the main indicators before and after implementing the 8Ds method. It is important to note that the total defects were reduced by 76.12%, while the customer complaints were reduced by 100%. Similarly, production, inspection, and packing times for the part number A were reduced by over 30%; and machines stoppages were reduced by over 77%. This reduction of time cycles allowed for increasing the production level by 34.22%.

Table 5. Comparison of the main indicators before and after applying the 8Ds method.

Indicator	Before Implementing the 8Ds Methodology	After Implementing the 8Ds Methodology	Difference
Total defects	67	16	−76.12%
Time for the production process of part number A	7.28 min	4.61 min	−36.68%
Time for the inspection and packing of part number A	6.5 min	4.28 min	−34.22%
Customer complaints	67	0	−100%
Machines stoppages	155 min/day	35 min/day	−77.42%
Production	850 products/day	1141 product/day	+34.22%

Moreover, the implementation of the 8Ds method had a positive impact on the company's competitiveness in terms of quality and safety. Furthermore, the 8Ds method had a significantly positive effect on employees and managerial responsibility, participation, and commitment, which streamlined and improved the company problem-solving process, especially by helping delegate equal responsibilities to the lowest organizational levels. Finally, the 8Ds method implementation allows collecting information concerning a problem in a quick manner, and reduces the communication time between the quality teamwork and operators.

When problems arise, a method, technique, or abstract tool ought to be implemented to find the best solution. On some occasions, the implementation process may require making small modifications in the organization, whereas in other cases, engineers must be more careful to spare the company losses. Additionally, in the implementation of any method, communication is a key element of success. A solid, rapid, and effective communication system encourages employees to be creative and be engaged in the problem-solving process and motivates employees to be prepared for any further change. In other words, the 8Ds method has a two-fold goal: to solve problems and to increase active employee participation in the problem-solving process. In order to achieve these goals, experts recommend the following strategies:

- Implement the 8Ds method to solve problems with other part numbers, and/or in other areas (purchase or sales, for instance).
- Always consider each employee's opinion, since it will make their work motivating.
- Engage customers' opinions and ideas to improve both the production processes and their satisfaction.

As future work and based on the findings obtained in the present case study, the authors of this research plan to implement the 8Ds method in some companies from the 914 manufacturing industries located in Baja California state to solve problems related to defective products and/or production process efficiency. Additionally, the authors plan to extend the 8Ds method implementation, as well as other industrial engineering tools (PDCA cycle, standardized work, poka-yoke, DMAIC, to mention few) not only to companies in the manufacturing sector, but also in another sectors, such as construction, education, agriculture, and food services.

Finally, the authors encourage researchers from the industrial engineering field to publish their case studies on the applications of different techniques, methods, or tools, supported by the CRA.

Author Contributions: Conceptualization, A.R.-V. and J.L.G.-A.; data curation, A.R.-V. and K.C.A.-S.; formal analysis, J.L.G.-A. and E.J.M.; funding acquisition, A.R.-V.; investigation, A.R.-V. and K.C.A.-S.; methodology, J.L.G.-A. and E.J.M.; project administration, A.R.-V. and J.L.G.-A.; validation, J.L.G.-A.; visualization, E.J.M.; writing—original draft, A.R.-V.; writing—review and editing, J.L.G.-A. and E.J.M. All authors have read and agreed to the published version of the manuscript.

Funding: This research has received no external funding.

Acknowledgments: The authors would like to acknowledge the manufacturing company where the 8Ds method was implemented. Additionally, the authors would like to thank the Tijuana Institute of Technology, the Autonomous University of Baja California, the Autonomous University of Ciudad Juarez, and the University of La Rioja for allowing the use of their facilities for this research. Finally, the authors would like to thank CONACYT and PRODEP for their constant support toward develop projects and research.

Conflicts of Interest: The authors declare that there is no conflict of interest.

Appendix A. Code for the Program on Visual Basic

The next step involved introducing the following command:

Command for the input variable:

Private Declare Function Inp Lib "inpout32.dll" _

Alias "Inp32" (ByVal PortAddress As Integer) As Integer

Command for the output variable:

Private Declare Sub Out Lib "inpout32.dll" _

Alias "Out32" (ByVal PortAddress As Integer, ByVal Value As Integer)

Command to tell the program that a delay function exists in milliseconds:

Private Declare Sub Sleep Lib "kernel32" (ByVal dwMilliseconds As Long)

The following instructions are given to the MOTOR TURNS TO THE LEFT button.

Private Sub Command2_Click()

MsgBox ("BE SURE THAT THE MOTOR IS TURNING COUNTER CLOCKWISE. PRESS OK TO START")

Dim x As Integer

For x = 10 To 500

Sleep 200

Out &H378, 6

Sleep 200

Out &H378, 5

Sleep 200

Out &H378, 9

Sleep 200

Out &H378, 10

Sleep 200

Next x

MsgBox ("END OF TEST TO THE LEFT")

End Sub

Now, instructions are given to the MOTOR TURNS TO THE RIGHT button.

Private Sub Command4_Click()

MsgBox ("BE SURE THAT THE MOTOR IS TURNING CLOCKWISE. PRESS OK TO START")

Dim x As Integer

For x = 10 To 500

Sleep 200

Out &H378, 10

Sleep 200

Out &H378, 9

Sleep 200

```
Out &H378, 5
Sleep 200
Out &H378, 6
Sleep 200
Next x
MsgBox ("END OF TEST TO THE RIGHT")
End Sub
```

Finally, instructions are given for the EXIT TEST button.

```
Private Sub Command3_Click()
MsgBox ("ARE YOU SURE YOU WANT TO EXIT?")
End
End Sub
```

Appendix B

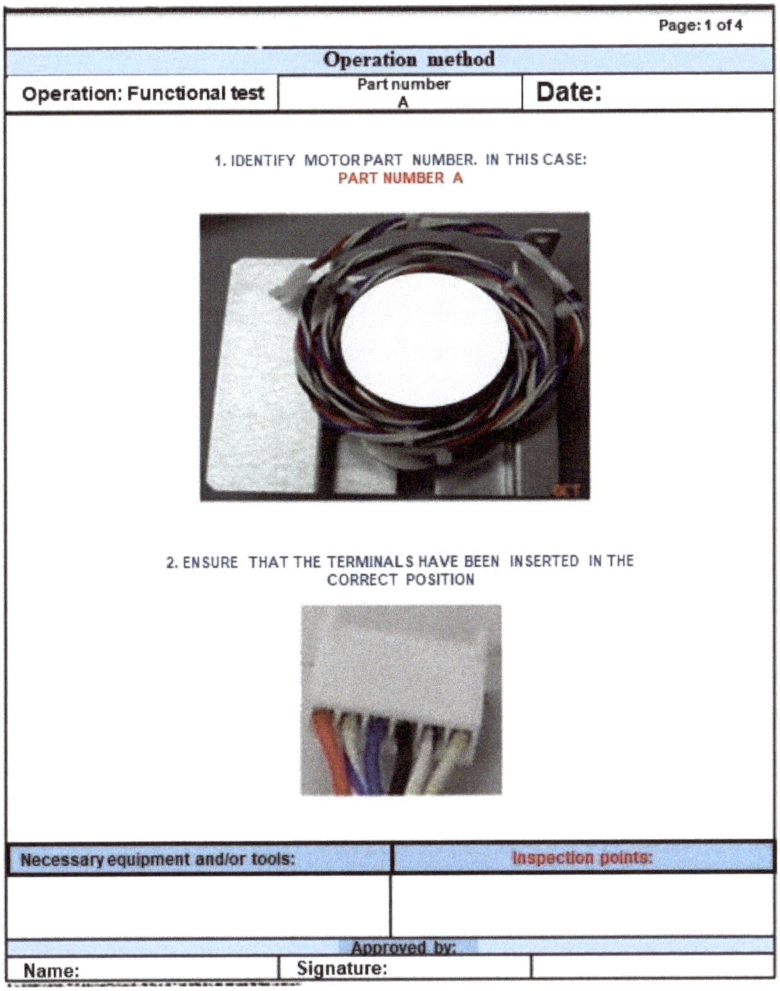

Page: 1 of 4

Operation method

Operation: Functional test	Part number A	Date:

1. IDENTIFY MOTOR PART NUMBER. IN THIS CASE: PART NUMBER A

2. ENSURE THAT THE TERMINALS HAVE BEEN INSERTED IN THE CORRECT POSITION

Necessary equipment and/or tools:	Inspection points:

Approved by:

Name:	Signature:

Figure A1. First Visual Aid to Conducting the Functional Test.

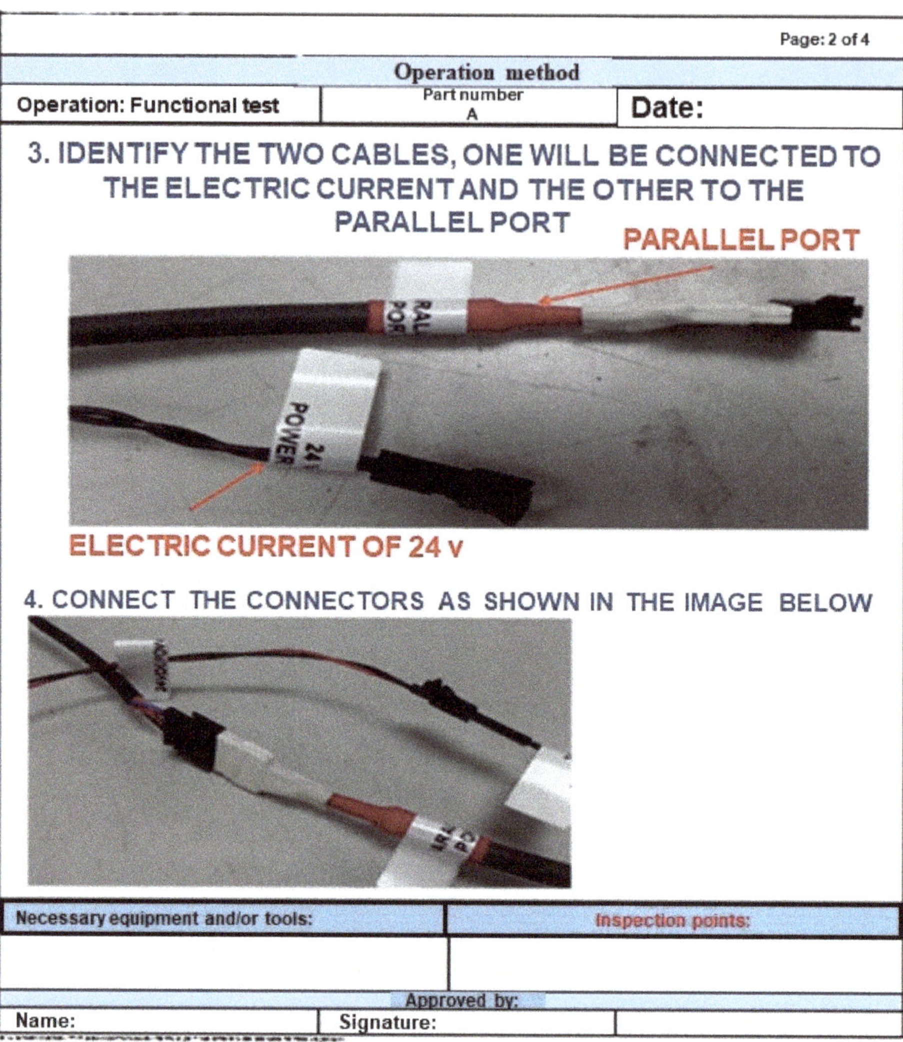

Figure A2. Second Visual Aid to Conducting the Functional Test.

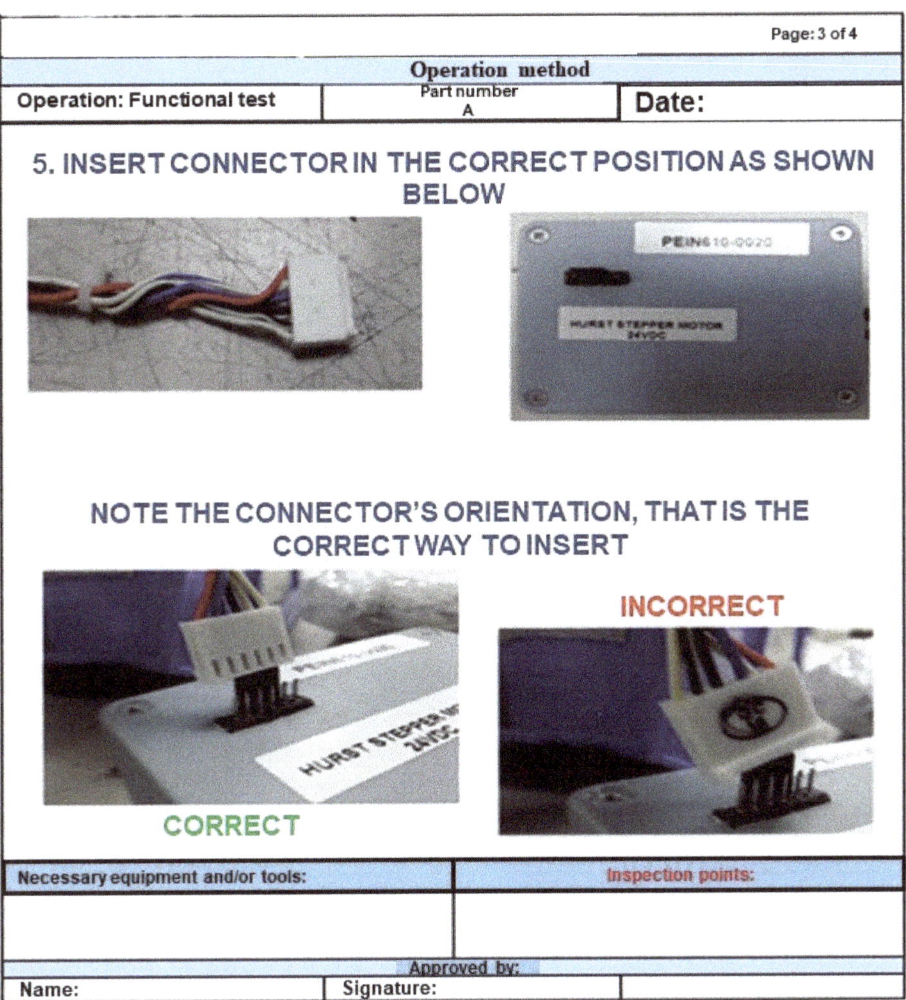

Figure A3. Third Visual Aid to Conducting the Functional Test.

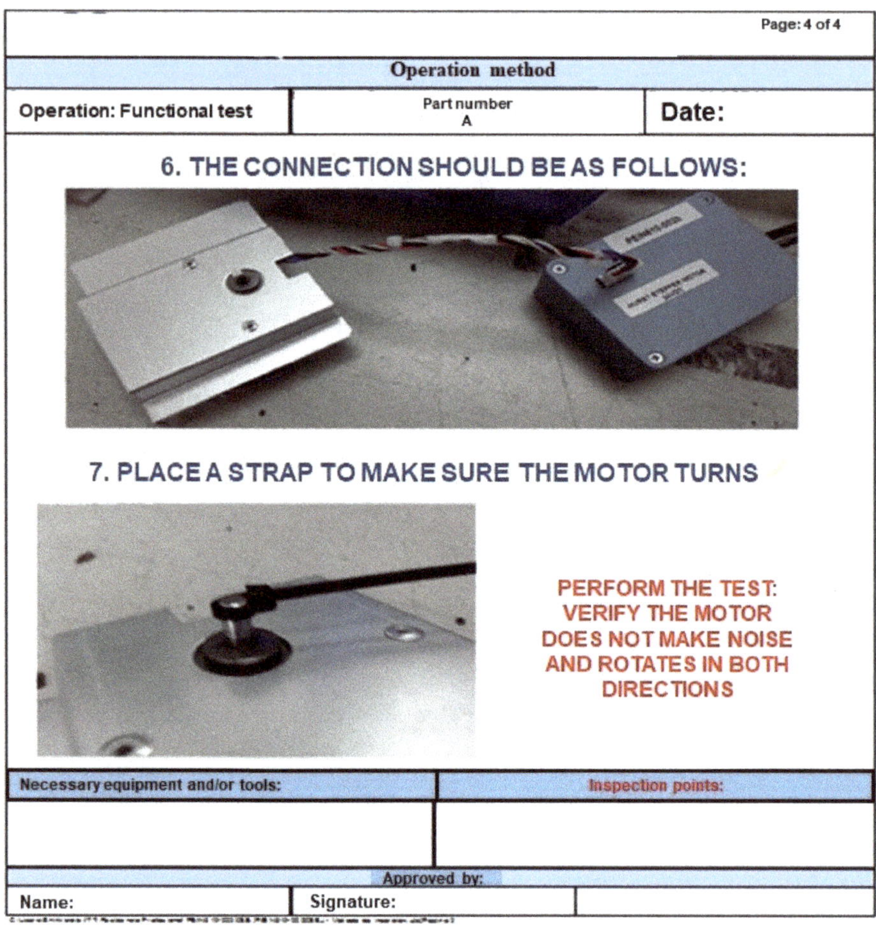

Figure A4. Fourth Visual Aid to Conducting the Functional Test.

References

1. Razali, N.M.; Rahman, M.N.A. Value Stream Mapping—A Tool to Detect and Reduce Waste for a Lean Manufacturing System. In *Proceedings of the 4th International Manufacturing Engineering Conference and the 5th Asia Pacific Conference on Manufacturing Systems*; Osman-Zahid, M., Abd.-Aziz, R., Yusoff, A., Mat-Yahya, N., Abdul-Aziz, F., Yazid-Abu, M., Eds.; Springer: Singapore, 2020; pp. 266–271. ISBN 978-981-15-0949-0.
2. Botti, L.; Mora, C.; Regattieri, A. Integrating ergonomics and lean manufacturing principles in a hybrid assembly line. *Comput. Ind. Eng.* **2017**, *111*, 481–491. [CrossRef]
3. Walder, J.; Karlin, J.; Kerk, C. Integrated lean thinking & ergonomics: Utilizing material handling assist device solutions for a productive workplace. In *MHIA White Paper: Material Handling Industry of America*; South Datota School of Mines: Charlote, NC, USA, 2007; pp. 1–18.
4. El-Namrouty, K.A.; Abushaaban, M.S. Seven wastes elimination targeted by lean manufacturing case study "gaza strip manufacturing firms". *Int. J. Econ. Financ. Manag. Sci.* **2013**, *1*, 68–80. [CrossRef]
5. Jadhav, P.K.; Nagare, M.R.; Konda, S. Implementing Lean Manufacturing Principle In Fabrication Process- A Case Study. *Int. Res. J. Eng. Technol.* **2018**, *5*, 1843–1847.
6. Guiras, Z.; Turki, S.; Rezg, N.; Dolgui, A. Optimization of Two-Level Disassembly/Remanufacturing/Assembly System with an Integrated Maintenance Strategy. *Appl. Sci.* **2018**, *8*, 666. [CrossRef]

7. Zhou, X.-Y.; Gosling, P.D. Influence of stochastic variations in manufacturing defects on the mechanical performance of textile composites. *Compos. Struct.* **2018**, *194*, 226–239. [CrossRef]
8. Buixansky, B.; Fleck, N.A. Compressive failure of fibre Composites. *J. Mech. Phys. Solids* **1993**, *41*, 183–211.
9. Pinho, S.T.; Iannucci, L.; Robinson, P. Physically-based failure models and criteria for laminated fibre-reinforced composites with emphasis on fibre kinking: Part I: Development. *Compos. Part A Appl. Sci. Manuf.* **2006**, *37*, 63–73. [CrossRef]
10. Gommer, F.; Endruweit, A.; Long, A.C. Quantification of micro-scale variability in fibre bundles. *Compos. Part A Appl. Sci. Manuf.* **2016**, *87*, 131–137. [CrossRef]
11. Sreedharan, R.; Rajasekar, S.; Kannan, S.S.; Arunprasad, P.; Trehan, R. Defect reduction in an electrical parts manufacturer: A case study. *TQM J.* **2018**, *30*, 650–678. [CrossRef]
12. González-Reséndiz, J.; Arredondo-Soto, K.C.; Realyvásquez-Vargas, A.; Híjar-Rivera, H.; Carrillo-Gutiérrez, T.; González-Reséndiz, J.; Arredondo-Soto, K.C.; Realyvásquez-Vargas, A.; Híjar-Rivera, H.; Carrillo-Gutiérrez, T. Integrating Simulation-Based Optimization for Lean Logistics: A Case Study. *Appl. Sci.* **2018**, *8*, 2448. [CrossRef]
13. Pérez-Domínguez, L.; Luviano-Cruz, D.; Valles-Rosales, D.; Hernández Hernández, J.; Rodríguez Borbón, M. Hesitant Fuzzy Linguistic Term and TOPSIS to Assess Lean Performance. *Appl. Sci.* **2019**, *9*, 873. [CrossRef]
14. Wang, C.-H. Incorporating customer satisfaction into the decision-making process of product configuration: A fuzzy Kano perspective. *Int. J. Prod. Res.* **2013**, *51*, 6651–6662. [CrossRef]
15. Abolhassani, A.; Harner, J.; Jaridi, M.; Gopalakrishnan, B. Productivity enhancement strategies in North American automotive industry. *Int. J. Prod. Res.* **2018**, *56*, 1–18. [CrossRef]
16. Saeidi, S.P.; Sofian, S.; Saeidi, P.; Saeidi, S.P.; Saaeidi, S.A. How does corporate social responsibility contribute to firm financial performance? The mediating role of competitive advantage, reputation, and customer satisfaction. *J. Bus. Res.* **2015**, *68*, 341–350. [CrossRef]
17. Galbreath, J. Twenty-first century management rules: The management of relationships as intangible assets. *Manag. Decis.* **2002**, *40*, 116–126. [CrossRef]
18. Liu, A.X.; Liu, Y.; Luo, T. What Drives a Firm's Choice of Product Recall Remedy? The Impact of Remedy Cost, Product Hazard, and the CEO. *J. Mark.* **2016**, *80*, 79–95. [CrossRef]
19. Wang, C.-N.; Chiu, P.-C.; Cheng, I.-F.; Huang, Y.-F. Contamination Improvement of Touch Panel and Color Filter Production Processes of Lean Six Sigma. *Appl. Sci.* **2019**, *9*, 1893. [CrossRef]
20. Bunce, M.M.; Wang, L.; Bidanda, B. Leveraging Six Sigma with industrial engineering tools in crateless retort production. *Int. J. Prod. Res.* **2008**, *46*, 6701–6719. [CrossRef]
21. Dudin, M.N.; Frolova, E.E.; Gryzunova, N.V.; Borisovna, E.S. The Deming Cycle (PDCA) Concept as an Efficient Tool for Continuous Quality Improvement in the Agribusiness. *Asian Soc. Sci.* **2015**, *11*, 239–246. [CrossRef]
22. Realyvásquez-Vargas, A.; Arredondo-Soto, K.C.; Carrillo-Gutiérrez, T.; Ravelo, G. Applying the Plan-Do-Check-Act (PDCA) Cycle to Reduce the Defects in the Manufacturing Industry. A Case Study. *Appl. Sci.* **2018**, *8*, 2181. [CrossRef]
23. Cheng, H.-R.; Chen, B.-W. A case study in solving customer complaints based on the 8Ds method and Kano model. *J. Chin. Inst. Ind. Eng.* **2010**, *27*, 339–350.
24. Kumar, S.; Adaveesh, B. Application of 8D Methodology for the Root Cause Analysis and Reduction of Valve Spring Rejection in a Valve Spring Manufacturing Company: A Case Study. *Indian J. Sci. Technol.* **2017**, *10*, 1–11. [CrossRef]
25. Mitreva, E.; Taskov, N.; Gjorshevski, H. Methodology for design and implementation of the TQM (Total Quality Management) system in automotive industry companies in Macedonia. In Proceedings of the Regionalna Naucno Strucna Konferencija—ERAZ 2015—Odrzivi Ekonomski Razvoj-Savremeni i Multidisciplinarni Pristup, Belgrade, Serbia, 9–11 June 2015; pp. 1–9.
26. Titu, M.A. Management of customers' complaints within SC COMPA SA Sibiu. Case study for a topical complaint. *J. Electr. Eng. Electron. Control Comput. Sci.* **2016**, *2*, 1–4.
27. Instituto Nacional de Estadística Geografía e Informática (INEGI) INEGI. Monthly Survey of Manufacturing (Emim). Available online: http://www.inegi.org.mx/sistemas/bie/default.aspx?idserPadre=10400100 (accessed on 18 March 2020). (In Spanish)

28. Easton, G. Critical realism in case study research. *Ind. Mark. Manag.* **2010**, *39*, 118–128. [CrossRef]
29. Tsang, E.W.K. Case studies and generalization in information systems research: A critical realist perspective. *J. Strateg. Inf. Syst.* **2014**, *23*, 174–186. [CrossRef]
30. Easton, G. *One Case Study is Enough*; The Department of Marketing: Lancaster, TX, USA, 2010.
31. Acero, R.; Torralba, M.; Pérez-Moya, R.; Pozo, J.A. Value Stream Analysis in Military Logistics: The Improvement in Order Processing Procedure. *Appl. Sci.* **2020**, *10*, 106. [CrossRef]
32. Pérez-Pucheta, C.E.; Olivares-Benitez, E.; Minor-Popocatl, H.; Pacheco-García, P.F.; Pérez-Pucheta, M.F. Implementation of Lean Manufacturing to Reduce the Delivery Time of a Replacement Part to Dealers: A Case Study. *Appl. Sci.* **2019**, *9*, 3932. [CrossRef]
33. Realyvásquez-Vargas, A.; Flor-Moltalvo, F.J.; Blanco-Fernández, J.; Sandoval-Quintanilla, J.D.; Jiménez-Macías, E.; García-Alcaraz, J.L. Implementation of Production Process Standardization—A Case Study of a Publishing Company from the SMEs Sector. *Processes* **2019**, *7*, 646. [CrossRef]
34. Saad, N.M.; Al-Ashaab, A.; Shehab, E.; Maksimovic, M. A3 Thinking Approach to Support Problem Solving in Lean Product and Process Development. In *Concurrent Engineering Approaches for Sustainable Product Development in a Multi-Disciplinary Environment*; Stjepandić, J., Rock, G., Bil, C., Eds.; Springer: London, UK, 2013; pp. 871–882. ISBN 978-1-4471-4425-0.
35. Joshuva, N.; Pinto, T. Study and Analysis of Process Capability of 4000 Ton Mechanical Press Using 8D Method. In Proceedings of the National Conference on Advances in Mechanical Engineering Science, Hangzhou, China, 9–10 April 2016; pp. 104–108.
36. Chelsom, J.V.; Payne, A.C.; Reavill, L.R.P. *Management for Engineers, Scientists, and Technologists*, 2nd ed.; John Wiley & Sons: Hoboken, NJ, USA, 2005; ISBN 0470021268.
37. Vargas, D.L. Resolução de problemas utilizando a metodologia 8D: Estudo de caso de uma indústria do setor sucroalcooleiro. In Proceedings of the Simpósio de Engenharia de Produção de Sergipe Anais do IX SIMPROD, São Cristóvão, Brazil, 28 Novembre–1 December 2017; pp. 464–477.
38. Quality-One Eight Disciplines of Problem Solving. Available online: https://quality-one.com/8d/ (accessed on 12 October 2018).
39. Bremmer, J. Update and Upgrade the Current Quality Assurance in the Global Supply Chain of a Heavy Truck Manufacturing Company: Managing Product Quality in the North Bound Flow (NBF) at Scania Production Zwolle. Master's Thesis, University of Twente, Snschede, The Netherlands, 2015.
40. Pacheco-Pacheco, V.H. Implementación de un Modelo de Mejora para Optimizar el Tiempo de Entrega de los Productos "Alto De Basta" y "Alto De Camisa" en el Proceso Confección, Arreglo y Modificación, mediante la Aplicación de las 8 Disciplinas. In *Caso: Symp—Sastrería*; Universidad Católica del Ecuador: Quito, Ecuador, 2018.
41. Zasadzień, M. Six Sigma methodology as a road to intelligent maintenance. *Prod. Eng. Arch.* **2017**, *15*, 45–48. [CrossRef]
42. Kumar, S.; Singh, D. Implementing 8D Model of Problem Solving in Employee Turnover: A Study of Selected Hotels in Delhi and Rajasthan. *J. Kashmir Tour. Cater. Technol.* **2015**, *2*, 1–10.
43. Zasadzień, M. Wykorzystanie metody 8D do doskonalenia procesu remontowego wagonów kolejowych. *Syst. Wspomagania w Inżynierii Prod.* **2016**, *z2*, 392–399.
44. Mitreva, E.; Zdrvkovska, L.; Taskov, N.; Metodijeski, D.; Gjorshevski, H. Quality management and practises in automotive parts production. In Proceedings of the XXIV International Scientific-Technical Conference on Transport, Road-Building, Agricultural, Hoisting & Hauling and Military Technics and Technologies, Sofia, Bulgaria, 29 June–2 July 2016; pp. 4–7.
45. Zhou, F.; Wang, X.; Mpshe, T.; Zhang, Y.; Yang, Y. Quality Improvement Procedure (QIP) based on 8D and Six Sigma Pilot Programs in Automotive Industry. *Adv. Econ. Bus. Manag. Res.* **2016**, *16*, 275–281.
46. Nicolae, V.; Ionescu, L.M.; Belu, N.; Elena Știrbu, L. Improvement of the 8D Analysis Through a System Based on the "Internet of Things" Concept Applied in Automotive Industry. In *CONAT 2016 International Congress of Automotive and Transport Engineering*; Chiru, A., Ispas, N., Eds.; Springer International Publishing: Berlin/Heidelberg, Germany, 2017; pp. 635–642.
47. Roque, R.; Berenice, E. Reengineering Process in the Manufacture of Dental Units Applying the Eight Discipline Methodology. Bachelor's Thesis, Universidad Tecnológica de Perú, Lima, Peru, 2017. (In Spanish)

48. Škůrková, K.L. Implementation of a System Quality Tool to Reduce tThe Costs of Scrap Loss in Industrial Enterprise. *Zesz. Nauk. Qual. Prod. Improv.* **2017**, *1*, 93–111. [CrossRef]

49. Wichawong, P.; Chongstitvatana, P. Knowledge management system for failure analysis in hard disk using case-based reasoning. In Proceedings of the 2017 18th IEEE/ACIS International Conference on Software Engineering, Artificial Intelligence, Networking and Parallel/Distributed Computing (SNPD), Kanazawa, Japan, 26–28 June 2017; pp. 1–6.

50. Chlpeková, A.; Večeřa, P.; Šurinová, Y. Enhancing the Effectiveness of Problem-Solving Processes through Employee Motivation and Involvement. *Int. J. Eng. Bus. Manag.* **2014**, *6*, 1–9. [CrossRef]

51. Slayter, E.; Higgins, L.M. Hands-On Learning: A Problem-Based Approach to Teaching Microsoft Excel. *Coll. Teach.* **2018**, *66*, 31–33. [CrossRef]

52. Hamad, M.M. *AutoCAD 2019: Beginning and Intermediate*; Stylus Publishing: Dulles, VA, USA, 2019; ISBN 9781683922599.

53. Garcés, K.; Casallas, R.; Álvarez, C.; Sandoval, E.; Salamanca, A.; Viera, F.; Melo, F.; Soto, J.M. White-box modernization of legacy applications: The oracle forms case study. *Comput. Stand. Interfaces* **2018**, *57*, 110–122. [CrossRef]

54. Xu, Q.; Jiao, R.J.; Yang, X.; Helander, M.; Khalid, H.M.; Opperud, A. An analytical Kano model for customer need analysis. *Des. Stud.* **2009**, *30*, 87–110. [CrossRef]

55. Meng, Q.; Zhou, N.; Tian, J.; Chen, Y.; Zhou, F. Analysis of Logistics Service Attributes Based on Quantitative Kano Model: A Case Study of Express Delivering Industries in China. *J. Serv. Sci. Manag.* **2011**, *4*, 42–51. [CrossRef]

56. Matzler, K.; Hinterhuber, H.H. How to make product development projects more successful by integrating Kano's model of customer satisfaction into quality function deployment. *Technovation* **1998**, *18*, 25–38. [CrossRef]

57. He, L.; Song, W.; Wu, Z.; Xu, Z.; Zheng, M.; Ming, X. Quantification and integration of an improved Kano model into QFD based on multi-population adaptive genetic algorithm. *Comput. Ind. Eng.* **2017**, *114*, 183–194. [CrossRef]

58. Patel, S. *The Global Quality Management System. Improvement through Systems Thinking*; Productivity Press: New York, NY, USA, 2016; ISBN 978-1-4987-3980-1.

59. Barsalou, M.A. *Root Cause Analysis, A Step-By-Step Guide to Using the Right Tool at the Right Time*; CRC Press: Boca Raton, FL, USA, 2014; ISBN 9781482258790.

60. Barsalou, M.A. In the Loop. Use 8D reports to track the status of custmer complaints. *Qual. Prog.* **2016**, *49*, 80.

61. Parker, F. *Strategy + Teamwork = Great Products. Management Techniques for Manufacturing Companies*, 1st ed.; CRC Press: Boca Raton, FL, USA, 2015; ISBN 978-1-4822-6011-3.

62. Camarillo, A.; Ríos, J.; Althoff, K.-D. Knowledge-based multi-agent system for manufacturing problem solving process in production plants. *J. Manuf. Syst.* **2018**, *47*, 115–127. [CrossRef]

63. Lira, L.H.; Hirai, F.E.; Oliveira, M.; Portellinha, W.; Nakano, E.M. Use of the Ishikawa diagram in a case-control analysis to assess the causes of a diffuse lamellar keratitis outbreak. *Arq. Bras. Oftalmol.* **2017**, *80*, 281–284. [CrossRef] [PubMed]

64. Silva, A.S.; Medeiros, C.F.; Vieira, R.K. Cleaner Production and PDCA cycle: Practical application for reducing the Cans Loss Index in a beverage company. *J. Clean. Prod.* **2017**, *150*, 324–338. [CrossRef]

65. Da Fonseca, C.M.; Leite, J.C.; De Oliveira Freitas, C.A.; Da Silva Vieira, A.; Fujiyama, R.T. Proposal for improvement the welding process of the micro-USB connector on the mother board on tablets. *J. Eng. Technol. Ind. Appl.* **2016**, *02*, 39–47. [CrossRef]

66. de Saeger, A.; Feys, B. *The Ishikawa Diagram for Risk Management: Anticipate and Solve Problems Within your Business*; Plurilingua Publishing: Brussels, Belgium, 2015; ISBN 9782806268426.

67. Joiner Associates, I. *Pareto Charts: Plain & Simple*; Reynard, S., Ed.; Oriel Incorporated: Madison, WI, USA, 1995; ISBN 9781884731044.

68. Beheshti, M.H.; Hajizadeh, R.; Dehghan, S.F.; Aghababaei, R.; Jafari, S.M.; Koohpaei, A. Investigation of the Accidents Recorded at an Emergency Management Center Using the Pareto Chart: A Cross-Sectional Study in Gonabad, Iran, During 2014–2016. *Health Emergencies Disasters* **2018**, *3*, 143–150. [CrossRef]

69. Webber, L.; Wallace, M. *Quality Control for Dummies*; John Wiley & Sons: Hoboken, NJ, USA, 2011; ISBN 9781118051030.
70. Dan, M.C.; Filip, A.M. Sorin Popescu mistakes in the application of 8d methodology and their impact on customer satisfaction in the automotive industry. In Proceedings of the 2016 International Conference on Production Research—Africa, Europe and the Middle East 4th International Conference on Quality and Innovation in Engineering and Management, Cluj-Napoca, Romania, 25–30 July 2016; pp. 298–303.

Article

Identification of Optimal Process Parameter Settings Based on Manufacturing Performance for Fused Filament Fabrication of CFR-PEEK

Kijung Park [1], Gayeon Kim [1], Heena No [1], Hyun Woo Jeon [2,*] and Gül E. Okudan Kremer [3]

[1] Department of Industrial and Management Engineering, Incheon National University, Incheon 22012, Korea; kjpark@inu.ac.kr (K.P.); g_y_kim@inu.ac.kr (G.K.); nhn@inu.ac.kr (H.N.)

[2] Department of Mechanical and Industrial Engineering, Louisiana State University, Baton Rouge, LA 70803, USA

[3] Department of Industrial and Manufacturing Systems Engineering, Iowa State University, Ames, IA 50011, USA; gkremer@iastate.edu

* Correspondence: hwjeon@lsu.edu; Tel.: +12-255-785-905

Received: 8 June 2020; Accepted: 1 July 2020; Published: 3 July 2020

Abstract: Fused filament fabrication (FFF) has been proven to be an effective additive manufacturing technique for carbon fiber reinforced polyether–ether–ketone (CFR-PEEK) due to its practicality in use. However, the relationships between the process parameters and their trade-offs in manufacturing performance have not been extensively studied for CFR-PEEK although they are essential to identify the optimal parameter settings. This study therefore investigates the impact of critical FFF parameters (i.e., layer thickness, build orientation, and printing speed) on the manufacturing performance (i.e., printing time, dimensional accuracy, and material cost) of CFR-PEEK outputs. A full factorial design of the experiments is performed for each of the three sample designs to identify the optimal parameter combinations for each performance measure. In addition, multiple response optimization was used to derive optimal parameter settings for the overall performance. The results show that the optimal parameter settings depend on the performance measures regardless of the designs, and that the layer thickness plays a critical role in the performance trade-offs. In addition, lower layer thickness, horizontal orientation, and higher speed form the optimal settings to maximize the overall performance. The findings from this study indicate that FFF parameter settings for CFR-PEEK should be identified through multi-objective decision making that involves conflicts between the operational objectives for the parameter settings.

Keywords: additive manufacturing; fused filament fabrication; CFR-PEEK; optimal process parameters; manufacturing performance; multiple response optimization

---

1. Introduction

Additive manufacturing has received increasing attention as industries have pursued new profit paths through the small volume production of more innovative, customized, and sustainable products with high competitiveness [1]. Additive manufacturing, defined as the process of building up materials layer by layer to make objects from 3D model data [2], initially emerged for rapid prototyping to create prototypes in a short time [3]. Additive manufacturing as a means of rapid prototyping has been extended to rapid manufacturing to take advantage of various materials and the design freedom provided by additive manufacturing [1,4,5]. Nowadays, additive manufacturing is employed for various application areas including patient-specific medical implants [6], lightweight parts in high-end engineering [7], artistic devices [8,9], and so on.

The emergence of additive manufacturing to replace traditional manufacturing processes has initiated the development of various additive manufacturing techniques. These include fused filament fabrication (FFF), stereolithography (SLA), selective laser sintering (SLS), laminated objective manufacturing (LOM), and three-dimensional printing (3DP) [10]. Among the various additive manufacturing techniques, FFF has become the most popular method commonly employed in a wide variety of application areas for polymer fabrication due to its cost-effectiveness and technological robustness [11]. In addition, FFF is able to accommodate various types of polymer-based materials. Common polymers for FFF are acrylonitrile–butadiene–styrene copolymers (ABS), polyamides (PA), polycarbonate (PC), and polylactide (PLA), which are placed at a commodity plastic level with low chemical and mechanical strength [11,12]. As the applications of additive manufacturing to advanced engineering and bio-medical devices have arisen simultaneously with the technological evolution of FFF, high-performance polymers such as polyetherimide (PEI) and polyether–ether–ketone (PEEK) have also been considered for FFF [13].

Carbon fiber reinforced PEEK (CFR-PEEK) is a newly emerging polymer, which is a semi-crystalline thermoplastic and a composite of PEEK with carbon fibers. CFR-PEEK has received a great deal of attention as an alternative material of metal for medical implants due to its high bio-compatibility [14,15]. CFR-PEEK provides more bio-compatibility advantages over normal PEEK due to chemical stability, and resistance to prolonged fatigue strain, the reduction in stress shielding and bone resorption, and manufacturability to realize the modulus of bone densities [16,17]. With the benefits in bio-mechanical and -chemical aspects, FFF can be more effective to fabricate CFR-PEEK than SLS due to the advantages of FFF in cost-effectiveness and easier material processing [18].

Despite the potential advantages of FFF for CFR-PEEK, CFR-PEEK has not been sufficiently discussed in the literature relevant to FFF applications. Most existing studies have considered low-end polymers such as PLA and ABS to identify the impact of variable process parameters for FFF mainly on mechanical properties [13]. Although Li, et al. [19] addressed the operational aspects of 3D printers, including the manufacturing cost, environmental impact, and surface quality, they focused on the general operational outcomes of PLA and ABS outputs through FFF with fixed process parameters. This research tendency brings the necessity of operational aspects to identify the effectiveness of FFF for CFR-PEEK to enhance the manufacturability of CFR-PEEK in practice. Since the process parameters of FFF that should be pre-determined can significantly affect additive manufacturing results [20], it is essential for practitioners to be able to determine optimal process parameter values by understanding the underlying trade-offs among various manufacturing performance variables. However, FFF process parameters are often determined in an ad hoc manner, causing unsatisfactory cost, time, and quality during the additive manufacturing process in practice. The negative impacts become even more serious problems for CFR-PEEK applications due to the higher material cost, longer processing time, and greater dimensional accuracy needs than other material applications.

Motivated by the above issues, this study aims to identify the dynamics of key FFF process parameters (i.e., layer thickness, build orientation, and printing speed) for CFR-PEEK on manufacturing performance measures (i.e., printing time, dimensional accuracy, and material cost) that are closely related to manufacturing time, quality, and cost. Herein, different design samples are considered to see whether the optimal combination of the process parameters varies depending on the design types. For each sample type, a design of experiments is repeatedly performed to identify the relationships between the process parameters and the performance measures through the analysis of variance (ANOVA) tests, and then a multiple response optimization model is built to look for the optimal process parameter settings that maximize the overall manufacturing performance. Findings from this study enable additive manufacturing practitioners to better understand the influence of the FFF parameters for CFR-PEEK on additive manufacturing performance that can lead to more cost-effective and reproducible applications using CFR-PEEK.

2. Literature Review

Various reviews relevant to additive manufacturing are available in the literature. For example, printing methods, materials, and recent developments for additive manufacturing are introduced in Wong and Hernandez [21] and Ngo, et al. [22]. Survey studies [23–26] are also available with a specific focus on application areas such as supply chain, aerospace engineering, dentistry, and medicine. Following the growing interest in additive manufacturing, many studies have investigated FFF and its various applications [27]. Since the additive manufacturing performance of FFF depends on the selection of process parameters, most studies have performed design of experiment (DOE) methods to investigate the effects of the process parameters on the performance measures of interest [20].

Table 1 summarizes the additive manufacturing studies for FFF based on DOE analysis. Input parameters commonly addressed in the existing additive manufacturing studies using DOE are layer thickness, build orientation, infill properties, and build temperature [20]. For the output parameters in the FFF experiments, mechanical properties have been mainly considered to optimize them by controlling the process parameters [13,20]. Typical response variables for mechanical properties include tensile strength, flexural strength, comprehensive strength, modulus of elasticity, residual stress bending strength, and angle of displacement [13,20]. Only a few studies, however, considered the impact of FFF process parameters on manufacturing performance. Sood, et al. [28] employed Taguchi's DOE to investigate the effects of layer thickness, build orientation, raster angle, air gap, and raster width on dimensional accuracy, and they observed that various conflicting factors distinctively affect the dimensional accuracy. Nancharaiah [29] identified that the highest levels in layer thickness and air gap are statistically significant to minimize printing time. Durgun and Ertan [30] considered different raster angles and build orientations to examine their effects on surface roughness and showed that build orientation affects the surface roughness more significantly than raster angles.

There are several research gaps in the current studies that should be scrutinized to boost the applicability of FFF in actual practice. Although some macroscopic operational performance measures (e.g., printing time, dimensional accuracy, and production cost) are considered in several studies, most studies in Table 1 focus on the mechanical properties as output variables. From a manufacturer's vantage point, operational parameters such as manufacturing cost, printing time, and dimensional accuracy are not ignorable since these parameters can significantly affect the total production cost. For example, the existing studies in Table 1 mostly disregard the cost factors in analyses, although the manufacturing cost of FFF outputs can be calculated from the material cost and printing time [19,30]. Moreover, manufacturing performance tends to be placed as a single performance measure in the existing studies, and therefore possible trade-offs among process parameter settings are not explicitly addressed in the literature. Since multiple input variables can have different effects on outputs in the FFF process [20], the DOE analysis using critical process parameters for FFF and operational performance measures is required to fully understand the dynamics among the relevant variables.

Table 1. Summary of additive manufacturing studies for fused filament fabrication (FFF) based on the design of experiment (DOE).

References	Material	Input Variables	Output Variables
Ahn, et al. [31]	ABS	Air gap, raster orientation, bead width, color, model temperature	Tensile and compressive strength
Lee, et al. [32]	ABS	Air gap, raster angle/width, layer thickness	Elasticity, flexibility
Lee, et al. [33]	ABS	Raster orientation, air gap, bead width, color, model temperature	Compressive strength
Sood, Ohdar and Mahapatra [28]	ABS	Print orientation, road width, layer thickness, air gap, raster angle	Dimensional accuracy
Masood, et al. [34]	PC	Air gap, raster angle/width	Tensile strength

Table 1. *Cont.*

References	Material	Input Variables	Output Variables
Nancharaiah [29]	ABS	Layer thickness, air gap, raster angle	Production time
Smith and Dean [35]	PC	Orientation	Elastic modulus, tensile strength
Lužanin, et al. [36]	PLA	Air gap, layer thickness, deposition angle	Flexural strength
Durgun and Ertan [30]	ABS	Orientations and raster angles	Surface roughness, tensile/flexural strength, production cost
Wu, et al. [37]	PEEK	Layer thickness, raster angle	Tensile, compressive and bending strength
Christiyan, et al. [38]	ABS	Layer thickness and printing speed	Flexural/tensile strength
Casavola, et al. [39]	ABS, PLA	Raster angle	Elastic/Poisson'/shear modulus
Chacón, et al. [40]	PLA	Build orientation, layer thickness, feed rate	Tensile/flexural strength
Webbe Kerekes, et al. [41]	ABS	Infill density, layer thickness	Ultimate strength, toughness/Young's modulus, initial yield stress, elongation at break
Han, et al. [42]	CFR-PEEK, PEEK	Material	Tensile/bending/compressive strength and modulus, surface characterization, cytotoxicity, cell adhesive and spreading

Acrylonitrile–butadiene–styrene copolymers (ABS); polycarbonate (PC); polylactide (PLA); polyether–ether–ketone (PEEK).

Furthermore, most materials for FFF examined in the existing studies are low-performance polymers such as ABS, PLA, and PC. There are a few studies relevant to FFF using high-performance polymers such as PEEK and CFR-PEEK, but the process parameters for the materials and their operational aspects have not been sufficiently discussed [37,42]. In particular, CFR-PEEK has been pointed out as a very promising material for 3D printing, since it can be used not only for various engineering applications, but also for medical applications due to its sturdy mechanical properties and low biological toxicity [15,17,42,43]. While the preliminary studies on CFR-PEEK are available in the literature, the breadth and depth of the relevant studies are less comprehensive than that of other common polymer materials.

In response to the above stated shortcomings, this study focuses on the FFF process of CFR-PEEK to identify the relationships between the FFF process parameters and manufacturing performance measures through a full factorial DOE, in which the information loss from the experiments is minimized. For this, the effects of important FFF process parameters (i.e., layer thickness, build orientation, and printing speed) [20] on the printing time, dimensional accuracy, and material cost for the experiments are investigated, respectively. Moreover, three different designs are considered to confirm whether identified relationships vary depending on design characteristics. Based on the DOE results, the optimal parameter settings considering all the manufacturing performance measures as well as the individual optimal parameter settings for each performance measure are suggested through the methodology proposed in the next section.

3. Methodology

This section illustrates the principal information of experimental design to identify the impact of the FFF process parameters for CFR-PEEK on the manufacturing performance of different product designs.

3.1. Preparation of Experiments

Experimental samples were fabricated by Apium P220 [44], which is a FFF-based 3D printer and compatible with a wide range of materials including high-performance polymers such as PEEK and CFR-PEEK. Table 2 summarizes the technical specifications of the machine. TECAPEEK CF30 [45], which has a 1.38 g/cm^3 density, 6000 MPa tensile modulus, and 112 MPa tensile strength, was used as the material for the experiments.

Table 2. Technical specifications of Apium P220 [44].

Specifications	Information
X/Y resolution	Product resolution: 0.5 mm, machine resolution: 0.0125 mm
Z resolution	Product resolution: 0.1 mm, machine resolution: 0.05 mm
Minimum/maximum layer thickness	0.1 mm/0.3 mm
Nozzle diameter	0.4 mm
Print head temperature	Heated up to 540 °C
Print bed temperature	Heated up to 160 °C
Build plate size	220 × 175 mm
Power consumption	Maximum 0.700 kW
Material types	PEEK, CFR PEEK, PEI 9085, PVDF (polyvinylidene fluoride), POM-C (polyoxymethylene), PP (polypropylene)

The experimental samples used for this study were the three specimen types based on ASTM D638 [46], ASTM D695 [47], and ASTM D3039 [48] (see Figure 1). The standard size of each specimen type was resized to have time efficiency in the experimental runs. Each sample design in Figure 1 was processed as follows: first, a pre-defined computer aided design (CAD) model of each design was created through SolidWorks [49] and then saved to a STL file. Since FFF deposits materials layer by layer, each CAD model needs a slicing process that transforms the designed CAD model into a series of layers to be printed. For this process, Simplify3D version 4.1 [50] was employed to transform each original CAD model into its G-code file which has all the operational commands for the additive manufacturing of the CAD model.

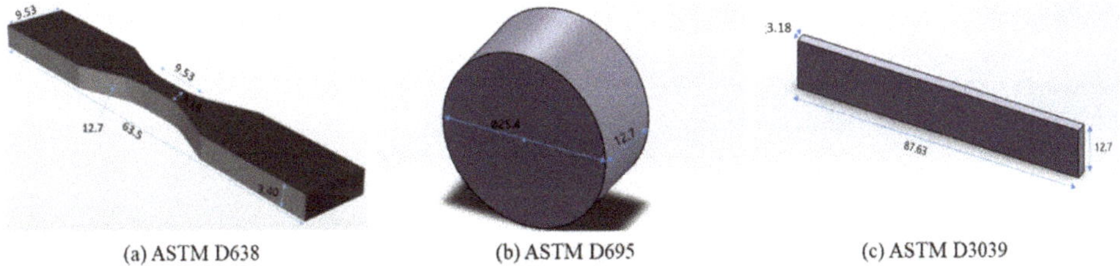

(a) ASTM D638   (b) ASTM D695   (c) ASTM D3039

Figure 1. Sample designs and their dimensional information used for the experiments (unit: mm). (**a**) ASTM D638; (**b**) ASTM D695; (**c**) ASTM D3039.

Figure 2 shows the examples of the 3D printing outputs simulated by Simplify3D. The areas in purple, blue, blue-green, and orange colors indicate the brim, outer perimeter, inner perimeter, and infill, respectively. Support structures were not generated to eliminate possible effects of support generation on the performance measures considered in experiments. Other process parameters, except for the input parameters, were fixed to the default settings for CFR-PEEK provided by the manufacturer through the parameter configuration of Simplify3D (see Table 3).

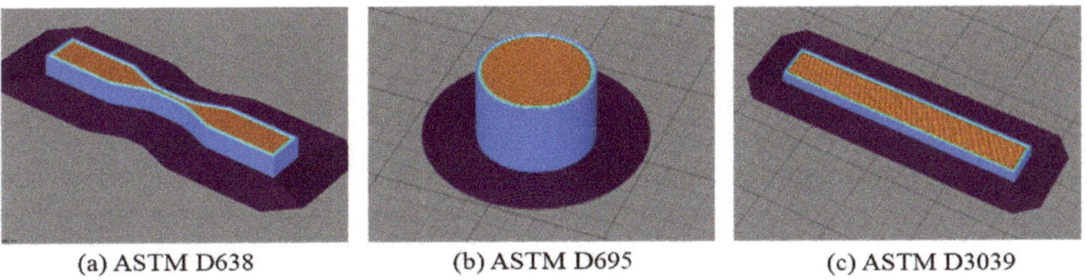

(a) ASTM D638   (b) ASTM D695   (c) ASTM D3039

Figure 2. Sliced models for the experiments. (**a**) ASTM D638; (**b**) ASTM D695; (**c**) ASTM D3039.

Table 3. Fixed process parameters.

Parameter	Unit	Value
Bed temperature	°C	120
Nozzle temperature	°C	510
Perimeters	Layers	3
Number of top layers	Layers	0
Number of bottom layers	Layers	0
Infill pattern	-	Rectilinear
Infill angle	°	+ 45/− 45
Infill rate	%	100
Extrusion with first layer	%	96

3.2. Design of Experiments

The DOE of this study was planned to statistically analyze the FFF parameters significantly affecting the manufacturing performance changes. In addition, it identified the process parameter settings that optimized the individual and overall manufacturing performance of the FFF for the CFR-PEEK. For this, three critical process parameters for FFF (i.e., layer thickness, build orientation, and printing speed) were considered as the input parameters for the experiments. First, layer thickness was the measure of each layer height deposited by a nozzle tip. Layer thickness determined the number of layers deposited for a printed part, and thus printing time and precision could be affected by this process parameter. Two values (i.e., 0.2 mm and 0.3 mm) were considered as the levels of layer thickness for the experimental design. It was noted that the 0.2 mm layer thickness was a reference parameter level recommended by the manufacturer for CFR-PEEK printing. Second, the build orientation represented the direction of a printed part that stood on a build plate. Since the movement directions of the material deposition were varied depending on the build orientation of a fabricated part, it could critically affect the operational performance of the outputs. Two main directions on the x axis (i.e., 0° and 90°) were considered for the build orientation of each design type (see Figure 3).

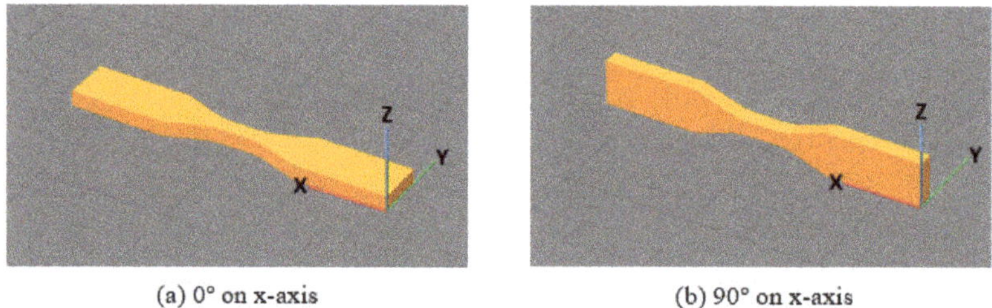

(a) 0° on x-axis   (b) 90° on x-axis

Figure 3. Illustration of the build orientation for the experiments. (**a**) 0° on x-axis; (**b**) 90° on x-axis.

The last input parameter was the printing speed defined as the nozzle's movement speed at which the material was deposited. Printing speed was another key process parameter for FFF that may have needed adjusting in practice to decrease the production lead-time. However, this may have led to poor product quality due to unstable polymer extrusion caused by fast nozzle movements. Based on the recommended printing speed of the machine (i.e., 1200 mm/min), 1000 mm/min and 1400 mm/min (a range of ±200 mm/min) were additionally considered to define three printing speed levels. Table 4 shows all the input parameters and their levels considered in the experimental design of this study. All the combinations of these parameters were applied to each design sample through Simplify3D.

Table 4. Input parameters for the DOE.

Parameter	Unit	Level 1	Level 2	Level 3
Layer thickness (L)	mm	0.2 (L1)	0.3 (L2)	-
Build orientation (O)	°	0 (O1)	90 (O2)	-
Printing speed (P)	mm/min	1000 (P1)	1200 (P2)	1400 (P3)

Printing time, dimensional accuracy, and material cost were selected for the response variables of the DOE. Each performance measurement is summarized in Table 5, and more detailed information is described below.

Table 5. Operational performance measurements for the DOE.

Response Variable	Unit	Description
Printing time	min	Total build time taken to finish fabrication
Dimensional accuracy	-	Mean squared error between the measured dimensions and the original CAD dimensions
Material cost	€	Filament cost calculated from consumed filament per printed sample

The printing time was measured by the duration in minutes between the start-time of the fabrication and the end-time of fabrication recorded by the machine. The start-time and end-time were recorded when the machine started fabrication after the completion of all the set-up processes and when the machine finished fabrication and started a cooling-down process, respectively. It was evident that an increase in printing speed led to a decrease in the total printing time. Moreover, the previous study reporting that layer thickness and build orientation were critical factors to minimize printing time [51] suggested that the different level combinations of the process parameters for CFR-PEEK may distinctively affect the printing time for each design.

Dimensional accuracy is measured by the mean squared error (MSE) between the measured specifications and the actual dimensional specifications of a fabricated part (see Equation (1)). A lower value of Equation (1) shows a lower dimensional error, indicating the better dimensional accuracy of the print. The dimensional specifications of each printed sample were measured multiple times by Mitutoyo NTD13-P15M, which is a digital vernier caliper with a ± 0.02 mm accuracy, as defined in Figure 4. The previous studies using ABS [28,52] observed that the dimensional accuracy of the fabricated parts of FFF were affected by the layer thickness, build orientation, and printing speed because they led to different deposition patterns and deformation effects:

$$Dimensional\ Accuacry = \sum_{i=1}^{n} (M_i - A_i)^2 / n, \tag{1}$$

where n is the number of measured dimensions for each design type, M_i is the measured value of the dimension i, and A_i is the actual CAD size of the dimension i.

The cost of the materials consumed during fabrication was calculated from the amount of consumed CFR-PEEK filament length. The CFR-PEEK used for the experiments costs EUR 450 per spool, and the total filament length of one spool is 150 m. Thus, the unit filament cost is EUR 3/m. Based on the unit filament cost, the total filament cost of each fabrication is estimated by the length of the used filament recorded by the machine. Since CFR-PEEK is relatively expensive compared to other polymer materials due to its chemical and mechanical advantages, the optimal process parameter settings that minimize the filament cost are essential to boost cost efficiency in additive manufacturing.

A total of 36 experiments for each design type were randomly ordered as a full factorial design with three replicates for all the possible combinations of the process parameters (2 levels × 2 levels × 3 levels × 3 replicates). Thus, a total of 108 experiments (36 experiments × 3 design types) were performed for all three design types. For each experiment of design type, the above performance measures were recorded to create a dataset for statistical analysis.

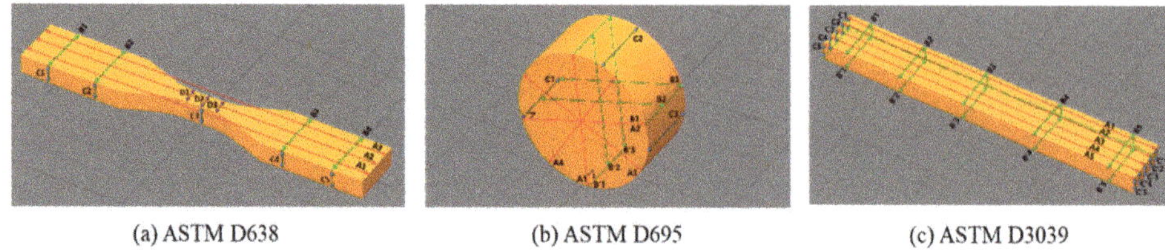

(a) ASTM D638	(b) ASTM D695	(c) ASTM D3039

Figure 4. Measured dimensional specifications of each design type. (**a**) ASTM D638; (**b**) ASTM D695; (**c**) ASTM D3039.

3.3. Analysis of Experiments

The analysis of variance (ANOVA) based on a general linear model (GLM) for the collected experimental data was performed by MINITAB 18 [53]. For each performance measure, three ANOVA tests for the individual design cases were separately performed to compare the effects of the process parameters. Equation (2) describes the simplified expression of the GLM considered in this study. Statistically significant terms (p-value ≤ 0.05) in the ANOVA results were identified to interpret the effects of the process parameters on the manufacturing performance. Then, each model was fitted again only with significant terms to identify optimal parameter levels:

$$y_{ijk} = \beta_0 + \sum_i \beta_i x_i + \sum_j \beta_j x_j + \sum_k \beta_k x_k + \sum_i \sum_j \beta_{ij} x_i x_j + \sum_i \sum_k \beta_{ik} x_i x_k \\ + \sum_j \sum_k \beta_{jk} x_j x_k + \sum_i \sum_j \sum_k \beta_{ijk} x_i x_j x_k + \varepsilon_{ijk},$$

(2)

where i is the factor level of layer thickness ($i = 1, 2$), j is the factor level of build orientation ($j = 1, 2$), k is the factor level of printing speed ($k = 1, ..., 3$), y_{ijk} is the value of a response variable (i.e., printing time, dimensional accuracy, and material cost), β_0 is an intercept, β_i, β_j, and β_k are the main effect coefficients, β_{ij}, β_{ik}, and β_{jk} are two-way interaction coefficients, β_{ijk} is a three-way interaction coefficient, x is the coded value ($-1, 0, +1$) of each factor level, and ε_{ijk} is an error term.

To derive parameter settings to simultaneously optimize all the manufacturing performance measures for each design case, the Derringer–Suich method for multi-response optimization [54] was used. Since all the manufacturing performance variables have the same optimality direction (i.e., minimization), the one-sided desirability function (d_i) expressed in Equation (3) was employed for each manufacturing performance measure:

$$d_i = \left(\frac{\hat{y}_i - U_i}{T_i - U_i}\right)^t,$$

(3)

where $T_i \leq \hat{y}_i \leq U_i$, $\hat{y}_i$ is a predicted value of performance measure i, T_i = a minimum value of i, U_i = an upper limit of i, and t = a weight to express the shape of the desirability function.

For all the predicted values of i, T_i and U_i that satisfy $\hat{y}_i \leq T_i$ and $U_i \leq \hat{y}_i$, respectively, were chosen to derive d_i ($0 \leq d_i \leq 1$). If $\hat{y}_i$ is at its goal T_i, then d_i becomes 1. Then, the desired parameter settings to satisfy the overall manufacturing performance were obtained to maximize the composite desirability (D) in Equation (4):

$$D = \left(\prod_{i=1}^{n} d_i\right)^{1/n},$$

(4)

For the derivation of D, the response optimizer tool provided in MINITAB 18 was employed in this study; MINITAB uses a reduced gradient algorithm to identify the optimal solution to maximize the composite desirability [55,56]. Based on the general linear regression model only including the significant factors in each design case, the optimal parameter settings to maximize composite desirability were derived by assuming the linearity of the individual desirability functions ($t = 1$).

4. Results

The following sub-sections show statistical results to identify the impact of the FFF process parameter combinations for the CFR-PEEK on the manufacturing performance measures, and the individual and overall optimal parameter settings for each design type are analyzed to derive their manufacturing implications.

4.1. Optimal Parameter Settings for the Individual Performance Measurements

4.1.1. Printing Time

For all the design types, the printing time is significantly affected by each process parameter itself along with its interactions with other process parameters (see Table 6). Figure 5 shows the main significant effects on printing time for each design case. The dotted line in Figure 5 represents the population mean of the printing time. The slope of each plot of the main effects indicates the impact of each parameter change; the steeper slope indicates the greater difference in the effect on the printing time. The greatest difference in printing time is observed in layer thickness for all the design cases; the 0.3 mm layer thickness leads to much a shorter mean printing time than the 0.2 mm layer thickness. In addition, printing time decreases when the build orientation is 0° regardless of the design cases; however, the impact of build orientation in the ASTM D695 case is relatively smaller than other design cases. Moreover, the fastest printing speed (i.e., at 1400 mm/min) results in the shortest mean printing time among the printing speed levels for all the design cases.

Table 6. ANOVA results for the printing time (* $\alpha < 0.05$).

ASTM D638	DF (Degrees of Freedom)	Sum of Square	Mean Square	F-Value	p-Value
Layer Thickness (L)	1	584.03	584.0280	3003.57	0.000 *
Build Orientation (O)	1	210.25	210.2500	1081.29	0.000 *
Printing Speed (P)	2	229.56	114.7780	590.29	0.000 *
L × O	1	20.25	20.2500	104.14	0.000 *
L × P	2	10.89	5.4440	28.00	0.000 *
O × P	2	8.67	4.3330	22.29	0.000 *
L × O × P	2	0.67	0.3330	1.71	0.201
Error	24	4.67	0.1940	-	-
Total	35	1068.97	-	-	-
ASTM D695	**DF**	**Sum of Square**	**Mean Square**	**F-Value**	**p-Value**
Layer Thickness (L)	1	5852.25	5852.2500	35,113.50	0.000 *
Build Orientation (O)	1	476.69	476.6900	2860.17	0.000 *
Printing Speed (P)	2	2355.50	1177.7500	7066.50	0.000 *
L × O	1	12.25	12.2500	73.50	0.000 *
L × P	2	108.50	54.2500	325.50	0.000 *
O × P	2	9.39	4.6900	28.17	0.000 *
L × O × P	2	0.17	0.0800	0.50	0.613
Error	24	4.00	0.1700	-	-
Total	35	8818.75	-	-	-
ASTM D3039	**DF**	**Sum of Square**	**Mean Square**	**F-Value**	**p-Value**
Layer Thickness (L)	1	2635.11	2635.11	18,972.80	0.000 *
Build Orientation (O)	1	2025.00	2025.00	14,580.00	0.000 *
Printing Speed (P)	2	992.39	496.19	3572.60	0.000 *
L × O	1	196.00	196.00	1411.20	0.000 *
L × P	2	51.39	25.69	185.00	0.000 *
O × P	2	45.50	22.75	163.80	0.000 *
L × O × P	2	3.17	1.58	11.40	0.000 *
Error	24	3.33	0.14	-	-
Total	35	5951.89	-	-	-

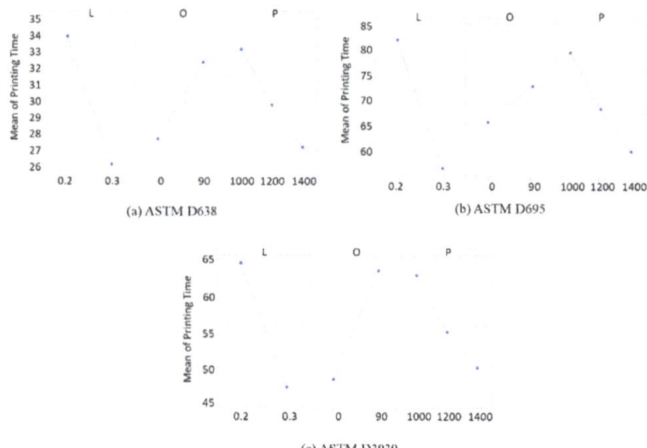

(a) ASTM D638

(b) ASTM D695

(c) ASTM D3039

Figure 5. Main effect plots of the significant factors for printing time. (**a**) ASTM D638; (**b**) ASTM D695; (**c**) ASTM D3039.

The non-parallel lines in Figure 6 indicate that all the design cases have similar interaction effects. For layer thickness, the 0° of build orientation and the 1400 mm/min of printing speed are associated with the shortest mean printing time if the 0.3 mm layer thickness is used. Similarly, the 0° build orientation is associated with the 0.3 mm layer thickness and the 1400 mm/min printing speed results in the shortest mean printing time for each case. This is also confirmed from the results showing that the 1400 mm/min printing speed was interacting with the 0.3 mm layer thickness and with the 0° build orientation which generates the shortest mean printing time.

It can be interpreted that one process parameter is less affected by another parameter if the lines on an interaction effect plot are close to parallel lines. Overall, the interaction effects existing in the ASTM D695 case are weaker than those of other design cases although the interaction effects are statistically significant. For example, the layer thickness for the ASTM D695 design affects less the relationship between the build orientation and the printing time, relatively, than the layer thickness for other designs.

The above main interaction effects of the process parameters on the printing time may result from the characteristic of FFF that stacks the material layer by layer. In the fabrication process of each layer, the nozzle moves back to the default position when the deposition of one layer is completed, and then the next layer is filled. In other words, the nozzle movement time increases as the number of layers increases. This can be a plausible reason for the impacts of the process parameters that increase printing time. When the layer thickness decreases, the total number of required layers for the print increases since more layers should be deposited for the same dimensions. Moreover, the number of layers increases when the 90° build orientation is used. For example, ASTM D638, ASTM D695, and ASTM D3039 have 16 layers, 64 layers, and 16 layers, respectively, when they are fabricated at the 0° build orientation; they increase to 48 layers, 127 layers, and 64 layers at the 90° build orientation. Consequently, the build orientation at 90° negatively affects the printing time.

Table 7 shows the printing time of each parameter combination for all the design cases, which are estimated by the prediction model only including the significant factors. In summary, the same process parameter settings are associated with a minimum printing time regardless of the design cases; the parameter combination of 0.3 mm (layer thickness), 0° (build orientation), and 1400 mm/min (printing speed) provides the minimum printing time for each design case. Therefore, these parameter settings can ensure the shortest printing time to fabricate outputs using CFR-PEEK if the operational objective of the additive manufacturing is only only minimize printing time.

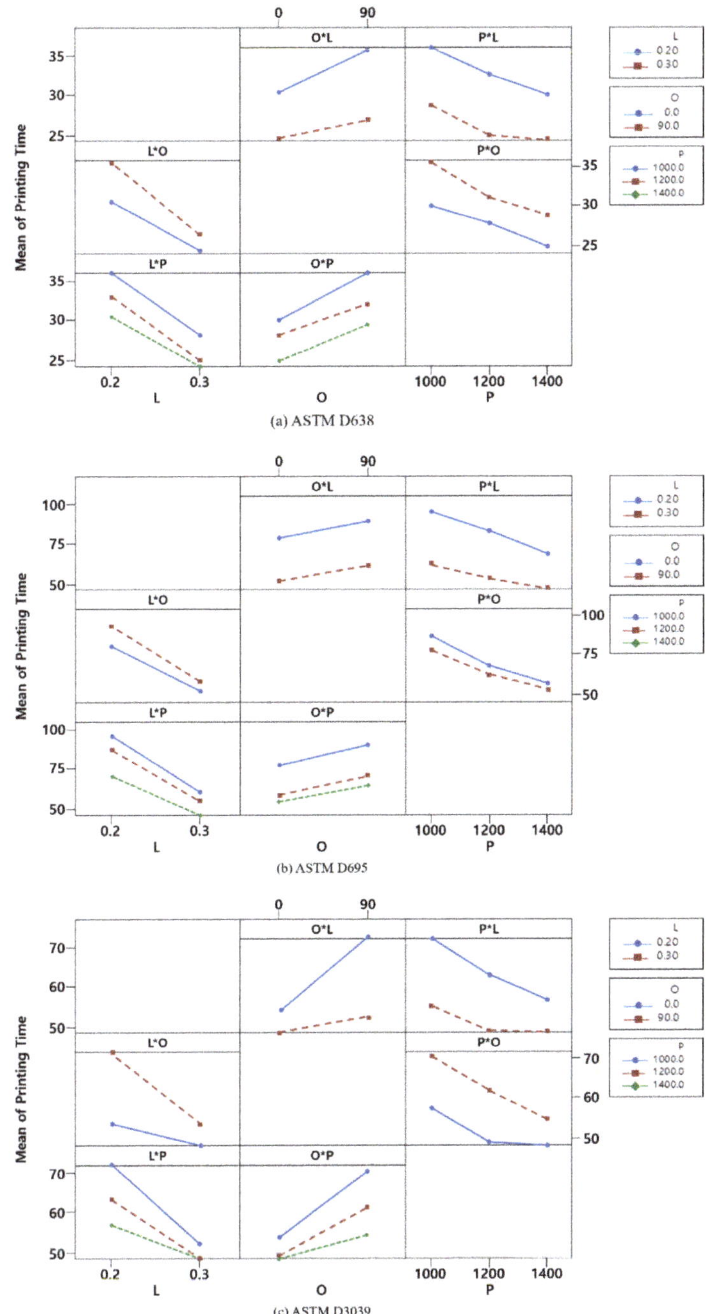

Figure 6. Interaction effect plots of the significant factors for printing time. (**a**) ASTM D638; (**b**) ASTM D695; (**c**) ASTM D3039.

Table 7. Estimated printing time of each parameter combination and the fitted regression models.

(a) Estimated Printing Time (min)

Sample	L1O1P1	L1O1P2	L1O1P3	L1O2P1	L1O2P2	L1O2P3	L2O1P1	L2O1P2	L2O1P3	L2O2P1	L2O2P2	L2O2P3
ASTM D638	34.17	30.67	27.83	41.83	36.67	33.17	26.17	24.33	22.5 *	30.83	27.33	24.83
ASTM D695	90.00	77.08	67.25	99.67	85.58	74.42	61.33	52.92	47.08 *	68.67	59.08	51.92
ASTM D3039	61.33	54.00	49.00	84.67	73.00	65.67	46.67	41.33	39.00 *	59.00	52.00	47.00

(b) Regression Model

Sample	Prediction Model	R-sq.
ASTM D638	$y = 30.03 + 4.03L1 - 4.03L2 - 2.42O1 + 2.42O2 + 3.22P1 - 0.28P2 - 2.94P3 - 0.75L1{\cdot}O1 + 0.75L1{\cdot}O2 + 0.75L2{\cdot}O1 - 0.75L2{\cdot}O2 + 0.72L1{\cdot}P1 - 0.11L1{\cdot}P2 - 0.61L1{\cdot}P3 - 0.72L3{\cdot}P1 + 0.11L2{\cdot}P2 + 0.61L2{\cdot}P3 - 0.67O1{\cdot}P1 + 0.16O1{\cdot}P2 + 0.50O1{\cdot}P3 + 0.67O2{\cdot}P1 - 0.17O2{\cdot}P2 - 0.50O2{\cdot}P3$	99.50%
ASTM D695	$y = 69.58 + 12.75L1 - 12.75L2 - 3.64O1 + 3.64O2 + 10.33P1 - 0.92P2 - 9.42P3 - 0.58L1{\cdot}O1 + 0.58L1{\cdot}O2 + 0.58L2{\cdot}O1 - 0.58L2{\cdot}O2 + 2.17 L1{\cdot}P1 - 0.08L1{\cdot}P2 - 2.08L1{\cdot}P3 - 2.17L2{\cdot}P1 + 0.08L2{\cdot}P2 + 2.08L2{\cdot}P3 - 0.61O1{\cdot}P1 - 0.03O1{\cdot}P2 + 0.64O1{\cdot}P3 + 0.61O2{\cdot}P1 + 0.03O2{\cdot}P2 - 0.64O2{\cdot}P3$	99.95%
ASTM D3039	$y = 56.06 + 8.56L1 - 8.56L2 - 7.50O1 + 7.50O2 + 6.86P1 - 0.97P2 - 5.89P3 - 2.33L1{\cdot}O1 + 2.33L1{\cdot}O2 + 2.33L2{\cdot}O1 - 2.33L2{\cdot}O2 + 1.53L1{\cdot}P1 - 0.14L1{\cdot}P2 - 1.39L1{\cdot}P3 - 1.53L2{\cdot}P1 + 0.14L2{\cdot}P2 + 1.39L2{\cdot}P3 - 1.42O1{\cdot}P1 + 0.08O1{\cdot}P2 + 1.33O1{\cdot}P3 + 1.42O2{\cdot}P1 - 0.08O2{\cdot}P2 - 1.33 O2{\cdot}P3 - 0.42L1{\cdot}O1{\cdot}P1 + 0.25L1{\cdot}O1{\cdot}P2 + 0.17L1{\cdot}O1{\cdot}P3 + 0.42L1{\cdot}O2{\cdot}P1 - 0.25L1{\cdot}O2{\cdot}P2 - 0.17L1{\cdot}O2{\cdot}P3 + 0.42L2{\cdot}O1{\cdot}P1 - 0.25L2{\cdot}O1{\cdot}P2 - 0.17L2{\cdot}O1{\cdot}P3 - 0.42L2{\cdot}O2{\cdot}P1 + 0.25L2{\cdot}O2{\cdot}P2 + 0.17L2{\cdot}O2{\cdot}P3$	99.94%

*: optimal value.

4.1.2. Dimensional Accuracy

Table 8 shows the statistically significant process parameters that affect the dimensional accuracy. It seems that dimensional accuracy is differently affected by the process parameters depending on the printed designs. Although dimensional accuracy in the ASTM D695 and ASTM D3039 cases is associated with a similar parameter effect, the ASTM D638 type has all the main and interaction terms as statistically significant factors on the dimensional accuracy except for the interaction effect between the layer thickness and build orientation and the three-way interaction effect.

The statistically significant main effects in Figure 7a show that each of 0.2 mm in layer thickness, 0° in build orientation, and 1200 mm/min in printing speed for the ASTM D638 case is associated with the lowest mean dimensional error. The ASTM D695 and ASTM D3039 cases, however, only have a layer thickness as a statistically significant factor in which the 0.2 mm layer thickness results in the minimum mean dimensional error (see Figure 7b,c). The interaction effects of the ASTM D638 case in Figure 8 support that the build orientation of the part design can play an important role in dimensional accuracy; the 0° build orientation significantly decreases the dimensional error at different layer thickness and printing speed levels.

Based on the above main interaction effects, it can be inferred that layer thickness is a critical factor that is closely related to dimensional accuracy regardless of the sample designs. The fact that the 0.2 mm layer thickness always offers lower dimensional error values in the experiments supports that the decrease in layer thickness can result in more sophisticated fabrication. In addition, a plausible explanation for the impact of build orientation in the ASTM D638 case can be found in the design characteristic of the fabricated part. If the build orientation becomes 90°, the ASTM D638 design has a bridge form that needs support structures for proper fabrication (see Figure 9). Since support structures were not created for the experiments to eliminate the possible impacts of support structure generation on the performance measurements, poor dimensional accuracy always occurs in the bridge structure of the ASTM D638 design at the 90° build orientation. However, the default printing speed (1200 mm/min), which is the recommended printing speed for CFR-PEEK from the machine provider, can decrease the negative impact of the 90° build orientation on dimensional accuracy. This seems to be caused by the deviations from the default printing speed that exacerbate the sagging problem on the bridge part as seen in Figure 9b.

Table 8. ANOVA results for dimensional accuracy (* $\alpha < 0.05$).

ASTM D638	DF	Sum of Square	Mean Square	*F*-Value	*p*-Value
Layer Thickness (L)	1	0.95570	0.95567	23.35	0.000 *
Build Orientation (O)	1	3.50630	3.50631	85.68	0.000 *
Printing Speed (P)	2	0.28090	0.14043	3.43	0.049 *
L × O	1	0.49120	0.49120	12.00	0.002 *
L × P	2	0.10290	0.05144	1.26	0.303
O × P	2	0.31940	0.15968	3.90	0.034 *
L × O × P	2	0.13450	0.06724	1.64	0.214
Error	24	0.98210	0.04092	-	-
Total	35	6.77290	-	-	-
ASTM D695	**DF**	**Sum of Square**	**Mean Square**	***F*-Value**	***p*-Value**
Layer Thickness (L)	1	0.41158	0.41158	42.90	0.000 *
Build Orientation (O)	1	0.03670	0.03670	3.83	0.062
Printing Speed (P)	2	0.00587	0.00293	0.31	0.739
L × O	1	0.00402	0.00402	0.42	0.523
L × P	2	0.00077	0.00039	0.04	0.961
O × P	2	0.00184	0.00092	0.10	0.909
L × O × P	2	0.00503	0.00251	0.26	0.772
Error	24	0.23027	0.00959	-	-
Total	35	0.69607	-	-	-
ASTM D3039	**DF**	**Sum of Square**	**Mean Square**	***F*-Value**	***p*-Value**
Layer Thickness (L)	1	0.32211	0.32211	34.70	0.000 *
Build Orientation (O)	1	0.00922	0.00922	0.99	0.329
Printing Speed (P)	2	0.01597	0.00799	0.86	0.436
L × O	1	0.00401	0.00401	0.43	0.517
L × P	2	0.01403	0.00701	0.76	0.481
O × P	2	0.00438	0.00219	0.24	0.791
L × O × P	2	0.00491	0.00246	0.26	0.770
Error	24	0.22278	0.00928	-	-
Total	35	0.59741	-	-	-

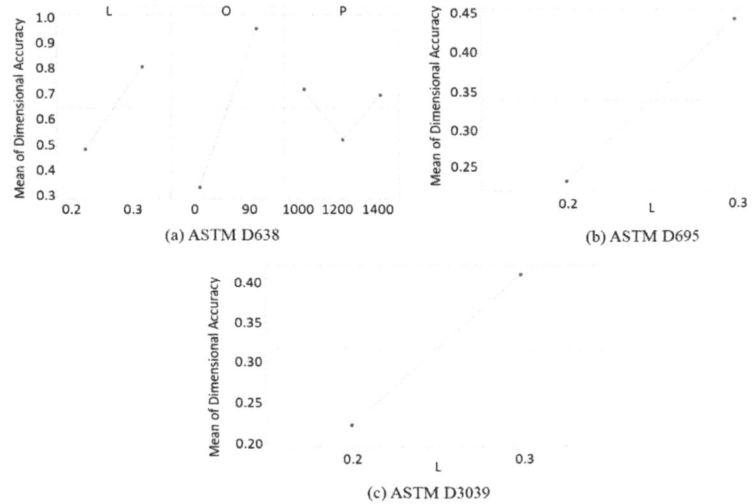

Figure 7. Main effect plots of the significant factors for dimensional accuracy. (**a**) ASTM D638; (**b**) ASTM D695; (**c**) ASTM D3039.

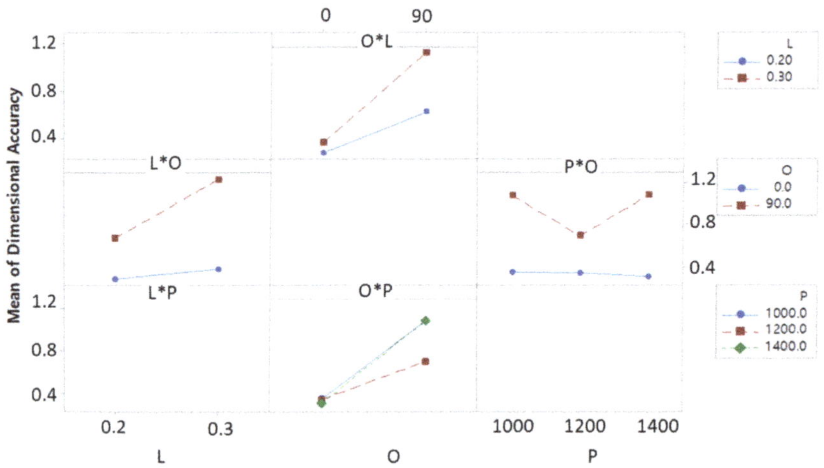

Figure 8. Interaction effect plots of the significant factors for the dimensional accuracy of ASTM D638.

Figure 9. Examples of the impact of build orientation on dimensional accuracy. (a) 0°; (b) 90°.

Table 9 shows the dimensional accuracy of each parameter combination for all the design cases, which is estimated by the prediction model only including the significant factors. Since layer thickness is the only significant factor existing in the regression models for the ASTM D695 and ASTM D3039 cases, the same predicted value is obtained for each parameter combination associated with the same layer thickness level regardless of other parameter levels. An interesting point is that the optimal parameter combination for the ASTM D638 consists of the 0.2 mm layer thickness, the 0° build orientation, and the 1400 mm/min printing speed although the main effect of the printing speed at 1200 mm/min is associated with the minimum mean dimensional error. This indicates that faster printing speed may be still good for dimensional accuracy due to its interaction effect with build orientation; a part design in which its build orientation is a critical factor due to the formation of a bridge structure may have a better dimensional accuracy at a printing speed higher than the default printing speed for CFR-PEEK when the build orientation becomes 0°. All the optimal parameter settings for the design cases in Table 9 show that the 0.2 mm layer thickness, the 0° build orientation, and the 1400 mm/min printing speed form the common optimal setting for all the design cases.

Table 9. Estimated dimensional accuracy of each parameter combination and the fitted regression models.

(a) Estimated Dimensional Accuracy

Sample	L1O1P1	L1O1P2	L1O1P3	L1O2P1	L1O2P2	L1O2P3	L2O1P1	L2O1P2	L2O1P3	L2O2P1	L2O2P2	L2O2P3
ASTM D638	0.31	0.30	0.26 *	0.81	0.42	0.81	0.40	0.39	0.35	1.36	0.98	1.37
ASTM D695	0.23 *	0.23 *	0.23 *	0.23 *	0.23 *	0.23*	0.45	0.45	0.45	0.45	0.45	0.45
ASTM D3039	0.22 *	0.22 *	0.22 *	0.22 *	0.22 *	0.22*	0.41	0.41	0.41	0.41	0.41	0.41

Table 9. *Cont.*

(b) Regression Model

Sample	Prediction Model	R-sq.
ASTM D638	y = 0.65 − 0.16L1 + 0.16L2 − 0.31O1 + 0.31O2 + 0.07P1 − 0.12P2 + 0.05P3 + 0.12L1·O1 − 0.12L1·O2 − 0.12L2·O1 + 0.12L2·O2 − 0.05O1·P1 + 0.13O1·P2 − 0.08O1·P3 + 0.05O2·P1 − 0.13O2·P2 + 0.08O2·P3	81.99%
ASTM D695	y = 0.34 − 0.11L1 + 0.11L2	59.13%
ASTM D3039	y = 0.32 − 0.09L1 + 0.09L2	53.92%

*: optimal value.

4.1.3. Material Cost

The ANOVA results in Table 10 show that the main terms for layer thickness and build orientation are only statistically significant to estimate the material cost in all the design cases. Since the same amount of the CFR-PEEK filament is used for the same parameter settings, it is noted that the calculated filament costs are the same for all the experimental replicates of the same parameter combination. Thus, the ANOVA table cannot calculate the statistics of interaction effects due to the lack of enough degrees of freedom for residual error, and the main effects are only presented in the result table in Table 10.

Table 10. ANOVA results for the material cost (* $\alpha < 0.05$).

ASTM D638	DF	Sum of Square	Mean Square	F-Value	p-Value
Layer Thickness (L)	1	0.499142	0.499142	908.58	0.000 *
Build Orientation (O)	1	0.017030	0.017030	31.00	0.000 *
Printing Speed (P)	2	0.000000	0.000000	0.00	1.000
Error	31	0.017030	0.000549	-	-
Total	35	0.533203	-	-	-
ASTM D695	**DF**	**Sum of Square**	**Mean Square**	**F-Value**	**p-Value**
Layer Thickness (L)	1	0.056882	0.056882	39.42	0.000 *
Build Orientation (O)	1	0.723350	0.723350	501.29	0.000 *
Printing Speed (P)	2	0.000000	0.000000	0.00	1.000
Error	31	0.044732	0.001443	-	-
Total	35	0.824965	-	-	-
ASTM D3039	**DF**	**Sum of Square**	**Mean Square**	**F-Value**	**p-Value**
Layer Thickness (L)	1	0.801920	0.801920	196.70	0.000 *
Build Orientation (O)	1	0.770010	0.770060	188.88	0.000 *
Printing Speed (P)	2	0.00000	0.00000	0.00	1.000
Error	31	0.126380	0.004077	-	-
Total	35	1.698310	-	-	-

The main effect plots in Figure 10 support that the material cost for CFR-PEEK depends on layer thickness and build orientation. The optimal settings that minimize the mean material cost are consistent across the design cases in which the 0.2 mm layer thickness and the 0° build orientation are optimal. However, the cost reduction effect of each process parameter is different depending on the design types; the cost reduction becomes the biggest at the 0.2 mm layer thickness for the ASTM D638 case, the 0° build orientation for the ASTM D695 case, and both factor levels for the ASTM D3039 case.

The material cost is related to the amount of the CFR-PEEK filament used for the fabrication of a final output. The average filament volumes consumed for the design cases fabricated at the 0.2 mm layer thickness are 19.99 cm^3 for ASTM D638, 67.03 cm^3 for ASTM D695, and 41.63 cm^3 for ASTM D3039. The amount of each filament volume increases to 21.90 cm^3 for ASTM D638, 67.67 cm^3 for ASTM D695, and 44.01 cm^3 for ASTM D3039 at the 0.3 mm layer thickness, respectively. The fact that the 0.2 mm layer thickness is associated with the lowest dimensional error indicates that the 0.2 mm layer thickness reduces the material cost due to its more precise fabrication. Moreover, it seems that the designs with a relatively thin dimension such as ASTM D638 and ASTM D3039 have a large impact of layer thickness on filament consumption and their material cost (see Figure 10a,c) since a lower layer thickness level can precisely deposit the filament to build the thin part. Similarly, the change in the build orientation from 0° to 90° increases the average amount of the filament consumption from 20.77 cm^3 to 21.12 cm^3 for ASTM D638, from 66.21 cm^3 to 68.48 cm^3 for ASTM D695, and from 41.65 cm^3 to 44.00 cm^3 for ASTM D3039, respectively. The greater impact of build orientation on material cost observed in the ASTM D695 and ASTM D3039 designs (see Figure 10b,c) seems to be caused by a brim generated for each design during the additive manufacturing process. Simplify3D automatically generates wider brim areas of the experiments for these design types to properly fix the fabricated parts than the ASTM D638 at the 90° build orientation, and thereby the experimental outputs consume a larger amount of the CFR-PEEK filament. This may result in the greater impact of build orientation on the material cost as seen in Figure 10b,c.

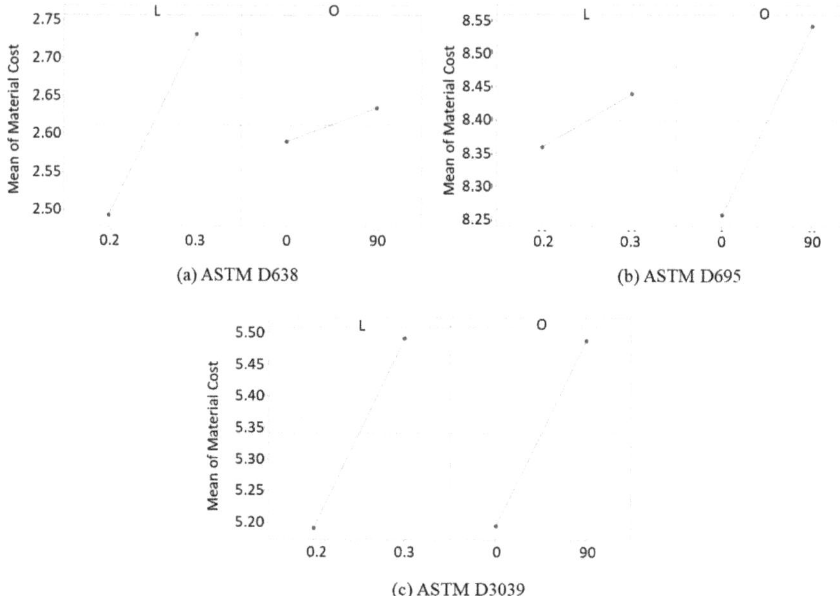

Figure 10. Main effect plots of the significant factors for material cost. (**a**) ASTM D638; (**b**) ASTM D695; (**c**) ASTM D3039.

The predicted material cost obtained from the regression model, only including statistically significant factors for each design, is shown in Table 11. It is noted that there are multiple parameter combinations that minimize the material cost of each design since layer thickness and build orientation only critically impact the material cost of each design. Thus, the 0.2 mm layer thickness and the 0° build orientation form the optimal parameter settings to minimize the material cost of each design regardless of its printing speed for the fabrication.

Table 11. Estimated material cost of each parameter combination and the fitted regression models.

(a) Estimated Material Cost (€)

Sample	L1O1P1	L1O1P2	L1O1P3	L1O2P1	L1O2P2	L1O2P3	L2O1P1	L2O1P2	L2O1P3	L2O2P1	L2O2P2	L2O2P3
ASTM D638	2.47 *	2.47 *	2.47 *	2.52	2.52	2.52	2.71	2.71	2.71	2.75	2.75	2.75
ASTM D695	8.22 *	8.22 *	8.22 *	8.50	8.50	8.50	8.30	8.30	8.30	8.58	8.58	8.58
ASTM D3039	5.05 *	5.05 *	5.05 *	5.34	5.34	5.34	5.34	5.34	5.34	5.64	5.64	5.64

Table 11. *Cont.*

(b) Regression Model		
Sample	**Prediction Model**	**R-Sq.**
ASTM D638	$y = 2.61 - 0.12L1 + 0.12L2 - 0.02O1 + 0.022O2$	99.95%
ASTM D695	$y = 8.40 - 0.04L1 + 0.04L2 - 0.14O1 + 0.14O2$	94.58%
ASTM D3039	$y = 5.34 - 0.15L1 + 0.15L2 - 0.15O1 + 0.15O2$	92.56%

*: optimal value.

4.2. Optimal Parameter Settings for Multiple Performance Measurements

As seen in the above results, the optimal parameter settings are varied depending on the performance measurement that is considered as an objective to be achieved for the additive manufacturing process. Thus, the determination of the optimal parameter settings for the FFF process of CFR-PEEK can be a multi-objective decision-making problem; trade-offs exist between the different performance measurements that are affected by the FFF process parameters. For example, the individual optimal setting results show that the 0.3 mm layer thickness minimizes printing time, but this layer thickness cannot achieve the minimized dimensional accuracy and material cost.

Table 12 shows the optimal performance settings under the multiple response optimization among the printing time, dimensional accuracy, and material cost based on the fitted regression models only with the significant factors in each design case. For all the design cases, the parameter combination of 0.2 mm in layer thickness, 0° in build orientation, and 1400 mm/min in printing speed maximizes the composite desirability among all the parameter combinations. The individual desirability less than 0.7 in Table 12 indicates that the optimal settings are less effective to the performance measurement; the optimal parameter settings involve trade-offs between the responses. The lowest individual desirability in the printing time is consistently observed in each design case, given the optimal parameter settings, although the dimensional accuracy and material cost have a relatively higher desirability at the optimal parameter settings regardless of the design cases. Since the equal importance of the responses was assumed to obtain the optimal settings in Table 12, it can be inferred that the current optimal settings compromise time reduction to improve the dimensional accuracy and material cost when the responses are equally important. Thus, the optimal settings can be varied if more importance is assigned to layer thickness. For example, the current optimal settings are very effective to individually minimize dimensional error and material cost for the ASTM D638 design ($d > 0.9$). However, the 0.3 mm layer thickness can be selected as an optimal setting to increase the desirability of printing time if the printing time has a much higher importance than other responses. Figure 11 shows the printed samples with the optimal parameter settings obtained from the multiple response optimization.

Table 12. Optimal parameter settings derived by multiple response optimization.

Design Type	Optimal Parameter Settings			Desirability (*d*)			Composite Desirability (*D*)
	Layer Thickness	**Build Orientation**	**Printing Speed**	**Printing Time**	**Dimensional Accuracy**	**Material Cost**	
ASTM D638	0.2 mm	0°	1400 mm/min	0.71	0.99	0.92	0.87
ASTM D695	0.2 mm	0°	1400 mm/min	0.62	0.74	1.00	0.77
ASTM D3039	0.2 mm	0°	1400 mm/min	0.78	0.80	0.90	0.83

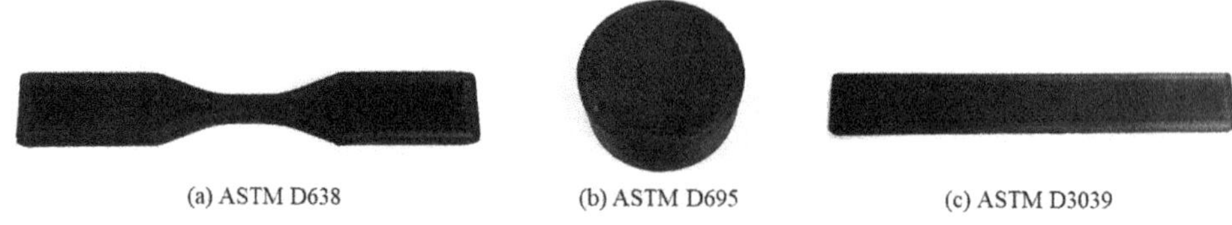

(a) ASTM D638   (b) ASTM D695   (c) ASTM D3039

Figure 11. Samples printed by the optimal parameter settings for the overall manufacturing performance. (**a**) ASTM D638; (**b**) ASTM D695; (**c**) ASTM D3039.

5. Conclusions and Discussion

Although many studies investigated the FFF process parameters, the majority of the existing works used common low-performance polymers to observe the effects of process parameters on the mechanical performance of fabricated outputs. Therefore, it has been hard to extract implications for other operational aspects of the FFF process using high-performance polymers. Since high-performance polymers are more expensive and should be carefully treated to be used for FFF, the relationships between the FFF process parameters for CFR-PEEK and manufacturing performance should be understood to achieve successful additive manufacturing operations for CFR-PEEK in practice. In this regard, this study focused on the impact of FFF process parameters for CFR-PEEK on manufacturing performance to investigate their dynamics and optimal parameter settings for different designs. For this, the layer thickness, build orientation, and printing speed were considered as key process parameters for FFF. Then, a full factorial experimental design of the parameter combinations with three replicates was planned for each of the three designs (i.e., ASTM D638, ASTM D695, and ASTM D3039) to measure the printing time, dimensional accuracy, and material cost of the fabricated outputs. The ANOVA results and regression models of each performance measure on the process parameters showed that there are common relationships observed across the three design cases. The minimum printing speed was related to greater layer thickness (0.3 mm), regular horizontal orientation (0°), and faster printing speed (1400 mm/min) in all the design cases. All the design types also had similar parameter effects that lead to the minimum dimensional accuracy at lower layer thickness (0.2 mm), but the 0° build orientation and the 1400 mm/min printing speed were significant parameters only for the ASTM D638 design case that formed a bridge structure at the vertical build orientation. Layer thickness and build orientation were statistically significant for the material cost in all the design cases, and the 0.2 mm layer thickness and the 0° build orientation resulted in the minimum cost.

The findings from this study show that the effects of the process parameters on the manufacturing performance measures are overall similar across the design cases. However, the dimensional accuracy is distinctively affected by the process parameters in the ASTM D638 case, in which the vertical orientation of the design can cause a sagging problem. This indicates that the parameter settings should be carefully determined for a design with complex shapes if the dimensional accuracy of the fabricated part is the most important factor for the additive manufacturing process since various parameters can simultaneously affect dimensional accuracy. Moreover, the optimal parameter settings separately obtained for the individual performance measures reveal that there are trade-offs in the performance measures caused by the layer thickness levels. That is, a greater layer thickness level decreases the printing time due to a decrease in the number of deposited layers, but it negatively affects the dimensional accuracy and material cost by causing over-deposition, due to a decrease in the printing resolution and an increase in the printed volume. However, such trade-offs in the performance measures are not observed for the build orientation and printing speed. This implies that the process parameter determination should be considered as a multi-objective decision-making problem that has conflicting manufacturing performance measures affected by the process parameter settings. Multiple response optimization was performed to consider the above trade-offs in optimal parameter determination, and the 0.2 mm layer thickness, the 0° build orientation, and the 1400 mm/min printing speed were identified as the parameter settings to optimize the overall manufacturing performance.

The manufacturing performance measures of each experiment are displayed in Figure 12. Since all the performance measures are desired to be minimized, a data point can be optimal as it becomes closer to the right lower corner of the performance space in Figure 12. The data points in red indicate the manufacturing performance measures of the optimal parameter settings from the multiple response optimization. They show that all three designs can properly achieve the overall manufacturing performance at the same parameter settings; universal parameter settings across designs to optimize the overall manufacturing performance can exist for the FFF process using CFR-PEEK. The optimal parameter settings for the overall manufacturing performance are obtained under the equal importance assumption among the performance measurements. Thus, the optimal settings can be varied if each performance measure has different importance. It indicates the necessity of an appropriate decision-making framework that enables the decision maker to reflect relative importance among performance measurements in finding optimal parameter settings for the overall performance improvement.

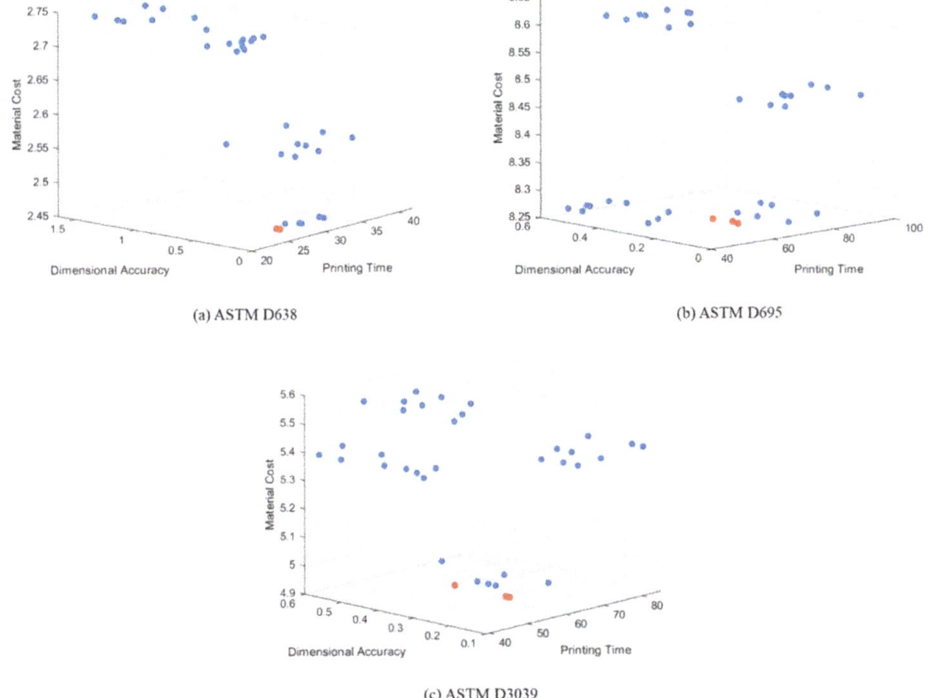

(a) ASTM D638

(b) ASTM D695

(c) ASTM D3039

Figure 12. Optimal parameter settings (in red) for the overall manufacturing performance. (**a**) ASTM D638; (**b**) ASTM D695; (**c**) ASTM D3039.

The primary contribution of this study is to establish a basis for additive manufacturing capabilities for CFR-PEEK applications from manufacturing performance perspectives. The findings from this study will provide useful information about the optimal parameter settings to enhance the manufacturing performance of fabricated products using CFR-PEEK. The approach of this study attempts a transition of prevailing mechanical performance viewpoints to manufacturing performance viewpoints. Thus, the approach will open potential research opportunities in understanding the complex dynamics among the different manufacturing performance measures and in addressing effective operational methods for additive manufacturing to improve the overall manufacturing performance. Nonetheless, the current study should be extended by considering several issues in future work. First of all, additional designs should be analyzed to generalize the findings of the optimal parameter settings since the current research compares three simple designs. Second, additional FFF process parameters and manufacturing performance measures that are critical for CFR-PEEK applications should be considered along with other advanced composite polymers to fully address the relationships between the process parameters and the manufacturing performance measures. Lastly, both the mechanical performance and manufacturing performance should be investigated together to identify the possible trade-offs between them depending on process parameters. Then, the determination of the optimal parameter settings will be formulated as a more complex decision-making problem in which various trade-offs exist between the mechanical and manufacturing performance.

Author Contributions: Conceptualization, K.P. and G.E.O.K.; methodology, K.P., G.K. and H.N.; formal analysis, K.P., G.K., H.N. and H.W.J.; investigation, K.P., G.K. and H.N.; writing—original draft preparation, K.P., G.K., H.N.; writing—review and editing, G.E.O.K. and H.W.J. All authors have read and agreed to the published version of the manuscript.

Funding: This work was supported by the National Research Foundation of Korea (NRF) grant funded by the Korea government (MIST) (No. 2018R1C1B5040256) for Kijung Park. This work was also supported by Research Assistance Program (2020) in the Incheon National University for Gayeon Kim.

Conflicts of Interest: The authors declare no conflict of interest.

References

1. Mellor, S.; Hao, L.; Zhang, D. Additive manufacturing: A framework for implementation. *Int. J. Prod. Econ.* **2014**, *149*, 194–201. [CrossRef]
2. ISO/ASTM52900-15. *Standard Terminology for Additive Manufacturing–General Principles–Terminology*; ASTM International: West Conshohocken, PA, USA, 2015; Available online: www.astm.org/Standards/ISOASTM52900 (accessed on 19 August 2019).
3. Campbell, I.; Bourell, D.; Gibson, I. Additive manufacturing: Rapid prototyping comes of age. *Rapid Prototyp. J.* **2012**, *18*, 255–258. [CrossRef]
4. Yang, S.; Zhao, Y.F. Additive manufacturing-enabled design theory and methodology: A critical review. *Int. J. Adv. Manuf. Technol.* **2015**, *80*, 327–342. [CrossRef]
5. Adam, G.A.O.; Zimmer, D. Design for Additive Manufacturing—Element transitions and aggregated structures. *CIRP J. Manuf. Sci. Technol.* **2014**, *7*, 20–28. [CrossRef]
6. Giannatsis, J.; Dedoussis, V. Additive fabrication technologies applied to medicine and health care: A review. *Int. J. Adv. Manuf. Technol.* **2009**, *40*, 116–127. [CrossRef]
7. Petrovic, V.; Vicente Haro Gonzalez, J.; Jordá Ferrando, O.; Delgado Gordillo, J.; Ramón Blasco Puchades, J.; Portolés Griñan, L. Additive layered manufacturing: Sectors of industrial application shown through case studies. *Int. J. Prod. Res.* **2011**, *49*, 1061–1079. [CrossRef]
8. Klein, J.; Stern, M.; Franchin, G.; Kayser, M.; Inamura, C.; Dave, S.; Weaver, J.C.; Houk, P.; Colombo, P.; Yang, M. Additive manufacturing of optically transparent glass. *3D Print. Addit. Manuf.* **2015**, *2*, 92–105. [CrossRef]
9. Vidim, K.; Wang, S.-P.; Ragan-Kelley, J.; Matusik, W. OpenFab: A programmable pipeline for multi-material fabrication. *ACM Trans. Gr.* **2013**, *32*, 1–12. [CrossRef]
10. Guo, N.; Leu, M.C. Additive manufacturing: Technology, applications and research needs. *Front. Mech. Eng.* **2013**, *8*, 215–243. [CrossRef]
11. Ligon, S.C.; Liska, R.; Stampfl, J.; Gurr, M.; Mülhaupt, R. Polymers for 3D printing and customized additive manufacturing. *Chem. Rev.* **2017**, *117*, 10212–10290. [CrossRef]
12. Apium. eGuide: Material Extrusion (FFF) 3D Printing with PEEK. Available online: https://apiumtec.com/download/eguide-material-extrusion-fff-3d-printing-with-peek (accessed on 5 June 2020).
13. Popescu, D.; Zapciu, A.; Amza, C.; Baciu, F.; Marinescu, R. FDM process parameters influence over the mechanical properties of polymer specimens: A review. *Polym. Test.* **2018**, *69*, 157–166. [CrossRef]
14. Bathala, L.; Majeti, V.; Rachuri, N.; Singh, N.; Gedela, S. The role of Polyether Ether Ketone (Peek) in dentistry—A review. *J. Med. Life* **2019**, *12*, 5. [CrossRef] [PubMed]

15. Panayotov, I.V.; Orti, V.; Cuisinier, F.; Yachouh, J. Polyetheretherketone (PEEK) for medical applications. *J. Mat. Sci. Mat. Med.* **2016**, *27*, 118. [CrossRef] [PubMed]

16. Green, S. Compounds and composite materials. In *PEEK Biomaterials Handbook*, 2nd ed.; Kurtz, S.M., Ed.; William Andrew Publishing: Cambridge, MA, USA, 2019; pp. 27–51. [CrossRef]

17. Li, C.S.; Vannabouathong, C.; Sprague, S.; Bhandari, M. The use of carbon-fiber-reinforced (CFR) PEEK material in orthopedic implants: A systematic review. *Clin. Med. Insights Arthritis Musculoskelet. Disord.* **2015**, *8*, 33–45. [CrossRef]

18. Singh, S.; Prakash, C.; Ramakrishna, S. 3D printing of polyether-ether-ketone for biomedical applications. *Eur. Polym. J.* **2019**, *114*, 234–248. [CrossRef]

19. Li, Y.; Linke, B.S.; Voet, H.; Falk, B.; Schmitt, R.; Lam, M. Cost, sustainability and surface roughness quality–A comprehensive analysis of products made with personal 3D printers. *CIRP J. Manuf. Sci. Technol.* **2017**, *16*, 1–11. [CrossRef]

20. Mohamed, O.A.; Masood, S.H.; Bhowmik, J.L. Optimization of fused deposition modeling process parameters: A review of current research and future prospects. *Adv. Manuf.* **2015**, *3*, 42–53. [CrossRef]

21. Wong, K.V.; Hernandez, A. A review of additive manufacturing. *ISRN Mech. Eng.* **2012**, *2012*, 1–10. [CrossRef]

22. Ngo, T.D.; Kashani, A.; Imbalzano, G.; Nguyen, K.T.; Hui, D. Additive manufacturing (3D printing): A review of materials, methods, applications and challenges. *Compos. Part B Eng.* **2018**, *143*, 172–196. [CrossRef]

23. Nuñez, J.; Ortiz, Á.; Ramírez, M.A.J.; González Bueno, J.A.; Briceño, M.L. Additive Manufacturing and Supply Chain: A Review and Bibliometric Analysis. In *Engineering Digital Transformation*; Ortiz, Á., Andrés Romano, C., Poler, R., García-Sabater, J.-P., Eds.; Springer International Publishing: Cham, Switzerland, 2019; pp. 323–331. [CrossRef]

24. Singamneni, S.; Yifan, L.; Hewitt, A.; Chalk, R.; Thomas, W. Additive manufacturing for the aircraft industry: A review. *J. Aeronaut. Aerosp. Eng.* **2019**, *8*, 1–13. [CrossRef]

25. Galante, R.; Figueiredo-Pina, C.G.; Serro, A.P. Additive manufacturing of ceramics for dental applications: A review. *Dent. Mat.* **2019**, *35*, 825–846. [CrossRef] [PubMed]

26. Culmone, C.; Smit, G.; Breedveld, P. Additive manufacturing of medical instruments: A state-of-the-art review. *Addit. Manuf.* **2019**, *27*, 461–473. [CrossRef]

27. Liu, Z.; Wang, Y.; Wu, B.; Cui, C.; Guo, Y.; Yan, C. A critical review of fused deposition modeling 3D printing technology in manufacturing polylactic acid parts. *Int. J. Adv. Manuf. Technol.* **2019**, *102*, 2877–2889. [CrossRef]

28. Sood, A.K.; Ohdar, R.K.; Mahapatra, S.S. Improving dimensional accuracy of Fused Deposition Modelling processed part using grey Taguchi method. *Mat. Des.* **2009**, *30*, 4243–4252. [CrossRef]

29. Nancharaiah, T. Optimization of process parameters in FDM process using design of experiments. *Int. J. Emerg. Technol.* **2011**, *2*, 100–102.

30. Durgun, I.; Ertan, R. Experimental investigation of FDM process for improvement of mechanical properties and production cost. *Rapid Prototyp. J.* **2014**, *20*, 228–235. [CrossRef]

31. Ahn, S.; Montero, M.; Odell, D.; Roundy, S.; Wright, P. Anisotropic material properties of fused deposition modeling ABS. *Rapid Prototyp. J.* **2002**, *8*, 248–257. [CrossRef]

32. Lee, B.H.; Abdullah, J.; Khan, Z.A. Optimization of rapid prototyping parameters for production of flexible ABS object. *J. Mat. Process. Technol.* **2005**, *169*, 54–61. [CrossRef]

33. Lee, C.S.; Kim, S.G.; Kim, H.J.; Ahn, S.H. Measurement of anisotropic compressive strength of rapid prototyping parts. *J. Mat. Process. Technol.* **2007**, *187–188*, 627–630. [CrossRef]

34. Masood, S.H.; Mau, K.; Song, W.Q. Tensile properties of processed FDM polycarbonate material. *Mat. Sci. Forum* **2010**, *654–656*, 2556–2559. [CrossRef]

35. Smith, W.C.; Dean, R.W. Structural characteristics of fused deposition modeling polycarbonate material. *Polym. Test.* **2013**, *32*, 1306–1312. [CrossRef]

36. Lužanin, O.; Movrin, D.; Plančak, M. Effect of layer thickness, deposition angle, and infill on maximum flexural force in FDM-built specimens. *J. Technol. Plast.* **2014**, *39*, 49–58.

37. Wu, W.; Geng, P.; Li, G.; Zhao, D.; Zhang, H.; Zhao, J. Influence of layer thickness and raster angle on the mechanical properties of 3D-printed PEEK and a comparative mechanical study between PEEK and ABS. *Materials* **2015**, *8*, 5834–5846. [CrossRef] [PubMed]

38. Christiyan, K.G.J.; Chandrasekhar, U.; Venkateswarlu, K. A study on the influence of process parameters on the mechanical properties of 3D printed ABS composite. *IOP Conf. Ser. Mat. Sci. Eng.* **2016**, *114*, 1–8. [CrossRef]

39. Casavola, C.; Cazzato, A.; Moramarco, V.; Pappalettere, C. Orthotropic mechanical properties of fused deposition modelling parts described by classical laminate theory. *Mat. Des.* **2016**, *90*, 453–458. [CrossRef]

40. Chacón, J.M.; Caminero, M.A.; García-Plaza, E.; Núñez, P.J. Additive manufacturing of PLA structures using fused deposition modelling: Effect of process parameters on mechanical properties and their optimal selection. *Mat. Des.* **2017**, *124*, 143–157. [CrossRef]

41. Webbe Kerekes, T.; Lim, H.; Joe, W.Y.; Yun, G.J. Characterization of process–deformation/damage property relationship of fused deposition modeling (FDM) 3D-printed specimens. *Addit. Manuf.* **2019**, *25*, 532–544. [CrossRef]

42. Han, X.; Yang, D.; Yang, C.; Spintzyk, S.; Scheideler, L.; Li, P.; Li, D.; Geis-Gerstorfer, J.; Rupp, F. Carbon fiber reinforced PEEK composites based on 3D-printing technology for orthopedic and dental applications. *J. Clin. Med.* **2019**, *8*, 240. [CrossRef]

43. Yao, S.-S.; Jin, F.-L.; Rhee, K.Y.; Hui, D.; Park, S.-J. Recent advances in carbon-fiber-reinforced thermoplastic composites: A review. *Compos. Part B Eng.* **2018**, *142*, 241–250. [CrossRef]

44. Apium. Apium Product Brochure. Available online: https://apiumtec.com/download/apium-product-brochure (accessed on 8 August 2019).

45. Apium. Apium CFR PEEK Data-Sheet. Available online: https://apiumtec.com/download/apium-cfr-peek-datasheet (accessed on 9 August 2019).

46. ASTM. ASTM D638-14: Standard Test Method for Tensile Properties of Plastics. Available online: https://www.astm.org/Standards/D638 (accessed on 17 March 2020).

47. ASTM. ASTM D695-15: Standard Test Method for Compressive Properties of Rigid Plastics. Available online: https://www.astm.org/Standards/D695 (accessed on 17 March 2020).

48. ASTM. ASTM D3039/D3039M-17: Standard Test Method for Tensile Properties of Polymer Matrix Composite Materials. Available online: https://www.astm.org/Standards/D3039 (accessed on 17 March 2020).

49. SolidWorks. Available online: https://www.solidworks.com/ (accessed on 8 August 2019).

50. Simplify3D. Available online: https://www.simplify3d.com/ (accessed on 8 August 2019).

51. Kumar, G.P.; Regalla, S.P. Optimization of support material and build time in fused deposition modeling (FDM). *Appl. Mech. Mat.* **2012**, *110–116*, 2245–2251. [CrossRef]

52. Zhang, J.W.; Peng, A.H. Process-parameter optimization for fused deposition modeling based on Taguchi method. *Adv. Mat. Res.* **2012**, *538–541*, 444–447. [CrossRef]

53. MINITAB. Minitab 18 Support. Available online: https://support.minitab.com/en-us/minitab/18/ (accessed on 19 August 2019).

54. Derringer, G.; Suich, R. Simultaneous optimization of several response variables. *J. Qual. Technol.* **1980**, *12*, 214–219. [CrossRef]

55. MINITAB. What is Response Optimization? Available online: https://support.minitab.com/en-us/minitab/18/help-and-how-to/modeling-statistics/using-fitted-models/supporting-topics/response-optimization/what-is-response-optimization (accessed on 19 August 2019).

56. Yalçınkaya, Ö.; Mirac Bayhan, G. Modelling and optimization of average travel time for a metro line by simulation and response surface methodology. *Eur. J. Oper. Res.* **2009**, *196*, 225–233. [CrossRef]

 *applied*
sciences

Article

Bootstrap Analysis of the Production Processes Capability Assessment

Patrycjusz Stoma [1], Monika Stoma [2,*], Agnieszka Dudziak [2] and Jacek Caban [2]

[1] Q&R Polska, Sp. z o.o., 20-806 Lublin, Ploand; stoma@qrpolska.pl
[2] Faculty of Production Engineering, University of Life Sciences in Lublin, 28 Głęboka Street, 20-612 Lublin, Poland; agnieszka.dudziak@up.lublin.pl (A.D.); jacek.caban@up.lublin.pl (J.C.)
* Correspondence: monika.stoma@up.lublin.pl; Tel.: +48815319726

Received: 19 November 2019; Accepted: 6 December 2019; Published: 8 December 2019

Abstract: The high customer requirements for appropriate product quality pose a challenge for manufacturers and suppliers and also cause them many problems related to ensuring a sufficiently high product quality throughout the entire production cycle. For the above reasons, it is so important to assess the capability of monitored processes, and shaping, analyzing and controlling the capability of processes is an important aspect of managing an organization that uses a process approach to management. The use of an appropriate method to analyze the course of production processes is a necessity imposed by quality standards, e.g., ISO 9001: 2015. That is why it is so important to propose a quick and low-cost method of assessing production processes. For this purpose, a method of assessing the capability of the manufacturing process using bootstrap analysis was used. The article presents the analysis of inherent properties of the production process based on the results of measurements of the characteristic features of the process or the characteristics of the manufactured products (process variables) for the shafts with grooves. The main goals of the work are to develop a procedure for determining process capability based on the bootstrap method, including criteria for the classification of production process capability; to develop the criterion values for confidence intervals of production process capability; as well as to demonstrate the practical application of bootstrap analysis in manufacturing. Moreover, comparative analyses of process capabilities using bootstrap and classic methods were carried out. They confirm both the narrowing of the confidence interval when using the bootstrap method and the possibility of determining a better estimator of the lower limit of this range compared to the results obtained using the classic method. The tests carried out for the unit production of shafts with grooves showed that the analysis of the process capability for measuring tests n = 10 is possible. Finally, new criterion values for the assessment of process capability for the bootstrap method were proposed. The model for assessing the capability of production processes presented in the paper was implemented in low-volume production in the defense industry.

Keywords: production process capability; product quality; monitoring of production processes; process variables

1. Introduction

In the modern globalized market, and in connection with the growing requirements of customers, one of the key problems and at the same time challenges of suppliers is the quality assurance of manufactured products throughout the entire production cycle. One of the aspects of product quality assurance is the quality of production processes and in particular their capability.

Process capability is an important factor in the cooperation between the supplier and the customer [1]. Currently, the recipients of components for the aviation, automotive or machine industry among others, in addition to product specifications and acceptable manufacturing defects, impose on their suppliers the required process capability, as the lack of process capability control can generate losses. To meet these requirements, the supplier should therefore monitor and measure the course of manufacturing processes so as to be able to correct and improve the quality of the product based on reliable information. That is why the problem of assessing the capability of monitored processes becomes so important, and shaping, analyzing and controlling the capability of processes is an important aspect of organization management, especially when it uses a process approach to management.

The problems of assessing the capability of production processes and maintaining the capability at the required level are a key element of business cooperation. When entering into contracts between interested parties, it is often impossible to sign such agreement. In most cases, this is a level set too high for the production process capability, i.e., the quality of the production. Elderly, time-worn machinery park and old technology mean low process capability; in turn, modern machinery park, automated production, precise control of production processes and modern technology allow achieving high capability of the production process [2]. That quality of processes, represented by their capability, is expected by recipients from manufacturers, especially from European Union countries. In the case of suppliers operating in the automotive sector having a quality management system certificate, continuous process capability analysis is even an obligatory action [3].

The results of the production process measurements depend on the complexity of the research methods used. An appropriate, structured research method is also a prerequisite for obtaining reliable information about the capability of the production process.

In the general case, the analysis of the capability of the production process consists in comparing the width of the tolerance range required with the distribution of results obtained in a selected range of the duration of this process [4]. Currently, two methods of testing process capabilities are used: classic and percentile analysis. Classic analysis is used for distributions that can be considered normal, while percentile analysis is used for distributions that deviate from the normal distribution. In both cases, the number of samples taken should exceed $n = 100$ measurement results [5]. Therefore, if in reality the tested process cannot be characterized by the required number of measurement results (less than 100 measurements), and in addition, their distribution deviates from the normal, the analysis of the process capability, carried out by classic or percentile methods, is not possible (assuming the correctness and reliability of the obtained capability values). Therefore, it seems that in the case of monitoring the course of such processes, the bootstrap method can be used to test their capability [6].

The bootstrap method involves drawing with return lots of small-scale bootstrap samples from a small number of results. Due to the size of the bootstrap sample set, this method is cumbersome to measure process capability in industrial practice. However, after using the appropriate software, it becomes accessible to operators who do not have extensive knowledge of statistical process control.

The basic use of bootstrap analysis to measure the capability of production processes can find place in industrial practice, mainly in cases where:

- This is job-lot or job production.
- The measurement methods used in the research of processes are in the form of destructive tests, and, at the same time, the value of the tested sample is relatively high.
- Process capability assessment is performed for a trial batch to verify that the capability value required in the contract with the recipient is achievable. In this case, the use of bootstrap analysis reduces the costs associated with the production and testing of a larger sample batch, necessary using the classic method.
- The capability of the production process is tested, where measurements are rarely done, e.g., due to the cost of testing and analysis.

The bootstrap method can be used for what-if studies. It is used in many different areas, including in simulation models analyzing medical data [7–12], in financial analyzes [13–15], in solving problems in the area of logistics and distribution [16–18], in environmental protection [19–22], safety sciences [23], automotive [24], risk management [25,26] and in classic queuing models [27].

The Aims of the Study

Taking the above into consideration, the following aims of the study were formulated:

1. Development of a process capability determination procedure based on the bootstrap method, including criteria for the classification of production process capability.
2. Using the developed procedure to analyze the capability of the production process.
3. Conducting a comparative analysis of process capability determined by bootstrap and classic methods.
4. Development of criterion values for confidence intervals of process capabilities.

The subject of consideration at work will be the production process. The analysis of inherent properties of the production process will be carried out on the basis of the results of measuring the characteristic features of the process or the characteristics of the products manufactured (process variables).

In the following, this paper discusses in detail the issues related to the capability of the production process and the description of the bootstrap method used in this study. Then, the research methodology, obtained research results and their discussion are presented. The last part of the work summarizes the conducted research and presents the conclusions of the paper and the possibilities of practical application of the bootstrap method.

2. Process Capability

The characteristics of the course of the process are best determined using appropriate statistical methods (PN-EN ISO 9004:2018-06) [28]. This involves the need to obtain quantitative results of measuring process variables (PN-EN ISO 9001:2015-10) [29] that will allow effective process monitoring.

The basic goals of process analysis are as follows [30]:

1. Formulating directions and priorities for improving their course.
2. Measurement of process improvement effects.
3. Regulation of the process flow.

An ideal, stationary process should be characterized by the lack of dispersion of results—any selected process variable characterizing its output during the process has a constant value. In reality (practice) there are no stationary processes. The result of each real process is in the form of a distribution of the values of the selected process variable—it is characterized by dispersion. Thus, the basic parameters for assessing the quality of processes are:

- A measure of the distribution dispersion of a selected process variable (process dispersion).
- Measure of the location of the distribution of the selected process variable (process centering).

The assessment of the quality of production processes is associated with the assumption that each product delivered to the customer is endowed with a defect (loss); the smaller it is, the higher is the quality of the product [31,32]. This loss is the higher the more the value of the product feature considered deviates from the target value, including within the tolerance range. This contradicts the view expressed by Taylor that the product quality is constant if the property under consideration falls within the tolerance range (Figure 1). The tolerance field is defined here by the lower LSL and upper USL tolerance limits.

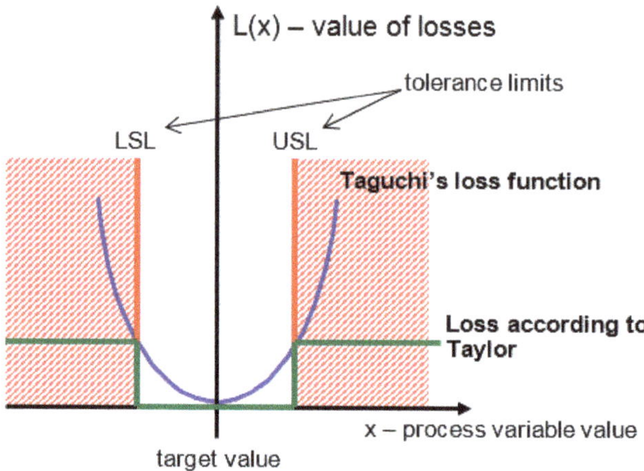

Figure 1. Taguchi and Taylor quality loss functions. Source: own study.

Therefore, given the approach presented by Taguchi, every product whose parameters deviate from the target value is characterized by a loss of quality [33,34].

To characterize the process with the capability specified for the selected process variable x, the width of the range in which we accept the obtained process results should be compared with the adopted limit of the process variable distribution, e.g., with the range of 6σ [35].

Indicators for process capability are increasingly used in industrial practice. Due to the introduction of a process approach in quality management in 2000 [29], it became necessary to monitor processes. "The organization should define the processes needed in the quality management system and their application in the organization and should (...) define and apply the criteria and methods (including monitoring, measurement and related performance indicators) needed to ensure the effective conduct and supervision of these processes". The point 8.5.1. of this standard refers to "the ability to achieve the planned results of production processes and services provided." In industrial practice, the planned result is the required width of the tested property range of the manufactured product.

The process capability is a special inherent feature of the process resulting from the statistical description of one of the outputs (adopted for the description of the process) carried out in a selected period of time. The process capability is the relationship between the required tolerance of the considered product property—process output—treated as a process variable and the obtained dispersion of the value of this variable—the result of the adopted method of limiting the distribution of variable values obtained in a given process duration. Graphic interpretation of the process capability assessment is presented in Figure 2.

The simplest form of the process capability indicator, denoted by C_P, is defined as follows:

$$C_P = \frac{\text{required tolerance}}{\text{process dispersion}},$$

(1)

This coefficient is used when the instantaneous average process value $\bar{x}$, obtained on the basis of measurements, is equal to the assumed—purposeful process value T. Arithmetic mean $\bar{x}$ and standard deviation s of measurements are given by the formula:

$$\bar{x} = \frac{1}{n}\sum x_i,$$

(2)

$$s = \sqrt{\frac{1}{n}\sum_{i=1}^{n}(x_i - \bar{x})^2} \quad \text{for n} > 30,$$

(3)

$$s = \sqrt{\frac{1}{n-1}\sum_{i=1}^{n}(x_i - \bar{x})^2} \quad \text{for n } \leq 30,$$

(4)

where:

x_i, measurement value;

n, sample size.

As follows from the above dependence, in order not to generate excessive losses associated with maintaining defective products, the process capability index should be [36]:

$$C_P \geq 1,$$

(5)

In the assessment of process capability in industrial practice, the distribution limited by six standard deviations is taken as the measure of the scatter of measurement results [37]. For the normal distribution of the process variable, the C_P process capability indicator takes the form:

$$C_p = \frac{USL - LSL}{6\sigma},$$

(6)

where:

USL, upper tolerance limit;

LSL, lower tolerance limit;

n, sample size;

σ, standard deviation of the general population.

Due to the growing requirements of customers, especially global concerns, the criteria for the minimum limit value of the C_P coefficient have been adopted for some industries [38]. According to Steinem et al., [39] the minimum values of C_P coefficient for selected industries are for the machinery industry $C_P = 1$, for the automotive industry $C_P = 1.33$, and in the aviation industry $C_P = 2$. Analyzing the capability of the production process, three ranges of the C_P value can be presented. When $C_P > 1$ the process dispersion is smaller than the width of the tolerance range, and this is the recommended process capability. For $C_P = 1$, the tolerance range is equal to the process dispersion, and this is a satisfactory process capability [40]. When $C_P < 1$, the tolerance range is narrower than the process dispersion—defective products are produced in excessive quantity. Then, the process capability is insufficient [41]. Another interpretation of the process capability coefficient value can be found in the literature, e.g., Kubera states that low process capability is $C_P < 1$, average $1 < C_P < 1.3$ and high $C_P > 1.33$ [41]. The interpretation may be different for different types of industry and processes, in this paper the capability at $C_P = 1$ level will be accepted as satisfactory.

Whether or not a production process to be executed is capable of achieving the assumed performance parameters depends, among others, on the reliability of the machines and technological devices that make up the system under design [42]. Layouts and temporal structure optimization of manufacturing requires application of a multi-criteria approach in designing production systems [43].

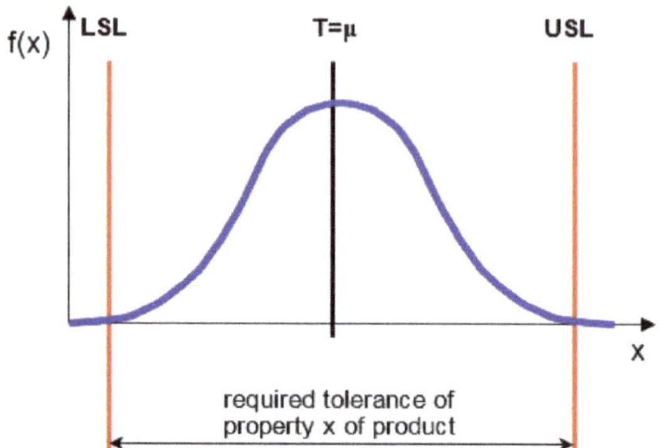

Figure 2. Graphic illustration of process capability. Source: own study.

3. Bootstrap Method

There are methods that simplify the procedure when assessing process capability using numerical indicators. One of such methods is bootstrap analysis based on the so-called bootstrap samples [44], which can be used when the sample size is not very large (show Figure 3). The purpose of this analysis is the possible verification of previously obtained results using factor methods.

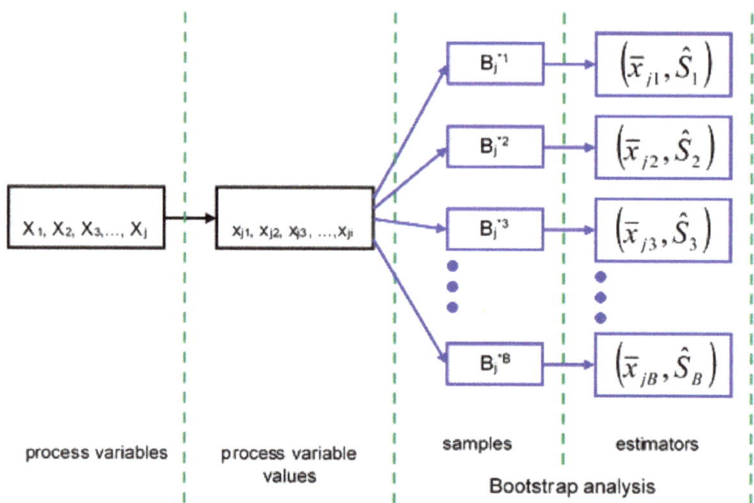

Figure 3. Scheme of bootstrap analysis. Source: own study.

Bootstrap methods have been known for over 20 years, but only in recent years they have been widely used, primarily in stochastic simulation models. The basis of this method is the assumption that the future is similar to the past. Therefore, instead of studying the past and trying to describe it using theoretical distributions, and then simulating the future using the selected distributions, you can generate simulation input data directly from historical data [45,46]. As a consequence, this means that since the observed sample of real data contains all the necessary information about the studied population, this sample can be treated as a population.

Bootstrap analysis consists of a draw with returning individual results from a random sample (from the original data) and then creating a new sample from the drawn samples for the study [47]. Thus, it is a method of estimating the distribution of estimation errors, using multiple random draws of observations with returns from the original sample. The draws take into account all possible combinations of elements from the sample, based on real data [48]. It allows checking the distribution of parameters in relation to the initial sample [49]. The bootstrap method is useful when the form of the distribution of the variable in the population is not known and when the quality or amount of information collected does not allow the use of classic statistical methods [50]. Due to the fact that it does not make assumptions about the distribution in the population, it is included in the non-parametric methods.

Let x_{j1}, x_{j2}, x_{j3}, ... x_{ji} denote the values of the process variable X_i of the production process under investigation. From a given set of x_{ji} values, we draw with returning n measurement values and in this way we get new bootstrap samples B_j*^1, B_j*^2, B_j*^3, ... B_j*^B.

Each bootstrap sample consists of exactly the same number of "n" elements as the number of values tested [30,51–53].

The number of bootstrap draws cannot be less than n^n. Based on empirical research, it has been shown that a sufficient number of draws for conducting tests is $B = 1000$ measurements [54]. After obtaining the set number of bootstrap attempts, based on them, inference is calculated by calculating the appropriate statistics ϕ. The empirical distribution obtained in this way is used to make inferences about the parameter θ [48,55].

The main advantage of this is that the process variable distribution is not studied but empirically constructed based on measuring a large number of samples of the process variable value [56]. The advantage of this is also the ability to assess process capabilities based on abnormal distributions, characterized by high skewness, flattening, drift, etc.

To sum up the above, the following analogy is crucial for the use of the bootstrap method in statistical inference: the bootstrap sample is for the sample drawn what the sample drawn for the entire population.

4. Subject and Research Method

The paper presents a method for assessing the manufacturing process capability of an shafts with grooves, a typical production process whose quality assessment is made on the basis of capability analysis and is carried out during the final product control. Based on the analysis of many industrial products it can be concluded that one half of all machine parts are rotational parts: shafts (over 40%), discs, sleeves, thin wall cylinders, rings, etc. [57].

In the case of production, there are two categories of processes when the capability analysis using the bootstrap method seems to be practically the best solution. These categories include processes in which:

- Measurements of the process variable are performed by the method destroying the sample and for economic reasons the sample size of the process capability becomes the sample size and
- The result of the course is a small collection of products (low-volume production).

The production process of an shafts with grooves belongs to the second of the mentioned categories of processes.

In the present case, due to the small size of the general population of products $n = 10$, the analysis of the process capability will be carried out using the bootstrap method, with the population size being the same as the sample size.

The shafts with grooves is manufactured by a manufacturing plant that provides complete reinforcement for defense purposes. The shafts is a necessary component of this armament.

The production of the shafts with the groove was chosen for the following reasons:

- The process is a typical example of a small batch process.
- It is necessary to make a cyclical assessment of the production process.

- Due to the number of shafts produced, it is not possible to assess the shape of the process variable distribution, so the assumption about the normality of the distribution of quality characteristics is rejected.

The results analyzed were obtained on the basis of measurements made using a CMM (coordinate-measuring machine). The geometry of this product defines 19 parameters (18 measurement parameters and total length), with parameter 12 selected as the process variable for analyzing the process capability.

To assess the process capability, a research procedure consisting of five activities was used (see Figure 4):

1. Measurement systems analysis (MSA).
2. Statistical process control (SPC).
3. Time series analysis.
4. Distribution studies of the selected process variable.
5. Process capability assessment, including capability assessment using three methods: classic, percentile, bootstrap.

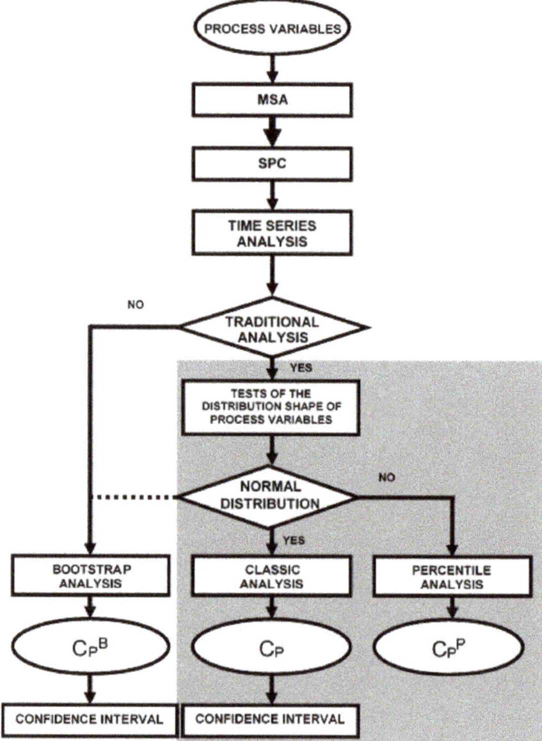

Figure 4. Diagram of the algorithm for assessing the quality of processes using the classic, percentile and bootstrap methods. Source: own study.

The principle of process capability assessment should be to take into account confidence intervals with a lower limit $C_{P\,MIN}$. This indicates that with some probability (the confidence interval was calculated at $P = 95\%$ in the work), the capacity of the analyzed process will not be lower than the lower limit of the confidence interval [58,59].

The detailed scope of research for individual activities is presented in Figure 5. Due to the large number of results (especially charts), the results are presented only for the final stage of the procedure, i.e., the assessment of the process capability C_P.

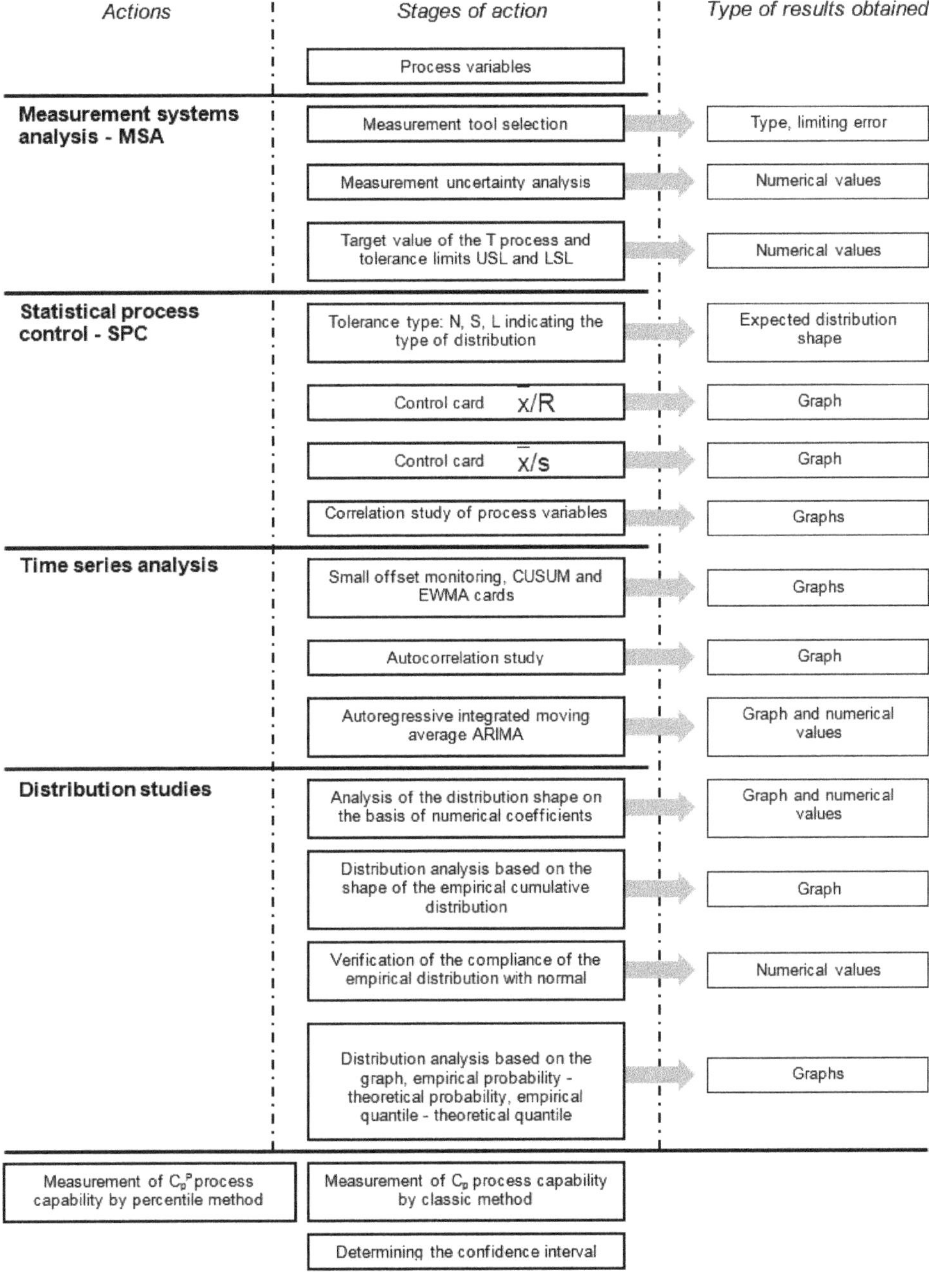

Figure 5. Sequence of actions in the process capability assessment procedure and presentation of individual results. Source: own study.

5. Results and Discussion

The results of the analyzes of process variable capacity 12, the unit production process of the shafts with the grooves are shown in Figure 6.

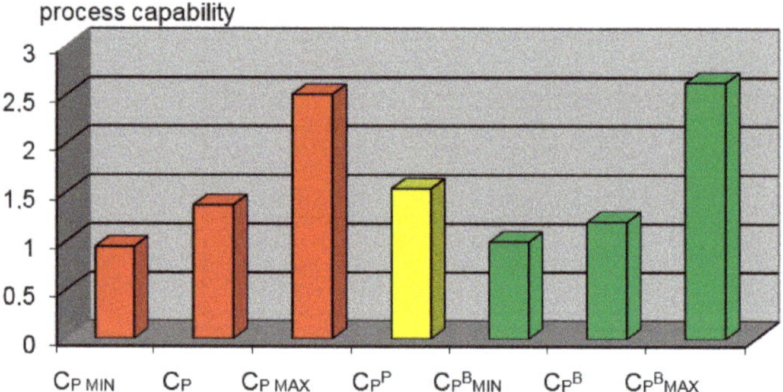

Figure 6. Comparison of the results of the value of the unit production process of the shafts with grooves, for the 12 process variable, by the classic method (red), percentile method (yellow) and bootstrap method (green). Source: own study.

For the unit production process of an shafts with grooves, the lowest value is taken by the process coefficient of the lower limit of the confidence interval of the classic analysis. The process capability coefficient determined by the percentile method C_P^P falls within the process capability range determined by the classic method as well as in the process range determined by the bootstrap method.

The tests carried out for the unit production of shafts with grooves have shown that the analysis of the process capability for ten measurements is possible. In both examined cases, the average value of process capability determined by the classic method C_P was greater than the lower bootstrap limit of the process capability. Whereas, the lower limit of the process capability determined by the classic method $C_{P\,MIN}$ was below the lower bootstrap limit $C_P^B{}_{MIN}$ (Figure 7).

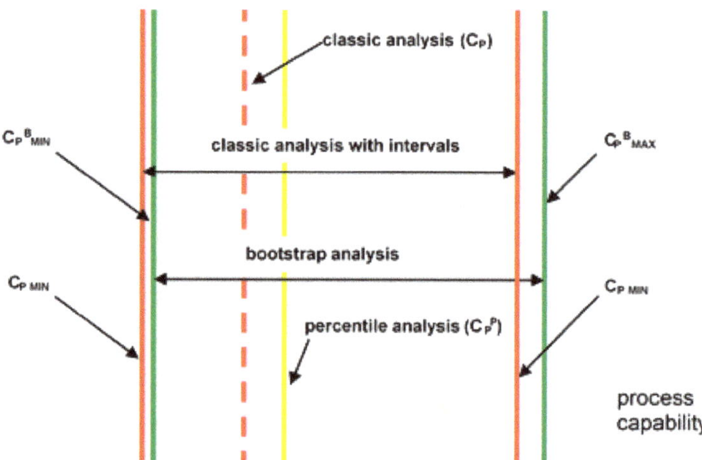

Figure 7. Graphic presentation of the results of the production process capability assessment of the shafts with grooves, process variable 12. Source: own study.

Based on the analyzes and results obtained, new criterion standards for the assessment of process capability can be introduced, resulting from the lower limit of process capability assessment. Analyzing the lower limits of process capability confidence intervals, processes can be assessed without the need to analyze the sample size, because interval estimation of process capability using the bootstrap method does not require this. The proposed new criterion values for process capability assessment for the bootstrap method are presented in Table 1.

Table 1. Sample criterion values of process capability assessment for the lower limit of bootstrap process capability in relation to the process capability determined by the traditional method for normal distribution. Source: own study.

C_P	$C_P{}^B{}_{MIN}$	
n = 100	n = 100	n = 10
1.0	0.86	0.78
1.33	1.15	1.05
1.67	1.44	1.31
2.0	1.72	1.57

6. Summary and Discussion

The article reviews and critically evaluates the methods used to determine process capability. The capability of production processes was treated here as the basic property determining their quality, conditioning control and, as a consequence, process management. In the case of a process-based approach to organization management, process capability can and should be one of the key instruments of organization management, in particular, quality management in the organization.

Against the background of the methods used so far to determine the capability of production processes, traditional and percentile, the article presents the assumptions of the bootstrap method and discusses the effects of using this method in relation to specific cases of process flows [60]. An attempt was also made to use the bootstrap method to analyze production processes.

In addition, a procedure for determining the capability of production processes based on the bootstrap method was developed and this procedure was used to analyze the process of manufacturing shafts with grooves (production process). The above model for assessing the capability of production processes using the bootstrap method for low-volume production has been implemented in the defense industry.

On the basis of the literature studies, research and analysis of the results obtained, the following conclusions can be drawn:

1. The procedure of process capability determination, developed on the basis of the bootstrap method, is a simplification compared to the classic method by omitting the assessment of the shape of the distribution of measured parameters.

2. An examined comparative analysis of the processes capabilities, determined by the bootstrap and classic methods, including those based on the simulation of results obtained using the classic method for a small sample size, allowed the determination of new criterion values of the process capability, analyzed by the bootstrap method, based on the small random sample size.

3. The use of bootstrap analysis results in a better approximation of confidence intervals of the production process capability factor than traditional analyses. The obtained test results confirm both the narrowing of the confidence interval when using the bootstrap method and the possibility of determining a better estimator of the lower limit of this range. The lower limit of process capability, obtained on the basis of the bootstrap method, is shifted upwards compared to the results obtained using the classic method. This is important when assessing process capability because the obtained value of the lower confidence interval limit means that at the selected confidence level the actual capability of the tested process will not be lower.

4. New criterion values for process capability determined by the bootstrap method were developed. The introduced criteria make it possible to assess the capabilities of processes taking place in various branches of the economy (different classes of processes), determined by the bootstrap method, and are equivalent to the criteria used so far resulting from the use of the classic method of determining capacity. The developed criteria allow the assessment of the capability of the processes examined also on the basis of small sample sizes ($n \geq 10$).

Based on the conducted research and literature study, the following possibilities of practical application of bootstrap analysis can be proposed:

1. The unit and low-volume production process capability (this type of process was described in the paper), as a result of which a few or tens of products are obtained, can not be assessed by traditional methods.
2. The production process capability assessment, carried out using the bootstrap method and realized on the base of destructive testing methods, shows an advantage over the classical method due to the reduction of the costs of performed tests.
3. The production process capability assessment of the trial batch, in the event that the production facility analyzes the commencement of production and the requirements imposed by the customer strictly specify the capability of the process, can also be performed with the help of bootstrap analysis, thereby reducing the sample batch size, and as a consequence reduction of project costs.
4. Bootstrap analysis of process capability assessment can also be used to determine required tolerance limits. In this case, based on the bootstrap analysis carried out and the assumed value of the bootstrap process capability coefficient, the searched tolerance limits sought can be determined.
5. Another application of bootstrap analysis is the ability of production process capability assessment when conducting a second party audit—a customer audit. On the basis of a small sample size and self-made measurements, auditors can easily and quickly form an opinion on the capability of the production processes in which they are interested. This is a very important application of bootstrap analysis, because the supplier can produce products in the production process with a different process capability than can be determined from the batch delivered to the customer.
6. The study of changes in process capability over time, conducted to determine the quality of a specific production machine and technological lines, can also be performed used bootstrap analysis. In this case, the process capability analysis gives a signal that the machine needs to be repaired or completely replaced. In the case of testing the capabilities of technological lines, this way it is possible to carry out an analysis of multimodal distributions obtained as a result of overlapping distribution of measured values for elements made on machines working in parallel in a technological line. Thus, the use of process capability assessment using bootstrap analysis makes it possible to assess numerically the state of the production equipment.
7. Bootstrap analysis of the production process capability can also be used as one of the possibilities to validate this process.

Author Contributions: Conceptualization, P.S. and M.S.; methodology, P.S., M.S. and A.D.; formal analysis, M.S., A.D. and J.C.; resources, M.S., A.D. and J.C.; writing—original draft preparation, P.S., M.S., A.D. and J.C.; writing—review and editing, M.S., A.D. and J.C.; visualization, M.S., A.D. and J.C.; project administration, M.S.

Funding: This research received no external funding.

Conflicts of Interest: The authors declare no conflict of interest.

References

1. Liao, M.Y. Process capability control chart for non-normal data–evidence of on-going capability assessment. *Qual. Technol. Quant. Manag.* **2016**, *13*, 165–181. [CrossRef]
2. Kashif, M.; Aslam, M.; Al-Marshadi, A.H.; Jun, C.H.; Khan, M.I. Evaluation of modified non-normal process capability index and its bootstrap confidence intervals. *IEEE Access* **2017**, *5*, 12135–12142. [CrossRef]
3. IATF 16949:2016. *Standard Systemu Zarządzania Jakością w Przemyśle Motoryzacyjnym*; Polski Komitet Normalizacyjny: Warszawa, Poland, 2016; p. 16949.
4. Saha, M.; Dey, S.; Maiti, S.S. Bootstrap confidence intervals of C_{pTk} for two parameter logistic exponential distribution with applications. *Int. J. Syst. Assur. Eng. Manag.* **2019**, *10*, 623–631. [CrossRef]
5. Franklin, L.A.; Wasserman, G.S. Bootstrap lower confidence limits for capability indices. *J. Qual. Technol.* **1992**, *24*, 196–202. [CrossRef]
6. Dwornicka, R.; Radek, N.; Pietraszek, J. The bootstrap method as a tool to improve the design of experiments. *Syst. Saf. Hum.-Tech. Facil.-Environ.* **2019**, *1*, 724–729. [CrossRef]
7. Bland, J.M.; Altman, D.G. Statistics Notes: Bootstrap resampling methods. *BMJ* **2015**, *350*, h2622. [CrossRef]
8. Grunkemeier, G.L.; Wu, Y. Bootstrap resampling methods: Something for nothing? *Ann. Thorac. Surg.* **2004**, *77*, 1142–1144. [CrossRef]
9. Lee, S.; Kim, C. Estimation of association between healthcare system efficiency and policy factors for public health. *Appl. Sci.* **2018**, *8*, 2674. [CrossRef]
10. Lipton, J.W.; Shaw, W.D.; Holmes, J.; Patterson, A. Short communication: Selecting input distributions for use in Monte Carlo simulations. *Regul. Toxicol. Pharmacol.* **1995**, *21*, 192–198. [CrossRef]
11. Schomaker, M.; Heumann, C. Bootstrap inference when using multiple imputation. *Stat. Med.* **2018**, *37*, 2252–2266. [CrossRef]
12. Vicente, E.; Jha, A.; Frey, E. A nonparametric sinogram-based bootstrap resampling method to investigate scan time reduction in nuclear medicine imaging. *J. Nucl. Med.* **2016**, *57*, 1872.
13. Boyle, P.; Broadie, M.; Glasserman, P. Monte Carlo methods for security pricing. *J. Econ. Dyn. Control* **1997**, *21*, 1267–1321. [CrossRef]
14. Hacker, S.; Hatemi, J.A. A bootstrap test for causality with endogenous lag length choice: Theory and application in finance. *J. Econ. Stud.* **2012**, *39*, 144–160. [CrossRef]
15. Du, K.; Worthington, A.C.; Zelenyuk, V. Data envelopment analysis, truncated regression and double-bootstrap for panel data with application to Chinese banking. *Eur. J. Oper. Res.* **2018**, *265*, 748–764. [CrossRef]
16. Alexandre, L.; da Silva, N.C.F.; da Silva, C.M. Bootstrap method in price analysis in reverse logistics of solid waste from commercial restaurants. *Int. J. Adv. Eng. Res. Sci. (IJAERS)* **2019**, *6*, 482–485. [CrossRef]
17. Ferrari, C.; Migliardi, A.; Tei, A. A bootstrap analysis to investigate the economic efficiency of the logistics industry in Italy. *Int. J. Logist. Res. Appl.* **2018**, *21*, 20–34. [CrossRef]
18. Lytras, M.D.; Visvizi, A. Who uses Smart City services and what to make of it: Toward interdisciplinary Smart Cities research. *Sustainability* **2018**, *10*, 1998. [CrossRef]
19. Barakat, H.M.; Nigm, E.M.; Khaled, O.M.; Momenkhan, F.A. Bootstrap method for order statistics and modeling study of the air pollution. *Commun. Stat. —Simul. Comput.* **2015**, *44*, 1477–1491. [CrossRef]
20. Keller, J.P.; Chang, H.H.; Strickland, M.J.; Szpiro, A.A. Measurement error correction for predicted spatiotemporal air pollution exposures. *Epidemiology* **2017**, *28*, 338–345. [CrossRef]
21. Koçak, E.; Şarkgüneşi, A. The impact of foreign direct investment on CO_2 emissions in Turkey: New evidence from cointegration and bootstrap causality analysis. *Environ. Sci. Pollut. Res.* **2018**, *25*, 790–804. [CrossRef]
22. Zhang, A.; Shi, H.; Li, T.; Fu, X. Analysis of the influence of rainfall spatial uncertainty on hydrological simulations using the Bootstrap method. *Atmosphere* **2018**, *9*, 71. [CrossRef]
23. Corral-De-Witt, D.; Carrera, E.V.; Munoz-Romero, S.; Tepe, K.; Rojo-Alvarez, J.L. Multiple correspondence analysis of emergencies attended by integrated security services. *Appl. Sci.* **2019**, *9*, 1396. [CrossRef]
24. Bąkowski, A.; Radziszewski, L.; Žmindak, M. Determining selected diesel engine combustion descriptors using the bootstrap method. *Procedia Eng.* **2016**, *157*, 451–456. [CrossRef]
25. Ardia, D.; Gatarek, L.T.; Hoogerheide, L.F. A new bootstrap test for multiple assets joint risk testing. *J. Risk* **2017**, *19*, 4. [CrossRef]

26. Valášková, K.; Spuchľáková, E.; Adamko, P. Non-parametric Bootstrap method in risk management. *Procedia Econ. Financ.* **2015**, *24*, 701–709. [CrossRef]

27. Barton, R.R.; Schruben, L.W. Resampling methods for input modeling. In Proceedings of the 2001 Winter Simulation Conference, Arlington, VA, USA, 9–12 December 2001; Peters, B.A., Smith, J.S., Medeiros, D.J., Rohrer, M.W., Eds.; pp. 372–378. [CrossRef]

28. PN-EN ISO 9004:2018-06. *Zarządzanie Jakością, Jakość Organizacji, Wytyczne Osiągnięcia Trwałego Sukcesu*; Polski Komitet Normalizacyjny: Warszawa, Poland, 2018; Volume 6, p. 9004.

29. PN-EN ISO 9001:2015-10. *Systemy Zarządzania Jakością Wymagania*; Polski Komitet Normalizacyjny: Warszawa, Poland, 2015; Volume 10, p. 9001.

30. Płaska, S. *Wprowadzenie do Statystycznego Sterowania Procesami Technologicznymi*; Wydawnictwo Politechniki Lubelskiej: Lublin, Poland, 2000.

31. Balamurali, S.; Usha, M. Determination of an efficient variables sampling system based on the Taguchi process capability index. *J. Oper. Res. Soc.* **2019**, *70*, 420–432. [CrossRef]

32. Taguchi, G.; Elsayed, E.A.; Hsiang, T.C. *Quality Engineering in Production Systems*; Mc-Graw-Hill College: New York, NY, USA, 1989.

33. Gavin, D.; Gallimore, K.; Brown, J. Does ISO 9000 give a quality emphasis advantage? A comparison of large service and manufacturing organizations. *Qual. Manag. J.* **2001**, *8*, 52–59.

34. Kuvaja, P.; Bicego, A. Bootstrap? A European assessment methodology. *Softw. Qual. J.* **1994**, *3*, 117–127. [CrossRef]

35. Dahlgaard, J.; Kristensen, K.; Kanji, G. *Podstawy Zarządzania Jakością*; Wydawnictwo Naukowe PWN: Warszawa, Poland, 2002.

36. Kapania, M. Measuring Your Process Capability. *Qual. Product. J.* **2000**, 1–13.

37. Wu, C.W.; Shu, M.; Pearn, W.; Liu, K. Bootstrap approach for supplier selection based on production yield. *Int. J. Prod. Res.* **2008**, *46*, 5211–5230. [CrossRef]

38. Suozzi, M. *Process Capability Studies*; Hughes Aircraft Company: Tucson, AZ, USA, 1990.

39. Steinem, S.; Bovas, A.; MacKay, J. *Understanding Process Capability Indices. Institute for Improvement of Quality and Productivity*; University of Waterloo: Waterloo, ON, Canada, 2003.

40. Jay, A. *Six Sigma Simplified Tools*; LifeStar: Denver, CO, USA, 2003.

41. Kubera, H. *Zachowanie Jakości Produktu*; Wydawnictwo Akademii Ekonomicznej w Poznaniu: Poznań, Polska, 2002.

42. Gola, A. Reliability analysis of reconfigurable manufacturing system structures using computer simulation methods. *Eksploatacja i Niezawodnosc—Maintenance and Reliability* **2019**, *21*, 90–102. [CrossRef]

43. Plinta, D.; Krajčovič, M. Production System Designing with the Use of Digital Factory and Augmented Reality Technologies. Progress in Automation, Robotics and Measuring Techniques. ICA 2015. In *Advances in Intelligent Systems and Computing*; Szewczyk, R., Zieliński, C., Kaliczyńska, M., Eds.; Springer: Cham, Germany, 2015; p. 350. [CrossRef]

44. Efron, B.; Tibshirani, R. Bootstrap methods for standard errors, confidence intervals, and other measures of statistical accuracy. *Stat. Sci.* **1986**, *1*, 54–75. [CrossRef]

45. Gentle, J.E. *Random Number Generation and Monte Carlo Methods*; Springer: New York, NY, USA, 2003.

46. Mielczarek, B. Metody próbkowania w symulacji Monte Carlo. *Prace Naukowe Instytutu Organizacji i Zarządzania Politechniki Wrocławskiej* **2007**, *83*, 187–199.

47. Clare, A. Machine Learning and Data Mining for Yeast Functional Genomics. Ph.D. Thesis, Departament of Computer Science University of Wales, Aberystwyth, UK, February 2003.

48. Zglińska-Pietrzak, A. Zastosowanie metody bootstrapowej w analizie portfelowej. *Przegląd Statystyczny* **2012**, *59*, 246–257.

49. Efron, B. Bootstrap methods: Another look at the jackknife. *Ann. Stat.* **1979**, *7*, 1–26. [CrossRef]

50. Rao, G.S.; Aslam, M.; Kantam, R. Bootstrap confidence intervals of C_{Npk} for inverse Rayleigh and log-logistic distributions. *J. Stat. Comput. Simul.* **2016**, *86*, 862–873. [CrossRef]

51. Domański, C.; Pruska, K. *Nieklasyczne Metody Statystyczne*; PWE: Warszawa, Poland, 2000.

52. Kozak, P.; Płaska, S.; Stoma, P. Ocena jakości procesów z wykorzystaniem metod bootstropowych, IX Konferencja Techniczna, pt. In *Metrologia w Technikach Wytwarzania Maszyn*; Wydawnictwo Politechniki Częstochowskiej: Częstochowa, Poland, 2001; pp. 129–134.

53. Saama, P. *Introduction to Resampling Methods—Based Methods*; UCLA Office of Academic Computing: Los Angeles, CA, USA, 1997.

54. Efron, B.; Tibshirani, R. *An Introduction to the Bootstrap*, 1st ed.; Chapman & Hall/CRC: Boca Raton, FL, USA; London, UK; New York, NY, USA; Washington, DC, USA, 1993; p. 456.

55. Kisielińska, J. Dokładna metoda bootstrapowa i jej zastosowanie do estymacji wariancji. *Przegląd Statystyczny* **2011**, *58*, 1–2, 60–73.

56. Davison, A.; Hinkley, D. *Bootstrap Methods and Their Application*; Cambridge Series in Statistical and Probabilistic Mathematics; Cambridge University Press: Cambridge, UK, 1997.

57. Świć, A.; Wołos, D.; Gola, A.; Šmidová, N. Accuracy control in the process of low-rigidity elastic deformable shafts turning. *Tehnicki Vjesnik-Technical Gazette* **2019**, *26*, 927–934. [CrossRef]

58. Dey, S.; Saha, M.; Maiti, S.S.; Jun, C.H. Bootstrap confidence intervals of generalized process capability index C_{pyk} for Lindley and power Lindley distributions. *Commun. Stat.-Simul. Comput.* **2018**, *47*, 249–262. [CrossRef]

59. Kashif, M.; Aslam, M.; Rao, G.S.; Al-Marshadi, A.H.; Jun, C.H. Bootstrap confidence intervals of the modified process capability index for Weibull distribution. *Arab. J. Sci. Eng.* **2017**, *42*, 4565–4573. [CrossRef]

60. Hall, P.; Härdle, W.; Simar, L. Iterated bootstrap with applications to frontier models. *J. Product. Anal.* **1995**, *6*, 63–76. [CrossRef]

Article

Modeling and Simulation of Processes in a Factory of the Future

Patrik Grznár [1,*], Milan Gregor [1], Martin Krajčovič [1], Štefan Mozol [1], Marek Schickerle [1], Vladimír Vavrík [1], Lukáš Ďurica [2], Martin Marschall [2] and Tomáš Bielik [2]

[1] Department of Industrial Engineering, Faculty of Mechanical Engineering, University of Žilina, Univerzitná 8215/1, 010 26 Žilina, Slovakia; milan.gregor@fstroj.uniza.sk (M.G.); martin.krajcovic@fstroj.uniza.sk (M.K.); stefan.mozol@fstroj.uniza.sk (Š.M.); schickerle@stud.uniza.sk (M.S.); vladimir.vavrik@fstroj.uniza.sk (V.V.)
[2] Institute of Competitiveness and Innovations, University of Žilina, Univerzitná 8215/1, 010 26 Žilina, Slovakia; lukas.durica@fstroj.uniza.sk (L.Ď.); martin.marschall@fstroj.uniza.sk (M.M.); tomas.bielik@fstroj.uniza.sk (T.B.)
* Correspondence: patrik.grznar@fstroj.uniza.sk; Tel.: +421-41-513 2733

Received: 4 May 2020; Accepted: 26 June 2020; Published: 29 June 2020

Featured Application: Application of article is mainly in the area of future manufacturing systems where the control system will use simulation for predicting future state and base on information carry out actions.

Abstract: Current trends in manufacturing, which are based on customisation and gradually customised production, are becoming the main initiator for the development of new manufacturing approaches. New manufacturing approaches are counted as the application of new behavioural management patterns that calculate the retained competencies of decision-making by the individual members of the system agent; the production becomes decentralised. The interaction of the members of such a system creates emergent behaviour, where the result cannot be accurately determined by ordinary methods and simulation must be applied. Modelling and simulation will, therefore, be an integral part of the planning and control of the processes of factories of the future. The purpose of the article is to describe the use of modelling and simulation processes in factories of the future. The first part of the article describes new manufacturing concepts that will be used in factories of the future, with a description of modelling and simulation routing in the frame of Industry 4.0. The next section describes how simulation is used for the control of manufacturing processes in factories of the future. The included subsection describes the implementation of this suggested pattern in the laboratory of ZIMS (Zilina Intelligent Manufacturing System), with an example of a metamodeling application and the results obtained.

Keywords: advanced industrial engineering; modelling and simulation; factory of the future; smart factory; manufacturing systems; production planning optimisation; decision support

---

1. Introduction

Future manufacturing systems will differ significantly from those of today. The changes will not only result in the pressure of customers on the variant of new products but also revolutionary changes in the impact of technological innovation. The most significant factor that affects the existing manufacturing environment is the customer. The factory must be able to produce the required product in the shortest possible time and at a reasonable cost. Future manufacturing will provide products that will be tailored to the requirements of a particular customer, highly sophisticated, complex, and capable of offering new functionality; therefore, it will require an entirely new manufacturing environment.

The customisation and personalisation of products are a complex problem that researchers are trying to tackle today. On the one hand, researchers have used the appropriate construction of new products, also known as modular, reconfigurable products. On the other hand, they have also tried to increase the flexibility of the manufacturing system, which we now refer to as reconfigurable manufacturing. However, future manufacturing systems will use completely new principles in their operation. Researchers have sought to develop and exploit new methods and approaches to product design and production due to the growing complexity of both products and manufacturing systems [1].

Industry 4.0 is a digital revolution being witnessed in the present generation, whereby the aim is to digitise the entire manufacturing process with minimal human or manual intervention [2]. We are in a time where every major breakthrough in technology changes the face of manufacturing industries. At present, we are in the era of Industry 4.0, which is hailed as the age of cyber-physical systems (CPSs) that has taken manufacturing and associated industry processes to an unforeseen level with flexible production, including manufacturing, supply chain, delivery, and maintenance [3]. The development of the Industry 4.0 concept was needed to develop new competitive business models. These business models need to be based on cooperation and better use of the available resources [4]. Industry 4.0 is based on digitalisation and application of exponential technologies. Digitisation and application of exponential technologies are directly linked to CPSs. CPSs presaturate physical devices with built-in tools for digital data collection, processing, and distribution, and, through the internet, are connected to each other online. CPSs form the basis for technology such as the Internet of Things and, in combination with the Internet of Services, form the base for Industry 4.0.

New factories, or their manufacturing systems, will have unique features that enable them to respond quickly and efficiently to frequently changing customer demands. These manufacturing systems will be designed as modular, reconfigurable, and intelligent holonic systems capable of rapidly changing their functions and capacities based on the auto diagnostic. The dynamism of complex manufacturing systems will no longer be possible to study using today's modelling and simulation techniques. The future dynamic manufacturing environment will require robust modelling and simulation tools that will be able to simulate complex phenomena and processes. New simulation systems must function as part of complex control systems, working in real-time and must be used to support decision-making and the creation of new knowledge. In this case, real-time work is seen as a rapid response to emerging events and time deterministic calculation of the trajectories of the development of future manufacturing system conditions [5].

Simulation has become the most essential tool for dynamic analysis of complex systems in recent decades. A high level of development has mainly seen a discreet simulation using the principles of the event orientation. The latest simulation tools have thus simplified the process of creating simulation models that today are being waived from the use of more straightforward analytical methods [6]. Today, artificial intelligence or virtual reality is the usual supportive technique used in simulations. The importance of simulation grows mainly with the increasing complexity of systems. They are mainly used where an erroneous decision can mean inefficient investment, long-term economic losses, and a weakening of its competitiveness.

In the growth of systems complexity and deployment of smart devices that decide on actions in factories of the future, it is, therefore, necessary to determine the outcome of the actions for management needs in a high emergence of processes. [7]. The requirement of frequent changes to the production base requires the rapid commissioning systems of automated manufacturing systems (Ramp Up), which will require new simulation tools. In the case of control, emulating technologies that are tied to the simulation may be used. One of the advantages of the emulation environment is that it can monitor the technical system (such as production, assembly, logistics) in real-time to evaluate the data collected and to update the model in question on a real-system basis and to carry out experiments on the simulation model simultaneously. In Industry 4.0, the introduction of the digital twinning of objects and processes is equally important [8].

The orientation of research into new manufacturing approaches is directed towards the area of intelligent manufacturing systems, using reconfigurable manufacturing systems, adaptive logistics, and the concept of competence islands. New simulation systems must also be adapted to this new requirement. They must possess the ability to simulate agent systems and model large networks. Modelling and simulation will, therefore, be an integral part of the planning and control of the processes of factories of the future.

Research on the principles of modelling and simulation and the development of factories of the future have been the long-term areas of research at the Department of Industrial Engineering, University of Žilina. The issue addressed is consistent with the strategy of Industry 4.0. Just by defining the characteristics of the systems used in factories of the future and their properties can be evaluated, as such systems can be modelled and simulated. The article, in its periphery, deals with the description of the manufacturing concepts that are potentially highly applicable for use in factories of the future, and the core descriptions of the use of modelling and simulation, mainly metamodeling, in the processes control of factories of the future. An example of using this approach is described in the processes control of laboratory ZIMS (Zilina Intelligent Manufacturing System).

2. Materials and Methods

2.1. Changing of Business Priorities of Factories

The changing demands of customers and emerging progressive technologies are revolutionising not only the existing manufacturing environment but, at the same time, bringing about a change in the main paradigm of business. Business priorities are dynamically changing. The technical level of production factories and systems for the planning and control of the activities of these factories fundamentally affect the productivity and efficiency of each factory and hence its competitiveness.

In the past, when world markets were not saturated, it was a priority for businesses to achieve high production capacity utilisation. Capacities accounted for capital invested, and businesses tried to assess it. The aim was to produce simple products in high production volumes (mass), which guaranteed the benefits from the economy of the quantity. Businesses were looking for low-wage territories, which triggered a mass shift to countries with cheap labour, known as offshoring. After the market was saturated, the customer's requirements and preferences gradually changed, resulting in increasing product variants and the combined growth of production complexity. Gradually, the priority of high capacity utilisation and low wages has been replaced by the priority of high flexibility and management of complexity in the production.

The flexibility of production is a prerequisite for the production of a wide range of different products. Thus, the priority of the high capacity utilisation was replaced by the requirement of the high flexibility of production and the ability to cope with the complexity of such production. Only this approach guarantees sustained productivity growth. The relationship between flexibility and productivity is shown in Figure 1.

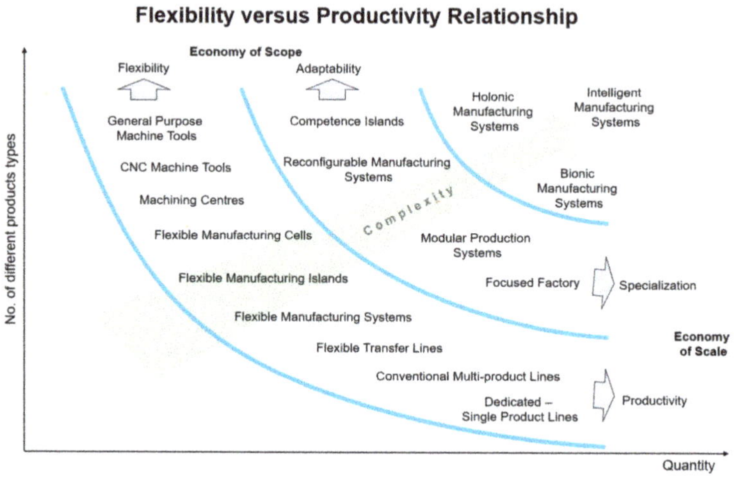

Flexibility versus Productivity Relationship

Figure 1. Flexibility versus productivity relationship [9].

Since the late 20th century, the latest information and communication technologies have begun to be mass-deployed in manufacturing companies, resulting in cheaper means of automation and industrial robots. The cost of automated production was lower than labour costs, and it brought a new phenomenon, now known as reshoring. Businesses began to move their production capacity from countries with cheap labour to their parent countries [10]. A good example is large corporations from the US (Intel, General Motors, General Electric).

The company is, therefore, transformed from production to services. New ways and forms of value creation are created. This brings significant changes to today's factories. The classic factories and their manufacturing systems are gradually transformed, as they end up with classic products we have perceived for centuries. Customers today do not need to possess physical product models when they look for a service that fills their requirements. The young generation no longer needs things to possess. They want to share them, thus stimulating the emergence of a so-called sharing economy. The importance of leasing has grown. Competition is shifting from products to business models. Today, businesses are already competing with their business models [9]. The price of the product is still the determining criterion by which most customers choose from the offerings. If several manufacturers offer similar products, the selling price is the only criterion in their decision-making.

The objective is the simple integrability of product variants in customer-oriented manufacturing systems and, consequently, the efficient implementation of mass production, so that the production of an affordable product is ensured through the economies of scale. It is known that the reduction of unit costs is due to the increasing volume of products produced. There are usually several global producers in the markets that offer similar products with a similar level of quality at relatively low prices. Small firms can compete in such an environment only by finding a more efficient production method, similar to that used by large producers. If the manufacturer wants to sell more expensive products, it must bring new value to the market, e.g., new products with different characteristics that customers will appreciate and buy. The new customer requirements are thus linked to the growth of product variants. The highest form of satisfaction of customer requirements is the personalisation of demand. This means that each product is tailor-made to the customer. The strategy of mass customisation may be appropriate for the economies of scope, and its effective implementation is not possible without advanced manufacturing systems capable of responding rapidly to changes [11]. These changes affect many of the long-established patterns of behaviour, known as paradigms.

2.2. Paradigm Changes

We live in a time when paradigm changes are underway, and their main drivers are new, emerging technologies. Advanced technologies affect the life of the whole society. Their most significant impact is reflected in the production sphere. Robotics is one of the areas that have, for several decades, undergone a technological revolution. Industrial robots have already become commonplace in production. The development and deployment of mobile robots in production and logistics are also on a similar path. Over the past five years, research labs have found their way into manufacturing workshops and cooperative robots (cobot). These represent an intervertebral stage in the transformation of manufacturing systems by integrating the activity of man and robot cooperatively.

At this stage, the man remains part of the production processes. Another development step is humanoid robots, which are gradually becoming a priority in research and quickly penetrating manufacturing practice and services. One of the main priorities for research is collective robotics, namely, the control and coordination of the target behaviour of a group of heterogeneous robots. Manufacturing systems and overall production are in the permanent transformation phase.

The current driving forces of the development of future enterprises by [9] can be classified into two independent groups, as shown in Figure 2.

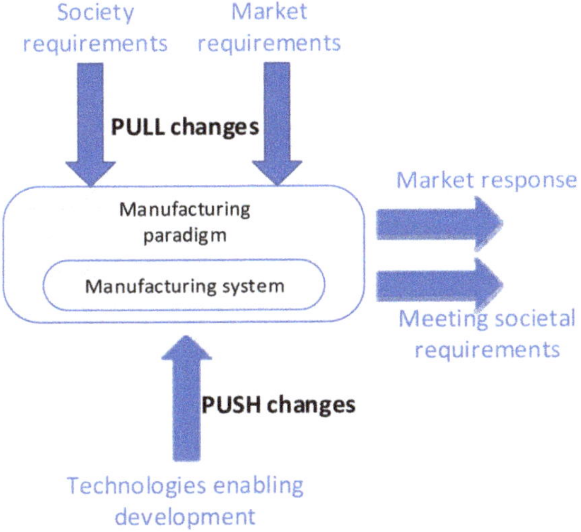

Figure 2. Illustration of PULL and PUSH changes [12].

Emerging Technologies (EmTe)—these are technologies that are often from the outset overlooked and, later, can transform entire industries or services sectors. These drivers are also called PUSH-changes because the changes are due to a technological push.

Market changes (changes in customer requirements)—these changes can often be tracked in long-term trends. Customers demand products (services) that better and more precisely satisfy their requirements, which manufacturers implement through the customisation and personalisation of products. As it is the customers who make this move, we call these changes "PULL changes".

In conventional manufacturing systems as a transfer line, lean manufacturing and flexible manufacturing systems may be advantageous to mass production.

In the mass production of standard products, the primary source of the competitive advantage is transfer lines for the production of one product type using fixed jigs and tools. The aim is to produce one product type in large production volumes and the required production quality, which means low production costs.

The concept of lean manufacturing was created at Toyota as an extension of the functions of mass manufacturing. It is also known as the Toyota Production System, and its primary purpose was to reduce the production lead time, linked with quality growth, cost reduction and loss-making and wastage.

Flexible manufacturing systems (FMSs) have been designed as solutions to ensure sufficient production of the family of similar products. FMSs use NC-controlled machines, automatic handling and transport systems and integrated production management. The deployment of information technologies with the possibility of rapid reprogramming has allowed the product range to be easily changed.

Conventional manufacturing systems used to control the production of classic PUSH approaches and, later, PULL-control. This manufacturing system was, at the time of origin, sufficient. However, now, some areas of markets in which conventional manufacturing systems are used have started to require a variety of products, with little time of product placement in the market. Hence, this growing demand of customers has created a complexity of production that has prompted the increasing experimentation of development of reconfigurable production lines or so-called competence islands.

The paradigm change that is currently occurring is characterised by the use of agents and the principles of multi-agent control in manufacturing. In practice, this means that the classic control systems will be progressively replaced by multi-agent control. Multi-agent control brings the manufacturing of emergence, which means that the characteristics of manufacturing systems are also changing as they become emergent.

The traditional manufacturing systems were complicated. The multi-agent control application in manufacturing represents a transformation of complicated systems into complex systems. For complex systems, the complexity of interrelationships between the various elements of the system is already so significant that it often tends to be very demanding, if not impossible, to use mathematical modelling for their studies. The dynamic behaviour of such systems can only be studied using the theory of complexity. Future manufacturing systems will operate as adaptive, dynamic manufacturing networks. New simulation systems must also be adapted to this new requirement. They must possess the ability to simulate agent systems and modelling large networks. Modelling and simulation will, therefore, be an integral part of the planning and control of the processes of factories of the future. In manufacturing, in addition to real objects, there will also be their virtual representatives, which we now refer to as digital twins. Such a dual representation of production is also known as virtual manufacturing. For the visualisation of future manufacturing systems, we can see a similarity to living organisms. Holonic production with multi-agent control will resemble more the emergence of the functioning of living organisms rather than a mechanical automaton.

A new trend in manufacturing systems development is reconfigurable manufacturing systems. Nature teaches us that when changing the environment, the living organism strives to adapt to changed conditions. It uses the change of internal structures and the number of elements and their composition. At the molecular level, it "stretches and recycles" unnecessary structures and reconfigures them into new, necessary structures. Recycling is a process of decomposition. Reconfiguration then represents the new use of existing structures [13].

Most of the activities in the industry of the future will be performed by intelligent robots. In order for the robots to be able to carry out their tasks, often in an unfamiliar environment, they have to possess autonomous capabilities, hence, the ability to adapt to their surroundings and the changing conditions of the surrounding area, collect and evaluate information about their internal state and environment (perception), predict future situations, make the necessary decisions and, of course, learn from the situations. Such tasks can now be tackled by the individual, advanced robotic systems.

The growing interest in mobile robotics applications has not only made changes to the part of users of robotic solutions but also to the part of their suppliers, i.e., manufacturers of mobile robots. Users increasingly prefer more complex mobile robotics solutions, with autonomous intelligent control, localisation and navigation.

The behaviour of future collective robots must resemble the behaviour of living organisms. From that point of view, we have to distinguish the concepts of robotics swarm and collective robotics.

Swarm robotics include a set of relatively simple, homogeneous robots. The behaviour of such robots imitates the behaviour of simple living organisms (we refer to them as swarms or flocks) such as bees, ants, or flying birds. For the collective behaviour of such a swarm, relatively simple rules apply. Each member of the swarm has a specified range of activities, which it carries out in favour of the whole swarm [14].

Collective robotics usually involves many, often very heterogeneous robots. Heterogeneous robots may include a whole set of autonomous robots, not requiring a human operation, from mobile robots, road robots, through to flying (drones) and floating robots. These robots possess strong autonomous functions, intelligence, and mobility capabilities. Such robots work with intelligent sensory networks and computer systems organised into cloud-based solutions. In the complex management of collective robots, it is no longer possible to use classical, centralised management. The results of the research in progress have shown that the management of collective robots will require a "proprietary" operating system [15].

The cooperation of collective robots differs significantly from the cooperation of simple swarm robots. In performing complex tasks, in a challenging and unfamiliar environment, collective robots must use distributed control mechanisms that can combine the behaviour of individual, autonomous robots into the complex behaviour of the entire group of robots. We refer to this behaviour as "holonic". For the control of the holonic systems, it is typical to use agent access and multi-agent systems (MAS). The process of cooperation of the group of individual and autonomous robots creates a higher level of collective intelligence, which we call emergence.

In the human body, we can change all the organs except the brain. Its change (disintegration structures and remastered) is blocked. Likewise, the company. Most structures change when reconfigured, but the central control system remains unchanged. It is possible, like the brain, only to expand its function (augmentation) through external expansion. Its architecture must be designed to reflect future changes. The custom control architecture remains to be maintained when reconfigured. A reconfigurable enterprise tries to behave like a living organism [16]. New manufacturing concepts are developed as a response to this paradigm.

2.3. New Manufacturing Concepts Designed for Factories of the Future

All new manufacturing concepts seek to meet one of the main objectives and, thus, adaptability, the ability to react immediately to rapid changes in the environment, is also referred to as turbulence. Adaptive manufacturing systems are, at present, a peak of scientists' efforts to formulate the contours of the future production environment. In order to meet the requirement of adaptability, it is possible to approach this in several ways, so scientists have developed and tested a whole group of new manufacturing concepts such as:

- Reconfigurable manufacturing systems
- Competence islands
- Multi-agent control systems

The manufactured product will behave in new manufacturing concepts as a smart entity, able to communicate with its surroundings and able to organise its processing entirely autonomously. Such a product will itself determine the sequence of its processing, allocate the required capacity in the relevant competence islands and sump a mobile robot to ensure its transport in production. To enable such a system of organisation to work safely and reliably and to fulfil the required tasks, it will require new ways of manufacturing planning and control. Next, the seemingly "chaotic" world of production will no longer operate current push control systems. With a vast number of smart elements (entities) in the manufacturing system, there will be complicated relationships and situations that are no longer able to deal effectively with today's hierarchical management. Complex relationships between individual entities cause a status called emergence, that is, the state in which it will no longer be challenging to predict the future behaviour of such complex systems. Therefore, researchers are experimenting with new management approaches based on the relative autonomy of the individual elements of the manufacturing system and their behaviour, which will resemble the behaviour of intelligent, living organisms. In production, in addition to real objects, there will also be their virtual representatives, which we now refer to as digital twins. Such a dual representation of production is also known as virtual manufacturing.

2.3.1. Reconfigurable Manufacturing Systems

The Reconfigurable Manufacturing System (RMS) is a production system, the structure of which is merely adjustable, with the possibility of scaling capacity and flexibility bounded by the selected product family [11]. Figure 3 illustrates the vision of reconfigurable manufacturing systems.

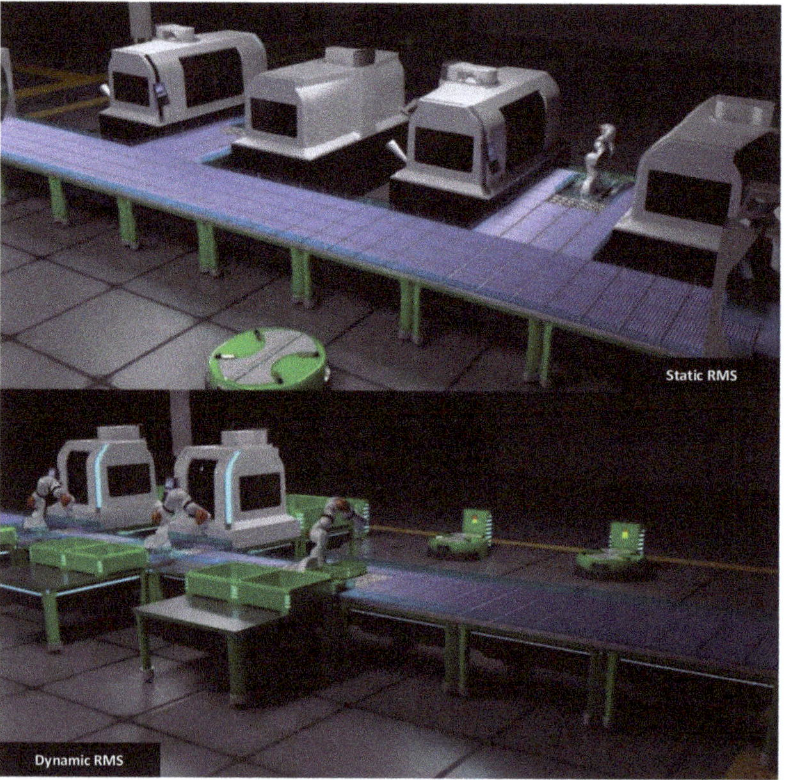

Figure 3. Comparison of the static reconfigurable manufacturing system and the dynamic reconfigurable manufacturing system [12].

Reconfigurable manufacturing systems represent the evolutionary phase of the development of manufacturing systems. Their application requires a new approach in which they play a dominant role in reconfigurable machines, jigs, tools, logistics, and reconfigurable control systems [11].

RMS is built to allow for easy and rapid conversion (reconfiguration). This feature pushes reconfigurable manufacturing systems into the adaptive systems area. Reconstructions enable the production system to be adapted to new product types (functionality) and new production quantities (capacity) [17]. Reconfigurability has thus become a new technology that can better meet market fluctuations and turbulence through the gradual rebuilding of the manufacturing system.

Reconfigurability represents the operational ability of the manufacturing system to adapt its functions and capacities to a particular product family.

It results in the desired flexibility of the manufacturing system. As opposed to reconfigurability in the manufacturing system, flexibility is firmly defined. Reconfigurability and elasticity make the adaptive ability of the manufacturing system, which is achieved through a change in its structure. Such a structural change makes it possible to adapt the functions and capacity of the manufacturing system to new requirements. The condition for effective reconfigurability is the requirement to minimise the effort undertaken and maximise the reduction in the time required for the implementation of the changes [11].

2.3.2. Competence Islands

The existing large-scale production method, organised rhythmically in production halls and working in the production cycle time, will no longer be able to respond to future customer requirements. Today's "static" production and assembly lines will be replaced by a set of autonomous workplaces called competence islands (Figure 4). It can imagine as virtual production lines, formed dynamically and virtually based on real needs. The competence islands will be equipped with technologies and cooperative robots capable of working safely and reliably with people [18]. Figure 4 illustrates the vision of the competence islands.

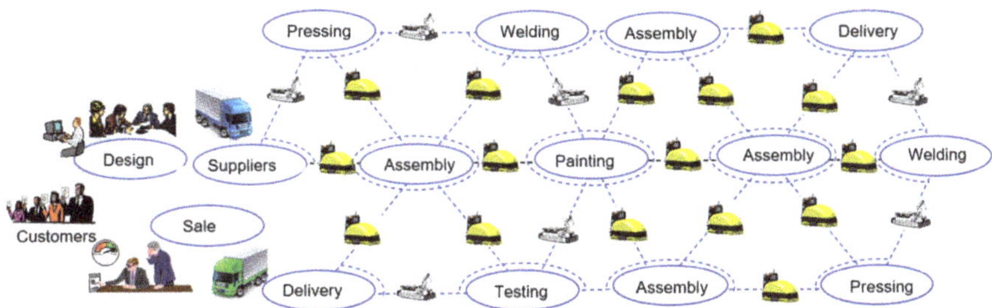

Figure 4. Concept of competence islands.

New manufacturing systems should, therefore, be designed as small, highly flexible production units, which will be deployed where there is sufficient real demand. Such manufacturing systems will be designed for the production of the selected product family, which requires that their concept be built on the principles of reconfigurable manufacturing systems.

The activities of future manufacturing systems will be organised differently. Classic production and assembly lines will only be maintained where it is still economically advantageous. Future production will seem to be complete chaos to the outside observer. It will seem that materials, intermediate and elaborate production, and mobile robots are moving unplanned and chaotically. However, each of them will be guided by a strict logic of the parent level, which will enable it to be relatively autonomous. It will, therefore, be organised chaos. For production management, the principles observed from nature, which offer an evolution of proven, optimal practices, will be used.

Holding a strong position in future factories will be intelligent mobile robots and mobile robotic systems and platforms. Thousands of such robots will ensure the movement of the worked products and their processing in a seemingly chaotic world.

The production will be organised as a living organism resembling an anthill, in which the ants appear messy but are strictly organized and specialized, and each of them performs precisely defined tasks that ensure the survival of the anthill.

The product, production equipment, technology and the entire production system will be changed. Manufactured products, manufacturing equipment and mobile logistics means will become intelligent and communicate with each other. In real-time, they will exchange and share all the necessary data and information.

Mobile robots, transporting a staged product, will move between the competence islands, while the product itself will determine the required operations and plan their order. The observer will not see the classic production line; what will be observed is the apparent physical chaos. However, there will be a hidden virtual line (its digital and virtual data model) made up of the competency islands required for the production of the customer product.

Future production will not be structured according to the production rhythm line, as is the case today, but according to the content of the work to be done. Functional relationships and not fixed cycle times will have a decisive role. This type of production environment will be suitable not only for small production companies but will be particularly advantageous for those types of products that work with high volumes, highly variants of production and which aim at high flexibility and efficiency. Such systems will be able to react more effectively to fluctuations in demand and rapid changes in the models produced by requiring different production technologies. The company Audi claims that the production islands will be much more efficient than today's linear concept.

2.3.3. Virtual Manufacturing and Intelligent Agents

The simulation model, detailed, hierarchical and more leveled, containing all the significant factors of the production process, will allow a new type of management, which will be built on dynamic analysis and prediction. If we link such a model to the information sources of production and its sensory system, it will operate as a human organism and will behave adaptively while using real-time data. It will work with its own "physical map", similar to the human body. Today's experiments with in-memory computing are about such future management systems.

In the area of virtual manufacturing and intelligent agents, we meet with the solution of syntactic, semantic and pragmatic boundaries [19]. The first aspect is syntax, which is important for the machine to machine communication. The communication capabilities of agents in multi-agent systems (MAS) are characterised by data exchange mechanisms based on proprietary messages in the form of Extensible Markup Language (XML) syntax and according to MAS standard communication models, for example, defined by the Foundation for Intelligent Physical Agents (FIPA). For establishing CPS in manufacturing environments, the usage of web services is inevitable for the realisation of scalable information exchange. Thus, in addition to a language that describes the information and provides data syntax and semantics, a common underlying mechanism for transfering the information from one entity to another or to perform interactions is needed [20], so the second aspect is semantics. In semantics, we find ontology, annotations, and definitions. The semantics give a mathematical meaning to formulas that, in theory, could be used to establish the truth of a logical formula by expanding all semantic definitions [21]. To provide a proper description of an agent that is readable, understandable and interpretable by other agents in an integrative manner, the description model of each agent needs to follow common design principles, e.g., by making use of a common ontology description, fixed namespaces for agent capabilities (among others) is needed [22]. According to [23], communication between agents can be realised if all agents can find and identify each other and all agents make use of a message system with a predefined ontology, which every agent can understand. In [22], one desired goal to deal with high amounts of raw data from the shop floor would be an automatic assignment of information from the lower levels of the factory. Automated annotation of production information with context information, such as metadata, would reach both machine-readable and interpretable information for autonomous process optimisation as well as a data basis understandable by humans. The third aspect is pragmatics, which means the question of how to use axiomatics to justify the syntactic renditions of the semantical concepts of interest. That is, how best to go about conducting a proof to justify the truth of a CPS conjecture [21]. That means that we must define how to use axiomatics to justify the truth.

New sensory systems allow for the end of the transition from static monitoring systems (sampling and data collection at an interval of one day) to dynamic monitoring (sampling in microseconds, as it is done today in the process industry). The average values of output parameters (statistics) must be replaced in the new generation monitoring systems with the immediate values and trends of changes in the last, most significant periods.

The monitoring system must include a watchdog, a function that will trace (seek) potential problems, an early warning system that notifies the occurrence of potential problems and an automatic correction mechanism that resolves the potential problem before its real emergence.

If we have enough data about the production system, we can, thanks to virtual reality, create a virtual image of production (its dynamic hologram) and then in such a "reality", virtually track the effect-change factors (visualise them), observe future status and decide on the changes that will be made. In long-enough time, such a system can gradually learn, with the support of a machine-learning system and knowledge system, how to adapt to changing surroundings. If a person makes a decision instead of using computers, manual management is applied. In direct management, in automatic mode, the direct control system decides, and manual interventions are replaced by automatic steering. In the case of manufacturing control, the virtual twin of each real object will be represented by an agent. We refer to a large group of such agents and their management as multi-agent systems (MAS) [24]. Future production will be represented by two worlds: the real world's and its virtual reflection, also called the virtual world. These worlds will be mutually integrated through data. Production data will be collected and processed in real-time. Almost immediately, information about each object in the production will be available—what it is doing, in what state is it located, what is further planned, and what is lacking. The status of each product, machine, tool, device, jig, robot, or person will be immediately scanned, and the processed information will be sent to the control centre. This information will be compared with the next step in the processing of the products, the sequence of future steps will be generated and the system will make the necessary decisions for further processing of the product. The virtual world will allow, if necessary, the simulation of future status and prediction of the effects of the necessary control actions.

2.4. Routing of Modelling and Simulation

The basic principle of simulation lies in the simplified representation of the real system of its simulation model, describing only those characteristics of the real system that interest us in terms of its study (simulation). Instead, it would be possible to say that a simulation is a supportive tool that allows the experiment to test the effects of its decisions on the simulation model. By this, we can obtain an answer to the question "what happens if". The great advantage of this approach is that it is possible to previsualise the future behaviour of the system and to realise the necessary interventions in the real system based on its knowledge [25].

In view of the future needs of the simulation, supporting strategic decision-making, new classes of simulation systems must be developed to enable work with aggregated data at different hierarchical levels of the systems being analysed. Such solutions will require the development of entirely new integrated, hierarchical simulation systems capable of modelling complex corporate systems and working with heterogeneous modelling approaches [26]. The main task of the creators of such systems will be to integrate heterogeneous environments into a single, holonic concept. The hierarchy will require integration at micro-, meso- and macrolevels. The simulation environment will provide modelling techniques and approaches for modelling of all corporate hierarchical structures.

Digital twin (DT) is the concept of the functioning of future production systems, based on the digital technology application currently promoted by Siemens. Although the principles of digital twins are known to be more distant, Siemens has stretched the development into a phase of products that are now offered on the market. The digital twin is now presented mainly at the product level, and its essence consists in the creation of a virtual (digital) model of a developed product, machine, or device. The virtual model thus created (digital twin) can be used in all phases of the development, operation, and improvement of the product. For example, the digital twin of a car allows the costs of developing and testing a car to be reduced. The entire development and most of the tests can be implemented through virtual testing and simulations, using the digital model. Physical tests are used only for the calibration of the test method [27].

The concept of the digital twin has been gradually expanded from product level to process levels, manufacturing systems to the enterprise level. The digital twin can be used in the performance of many business processes, whether it is logistics, manufacturing, assembly, and machining [28]. Industry 4.0 requires phenomenon twins to functionalised the relevant systems (e.g., cyber-physical systems). A phenomenon twin means the computable virtual abstraction of a real phenomenon [29]. The digital factory includes digitisation of the three most important business areas: products, processes and resources. Thus, the era is launched, in which all critical physical production entities are represented by their digital copies and digital models, also called the digital mock-ups (DMUs). In addition to the real manufacturing system, a digital manufacturing system, which is represented by a set of static, kinematic and dynamic digital models, which is integrated into a single digital development environment, i.e., a digital factory, will also be available to all companies.

This has allowed us to study and analyse the efficiency and performance of production before putting it into real operation. Decision-making has begun to become more and more algorithmised, with the database for decision making being the results of dynamic computer simulations. Hence, the beginning of 21st-century enterprises are confronted with two parallel worlds, real and digital (a real-digital world) [27].

Sensor hardening, the rapid development of new communication equipment and systems, have enabled the virtualisation of the world of manufacturing. Such a virtual manufacturing world has generated vast amounts of data that businesses have kept, analysed and started to use for predicting the future behaviour of manufacturing systems. Virtualisation, in this case, means that the managers obtain information about the immediate state of the manufacturing system through sensors. Data from sensors, processed by intelligent algorithms, create a dynamic, virtual image of a real production, which is named "virtual factory", and represent the duality of the real–virtual world (Figure 5). By linking digital, real and virtual worlds, this new quality is now known as the digital twin.

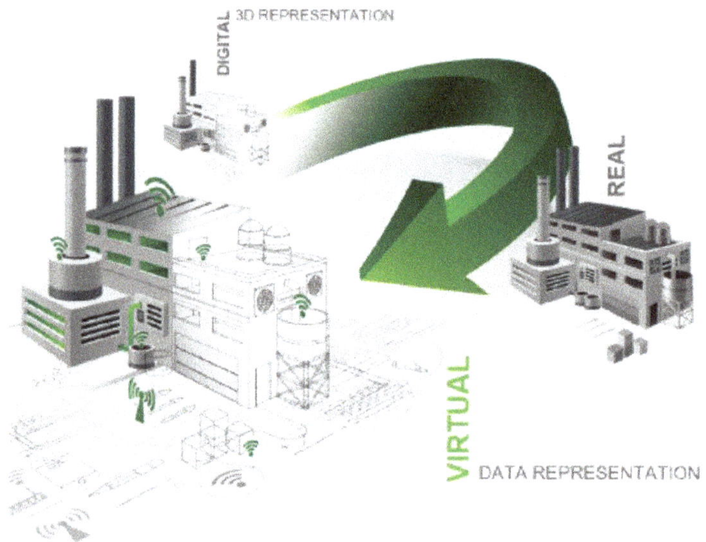

Figure 5. Combining three worlds—digital, real and virtual [30].

3. Results

3.1. Control and Simulation in the Processes of Factories of the Future

The manufacturing system is a multi-factor system. Its model is dynamic, not static. Therefore, it is not possible to say that the efficiency of the manufacturing system is a function of low stock or short, intermediate periods. The efficiency of manufacturing depends on a set of (huge) factors that are dynamically changing over time and are different for each manufacturing system. Although we do not have to know in detail the functioning of each element of manufacturing and we do not have to understand it fully, we can control it. However, we only apply its effectiveness to a very narrow range of criteria (most significant parameters) [30].

Correlation is a statistical characteristic of the statistical dependency rate of two (or more) statistical variables (random quantities). If we consider only two variables, we can easily interpret the dependencies. However, if we move in n-dimensional space with hundreds of variables, relationships will begin between variables (statistical dependencies) to acquire an often meaningless character. In manufacturing systems, we work in reality with an almost infinite number of variables (factors). Therefore, it is very complex (if not impossible) to compile an exhaustive mathematical model of the manufacturing system that would faithfully and accurately represent its dynamism. In this case, the approximate method of computer simulation will help. A cause and its effect, represented by correlation, does not always reveal the causes of the latter, and, rather, may reveal the consequences. Too much data brings the so-called "elusive correlation". A lot of data is used for many different estimates and predictions [31].

The control concept that uses virtualisation contains predictive mechanisms that enable the control system to "see potential scenarios for the future" [31]. The data structure for such a control concept is illustrated in Figure 6.

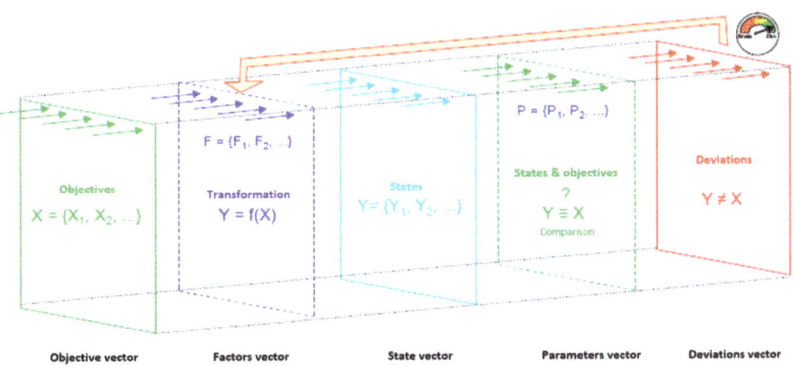

Figure 6. The data structure of the factory control system.

The use of a multi-agent control system is for the distribution of tasks and the hierarchical behaviour of the members of the system. In such a system, the holonic system works; it can be seen as a system consisting of subsystems, but at the same time, the system is part of a larger whole (system). A set of holons with their characteristics creates a holonic organisation called holarchy, which is characterised by the fulfillment of common objectives. Holarchy allows the creation of structures and representations of the behaviour of complex systems, often referred to as social systems. The functioning of the holonic systems is based on the use of the ability of autonomous agents. An agent is a system entity that has a specific degree of independence, allowing it to autonomously address tasks within a defined level of action. Agents accept tasks from the parent level of the holarchy, but their solution is carried out autonomously.

An intelligent agent is a computational or natural system capable of perceiving its surroundings and, based on its monitoring, performing actions that result in the extreme of its objective function (minimum, maximum), thereby fulfilling the global objectives of the system. In the agent systems, in the vast amount of interactions that occur between individual, autonomous agents (for example, in social systems), we are no longer able to predict the future behaviour of such a system [32]. If we were modelling such a system, it would be better to define the behaviour of individual parts of the system (agents). An agent can use the services of holon, which is used for simulation of varying inputs and to see the outcomes of actions. The use of simulation metamodelling within the holon simulation is illustrated in Figure 7.

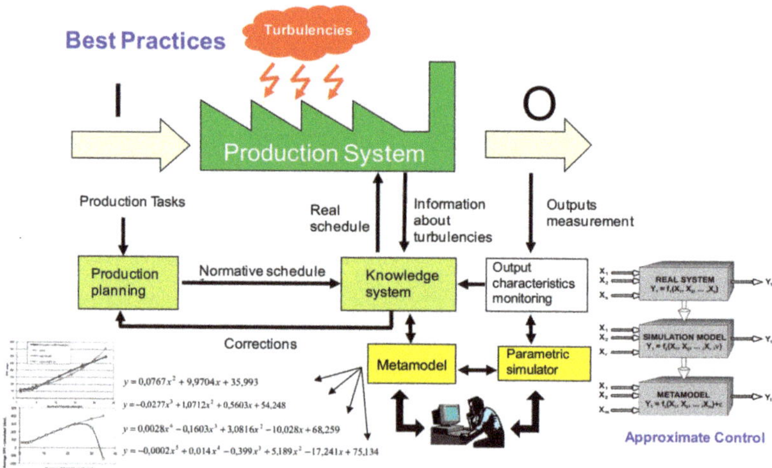

Figure 7. Use of simulation metamodelling in the manufacturing control of complex manufacturing systems.

Figure 8 also illustrates the principle of application of digital factory instruments to changes in the product range, the exchange of technology and the change of layout. As seen, the entire management concept is first developed and tested offline in the virtual environment of the digital factory. Agents that represent the physical elements of the system use the knowledge of previous actions as well as existing models and, on the virtual model of the manufacturing system, carry out experiments in which scenarios are tested. Then, it selects the appropriate scenario that matches the target characteristics of the system. After completion of the development, the validated control concept is transferred to the real production system. Therefore, the simulation becomes an emulation when the startup point of a real-element agent that is recorded in a specific position predicts future statuses. The principle of the knowledge-based environment will support the system in the form of learning from process activities.

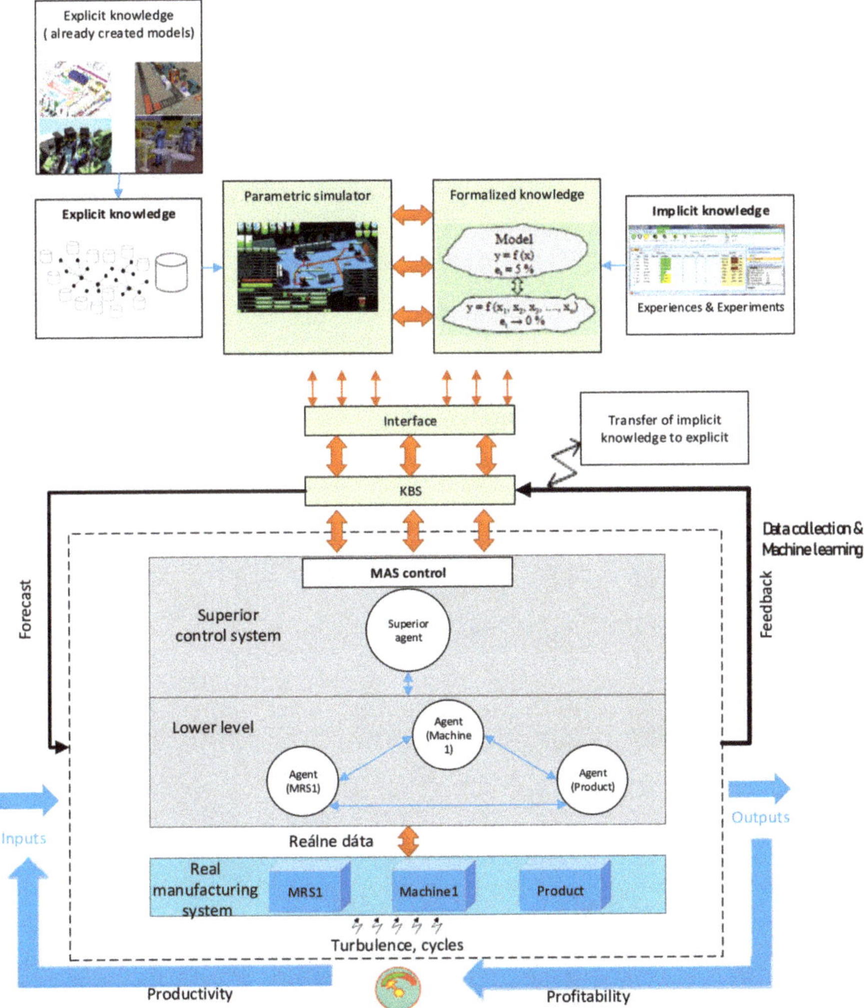

Figure 8. Linking simulations in continuity to control and learning from processes.

3.2. Application to the Zilina Intelligent Manufacturing System

The question involved was applied to the Zilina Intelligent Manufacturing System (ZIMS), whose structure is illustrated in Figure 9. The Zilina Intelligent Manufacturing System (ZIMS) has been built to faithfully represent advanced manufacturing systems with its practical design, while also enabling experimentation and further research in the field of intelligent manufacturing systems. The whole concept of ZIMS was designed as a holistic system. Individual holons represent the main subsystems and elements of advanced business systems. As part of the simulation application design, a control system based on simulations, emulations and metamodelling was applied.

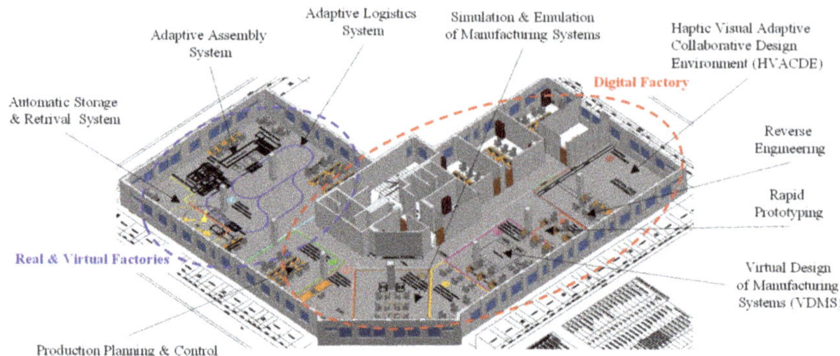

Figure 9. Structure of the Zilina Intelligent Manufacturing System (ZIMS) laboratory.

The logic of a more detailed subdivision of the holons is shown in Figure 10, in which the modelling and simulation are under holon manufacturing.

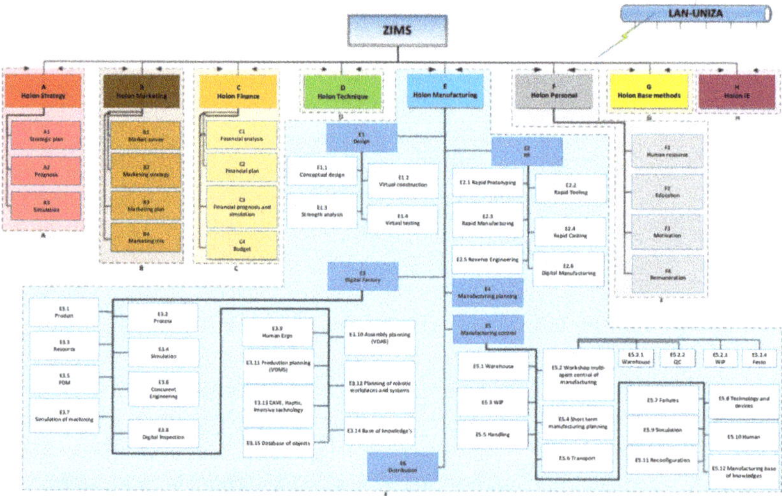

Figure 10. A more detailed breakdown of the holons of the ZIMS laboratory.

As seen from Figure 11, the structure of the holonic control respects the functional requirements of the enterprise control system. Different data connections and standards are used at the level of the individual holons: STEP, IGES, PNG, RAW, RGB, VRML, DXF.

Communication within the holons is illustrated by the example of manufacturing (Figure 12).

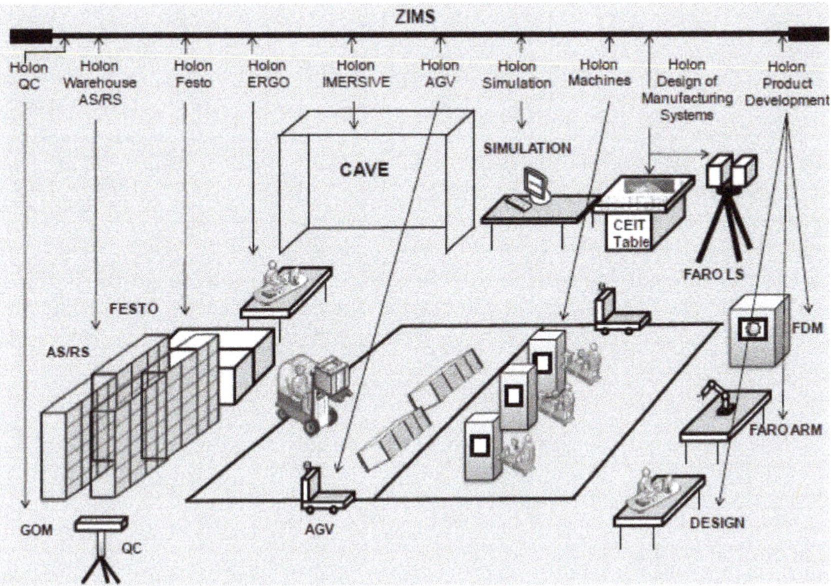

Figure 11. Structure of the holon manufacturing in ZIMS.

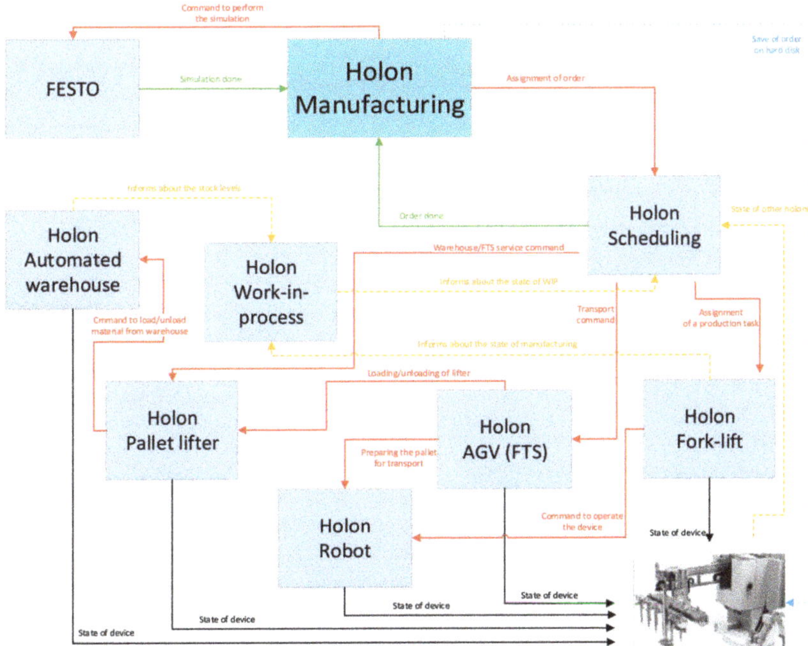

Figure 12. Communication of the holon manufacturing control in ZIMS.

3.3. Use of Metamodelling in Laboratory ZIMS

In the laboratory, ZIMS is used for the determination of individual holon action metamodelling. An essential part of using the simulation as support for control is to train the simulation network with data. For the verification of metamodeling, a model of manufacturing cell of the ZIMS concept is created, which consists of three machines. They gradually work on intermediate A, which enters the system at regular intervals, every 3 minutes. It is then transported by a conveyor belt to the buffer of the first machine, S1, from which it is taken and subsequently worked on if the machine is free. When the operation is completed, the semi-finished goods are transported again using the conveyor belt into the buffer of the next machine, S2, and the procedure is repeated. The transport between the last machine, S3, and the "output" is not considered.

The operational times of the machinery and the times of transport between the workplaces (Table 1) are the same at a purely theoretical level; in the event of a failure to run the system, the machines work at 100 (which is unrealistic, but it is only an explanation of the process of working in the formation of the metamodel). However, overall system productivity is affected by the failure of the second machine, S2, which occurs at particular time intervals $X1 = \{x11,x12,...,x17\} = \{15,20,25,30,40,50,60\}$ and the repair time is defined by a set of $X2 = \{x21,x22,...,x26\} = \{5,8,10,12,15,20\}$.

Table 1. Times, transport times and the interval of arrivals of intermediate products into the system.

Machine	Operation Time (min)	Input	Interval of Arrivals into System (min)	Ways	Transport Time (min)
S1	3			c1	1
S2	3	A	3	c2	1
S3	3			c3	1

The denotation of variables is X1—time between failures; X2—repair time; Y—lead time of production.

Then we selected (based on short pilot runs) time simulation, namely, one working week with a single-shift 7.5-hour operation (i.e., 2250 min) and a time of production of 50 min. After completing all these steps, we could proceed to the implementation of simulation experiments.

These input data were performed for all combinations of the levels of factors X1 and X2 mentioned above, which totals 42 simulation runs (Table 2).

Table 2. Results of simulation experiments.

X1	X2	Y1	X1	X2	Y1
	5	294.17		12	333.62
	8	403.55	30	15	386.73
15	10	461.68		20	461.09
	12	511.68		5	137.65
	15	574.15		8	200.25
	20	654.48	40	10	237.61
	5	237.62		12	271.98
	8	334.37		15	318.84
20	10	388.11		20	386.54
	12	434.56		5	114.79
	15	494.27		8	167.78
	20	573.89	50	10	199.99
	5	200.62		12	230.04
	8	286.01		15	271.60
25	10	334.50		20	333.50
	12	377.58		5	98.88
	15	434.15		8	144.76
	20	511.66	60	10	173.01
	5	173.45		12	199.86
30	8	249.62		15	237.46
	10	293.80		20	293.87

Subsequently, after verifying the data from the simulation, we determined the ones that will serve to train the network and those that will be test data. Artificial neural networks (ANNs) were also tested during the learning process, and validation was not necessary [33]. For training, we selected a set of 35 combinations of data obtained from simulation runs Table 3. The training set Table 4 modified the scales, and a generating error was detected using the test set. The entire process of creation, training, testing and validation took place in the Matlab program environment [34].

Table 3. Training data (inputs and outputs) for the artificial neural network (ANN).

X1	X2	Y1	X1	X2	Y1
15	5	294.17	30	15	386.73
15	10	461.68	30	20	461.09
15	12	511.68	40	5	137.65
15	15	574.15	40	8	200.25
15	20	654.48	40	10	237.61
20	5	237.62	40	15	318.84
20	8	334.37	40	20	386.54
20	12	434.56	50	5	114.79
20	15	494.27	50	8	167.78
20	20	573.89	50	10	199.99
25	5	200.62	50	12	230.04
25	8	286.01	50	20	333.50
25	10	334.50	60	5	98.88
25	12	377.58	60	8	144.76
25	15	434.15	60	10	173.01
30	8	249.62	60	15	237.46
30	10	293.80	60	20	293.87
30	12	333.62			

Table 4. Test data for the ANN.

XT1	XT2	YT
15	8	403.55
20	10	388.11
25	20	511.66
30	5	173.45
40	12	271.98
50	15	271.60
60	12	199.86

After we enter all the input factors and commands to display an error between the outputs of the ANN and the specified Y results, rendering the ANN output differences for the test data and the actual output of YT, a network training order with the training data is entered. When starting the learning process, a Figure 13 window appears, which can be followed by the training process, the number of running eras, the duration of learning, and a shrinking/increasing error [35].

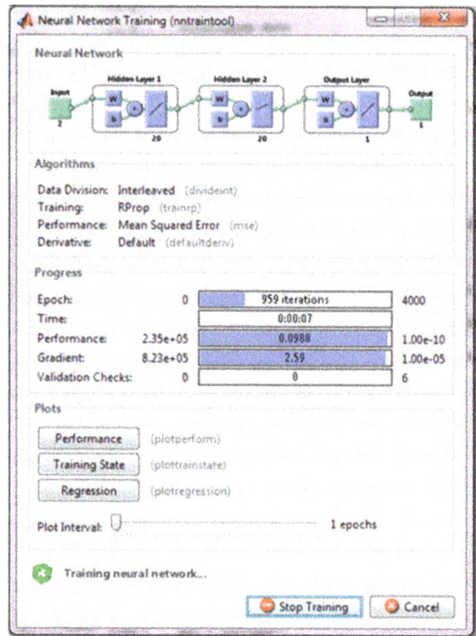

Figure 13. Example of the ANN training process.

Once the ANN has been trained and reaches a satisfactory result, it is possible to use the ANN to address specific problems by putting new values into ANN for which no simulation runs have been made, but the responses to them are of interest to us. The validation itself does not need to be carried out since test data have already been used. However, for a demonstration, in Table 5, we compare outputs generated by our ANN for specified input data with simulation outputs.

Table 5. Comparison of the results of the simulation and trained ANN.

X1	X2	Y1-Simulation	Y1-ANN
17	5	268.51	268.6991
18	9	387.27	387.7591
22	4	185.78	196.6634
27	11	338.75	338.5108
29	14	379.01	378.5756
35	10	262.80	262.2215
38	16	345.35	344.4772
42	6	153.04	154.3730
46	18	328.02	327.5044
51	13	240.94	241.3330
58	17	267.40	266.8995
63	7	124.98	125.6077

4. Discussion

Based on the knowledge learned from the long-term research in the field and the practical experience gained in dealing with the projects in the industry, we can anticipate the development of simulation environment requirements for the factories of the future.

Due to the fast onset of solutions of Industry 4.0 and the extensive use of sensors the main task for future simulation environments is the ability to model and simulate the behaviour of complex systems. When using a large number of sensors, processing data in real-time and the autonomous behaviour of the elements of the manufacturing system, the factories of the future will experience emergent phenomena.

This change will cause the simulation systems today to be used to simulate an emerging complexity. Therefore, one of the crucial tasks for the creators of simulation systems will be to develop solutions to simulate complexity in manufacturing systems. For the dynamism of the autonomous behaviour of the elements of manufacturing systems, the principles of multi-agent systems can be used in simulations, which today represents the agent simulation. Another suggested development will be the effort to "simplify" complex problems, in which way, one of the routes can be the use of simulation metamodelling. Several types of research work addressed in our department declare this development trend.

One of the crucial requirements for a new simulation environment will be its ability to offer the functionality of the emulatory environment. The integration of the real manufacturing system with its digital and virtual models will enable both offline and online optimisation, and the simulation will become part of real-time control systems.

The future simulation environment will naturally reflect the requirements of the factory of the future. In its creation, all modelling and statistical support tools, which are now commonly used in the simulation, will be used.

However, this will fundamentally change the way the simulation is implemented. Three main approaches to simulations (event orientations, process orientation and activity orientation) have traditionally been used, while new simulation algorithms will be built on distributed, autonomous principles. Due to the requirement for the reconfigurability of manufacturing systems, new simulation systems will have to offer entirely new functionalities and thematic templates, as shown in the example of the research and development of the agent simulator for future hospitals or the development of a multi-agent control (and simulation) system of complex logistics systems.

Methods and tools supporting the transformation of physical systems into virtual ones are evolving. The dynamics of the development of such systems are displayed in Figure 14.

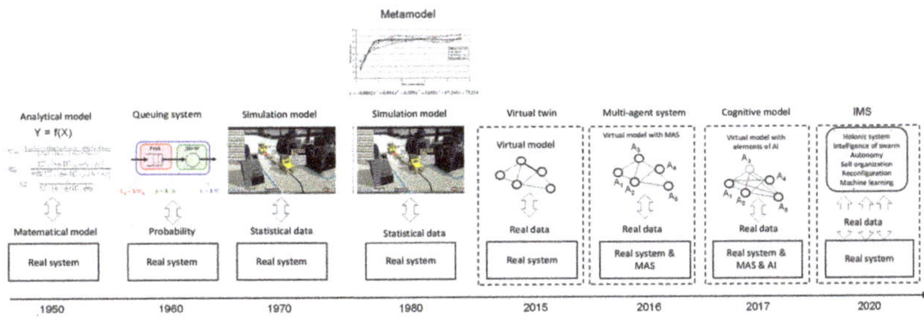

Figure 14. Evolution of real system to virtual model [36].

As to further perspectives for the development of modelling and simulation in factories of the future, the Department of Industrial Engineering sees, in the following period, further integration of the various industrial engineering methods and tools used in the industry into computer simulation software tools. The concept of planning and control of future factories, through the use of computer simulations, forecasting of demand and monitoring of enterprise performance indicators (e.g., productivity), must also be developed within the framework of the approach mentioned above. These themes enter into a new dimension because the coordination of the production chain is becoming a prevalent task for all stakeholders, and the aim is to achieve a common synergy effect.

The research conducted has clearly shown that the theory of complex systems, on the basis of the requirements of factories of the future, has progressed considerably and is already providing practical tools for designers of future manufacturing systems. Key contributions of the research base on case applications are:

- Growing performance
- Thanks to metamodelling applications, a quick prediction of emergent properties of the analysed system is achieved
- Identification of bottlenecks
- Identification of obsolete and nonturnover stocks
- Reduction of stocks of finished products
- Increasing key indicators of the enterprise (such as performance, productivity)

In the frame of research limitation, it can be said that we mainly focused on an application on the reconfigurable manufacturing system and the processes within (e.g., manufacturing, logistics). However, the model can be used on conventional manufacturing systems that have a certain level of communication ability and self-awareness.

We focused our current efforts on researching new approaches to simulating complex systems using agent simulation. In the future, we want to focus our research work mainly on the area of intelligent manufacturing systems, which using reconfigurable manufacturing systems, adaptive logistics and the concept of competency islands.

5. Conclusions

Manufacturing systems of the factories of the future will have new features that will enable them to respond quickly and efficiently to frequently changing customer demands. These manufacturing systems will be designed as modular, reconfigurable and intelligent holonic systems, capable of rapidly changing their functions and capacities based on auto diagnostics. Designing and analysing the behaviour of future manufacturing systems will require heterogeneous simulation models and new simulation tools, allowing for rapid, comprehensive analysis and interpretation of the results obtained. The future simulation environment will naturally reflect the requirements of the future factory. In its creation, all modelling and statistical support tools, which are now commonly used in the simulation, will be used. However, this will fundamentally change the way the simulation is implemented. Three main approaches to simulations (event orientations, process orientation and activity orientation) have traditionally been used, while new simulation algorithms will be built on distributed, autonomous principles. In view of the requirement for reconfigurability of manufacturing systems, new simulation systems will have to offer entirely new functionalities and thematic templates. By engaging the simulation, it is possible to at least partially estimate the results of the interactions at emergence and to control the manufacturing system. The purpose of the article is to outline the use of modelling and simulation in control of processes in a factory of the future. An application of metamodelling inside a manufacturing holon in laboratory ZIMS was presented as an example. The article in the periphery also describes the developed manufacturing concepts that will be used in the factories of the future, which will meet the demands of the paradigm of mass customisation and personalisation.

Author Contributions: All authors contributed to writing the paper, documented the literature review, analysed the data and wrote the paper. All authors were involved in the finalisation of the submitted manuscript. All authors have read and agreed to the published version of the manuscript.

Funding: Slovak Research and Development Agency under contract no. APVV-18-0522.

Acknowledgments: This work was supported by the Slovak Research and Development Agency under contract no. APVV-18-0522.

Conflicts of Interest: The authors declare no conflict of interest.

References

1. Mleczko, J.; Dulina, L. Manufacturing Documentation for the High-Variety Products. *Manag. Prod. Eng. Rev.* **2014**, *5*. [CrossRef]
2. Kumar, A.; Nayyar, A. *Si3-Industry: A Sustainable, Intelligent, Innovative, Internet-of-Things Industry*; Springer: Cham, Switzerland, 2020; pp. 1–21. [CrossRef]

3. Sharma, A.; Jain, D. *Development of Industry 4.0*; Springer: Cham, Switzerland, 2020; pp. 23–38. [CrossRef]

4. Grabowska, S.; Gajdzik, B.; Saniuk, S. The Role and Impact of Industry 4.0 on Business Models. In *Sustainable Logistics and Production in Industry 4.0: New Opportunities and Challenges*; Grzybowska, K., Awasthi, A., Sawhney, R., Eds.; EcoProduction; Springer International Publishing: Cham, Switzerland, 2020; pp. 31–49. [CrossRef]

5. Kolarovszki, P.; Vaculik, J. *Middleware—Software Support in Items Identification by Using the UHF RFID Technology*; Springer: Cham, Switzerland, 2014; pp. 358–369. [CrossRef]

6. Sobaszek, Ł.; Gola, A.; Kozłowski, E. Application of Survival Function in Robust Scheduling of Production Jobs. In Proceedings of the 2017 Federated Conference on Computer Science and Information Systems, Prague, Czech Republic, 3–6 September 2017; pp. 575–578. [CrossRef]

7. Bubeník, P.; Horák, F. Proactive Approach to Manufacturing Planning. *Qual. Innov. Prosper.* **2014**, *18*. [CrossRef]

8. Danilczuk, W. The use of simulation environment for solving the assembly line balancing problem. *Appl. Comput. Sci.* **2018**, *14*, 42–52. [CrossRef]

9. Gregor, M.; Haluška, M.; Grznár, P.; Hoč, M.; Ďurica, L.; Gregor, M.; Marschall, M.; Vavrík, V.; Gregor, T.; Krkoška, L. *Budúce Továrne. Technologické Zmeny a Ich Vplyv na Budúce Výrobné Systémy. Žilinská Univerzita*; Study CEIT-Š001-05-2017; CEIT: Žilina, Slovakia, 2017; p. 47, ISBN 978-80-971684-6-9.

10. Pekarčíková, M.; Trebuňa, P.; Markovič, J. Case Study of Modelling the Logistics Chain in Production. *Procedia Eng.* **2014**, *96*, 355–361. [CrossRef]

11. Haluška, M. Rekonfigurovateľné Výrobné Systémy. Písomná Práca k Dizertačnej Skúške. Ph.D Thesis, Žilinská Univerzita v Žiline, Žiline, Slovakia, 2014; p. 65.

12. Vavrík, V. Projektovanie Produkčných Liniek s Využitím Princípov Rekonfigurácie. Ph.D Thesis, Žilinská Univerzita v Žiline, Žilina, Slovakia, 2019; p. 190.

13. Koren, Y. General RMS Characteristics. Comparison with Dedicated and Flexible Systems. In *Reconfigurable Manufacturing Systems and Transformable Factories*; Dashchenko, A.I., Ed.; Springer: Berlin/Heidelberg, Germany, 2006; pp. 27–45. [CrossRef]

14. Yang, X.; Dai, H.; Yi, X.; Wang, Y.; Yang, S.; Zhang, B.; Wang, Z.; Zhou, Y.; Peng, X. micROS: A morphable, intelligent and collective robot operating system. *Robot. Biomim.* **2016**, *3*, 21. [CrossRef] [PubMed]

15. Michulek, T.; Capák, J. *Výskum a Vývoj Kráčajúceho Robotického Systému*; Realization Study; CEIT: Žilina, Slovakia, September 2007; p. 18.

16. Gregor, M.; Mačuš, P. Digitálne dvojča & Factory Twin. March 2019. Available online: https://www.mmspektrum.com/clanek/digitalne-dvojca-factory-twin.html (accessed on 29 March 2020).

17. Westkämper, E. Digital Manufacturing in the global era. In Proceedings of the 3rd International CIRP Conference on Digital Enterprise Technology, Setúbal, Portugal, 18–20 September 2006.

18. Carlile, P.R. Apragmatic viewof knowledge and boundaries: Boundary objects in new product development. *Organ. Sci.* **2002**, *13*, 442–455. [CrossRef]

19. Zhao, Q. *Presents the Technology, Protocols, and New Innovations in Industrial Internet of Things (IIoT)*; Springer: Cham, Switzerland, 2020; pp. 39–56. [CrossRef]

20. Platzer, A. (Ed.) Cyber-Physical Systems: Overview. In *Logical Foundations of Cyber-Physical Systems*; Springer International Publishing: Cham, Switzerland, 2018; pp. 1–24. [CrossRef]

21. Hoffmann, M. *Smart Agents for the Industry 4.0: Enabling Machine Learning in Industrial Production*; Springer: Cham, Switzerland, 2019. [CrossRef]

22. Wooldridge, M.A. *Introduction to MultiAgent Systems*, 2nd ed.; Wiley: Chichester, UK, 2009.

23. Bauernhansl, T. Die Vierte Industrielle Revolution—Der Weg in ein wertschaffendes Produktionsparadigma. In *Industrie 4.0 in Produktion, Automatisierung und Logistik: Anwendung Technologien Migration*; Bauernhansl, T., Ten Hompel, M., Vogel-Heuser, B., Eds.; Springer Fachmedien: Wiesbaden, Germany, 2014; pp. 5–35. [CrossRef]

24. Trebuňa, P.; Kliment, M.; Edl, M.; Petrik, M. Creation of Simulation Model of Expansion of Production in Manufacturing Companies. *Procedia Eng.* **2014**, *96*, 477–482. [CrossRef]

25. Gola, A.; Wiechetek, Ł. Modelling and simulation of production flow in job-shop production system with enterprise dynamics software. *Appl. Comput. Sci.* **2017**, *13*. [CrossRef]

26. Osinga, F.P.B. *Science, Strategy and War: The Strategic Theory of John Boyd*, 1st ed.; Routledge: London, UK, 2007.

27. Ghosh, A.; Ullah, S.; Kubo, A. Hidden Markov Model-Based Digital Twin Construction for Futuristic Manufacturing Systems. *Artif. Intell. Eng. Des. Anal. Manuf.* **2019**, *33*, 317–331. [CrossRef]

28. Ghosh, U.; Kubo, A.; D'Addona, D.M. Machining Phenomenon Twin Construction for Industry 4.0: A Case of Surface Roughness. *J. Manuf. Mater. Process.* **2020**, *4*, 11. [CrossRef]

29. Ristvej, J.; Holla, K.; Simak, L.; Titko, M.; Zagorecki, A. Modelling, simulation and information systems as a tool to support decision-making process in crisis management. In *Modelling and Simulation 2013—European Simulation and Modelling Conference, Proceedings of the ESM 2013, Lancaster, UK, 23–25 October 2013*; EUROSIS: Ostend, Belgium, 2013; pp. 71–76.

30. Gregor, M.; Haluška, M.; Grznár, P. *Komplexné systémy. Využitie teórie komplexných systémov pri analýze a navrhovaní produkčných systémov. Žilinská univerzita*; CEIT: Žilina, Slovakia, 2018; p. 107, ISBN 978-80-89865-10-9.

31. Krajčovič, M.; Rakyta, M.; Dulina, Ľ.; Grznár, P.; Gašo, M. *Zásobovacia a Distribučná Logistika 1*; EDIS: Žilina, Slovakia, 2018; p. 492, ISBN 978-80-554-1490-4.

32. Gregor, M. *CEIT_MRS_2020. Smerovanie Vývoja Mobilnej Robotiky v CEIT do Roku 2020. Study*; CEIT-Š010-09-2015; CEIT: Žilina, Slovakia, 2015; p. 66.

33. Grznár, P.; Furtáková, S.; Hnát, J. Analýza virtuálneho prototypovania. *Časopis Slovenskej Spoločnosti pre Systémovú Integráciu* **2013**, *1*, 1–4, ISSN 1336-5916.

34. Gregor, M.; Palajová, S.; Gregor, M. Simulation metamodeling of manufacturing systems with the use of artificial neural networks. In Proceedings of the MITIP 2012, 14th International Conference on Modern Information Technology in the Innovation Processes of the Industrial Enterprises, Budapest, Hungary, 24–26 October 2012; Hungarian Academy of Science: Budapest, Hungary, 2012; pp. 178–190, ISBN 978-963-311-3738.

35. Palajová, S. Simulačné Metamodelovanie Výrobných Systémov. Ph.D Thesis, Žilinská Univerzita v Žiline, Žilina, Slovkia, 2012; p. 138.

36. Gregor, M.; Krajčovič, M.; Gregor, T.; Grznár, P.; Haluška, M.; Ďurica, L. *Smart Logistics. Nové technológie pre logistiku. Štúdia č. 04-2015 APVV*; CEIT, Žilinská Univerzita v Žiline: Žilina, Slovkia, 2015; p. 68, ISBN 978-80-971684-8-3.

 *applied sciences*

Article

Predictive Scheduling with Markov Chains and ARIMA Models

Łukasz Sobaszek [1,*], Arkadiusz Gola [1] and Edward Kozłowski [2]

[1] Faculty of Mechanical Engineering, Lublin University of Technology, 20-618 Lublin, Poland; a.gola@pollub.pl
[2] Faculty of Management, Lublin University of Technology, 20-618 Lublin, Poland; e.kozlovski@pollub.pl
* Correspondence: l.sobaszek@pollub.pl; Tel.: +48-81-538-45-85

Received: 13 July 2020; Accepted: 29 August 2020; Published: 3 September 2020

Abstract: Production scheduling is attracting considerable scientific interest. Effective scheduling of production jobs is a critical element of smooth organization of the work in an enterprise and, therefore, a key issue in production. The investigations focus on improving job scheduling effectiveness and methodology. Due to simplifying assumptions, most of the current solutions are not fit for industrial applications. Disruptions are inherent elements of the production process and yet, for reasons of simplicity, they tend to be rarely considered in the current scheduling models. This work presents the framework of a predictive job scheduling technique for application in the job-shop environment under the machine failure constraint. The prediction methods implemented in our work examine the nature of the machine failure uncertainty factor. The first section of this paper presents robust scheduling of production processes and reviews current solutions in the field of technological machine failure analysis. Next, elements of the Markov processes theory and ARIMA (auto-regressive integrated moving average) models are introduced to describe the parameters of machine failures. The effectiveness of our solutions is verified against real production data. The data derived from the strategic machine failure prediction model, employed at the preliminary stage, serve to develop the robust schedules using selected dispatching rules. The key stage of the verification process concerns the simulation testing that allows us to assess the execution of the production schedules obtained from the proposed model.

Keywords: robust scheduling; predictive scheduling; machine failure; failure prediction

---

1. Introduction

While conducting their activities, manufacturing enterprises establish a range of various goals. Certainly, one of the common strategic business objectives is to strengthen the market position. An enterprise that aims to broaden the group of clients, as well as foster the already existing business relations, must first and foremost be reliable and deliver quality goods within contractual deadlines [1,2]. Therefore, proper planning of works becomes central to sound execution of production processes. Production scheduling is the solution that can boost the capacity of manufacturers, hence there are numerous scientific publications in the field [3]. Researchers are still taking active efforts to optimise the effectiveness of production jobs scheduling in order to streamline the production planning process [2,4].

Unfortunately, most of the proposed solutions display numerous limitations [5]. It is common practice that the job scheduling algorithms build schedules for idealized production environments, i.e., assuming a static and stable production flow [5,6]. Thereby, a number of disruptive factors are excluded, which would bring the production to a halt in case they occur [7,8]. As a consequence, contractual deadlines would be missed, penalties would be imposed and the manufacturer's credibility would diminish. Among the various uncertainty factors, we can highlight the following [5,9]:

- disruptions of resource availability (machine or robot failure)

- disruptions of orders (placement of new orders)
- disruptions of processes (material shortage, poor product quality)
- disruptions associated with misestimation of the ongoing process parameters (incorrect estimation of operation times)
- disruptions related to the change in the duration of the operation (employee absence or malaise, shorter or extended operation times)

Scheduling production in real manufacturing systems cannot afford to pretend to be disruption-free. It is, therefore, of the essence that scheduling endeavors should consider production problems under uncertainty, which is capable of having a colossal effect on the timeliness of production [10,11]. Given that the more the process changes, the greater its disorganization, the scientific literature in the field of scheduling has recently turned towards robust scheduling [5,9].

The predictive scheduling method proposed in this work employs Markov chains and ARIMA (auto-regressive integrated moving average) models whose combination enables determining the values of the machine failure parameters (time to failure and repair time of the machine). In the next step time buffers are directly integrated into the scheduling process and determine the completion time of the production, which corresponds to the delivery date agreed with the customer.

Section 2 summarizes the essential information regarding robust production scheduling and reviews existing literature. The new methodology for scheduling under machine failure and failure prediction is described in Section 3, and the proposed solutions and results are discussed in the subsequent section. Conclusions and plans for further research work are presented in the last section of the work.

2. Existing Work on Robust Production Scheduling

2.1. Essentials of Robust Scheduling

The purpose of a robust production schedule is predominantly to absorb potential disruptions, by allowing variability to the production system parameters.

Two phases of scheduling are distinguished [9,12]:

1. Predictive scheduling-related to the planning stage.
2. Reactive scheduling-related to the production stage.

A well-executed process of scheduling production jobs must pertain to the first of the phases (also referred to as the offline phase) when the available production data give the foundation for creating [3]:

- a nominal schedule-based on the current system parameters,
- a robust schedule-based on the assumption of uncertainty and variability of production.

Reasonable scheduling in this phase requires not only implementing appropriate tools but also suitable methods for determining uncertainty factors [6]. Unfortunately, there is a distinct paucity of solutions that consider the impact of process disruptions [5,6].

2.2. Existing Literature on Robust Production Scheduling

Due to the practical nature of the problem, robust production scheduling solutions are mainly developed for flow-shop and job-shop systems, which are the prevailing forms of organization in real production systems.

Although robust task scheduling in a flow-shop environment is rather neglected in the literature, a certain number of publications on this issue can be found [13–15]. Various approaches have been applied to building robust schedules in a flow-shop environment—from classical local search algorithms [16] to genetic algorithms [17], integer programming applications [18] and dynamic programming [19]. However, in an overwhelming majority, the publications are concerned with predictive-reactive scheduling, thus, tend to focus on the investigation of effective re-scheduling methods [19], and not on the analysis of uncertainty factors itself.

The second most-investigated scheduling problem is robust production scheduling in job-shop systems [20]. This system is a close reflection of a typical production environment, where the order of operations is imposed by the technological routes of their jobs. Researchers have long highlighted that the real job-shop problem requires a distinctly different approach than the shown by the prevailing theoretical tendencies [21], however, to date no clear trend has emerged. Robust scheduling solutions proposed in the aspects of job-shop production processes resemble the solutions for flow-shop systems inasmuch as they are mainly dedicated to the predictive-reactive approach. Standard approaches are shown to draw from various methods, such as genetic algorithms and their hybrids [20,22], immunological algorithms [23,24] and stochastic programming [25]. Other authors propose robust scheduling methods using expert systems [26].

2.3. Machine Failure as the Major Uncertainty Factor

Although many uncertainty factors can be named, the failure of technological machinery is still considered to be the central problem in manufacturing. This disturbance is regarded to have the greatest impact on performed processes. Failure will not only halt the production but its consequences will linger throughout the remaining production process [5,7].

From the analyzed scientific papers dealing with the topic of machine failure in scheduling, it can be seen that the search for methods that will enable approaching the problem of machine failure and predicting its occurrence are very much in place. Developing effective prediction methods is extremely important from the perspective of robust scheduling [18,19,26].

To this end, a probability distribution is among the most widely used approaches in the field of failure analysis. Researchers employ typical distributions and their combinations. The failure description proposed by Jensen [27] applies a uniform distribution. A similar solution is proposed by Al-Hinai and ElMekkawy [28], who, however, assume that the probability of failure is constant. In contrast, in their description of production process disturbances, Davenport et al. [29] implement a normal distribution, while Mehta and Uzsoy [21] utilize an exponential distribution. The authors propose the use of interesting approaches, such as the methods based on combinations of various distributions. The latter is used by Gürel et al. [4], who combine normal, triangular and exponential distributions.

Recently, researchers have also investigated the application of typical key performance indicators (KPIs) used in maintenance, e.g., MTTF (mean time to failure), MTBF (mean time between failures) and MTTR (mean time to repair). In their direct application of the indicators, Deepu [9] and Gao [5] analyze specially prepared scenarios that assume a certain frequency of machine failure, i.e., high, medium or low, to study the consequences of machine downtime and propose solutions to absorb the emerging disruptions to the schedule. With respect to the indirect use of the indicators, Kempa et al. [30,31] propose the use of the aforementioned reliability indicators indirectly for the purpose of estimating Weibull distribution parameters, while Rosmaini and Shahrul [32] in their study, couple the said indicators with statistical methods. These studies, however, suffer from the major drawback—the acquisition and use of the respective quantities is treated quite theoretically and lacks practical verification on real data of machine failure rates [9,31].

In addition to the methods referenced in the preceding paragraphs, a range of alternative failure prediction methods can be found in the literature. Jian et al. [20] propose accumulating failures to a single occurrence, describing it by means of the MTTR parameter and their original indicator, MBL (machine breakdown level). In turn, Rawat and Lad [33] determine failure rates from the analysis of machine load time distributions, and Baptista et al. in [34] use artificial neural networks for failure analysis.

Although constituting an interesting and important voice in the robust production scheduling studies, these models are associated with certain limitations. Their verification is often carried out on test data, which may not be the most accurate representation of actual problems in manufacturing systems. Secondly, the questions arise as to insufficient argumentation regarding the selection of the solutions. Consequently, the key aspect of implementing historical data in studies of the failure rate of machines is omitted.

The need to use real data on uncertainty factors is also emphasized by Davenport et al. [29] and Kalinowski et al. [35]. Only real knowledge on process disruptions can actually solve actual the problems that result from their occurrence. The issue was addressed in our previous work [36], where a model for the prediction of technological operation times in the framework of an intelligent job scheduling system was conceptualized. The study in question considered the impact of real processing time uncertainty on the production schedule and the developed intelligent module also implemented ARMA/ARIMA time series models, however, a problem of a different size was concerned and the verification was carried out for different production data. While such solutions can be found in the literature, the body of knowledge in the field still appears to lack proper depth [6].

3. Production Scheduling under Technological Machine Failure Constraint

3.1. Objectives

This paper formulates a predictive production scheduling process model in the job-shop environment under technological machine failure established with the help of Markov chains and ARIMA models. Our solutions predict the time of machine failure, as well as the time of repair, and constitute an alternative method to the models proposed in the literature. The objective function of our predictive production scheduling is to minimize the makespan, i.e., to produce a schedule with a minimum completion time of all jobs.

3.2. Basic Mathematical Notation of the Problem

Prior to formulating the problem of robust job scheduling under uncertainty in the job-shop system, we need to define the elements of the production process:

- Set M is a set of m machines (workstations) processing jobs:

$$M = \{M_1, M_2, \ldots, M_m\}. \tag{1}$$

- Set J is a set of n jobs (tasks) to process

$$J = \{J_1, J_2, \ldots, J_n\}. \tag{2}$$

Processing job J_i on machine M_j constitutes an operation, which is called operation j of job i in the following. Therefore, it is necessary to define:

- MO—a matrix of m columns and n rows describing the technology (the job order):

$$MO = \left[o_{ij}\right], \tag{3}$$

where o_{ij}—the order position of the operation j of job i, which is $o_{ij} = 0$ when the job is not processed on the machine; and $o_{ij} = \{1, \ldots, m\}$, when it is.
- Matrix PT—a matrix describing processing times of operations:

$$PT = \left[pt_{ij}\right], \tag{4}$$

where pt_{ij}—the processing time of operations j of job i; for each $o_{ij} = 0$, $pt_{ij} = 0$.

- Set FT_{Ml} of potential machine failure times:

$$FT_{Ml} = \left\{ ft_{Ml_1}, ft_{Ml_2}, \ldots, ft_{Ml_z} \right\}, \; l \in \;<1; m>, \tag{5}$$

where ft_{Ml_z}—time to failure of the machine l; where z is a natural number representing the z-th machine failure.

- Set TB_{Ml} of time buffers to include in the nominal schedule (for machine l) to obtain a robust schedule:

$$TB_{Ml} = \left\{ tb_{Ml_1}, tb_{Ml_2}, \ldots, tb_{Ml_z} \right\}, \tag{6}$$

where tb_{Ml_z}—the size of the time buffer in the schedule at the failure time ft_{Ml_z}; where for $ft_{Ml_z} \neq 0$, $tb_{Ml_z} \neq 0$.

3.3. Prediction of Failure and Machine Repair Times

In the paper, we analyze the system describing the shift on which a failure occurs. Additionally, the repair time [37] required for failure removal is analyzed. Let $(\Omega, \mathcal{F}, P)$ be a probabilistic space: Ω—sample space (set of elementary events, outcomes), field $\mathcal{F}$ is a family of sample space Ω (set of all subsets of sample space Ω), P—probability measure (function that assigns each element from field $\mathcal{F}$ the probability, the value between 0 and 1), N—a set of natural numbers, R—a set of real numbers, $S = \{s_1, s_2, \ldots, s_k\}$—a set of possible shifts, $k \in N$, $k < \infty$—the number of possible shifts.

Definition 1. *A family $\{X_t\}_{t \in N}$ of random variables $X_t : \Omega \to S$ for any $t \in N$ is called a stochastic process with discrete time [38,39].*

At any $t \in N$ time, the system can take one of the possible states denoted as $X_t(\omega) = x_t \in S$ and $P(X_t(\omega) = s_i) = p_i(t)$ value means the probability that the system is in a state $s_i \in S, 1 \leq i \leq k$ at a moment $t \in N$, and $\sum_{i=1}^{k} p_i(t) = 1$.

Definition 2. *A stochastic process $\{X_t\}_{t \in N}$ with a discrete time is called a Markov chain [38,39]. If for each $n \in N$, moments $t_1, t_2, \ldots, t_n \in N$ satisfying the condition $t_1 < t_2 < \ldots < t_n$ and any $x_1, x_2, \ldots, x_n \in S$, the equality:*

$$P(X_{t_n} = x_n | X_{t_{n-1}} = x_{n-1}, X_{t_{n-2}} = x_{n-2}, \ldots, X_{t_1} = x_1) = P(X_{t_n} = x_n | X_{t_{n-1}} = x_{n-1}), \tag{7}$$

holds.

Below we assume $t_n = n \in N$. If $\{X_t\}_{t \in N}$ is a heterogeneous Markov chain, then for any $t \in N$ and $1 \leq i, j \leq k$, the value:

$$P\left(X_t = s_j | X_{t-1} = s_i\right) = p_{ij}(t), \tag{8}$$

is the transition probability from s_i state at the moment $t - 1$ to s_j state at moment t. From Markov property (7), the conditional probability distribution of the future process state depends only on the current state at moment t, regardless of the past. The matrix $P(t) = \left[p_{ij}(t) \right]_{1 \leq i, j \leq k}$ is called the transition probabilities matrix at the moment t and the elements of the $P(t)$ matrix satisfy the condition $\sum_{j=1}^{k} p_{ij}(t) = 1$ for $t \in N$ and $1 \leq i \leq k$.

Definition 3. *The Markov chain $\{X_t\}_{t \in N}$ is homogeneous, if the probabilities of transition $p_{ij}(t)$ do not depend on the moment $t \in N$.*

Thus, if for a homogeneous Markov chain [39,40] the matrix $P = \left[p_{ij} \right]_{1 \leq i, j \leq k}$ satisfies the condition $\sum_{j=1}^{k} p_{ij} = 1, 1 \leq i \leq k$, then it is known as the one-step transition probability matrix. From the above, for a homogeneous Markov chain, the transition probability from s_i state at t moment to the s_j state at $t + n$ moment is calculated as follows [38,40]:

$$P\big(X_{t+n} = s_j \big| X_t = s_i\big) = p_{ij}{}^{(n)} \tag{9}$$

where $\left[p_{ij}^{(n)}\right]_{1 \le i,j \le k} = P^n$, $n \in N$ is the transition probability matrix in n steps.

Definition 4. *If* $\{X_t\}_{t \in N}$ *is a homogeneous Markov chain and there is a distribution* $\pi = (\pi_1, \pi_2, \ldots, \pi_k)$ *where* $\pi_i \ge 0$, $1 \le i \le k$ *and* $\sum_{i=1}^{k} \pi_i = 1$ *satisfying the equation:*

$$\pi P = \pi \tag{10}$$

then the distribution π *is called the stationary distribution of the homogeneous Markov chain.*

This property means that if at some $n \in N$ moment the chain reaches a stationary distribution, then for each subsequent moment greater than n, the distribution will remain the same. To determine the stationary distribution, we solve Equation (10).

Let $\{x_t\}_{0 \le t \le n}$ be the realization of Markov chain, where $n_i = \#\{t : x_t = s_i, 0 \le t \le n\}$ is the number of moments for which the system was in s_i state, $1 \le i \le k$ and $\sum_{i=1}^{k} n_i = n$. The value $n_{ij} = \#\{t : x_t = s_i, x_{t+1} = s_j, 0 \le t \le n-1\}$ represents the number of transitions from the state s_i to the state s_j for $1 \le i, j \le k$ and $\sum_{j=1}^{k} n_{ij} = n_i$. We calculate the estimator of transition probability from s_i state to s_j state as $\hat{p}_{ij} = \frac{n_{ij}}{n_i}$ for $1 \le i, j \le k$.

In this work, the goodness of fit test is used to verify Markov property χ^2 [40,41]. At the significance level $\alpha \in (0, 1)$, we create a working hypothesis: $H_0 : P(X_t = x | X_{t-1} = y, X_{t-2} = z) = P(X_t = x | X_{t-1} = y)$ (the chain $\{X_t\}_{t \in N}$ meets Markov property) and an alternative hypothesis: $H_1 : P(X_t = x | X_{t-1} = y, X_{t-2} = z) \ne P(X_t = x | X_{t-1} = y)$ (the chain $\{X_t\}_{t \in N}$ does not meet Markov property), where $x, y, z \in S$.

To verify the hypothesis H_0, we calculate the test statistics:

$$\chi_e^2 = \sum_{i=1}^{k} \sum_{j=1}^{k} \sum_{v=1}^{k} \frac{\big(n_{ijv} - n_{ij}\hat{p}_{jv}\big)^2}{n_{ij}\hat{p}_{jv}}, \tag{11}$$

which has a χ^2 distribution with k^3 degrees of freedom and $n_{ijv} = \#\{t : x_t = s_i, x_{t+1} = s_j, x_{t+2} = s_v, 0 \le t \le n-2\}$ is the number of transitions from state s_i to state s_j and next to state s_v for $1 \le i, j, v \le k$. The critical value is a quantile of order $1 - \alpha$ for χ^2 distribution with k^3 degrees of freedom. We denote as $\chi^2(1 - \alpha, k^3)$. If $\chi_e^2 < \chi^2(1 - \alpha, k^3)$, then at the significance level α, there are no grounds for rejecting the working hypothesis H_0. So, we can assume that the chain $\{X_t\}_{t \in N}$ meets Markov property. When $\chi_e^2 \ge \chi^2(1 - \alpha, k^3)$, then at the significance level α we reject the working hypothesis H_0 in favor of the alternative hypothesis. Thus, the chain $\{X_t\}_{t \in N}$ does not meet Markov property.

An ARIMA model, which usually correlates historical values in a time series, is applied to forecast the repair time. The behavior of the considered time series can be predicted (i.e., forecast with appropriate probability) based on current observation and historical data (dataset). Let $\{rt_t\}_{t \in N}$ denote the sequence of times needed to repair a plant. Because the times needed do remove the failures can take only positive values, the variance-stabilizing transformation

$$\varepsilon_t = ln\big(rt_t\big) \tag{12}$$

can be applied.

The series $\{\varepsilon_t\}_{t \in N}$ is identified using $ARIMA(p,r,q)$ models, $p, r, \ q \in N$ (auto-regressive integrated moving average) [42–45]. In this paper, the logarithm of repair time is modelled as follows:

$$\Delta^r \varepsilon_t = \alpha_0 + \alpha_1 \, \Delta^r \varepsilon_{t-1} + \ldots + \alpha_p \Delta^r \varepsilon_{t-p} + \epsilon_t - \theta_1 \epsilon_{t-1} - \ldots - \theta_q \epsilon_{t-q} \tag{13}$$

where $\{\epsilon_t\}_{t \in N}$ is a sequence of independent random variables with distribution $N\left(0, \ \sigma^2\right)$. To estimate the integration degree, Augmented Dickey-Fuller (ADF) and the Kwiatkowski-Phillips-Schmidt-Shin (KPSS) [42,46,47] tests are applied.

The Markov chains and ARIMA models are implemented to determine the values of elements of sets FT_{MI} and TB_{MI}. The analysis of historical machine failure data leads to determining important failure parameters, which can subsequently help establish buffer time periods in the predictive production scheduling method.

4. Experimental Verification of the Proposed Solution

4.1. Historical Data

In order to verify the solutions proposed in this publication, a process of robust job scheduling is performed on the production data and on historical data of technological machine failure. The production data describes the execution of 9 production jobs processed by 12 machines, constituting a manufacturing cell (Table 1). The parts are produced in batches of 50 elements ($b = 50$) and the setup times of individual operations are not taken into account in the production scheduling process (uncertainty of setup times is a different factor and requires additional research). Therefore:

$$pt_{ij} = b \cdot to_{ij}, \tag{14}$$

where b—quantity of elements in the production batch, to_{ij}—operation time.

Table 1. Production data implemented in the robust scheduling solution.

Job	Operation		Machine	Type of Operation	ts_{ij} * [min]	ts_{ij} * [h]	to_{ij} * [min]	to_{ij} * [h]
	10	M_1	Laser1	Laser-cutting sheets	22	0.367	4	0.067
	20	M_4	CNC saw	Band-saw cutting	6	0.100	0.5	0.008
	30	M_3	CNC press	Edge bending	16	0.267	3	0.050
2	40	M_8	Drill	Drilling holes and threading	12	0.200	1	0.017
	50	M_5	Metalworking	Metalworking	5	0.083	1	0.017
	60	M_6	MIG welder	MIG welding	8	0.133	5.5	0.092
	10	M_1	Laser1	Laser-cutting sheets	12	0.200	0.3	0.005
	20	M_2	Laser2	Laser-cutting profiles	14	0.233	1	0.017
6	30	M_5	Metalworking	Metalworking	5	0.083	1	0.017
	40	M_6	MIG welder	MIG welding	8	0.133	1	0.017
	50	M_{10}	Turning lathe	Turning	11	0.183	2	0.033
	10	M_1	Laser1	Laser-cutting sheets	20	0.333	5	0.083
	20	M_2	Laser2	Laser-cutting pipes and profiles	12	0.200	2	0.033
	30	M_4	CNC saw	Band-saw cutting	6	0.100	1	0.017
9	40	M_3	CNC press	Edge bending	25	0.471	6.5	0.108
	50	M_8	Drill	Drilling holes and threading	12	0.200	7	0.117
	60	M_5	Metalworking	Metalworking	5	0.083	2	0.033
	70	M_6	MIG welder	MIG welding	8	0.133	7.5	0.125

* ts_{ij}—setup time, to_{ij}—operation time.

The data describing the failure rate of machines in the scheduled production process have been obtained from the computer records of machine operation from a maintenance department (Table 2). The collected data describe the failure rate of six technological machines that are crucial for the performed production process. The numbers of observations are as follows:

- Machine M_1—197 observations
- Machine M_2—166 observations

- Machine M_3—180 observations
- Machine M_6—157 observations
- Machine M_7—208 observations
- Machine M_8—97 observations

Table 2. Machine failure and repair time data implemented in the scheduling solution.

Machine M_1		Machine M_2		Machine M_3		Machine M_6		Machine M_7		Machine M_8	
Failure –Shift [–]	Repair Time [min]	Failure –Shift [–]	Repair Time [min]	Failure –Shift [–]	Repair Time [min]	Failure –Shift [–]	Repair Time [min]	Failure –Shift [–]	Repair Time [min]	Failure –Shift [–]	Repair Time [min]
3	230	2	50	3	70	1	10	2	20	2	235
2	120	1	15	3	30	1	50	1	20	1	30
1	15	2	20	1	35	1	15	1	40	1	15
2	95	2	20	3	190	3	110	1	20	2	215
1	80	1	15	2	125	1	120	2	20	2	100
2	30	3	250	2	30	2	130	3	80	2	10
3	130	2	15	3	15	1	30	2	10	1	40

The nominal and robust schedules are subsequently built based on the data presented above. This, however, must be preceded by the prediction of failure parameters with the application of Markov chain and ARIMA models, which is described in the next section.

4.2. Prediction of Machine Failure Parameters

The failure prediction process is performed using appropriate scripts formulated in the RStudio computing environment [48] that enable the analysis in the range described in Section 3.3.

Modelling the machine failure rates using the Markov chain (containing information on changes in production) is performed with the use of the *markovchain* library. Initially, the collected empirical data is verified to check whether they fulfil the properties of the Markov process. The analyses confirm that the data from the analyzed machines meet the required properties, which is further evidenced by the p-value index (Table 3). The p-value is the probability of obtaining hypothesis test results as extreme as the observed results, assuming that the null hypothesis is correct (data chain has a Markov property).

Table 3. Markov process identification results.

Machine No.	p-Value [–]
M_1	0.8922
M_2	0.9051
M_3	0.9510
M_6	0.7361
M_7	0.9684
M_8	0.5618

In the next stage of the machine failure prediction analysis, the transition rate matrices are generated from the collected data. From the obtained information, we determine, at a given probability level, the occurrence of subsequent chain elements, which in this study is the probability of machine failure occurrence at subsequent production shifts (Table A1).

For clarity of presentation, the results from calculations are given in the form of transition diagrams (Figure 1), which additionally enable determining the probability of failure not only on subsequent shifts but also during the current shift (the arrow returning to the node). In Figure 1, knots represent shifts and the probability of machine failure during a given shift is given at the beginning of each arrow (next to the knot), e.g., for machine M_6, the probability that the machine failure will occur after shift 1 during shift 2 is 0.593.

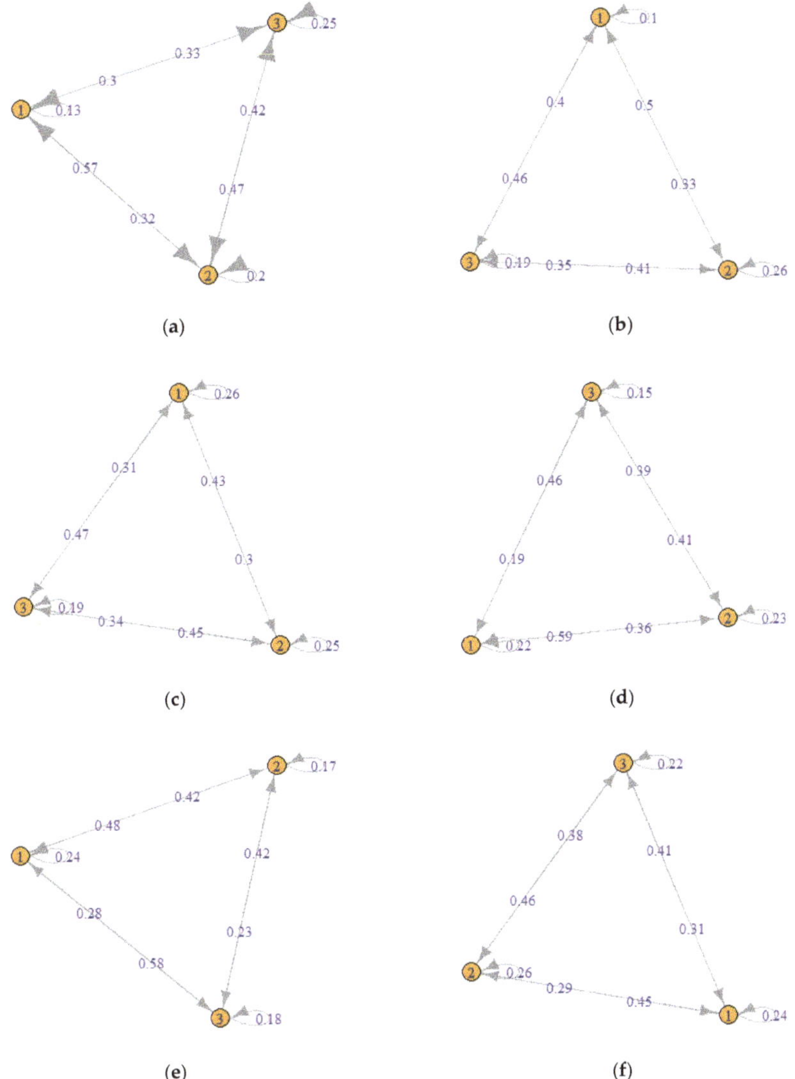

Figure 1. Markov chains with transition diagrams for the considered machines: (**a**) Machine M_1; (**b**) Machine M_2; (**c**) Machine M_3; (**d**) Machine M_6; (**e**) Machine M_7; (**f**) Machine M_8.

The second key parameter of failure is its time. To this end, the *forecast* package is used, which enables identifying the elements of ARIMA time series. As a result, the predicted machine failure repair times are determined. Before forecasting, each machine is tested to verify whether the process is stationary, also the component models are established (autoregression, moving average and the integration). In the subsequent step, future repair times are predicted. Due to the fact that in ARIMA models, the forecasts may also take negative values, the collected observations have been first subjected to variance-stabilizing transformation, and after the prediction, the process is completed and their original sets of values are returned. The results from the model identifying exemplary predicted times for the 5 subsequent steps of the time series are presented in Table 4.

Table 4. Predicted machine repair times.

Machine No.	ARIMA Model	Predicted Repair Times [min]				
		1	2	3	4	5
M_1	ARIMA(1,0,0)	38.77	42.11	41.90	41.91	41.91
M_2	ARIMA(0,0,0)	39.79	39.79	39.79	39.79	39.79
M_3	ARIMA(1,1,2)	36.78	40.80	40.02	40.16	40.14
M_6	ARIMA(2,0,1)	48.12	37.57	43.20	37.78	42.13
M_7	ARIMA(1,0,1)	54.80	54.20	53.72	53.35	53.06
M_8	ARIMA(0,0,1)	51.83	49.85	49.85	49.85	49.85

The data below display the disparity between individual machine repair times. Each machine is coupled with a different ARIMA model. From the set of possible ARIMA models, we select a model that has a smaller AIC (Akaike Information Criterion) value. The analytical process is carried out with the exclusive use of autoregression (Machine M_1), or the moving average (Machine M_8), or a combination thereof (Machine M_2, M_6, M_7). In a single case (Machine M_2), time variability is a series of independent random variables; therefore, the forecast repair times are established on the basis of mean observations.

The results of the prediction of machine failure parameters are perfectly applicable to robust production scheduling. Expressing the time of failure through production shifts allows us to determine the intervals in which machine failures are likely to occur. In turn, the forecasted repair times could be further employed to the determination of machine inspection and maintenance times. Therefore, in the next stage, the obtained analysis results are used to formulate a robust production schedule, whose effectiveness is subsequently verified.

4.3. Production Process Modelling and Scheduling

Before the obtained prediction results could be subjected to further processing, nominal schedules are generated from the real data. Let us assume that the product is manufactured in batches of 50 pieces and the objective function of the schedule is to minimize the makespan ($C_{\max}$).

The schedules are developed using LiSA (A Library of Scheduling Algorithms) software [49], for the analysis of scheduling problems in various environments. The production data serve to represent the production system: a set of machines M and jobs J, the technology matrix MO and the matrix of processing times PT. To test the alternative versions of scheduling, the choice of the next operation is determined by two dispatching rules [50]:

- LPT (longest processing time)
- SPT (shortest processing time)

The robustness of the production schedules is to be provided by the inclusion of the results from the predictions of failure parameters using the data describing the states of production shifts set S and predicted repair times rt_t. As a result, we have managed to determine the elements from the set of predicted machine failure times FT_{MI} and the service time buffer set TB_{MI}. As noted in the introduction, the data obtained for strategic machines are analyzed from the perspective of executed production processes, hence the discrepancies in the designations in the technology records and the schedule. All the data that serve to generate the robust schedule are presented in Table 5.

Table 5. The data implemented into the robust production schedule.

Machine No.	Elements of Set FT_{MI} [h]	Elements of Set TB_{MI} [h]
M_1	$FT_{M1} = \{8\}$	$TB_{M1} = \{0.646, 0.702, 0.698, 0.699, 0.699\}$
M_2	$FT_{M2} = \{8\}$	$TB_{M2} = \{0.663, 0.663, 0.663, 0.663, 0.663\}$
M_3	$FT_{M3} = \{8\}$	$TB_{M3} = \{0.613, 0.680, 0.667, 0.669, 0.669\}$
M_6	$FT_{M6} = \{8\}$	$TB_{M6} = \{0.802, 0.626, 0.720, 0.630, 0.702\}$
M_7	$FT_{M7} = \{8\}$	$TB_{M7} = \{0.913, 0.903, 0.895, 0.889, 0.884\}$
M_8	$FT_{M8} = \{8\}$	$TB_{M8} = \{0.864, 0.831, 0.831, 0.831, 0.831\}$

Since it is built on the data above, the obtained schedule is robust to machine failure disturbances. The procedure for generating the robust schedule is rather straightforward: service time buffers TB_{MI} are implemented into the nominal schedule in the slots indicated by the set of machine failure times FT_{MI}. The time-to-failure is counted only for the machines processing jobs (idle time was disregarded). In the case when a service time buffer is required during the operation, any interfering operation was shifted right in the order of jobs.

4.4. Evaluation Criteria

To verify the effectiveness of the robust scheduling solutions, as well as for the sake of comparative analysis against the nominal schedules, the following assessment criteria are applied:

- makespan $C_{\max}$—total production time,
- mean completion time $\overline{C}$ given by:

$$\overline{C} = \frac{1}{n} \sum_{i=1}^{n} (C_i), \tag{15}$$

where C_i—the completion time of job i.

- mean flow time $\overline{F}$ given by:

$$\overline{F} = \frac{1}{n} \sum_{i=1}^{n} (F_i), \tag{16}$$

where F_i—the flow time of job i.

- the number of critical operations Y_K is derived from:

$$Y_K = \sum_{i=1}^{n} \sum_{j=1}^{m} (y_{ij}), \tag{17}$$

$$y_{ij} = \begin{cases} 1, & \text{when } (tz_{ij} - tr_{ij+1}) = 0) \\ 0, & \text{when } (tz_{ij} - tr_{ij+1}) \neq 0) \end{cases}, \tag{18}$$

where Y_K—the number of critical operations, tz_{ij}—the completion time of operation o_{ij} (current), tr_{ij+1}—the start time of operation $o_{ij} + 1$ (subsequent).

The verification of the obtained schedules is performed during the online stage (production execution), modelled with the Enterprise Dynamics software in a series of simulation tests. The computations serve to determine total completion times of production jobs under strategic machinery failure. The modelling tool used in the study allows detecting machine failure times by setting the MTTF and MTTR indicators and selected probability distributions (Table 6). The MTTF values are specified for the uniform probability distribution (i.e., the machine failure can occur at any time—from the start of the job on the machine until its completion). On the other hand, the MTTR values are determined using the Weibull distribution, obtained for machine repair times from the Cullen–Frey graph [6].

Table 6. Machine failure parameters defined in the simulation environment.

Machine No. (Technology)	MTTF *	MTTR *
M_1	Uniform(0, 16.763)	Weibull(0.88, 1.28)
M_2	Uniform(0, 8.673)	Weibull(0.75, 1.51)
M_3	Uniform(0, 15.247)	Weibull(0.679, 1.72)
M_6	Uniform(0, 22.083)	Weibull(0.769, 1.43)
M_7	Uniform(0, 8.34)	Weibull(0.973, 1.58)
M_8	Uniform(0, 19.24)	Weibull(0.877, 1.45)

* the parameters are expressed in hours.

Twenty-five simulations of the production process are performed for each of the LPT or SPT schedules. The indicators employed in the assessment of the results from simulations are:

- Increase of completion time of all jobs ΔC_{max} given by:

$$\Delta C_{max} = C_{max} - C'_{max} \, , \tag{19}$$

where ΔC_{max}—increase of completion time of all jobs, C_{max}—nominal schedule makespan, C'_{max}—actual (executed) schedule makespan.

- Relative increase of makespan $E_{C_{max}}$ given by:

$$E_{C_{max}} = \frac{C_{max}}{C'_{max}}, \tag{20}$$

where $E_{C_{max}}$—relative increase of makespan, C_{max}—nominal schedule makespan, C'_{max}—actual (executed) schedule makespan.

4.5. Experimental Results

The first of the verification objectives is to compare the nominal and robust production schedules in terms of evaluation criteria. The values of the evaluation indicators of the schedules have been determined and are summarized below (Table 7).

Table 7. Evaluation criteria in the nominal and robust schedules.

Dispatching Rule	Evaluation Criterion [h]								
	$\bar{F}$			$\bar{C}$			C_{max}		
	Nominal Sched	Robust Sched	Elong. [%]	Nominal Sched	Robust Sched	Elong. [%]	Nominal Sched	Robust Sched	Elong. [%]
LPT	23.34	23.86	2.3%	31.94	36.49	14.2%	46.93	53.14	13.2%
SPT	18.33	19.94	8.8%	20.95	23.42	11.8%	47.26	55.68	17.8%

From the presented data, it can be seen that the implementation of service time buffers increases the completion times of all jobs. As a consequence, in each of the analyzed cases, one additional shift is required to complete the production process. This effect is not at all unexpected, given that incorporating service time buffers is inseparably connected with elongation. It should be noted, however, that in the robust schedule the time spent in the production system is not extended owing to the fact that the mean flow time is subject to slight elongation.

Figure 2 presents the visual interpretation of the nominal and robust SPT schedules. Service time buffers are represented by crossed white blocks.

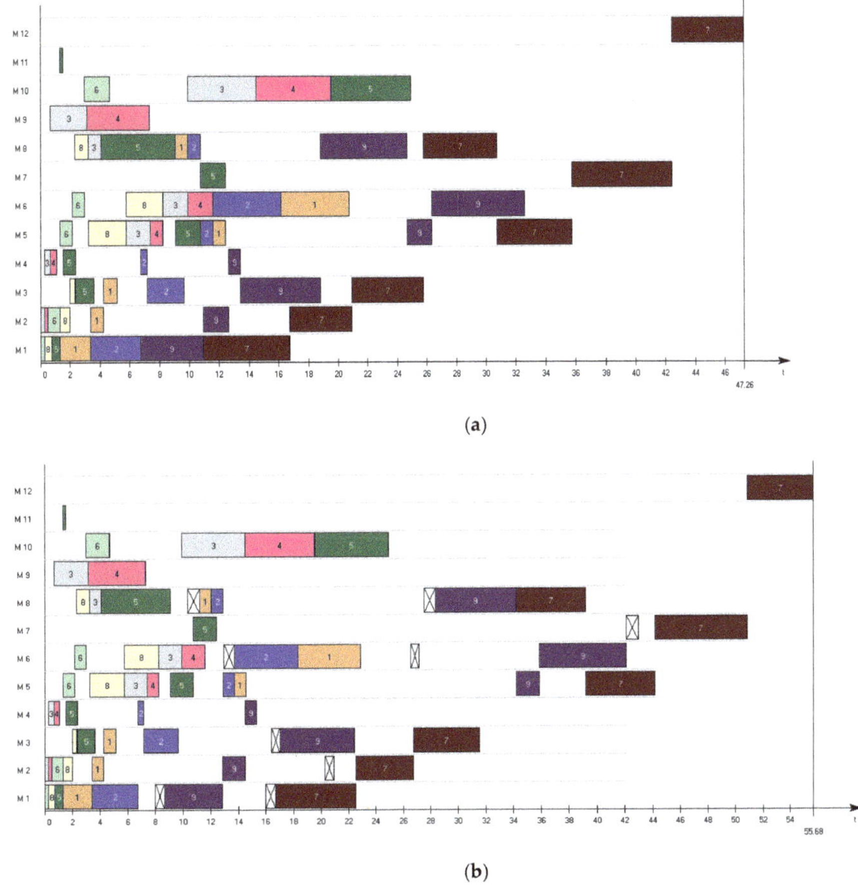

Figure 2. Schedules obtained using the SPT (shortest processing time) rule: (**a**) nominal, (**b**) robust.

A further indicator of robustness that is of great importance in scheduling is the number of critical operations. Scheduling should minimize its value because the stability of the executed process is compromised with the rising number of critical operations. In the analyzed example, the number of critical operations is considered in relation to individual jobs (Y_{KJ}) and machines (Y_{KM}). The robust scheduling results with respect to the number of critical operations are presented in Table 8.

Table 8. The number of critical operations in the nominal and robust schedules.

Dispatching Rule	Number of Critical Operations [–]					
	Y_{KJ}			Y_{KM}		
	Nominal Sched.	Robust Sched.	Reduction [%]	Nominal Schedule	Robust Sched.	Reduction [%]
LPT	30	24	−20.0%	26	21	−19.2%
SPT	32	27	−15.6%	25	21	−16.0%

The incorporation of service time buffers is shown to have a positive effect on the considered parameters. The number of critical operations is reduced by up to 20%. This confirms the legitimacy of implementing service time buffers, which generate additional space in the schedule and thus can prove to be beneficial in the event of machinery failure or other process disruptions.

Simulation tests are conducted to indicate which of the schedules features a makespan closer to the simulated completion time of all jobs. The tests follow the procedure presented in the preceding section and their results are given below, in Tables 9 and 10.

Table 9. Results from simulation: nominal and robust schedules (LPT—longest processing time).

Sim. No.	Executed Schedule (Simulation) C'_{max} [h]	Increase of Makespan and Relative Increase of Makespan					
		Nominal Schedule			Robust Schedule		
		C_{max} [h]	ΔC_{max} [h]	E_{Cmax} [–]	C_{max} [h]	ΔC_{max} [h]	E_{Cmax} [–]
1	52.15		−5.22	0.90		0.99	1.02
2	49.75		−2.82	0.94		3.39	1.07
3	50.93		−4.00	0.92		2.21	1.04
4	57.57		−10.64	0.82		−4.43	0.92
5	52.79		−5.86	0.89		0.35	1.01
6	52.62		−5.69	0.89		0.52	1.01
7	50.01		−3.08	0.94		3.13	1.06
8	55.23		−8.30	0.85		−2.09	0.96
9	50.69		−3.76	0.93		2.45	1.05
10	53.73		−6.80	0.87		−0.59	0.99
11	50.62		−3.69	0.93		2.52	1.05
12	49.26	46.93	−2.33	0.95	53.14	3.88	1.08
13	51.98		−5.05	0.90		1.16	1.02
14	51.73		−4.80	0.91		1.41	1.03
15	50.20		−3.27	0.93		2.94	1.06
16	52.17		−5.24	0.90		0.97	1.02
17	50.71		−3.78	0.93		2.43	1.05
18	51.01		−4.08	0.92		2.13	1.04
19	50.61		−3.68	0.93		2.53	1.05
20	50.65		−3.72	0.93		2.49	1.05
21	49.95		−3.02	0.94		3.19	1.06
22	50.22		−3.29	0.93		2.92	1.06
23	51.83		−4.90	0.91		1.31	1.03
24	52.21		−5.28	0.90		0.93	1.02
25	50.79		−3.86	0.92		2.35	1.05

The schedules generated with the LPT and SPT dispatching rules are shown to outperform the nominal schedule. Their accuracy of predictions is closer to the production data established in simulations. At a closer investigation, the LPT schedules (Table 9) exhibit good compliance of robust and simulated makespans. The schedule is robust for an average of 1.56 h longer than the simulated process; however, considering the nominal schedule, the completion time of all jobs is on average 4.65 h shorter. The comparable makespan length of the robust schedule and the executed production schedule is further confirmed by the mean value of indicator E_{Cmax}, which amounts to 1.03 for the robust schedule, and 0.91 for the nominal schedule.

A similarity of a comparable magnitude is also shown to occur in the production process simulations conducted according to the SPT schedules (Table 10). The mean makespan of the nominal schedule is −5.75 h, while of the robust schedule 2.67 h. In the same case, the mean relative increase is 0.89 for the nominal schedule and 1.05 for the robust schedule.

Nevertheless, it should be noted that in several simulations (for both the LPT and SPT rules), the nominal schedules display a closer resemblance to the executed production process; still, the robust and the executed schedules also show a good fit (e.g., simulation 2 for the LPT rule, or simulation 18 for the SPT rule).

To summarize, the data obtained in the study clearly indicate that the schedule with service time buffers achieves a closer resemblance to the simulated makespan.

Table 10. Results from simulation: nominal and robust schedules (SPT).

Sim. No.	Executed Schedule (Simulation) C'_{max} [h]	Increase of Makespan and Relative Increase of Makespan					
		Nominal Schedule			Nominal Schedule		
		C_{max} [h]	ΔC_{max} [h]	E_{Cmax} [−]	C_{max} [h]	ΔC_{max} [h]	E_{Cmax} [−]
1	51.86		−4.60	0.91		3.82	1.07
2	53.32		−6.06	0.89		2.36	1.04
3	52.11		−4.85	0.91		3.57	1.07
4	55.09		−7.83	0.86		0.59	1.01
5	54.27		−7.01	0.87		1.41	1.03
6	55.36		−8.10	0.85		0.32	1.01
7	52.55		−5.29	0.90		3.13	1.06
8	52.65		−5.39	0.90		3.03	1.06
9	51.60		−4.34	0.92		4.08	1.08
10	53.19		−5.93	0.89		2.49	1.05
11	53.99		−6.73	0.88		1.69	1.03
12	51.07	47.26	−3.81	0.93	55.68	4.61	1.09
13	53.76		−6.50	0.88		1.92	1.04
14	51.54		−4.28	0.92		4.14	1.08
15	55.85		−8.59	0.85		−0.17	1.00
16	54.55		−7.29	0.87		1.13	1.02
17	53.95		−6.69	0.88		1.73	1.03
18	51.47		−4.21	0.92		4.21	1.08
19	51.69		−4.43	0.91		3.99	1.08
20	50.71		−3.45	0.93		4.97	1.10
21	51.75		−4.49	0.91		3.93	1.08
22	53.29		−6.03	0.89		2.39	1.04
23	54.03		−6.77	0.87		1.65	1.03
24	53.47		−6.21	0.88		2.21	1.04
25	52.11		−4.85	0.91		3.57	1.07

Figures 3 and 4 display the results for makespan increase indicators, which provide further evidence confirming the legitimacy of our solutions. The proximity of the robust schedules to the simulated schedules is again highlighted by their being situated close to the dashed line.

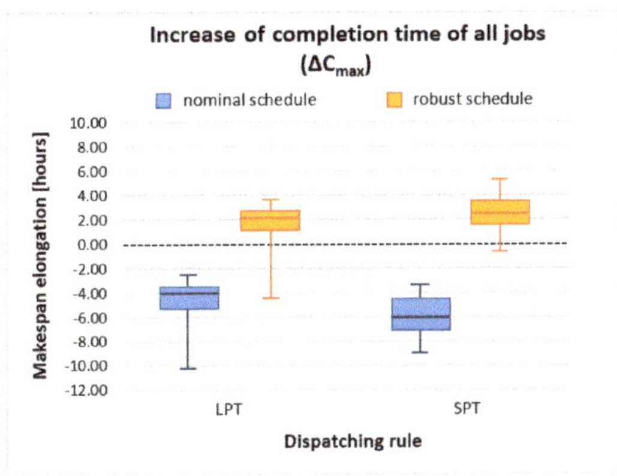

Figure 3. Increase of makespan in LPT and SPT schedules.

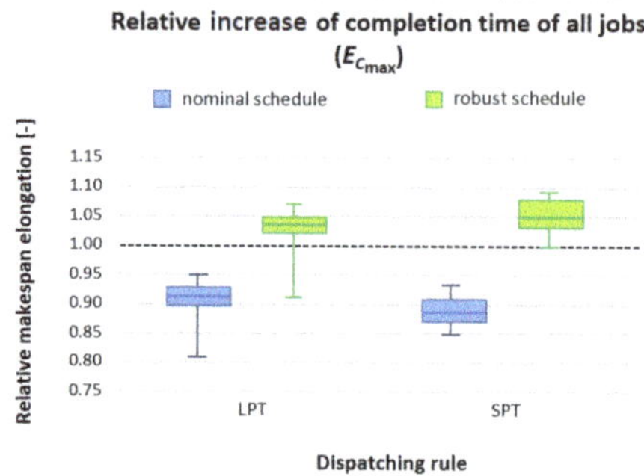

Figure 4. Relative increase of makespan in LPT and SPT schedules.

Robust production schedules generated with the application of our solutions determine makespans closer to the simulated completion times of all jobs in the simulated production conditions under technological machinery failure uncertainty. This is evidenced by several indications, e.g., the fact that for the robust schedules, the values of ΔC_{max} tend to be close to 0 (Figure 3), whereas in the case of $E_{C_{max}}$ around 1 (Figure 4). Values 0 and 1 of the considered indicators denote igh compliance of the robust schedule with the production execution (simulation).

5. Summary and Conclusions

The execution of production processes is associated with the occurrence of various uncertainty factors. Disruptions generate problems that may have a marked effect on production schedules. Therefore, more effort is required in developing techniques and methods that affirm the relevance of uncertainty factors in manufacturing and propose viable solutions. Robust scheduling exhibits the required potential to cope with disruptions and, thus, should be studied further.

In this investigation, the aim was to design a robust production scheduling method with the implementation of Markov chain theory and ARIMA models that will provide for the negative effects of technological machinery failure. The analyses reveal that the inclusion of machine failure in the production schedule results in the extension of the performance indicators, mean flow time, mean job completion time, as well as the central criterion describing the performance of the production system—the completion time of all jobs (makespan). However, the elongation remains within the reasonable limits given that the production is carried out according to failure-inclusive schedules. The simulations evidence that the robust schedules bear a closer similarity to the simulated production process than their nominal equivalents. In other words, the proposed model generates high-accuracy makespan while increasing the robustness and stability of the schedule.

To extend our research in the future, we intend to develop improved models that will: provide for the management of other uncertainty factors in production scheduling (e.g., disruptions related to transport, availability of materials or employee absence), enable reactive scheduling of production jobs or extend the versatility of the proposed solutions over other manufacturing systems. Our current findings and methodologies should make a noteworthy contribution to the theory of production scheduling, as well as appeal to practitioners representing various manufacturing industries and different-sized enterprises.

Author Contributions: A.G. gave the theoretical and substantive background for the developed solution and conceived and designed the experiments, E.K. prepared and provided mathematical description of the method, Ł.S. prepared conception of proposed method and conducted experimental verification of the solution. All authors have read and agreed to the published version of the manuscript.

Funding: The project/research was financed from the Lublin University of Technology Project—Regional Initiative of Excellence from the funds of the Ministry of Science and Higher Education on the basis of a contract No. 030/RID/2018/19.

Conflicts of Interest: The authors declare no conflict of interest.

Appendix A

In the table below, the rows and columns describe individual production shifts. The probability of the machine failure during a given shift is derived from the matrix by setting the shift number in a given row against an appropriate column, e.g., for Machine M_6, the probability that the machine failure will occur after shift 1 during shift 2 is 0.593 (first row, second column). The procedure of machine failure probability calculation is described in detail in Section 3.3 (Definition 4).

Table A1. The transition rate matrix for each machine in the production system. Bold numbers represent the highest probability of the machine failure.

	Transition Rate Matrix										
	Machine M_1				Machine M_2				Machine M_3		
shift	1	2	3	shift	1	2	3	shift	1	2	3
1	0.132	**0.566**	0.302	1	0.100	**0.500**	0.400	1	0.262	**0.426**	0.311
2	0.324	0.203	**0.473**	2	0.328	0.262	**0.410**	2	0.300	0.250	**0.450**
3	0.333	**0.420**	0.246	3	**0.463**	0.352	0.185	3	**0.466**	0.345	0.190
	Machine M_6				Machine M_7				Machine M_8		
shift	1	2	3	shift	1	2	3	shift	1	2	3
1	0.222	**0.593**	0.185	1	0.244	**0.476**	0.280	1	0.241	**0.448**	0.310
2	0.361	0.230	**0.410**	2	0.415	0.169	**0.415**	2	0.286	0.257	**0.457**
3	**0.463**	0.390	0.146	3	**0.583**	0.233	0.183	3	**0.406**	0.375	0.219

References

1. Perłowski, R.; Antosz, K.; Zielecki, W. Optimization of the Medium-Term Production Planning in the Company—Case Study. *Lect. Notes Electr. Eng.* **2018**, *505*, 369–376.
2. Burduk, A.; Musial, K.; Kochanska, J. Tabu Search and Genetic Algorithm for Production Process Scheduling Problem. *LogForum* **2019**, *15*, 181–189. [CrossRef]
3. Sobaszek, Ł.; Gola, A.; Kozłowski, E. Application of survival function in robust scheduling of production jobs. In Proceedings of the 2018 Federated Conference on Computer Science and Information Systems (FEDCSIS), Prague, Czech Republic, 3–6 September 2017; Ganzha, M., Maciaszek, M., Paprzycki, M., Eds.; IEEE: New York, NY, USA, 2017; Volume 11, pp. 575–578.
4. Gürel, S.; Körpeoğlu, E.; Aktürk, M.S. An Anticipative Scheduling Approach with Controllable Processing Times. *Comput. Oper. Res.* **2010**, *37*, 1002–1013. [CrossRef]
5. Gao, H. Building Robust Schedules using Temporal Potection—An Empirical Study of Constraint Based Scheduling Under Machine Failure Uncertainty. Ph.D. Thesis, Univeristy of Toronto, Toronto, ON, Canada, 1996.
6. Sobaszek, Ł.; Gola, A.; Kozlowski, E. Job-shop scheduling with machine breakdown prediction under completion time constraint. *Adv. Intell. Syst. Comput.* **2018**, *637*, 358–367.
7. Daniewski, K.; Kosicka, E.; Mazurkiewicz, D. Analysis of the correctness of determination of the effectiveness of maintenance service actions. *Manag. Prod. Eng. Rev.* **2018**, *9*, 20–25.
8. Janardhanan, M.N.; Li, Z.; Bocewicz, G.; Banaszak, Z.; Nielsen, P. Metaheuristic algorithms for balancing robotic assembly lines with sequence-dependent robot setup times. *Appl. Math. Model.* **2019**, *65*, 256–270. [CrossRef]
9. Deepu, P. Robust Schedules and Disruption Management for Job Shops. Ph.D. Thesis, Montana State Univeristy, Bozeman, MT, USA, 2008.
10. Jasiulewicz-Kaczmarek, M.; Gola, A. Maintenance 4.0 Technologies for Sustainable Manufacturing—An Overview. *IFAC PapersOnLine* **2019**, *52*, 91–96. [CrossRef]

11. Gola, A.; Kłosowski, G. Development of computer-controlled material handling model by means of fuzzy logic and genetic algorithms. *Neurocomputing* **2019**, *338*, 381–392. [CrossRef]
12. Klimek, M. Techniques of Generating Schedules for the Problem of Financial Optimization of Multi-Stage Project. *Appl. Comput. Sci.* **2017**, *15*, 20–34.
13. Rahman, H.F.; Sarker, R.; Essam, D. A Real-Time Order Acceptance and Scheduling Approach for Permutation Flow Shop Problems. *Eur. J. Oper. Res.* **2015**, *247*, 488–503. [CrossRef]
14. Choi, S.H.; Wang, K. Flexible Flow Shop Scheduling with Stochastic Processing Times: A Decomposition-Based Approach. *Comput. Ind. Eng.* **2012**, *63*, 362–373. [CrossRef]
15. Kianfar, K.; Fatemi, G.S.M.T.; Oroojlooy, J.A. Study of Stochastic Sequence-Dependent Flexible Flow Shop via Developing a Dispatching Rule and a Hybrid GA. *Eng. Appl. Artif. Intell.* **2012**, *25*, 494–506. [CrossRef]
16. Almeder, C.; Hartl, R.F. A Metaheuristic Optimization Approach for a Real-World Stochastic Flexible Flow Shop Problem with Limited Buffer. *Int. J. Prod. Econ.* **2013**, *145*, 88–95. [CrossRef]
17. Rahman, H.F.; Sarker, R.; Essam, D. A Genetic Algorithm for Permutation Flow Shop Scheduling Under Make to Stock Production System. *Comput. Ind. Eng.* **2015**, *90*, 12–24. [CrossRef]
18. Chung-Cheng, L.; Kuo-Ching, Y.; Shih-Wei, L. Robust Single Machine Scheduling for Minimizing Total Flow Time in the Presence of Uncertain Processing Times. *Comput. Ind. Eng.* **2014**, *74*, 102–110.
19. Bibo, Y.; Geunes, J. Predictive–reactive scheduling on a single resource with Uncertain Future Jobs. *Eur. J. Oper. Res.* **2008**, *189*, 1267–1283.
20. Jian, X.; Li-Ning, X.; Ying-Wu, C. Robust Scheduling for Multi-Objective Flexible Job-Shop Problems with Random Machine Breakdowns. *Int. J. Prod. Econ.* **2013**, *141*, 112–126.
21. Mehta, S.V.; Uzsoy, R.M. Predictable Scheduling of a Job Shop Subject to Breakdowns. *IEEE Trans. Robot. Autom.* **1998**, *14*, 365–378. [CrossRef]
22. Bierwirth, C.; Mattfeld, D.C. Production Scheduling and Rescheduling with Genetic Algorithms. *Evol. Comput.* **1999**, *7*, 1–17. [CrossRef]
23. Xingquan, Z.; Hongwei, M.; Jianping, W. A robust scheduling method based on a multi-objective immune algorithm. *Inf. Sci.* **2009**, *179*, 3359–3369.
24. Jensen, M.T. Robust and Flexible Scheduling with Evolutionary Computation. Ph.D. Thesis, University of Aarhus, Aarhus, Denmark, 2001.
25. Janak, S.L.; Lin, X.; Floudas, C.A. A New Robust Optimization Approach for Scheduling Under Uncertainty—II. Uncertainty with Known Probability Distribution. *Comput. Chem. Eng.* **2007**, *31*, 171–195. [CrossRef]
26. Henning, G.P.; Cerda, J. Knowledge-based predictive and reactive scheduling in industrial environments. *Comput. Chem. Eng.* **2000**, *24*, 2315–2338. [CrossRef]
27. Jensen, M.T. Improving robustness and flexibility of tardiness and total flow-time job shops using robustness measures. *Appl. Soft Comput.* **2001**, *1*, 35–52. [CrossRef]
28. Al-Hinai, N.; ElMekkawy, T.Y. Robust and Stable Flexible Job Shop Scheduling with Random Machine Breakdowns Using a Hybrid Genetic Algorithm. *Int. J. Prod. Econ.* **2011**, *132*, 279–291. [CrossRef]
29. Davenport, A.; Gefflot, C.; Beck, C. Slack-based Techniques for Robust Schedules. In Proceedings of the Sixth European Conference on Planning, Toledo, Spain, 12–14 September 2001.
30. Kempa, W.; Paprocka, I.; Kalinowski, K.; Grabowik, C. Estimation of reliability characteristics in a production scheduling model with failures and time-changing parameters described by Gamma and exponential distributions. *Adv. Mater. Res.* **2014**, *837*, 116–121. [CrossRef]
31. Kempa, W.; Wosik, I.; Skołud, B. Estimation of Reliability Characteristics in a Production Scheduling Model with Time-Changing Parameters—First Part, Theory. *Manag. Control Manuf. Process.* **2011**, *1*, 7–18.
32. Rosmaini, A.; Shahrul, K. An overview of time-based and condition-based maintenance in industrial application. *Comput. Ind. Eng.* **2012**, *63*, 135–149.
33. Rawat, M.; Lad, B.K. Novel approach for machine tool maintenance modelling and optimization using fleet system architecture. *Comput. Ind. Eng.* **2018**, *126*, 47–62. [CrossRef]
34. Baptista, M.; Sankararaman, S.; De Medeiros, I.P.; Nascimento, C.; Prendinger, H.; Henriques, E.M.P. Forecasting fault events for predictive maintenance using data-driven techniques and ARMA modeling. *Comput. Ind. Eng.* **2018**, *115*, 41–53. [CrossRef]
35. Kalinowski, K.; Krenczyk, D.; Grabowik, C. Predictive-reactive strategy for real time scheduling of manufacturing systems. *Appl. Mech. Mater.* **2013**, *307*, 470–473. [CrossRef]

36. Sobaszek, Ł.; Gola, A.; Kozłowski, E. Module for prediction of technological operation times in an intelligent job scheduling system. In *Intelligent Systems in Production Engineering and Maintenance—ISPEM 2018: International Conference on Intelligent Systems in Production Engineering and Maintenance*; Burduk, A., Chlebus, E., Nowakowski, T., Tubis, A., Eds.; Springer: Cham, Switzerland, 2018; pp. 234–243.

37. Knopik, L.; Migawa, K. Semi-Markov system model for minimal repair maintenance. *Eksploat. I Niezawodn. Maint. Reliab.* **2019**, *21*, 256–260. [CrossRef]

38. Stewart, W.J. *Probability, Markov Chains, Queues, and Simulation*; Princeton University Press: Princeton, NJ, USA, 2009.

39. Ross, S. *Introduction to Probability Models*, 6th ed.; Academic Press: San Diego, CA, USA, 1997.

40. Kozłowski, E.; Borucka, A.; Świderski, A. Application of the logistic regression for determining transition probability matrix of operating states in the transport systems. *Eksploat. I Niezawodn. Maint. Reliab.* **2020**, *22*, 192–200. [CrossRef]

41. Chow, G.C. *Ekonometria*; PWN: Warszawa, Poland, 1995.

42. Kozłowski, E. *Analiza i Identyfikacja Szeregów Czasowych*; Politechnika Lubelska: Lublin, Poland, 2015.

43. Box, G.E.P.; Jenkins, G.M. *Time Series Analysis: Forecasting and Control. Holden-Day*; Wiley: San Francisco, CA, USA, 1970.

44. Rymarczyk, T. Characterizations of the shape of unknown objects by inverse numerical methods. *Prz. Elektrotechniczny* **2020**, *88*, 138–140.

45. Kozłowski, E.; Mazurkiewicz, D.; Żabiński, T.; Prucnal, S.; Sęp, J. Machining sensor data management for operation-level predictive model. *Expert Syst. Appl.* **2020**, *159*, 113600. [CrossRef]

46. Shumway, R.H.; Stoffer, D.S. *Time Series Analysis and Its Applications with R Examples*; Springer: Cham, Switzerland, 2017.

47. Kosicka, E.; Kozłowski, E.; Mazurkiewicz, D. The use of stationary tests for analysis of monitored residual processes. *Eksploat. I Niezawodn. Maint. Reliab.* **2015**, *17*, 604–609. [CrossRef]

48. RStudio Team. *RStudio: Integrated Development for R*; PBC: Boston, MA, USA, 2020. Available online: http://www.rstudio.com (accessed on 10 July 2020).

49. Bräsel, H.; Dornheim, L.; Kutz, S.; Mörig, M.; Rössling, I. *LiSA—A Library of Scheduling Algorithms*; Magdeburg University: Magdeburg, Germany, 2001.

50. Chiang, T.C.; Fu, L.C. Using Dispatching Rules for Job Shop Scheduling with Due Date-Based Objectives. *Int. J. Prod. Res.* **2007**, *45*, 1–28. [CrossRef]

Article

Simulation-Based Analysis on Operational Control of Batch Processors in Wafer Fabrication

Pyung-Hoi Koo [1,*] **and Rubén Ruiz** [2]

1   Department of Systems Management & Engineering, Pukyong National University, Busan 48513, Korea
2   Instituto Tecnológico de Informática, Universitat Politècnica de València, 46021 València, Spain; rruiz@eio.upv.es
*   Correspondence: phkoo@pknu.ac.kr; Tel.: +82-51-629-6485

Received: 13 August 2020; Accepted: 25 August 2020; Published: 27 August 2020

Abstract: In semiconductor wafer fabrication (wafer fab), wafers go through hundreds of process steps on a variety of processing machines for electrical circuit building operations. One of the special features in the wafer fabs is that there exist batch processors (BPs) where several wafer lots are processed at the same time as a batch. The batch processors have a significant influence on system performance because the repetitive batching and de-batching activities in a reentrant product flow system lead to non-smooth product flows with high variability. Existing research on the BP control problems has mostly focused on the local performance, such as waiting time at the BP stations. This paper attempts to examine how much BP control policies affect the system-wide behavior of the wafer fabs. A simulation model is constructed with which experiments are performed to analyze the performance of BP control rules under various production environments. Some meaningful insights on BP control decisions are identified through simulation results.

Keywords: batch processors; real-time control; dispatching; wafer fabrication; semiconductor manufacturing; system-wide performance

---

1. Introduction

In wafer fabrication facilities, semiconductor chips are made out of silicon wafers, thin and round slices of semiconductor material, by building electrical circuits on wafers layer by layer. Each layer requires a number of different processes, including oxidation, photolithography, etching, diffusion, deposition, ion implantation, etc. Wafers in the wafer fab move through these processes in lots generally consisting of 20–25 individual wafers. In general, a wafer lot goes through 500–700 process steps on more than one hundred machines [1]. Since semiconductor processing equipment is very expensive, it is shared by the jobs during different process steps, leading to reentrant product flow. (In this paper, the jobs refer to wafer lots.) Because of the long sequences of operations and resource sharing caused by reentrant product flow, most wafer fabs suffer from high work-in-process (WIP) inventories and long lead times (often more than one month). Since the cost of building a wafer fab is enormous, often more than ten billion dollars [2], the system capacity cannot be easily expanded. Hence, the wafer fab should be efficiently operated with a given facility capacity.

The process equipment in the wafer fabs is often classified into two types: Discrete processors (DPs) and batch processors (BPs). The DPs process wafers one at a time while the BPs process several wafer lots simultaneously as a batch. Diffusion furnace is a typical example of the batch processor where several wafer lots are placed in a reactor, which is then sealed, heated and filled with carrier gas for changes of their electrical and chemical characteristics [3]. The wafer lots arriving at the BP station are formed as a batch before being served by one of the batch processors. After a batch of wafer lots receive a process service, the batch is split into individual wafer lots before they are transferred to a downstream station. Due to process or facility constraints, there is a limitation on the number of wafer lots in a batch that a batch processor can accommodate, which is referred to as batch capacity. Once a processing cycle begins, additional lots cannot join the batch being processed and must wait for a batch processor to become free. Due to the chemical nature of the process, it is often impossible to include the wafer lots with different recipes together in the same batch [4]. The wafer lots with the same recipe can be viewed as a job family. The job families are incompatible in that the jobs of different families cannot be processed together. In wafer fabs with reentrant characteristics, the wafers of the different processing steps need different processing recipes. Hence, even for the same wafer lots, the job family of a wafer lot depends on its current operation step. Figure 1 shows a schematic representation of a BP station with three batch processors whose batch capacity is six. In the figure, two BPs are under operation while one BP is idle because there are no job families with a full load currently waiting in the queue. (Here, a full load operation is assumed.) The processing time at the batch processors is very long compared with discrete processing times. When a BP finishes processing a batch of wafers, it releases multiple wafer lots to its downstream workstations. The bunched flow of wafer lots leads to the formation of high WIP inventories at the downstream stations, and so the manufacturing systems with BPs suffer from high flow variability. The distinct characteristics of batch processors have a significant influence on the overall wafer fab performance.

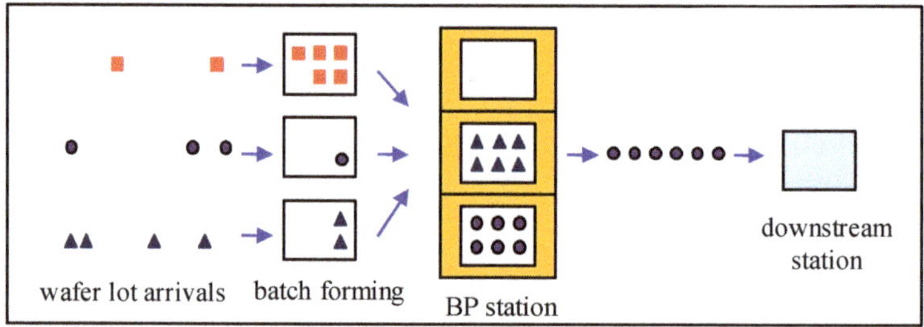

Figure 1. Schematic representation of a batch processor (BP) station with multiple batch processors.

A batch of wafers should be loaded first before they have processing service on a BP. The loading decision at a BP station is made at two instances [5]: (i) When wafer lots arrive at the BP station (a push decision); or (ii) when a batch processor becomes available (a pull decision). If a batch with a full load and a batch processor are both available, the batch is loaded and processed immediately. However, if a partial load of a batch is waiting and a batch processor is available, a loading decision should be made as to whether to load the batch immediately or to wait for more lots to arrive to form a larger batch size. Starting a batch immediately with a partial batch size undermines the BP capacity, while delayed batch loading increases the waiting time for the jobs currently waiting at the BP station, and therefore, potentially increases lead times.

Some research studies have addressed the BP control problem in semiconductor wafer fabs. Most of the BP control rules developed so far have focused on optimizing the local performance of batch processors. However, their application in multi-stage reentrant production systems may not result in as good a system-wide performance as expected [6]. For example, waiting time minimization at the BP stations may deteriorate the system-wide performance: It may lead to an unbalanced and excessive WIP level throughout the manufacturing system, resulting in increased system lead time. Hence, the decision making at batch machines should consider overall system performance, not only local performance. This paper examines the impact of BP control decisions on the system-wide performance of wafer fabs. In this paper, two system-wide performance measures are considered to examine the system performance: Throughput rate and lead times. The throughput rate may be the most important performance measure when production managers want to have a maximum throughput with given manufacturing facilities. Lead time is defined as the time needed for a wafer lot to go all the way through a wafer fab. Lead time is widely recognized as a key performance indicator in today's lean manufacturing environments [7], whose major objective is to keep high on-time delivery with minimum WIP levels while achieving a target throughput. Little's law describes a relationship among throughput rate, lead time, and WIP inventory: $L = \lambda W$ where L is the WIP inventory, λ is the throughput rate, and W is the lead time. This law states that given a specific production rate, the lead time is proportional to the WIP level.

The objectives of this study include: (1) Identifying how much BP stations affect the behavior of the system when compared to discrete processor stations; (2) comparing the performance of the loading decisions with and without system status; and (3) examining the relationship between the local performance of the BP station and system-wide performance. A simulation model is constructed for a wafer fab, and the performance of BP operational controls is examined with simulation experiments. This paper is organized as follows: In the next section, existing BP control rules are reviewed. Section 3 describes the operational control rules for the batch processors that will be examined through simulation experiments. The simulation model and experimental results are presented in Section 4. Finally, in Section 5, conclusions are drawn, and some useful insights are discussed on BP control issues.

2. Literature Review on BP Control Rules

The BP control rules involve an event-based decision procedure for loading a batch of wafer lots on batch processors in real-time. When a batch processor completes the processing service of a batch and there exist products waiting in the queue, decisions should be made whether to start loading a batch of wafer lots right away or wait for wafer lots to arrive, and which job type to be loaded. One possible loading alternative is to load jobs immediately regardless of how many jobs exist in the waiting queue. In this case, only a single wafer lot might be loaded and processed on the batch processor. Another possible control rule is that the processing cycle is initiated only when the batch size reaches the batch capacity. In this case, the jobs which have arrived earlier should wait for upcoming jobs to form a full batch load. An in-between strategy is a threshold strategy. A commonly used threshold strategy is MBS (minimum batch size) [8]. The MBS loading strategy works as follows: Let q be the number of wafer lots waiting in the queue at a BP station, B be the predetermined MBS value (or threshold value), and C be the batch capacity. When a batch processor becomes idle, a BP control decision is initiated. If $q < B$, the BP waits until there are B jobs present at the queue—at which point, it starts serving them together. If $B \leq q \leq C$, all the q jobs are immediately loaded as a batch for having a BP process service. If $q > C$, only C wafer lots are loaded immediately for processing, and the others must wait. The determination of the optimal MBS is a main research topic in the MBS-based control strategy. A stochastic dynamic program is provided in Reference [9] to determine the optimal MBS value for a BP station. It is proved that the MBS control policy is optimal for minimizing mean waiting time when the jobs arrive with a Poisson process, and the processing times are independent and identically distributed. An MBS-based BP control problem is examined in Reference [10] for a reentrant two-stage $\delta \rightarrow \beta$ system where the first stage is a DP station with multiple discrete processors and the second is a BP station with only one batch processor. (In this paper, notations δ and β will be used for discrete and batch processes, respectively.) The discrete processors are subject to failure, which implies that the arrivals of jobs to the BP station are uncertain. Since the system is reentrant, the jobs leaving the BP station visit the DP station again. It is claimed that the control decision at the BP station affects the availability of material at the downstream station, which is especially important when the downstream DP station is a bottleneck resource. An MBS-based control algorithm is introduced in Reference [11] to determine the optimal threshold value with the objective of minimizing the number of customers in the system. It is found that at high BP traffic intensities, the system performance is relatively insensitive to the threshold value because there are always enough customers in the queue at a service completion, while at low BP traffic intensities, it is better to set the MBS value low because the time spent waiting for jobs to arrive is long. A theory-based queueing model is presented in Reference [12] to determine the threshold value for a BP station with multiple machines where multiple types of products are processed. A closed-form formula is presented to approximate the steady-state performance measures, including lead time and WIP levels. A genetic algorithm (GA) based batching decision is proposed to find the near-optimal batch sizes for different product families.

The MBS-based strategies discussed above only consider WIPs in the queue for BP loading decisions. In modern wafer fabs with advanced shop-floor information systems, real-time manufacturing data can be collected with which smart operational controls become possible. With information on product arrivals and the state of the manufacturing system, the batch processors may be more efficiently controlled. For example, if we have information about future product arrivals at the time that a machine becomes idle, and there are wafer lots in the queue smaller than the batch capacity, we may use the information to determine whether to start the process right away or wait for future job arrivals to form a batch with more jobs. A dynamic batching heuristic, called DBH, is presented in Reference [13] for a BP station with a single BP and a single job family where information about upcoming job arrivals is considered. In DBH, the planning horizon is the BP processing time T and the number of forecasted arrivals, L, is smaller than or equal to $C - 1$ where C is the batch capacity. DBH is activated when a batch processor becomes idle, and $q \geq 1$ wafer lots are waiting in the queue. At the time epoch t_0 (current decision time), DBH first examines q against the capacity of the batch processor. If $q \geq C$, C lots are loaded and processed immediately. If $q < C$, DBH makes a batching decision to minimize the total delay time, given the forecasted job arrivals. The amount of additional delay time for products currently waiting in the queue is compared to the amount of saved delay time for the future arrivals by waiting until the arrivals occur. Figure 2 shows a waiting time change when the BP station waits for one incoming lot and load $q + 1$ lots at time epoch t_1 where t_i represents the arriving time epoch of the next ith lot after t_0. The saved wait time can be calculated as $\text{Area}_2(t_1) - \text{Area}_1(t_1)$, compared to the wait time in the immediate loading case. From Figure 2, $\text{Area}_1(t_1) = q(t_1 - t_0)$ and $\text{Area}_2(t_1) = T - (t_1 - t_0)$. The saved wait time is calculated for all the time epochs of L, and the one with the largest positive value is selected for the loading time. If all the saved wait times are negative (this means that the delayed loading leads to more waiting time), the machine starts processing jobs with the partial batch immediately.

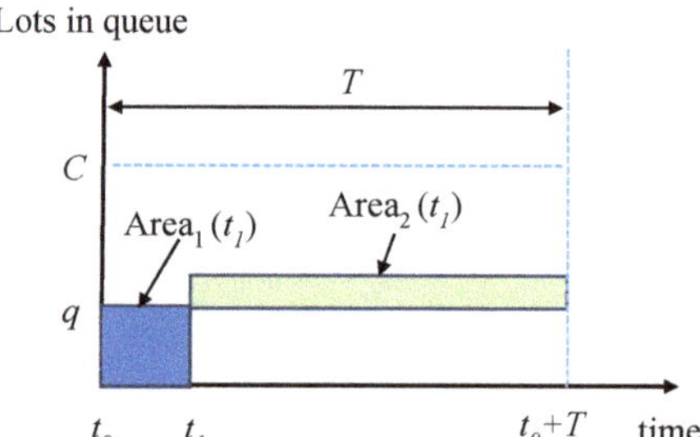

Figure 2. Changes in waiting time when the lots are loaded at time t_1. The saved wait time is $\text{Area}_2(t_1) - \text{Area}_1(t_1)$.

A BP control heuristic named the NACH (Next Arrival Control Heuristic) is presented in Reference [14] for both a single product family and multiple product families. The heuristic only considers the next job arrival and determines if it is more efficient to start the batch process immediately or at the next arrival time. When the loading is postponed until the next arrival time, the decision process is repeated at the next arrival time. It is shown that their heuristic is robust in the sense that it performs well even with prediction errors. The NACHM (NACH with multiple processors) is an extended version of NACH for a BP station with parallel batch processors [5]. Extensive review work on BP control problems with real-time job arrival data is carried out in References [15,16]. Even though BP control decisions with real-time shop-floor information mostly provide better performance than threshold policies, most real-world wafer fabs use MBS-based policies for batching decisions [17]. This is because threshold policies are easy to implement and the future information used in look-ahead control policies is often incorrect, due to unpredictable problems, such as equipment malfunctions, product quality problems, urgent orders, etc.

Most research studies on BP control problems have considered stand-alone BP stations. A few pieces of research handle discrete processors in the downstream, upstream, and both upstream/downstream together with batch processors. A loading decision procedure is proposed in Reference [18] for a $\beta \rightarrow \delta$ network to minimize the production lead time. Experimental results show that the use of upstream and downstream information in batching decisions may lead to additional improvements at the light to moderate traffic intensity, but the improvement vanishes under high traffic conditions. A BP control heuristic for a $\delta \rightarrow \beta \rightarrow \delta$ manufacturing network is presented in Reference [19], in which the states of both upstream and downstream machines are considered. It is claimed through simulation experiments that the benefit of utilizing information about the state of an upstream discrete machine appears to be an order of magnitude larger than that of utilizing information about the state of a downstream discrete machine. An extension of the NACH policy is developed in Reference [20] for a two-stage $\beta \rightarrow \delta$ system that incorporates knowledge about future arrivals and the status of critical machines in downstream processing in the control decision process. The idea is to balance the time for the lots to spend waiting at a BP station with the time spent in the setup at a downstream DP station, thus improve the overall lead time. A rolling horizon BP control strategy is proposed in Reference [21] for a $\delta \rightarrow \beta$ network which incorporates the sequence decisions on the upstream processor through a re-sequencing approach to improve the mean lead time performance of the batch processor. Simulation experiments show that the re-sequencing approach improves the lead time performance of the $\delta \rightarrow \beta$ network as compared to the NACH and MBS-based policies, especially when the number of product types is large, and the BP traffic intensity is low or moderate.

Very few pieces of research work deal with the control problems of the batch processors from a systems perspective. The effect of multiple operational decisions, including job release, mask scheduling, and batch loading on wafer fab performance (throughput and lead time) is examined in Reference [22]. For a batch loading decision, BFQL (back and front queue leveling) heuristic is proposed in which possible starvation of the immediate downstream station is incorporated in an MBS-based loading policy. Appropriate combinations of the lot release, lot dispatching, and batching decisions are examined through simulation experiments. One possible drawback of the BFQL rule is that the control decision only considers the immediate downstream station regardless of its scheduled workload. If the scheduled workload of the downstream station is low, the system performance is not much affected by the downstream station so that the downstream status is not that important to be included in the batching decision process. The effect of dispatching rules and job release policies on system performance in wafer fabs is examined in References [23,24]. Several composite dispatching rules have been suggested, and the effect of the combination of dispatching rules and job release policies is compared through simulation experiments. However, in these studies, the effect of batch processors on system performance is not explicitly examined as only an MBS-based loading rule, or a full-batch loading rule is assumed.

To the best of our knowledge, little work has been done to use simulation analysis to examine the effect of different BP operational decisions on system-wide performance where the BP processors are explicitly considered. Our study in this paper is to examine the impact of BP control decisions on system-wide performance through simulation experiments with different BB control rules.

3. Operational Control Rules for Simulation Analysis

The control problems in wafer fab can be classified into two major problems, lot release and dispatching [24]. Lot release involves a fab-level decision that specifies when and how many wafer lots should be released to the wafer fab for the first process, while dispatching is a workstation-level decision that determines which job should be loaded next on an idle machine. There are two typical control strategies in terms of lot release: Open-loop release and closed-loop release. The open-loop release policies release wafer lots into the fab based on static production scheduling regardless of the current system status, while the closed-loop release policies consider wafer fab status in release decision making. It is widely recognized that the performance of closed-loop release policies is better compared with that of open-loop release policies [24]. One of the simple open-loop release policies is CONTIME (CONstant TIME) where wafer lots are released into the fab at a constant time interval (for example one wafer lot is released every 2 h). With the CONTIME lot release rule, the production rate can be anticipated by the release interval, while the WIP inventories are variable depending on the operational control decisions at the workstations. A typical closed-loop release policy is CONWIP (CONstant WIP) where a constant number of WIPs are maintained in the fab [25]. In the CONWIP release rule, whenever a wafer lot leaves the wafer fab, a new wafer lot is introduced into the system. Contrary to the open-loop control systems, the throughput rates are variable depending on the operational control decisions at the workstations, while WIP levels are constantly maintained. The other widely known closed-loop release policies include the workload regulating (WR) rule [26] and the starvation avoidance (SA) rule [27] where a new lot is released to the wafer fab when the capacity load of the bottleneck station drops to a specified level. In both the WR and SA rules, the throughput is controlled by changing the critical values. Various closed-loop release rules have been proposed subsequently, which are mostly variations of the WR and SA rules [24,28].

Once the wafer lots are released into the wafer fab, they visit a number of workstations where a dispatching decision should be made as to which lot should be loaded to be processed next. Most of the existing dispatching approaches are based on priorities that are set by using product information, such as FCFS (first come first served), SRPT (shortest remaining processing time), and EDD (earliest due date), as well as system status including workload and WIP levels in downstream and/or upstream workstations. A wide variety of dispatching policies are reviewed extensively in Reference [15]. For the batch processors, the BP control decisions on batch forming and batch loading time should be made. In this paper, three control schemes, MBS-based rule, Look-upstream rule, and Look-downstream rule are examined: A control rule only with local information, a control rule with upstream information, and a control rule with downstream bottleneck station information, respectively (See Figure 3). The BP control rules are based on the existing BP control policies because our objective is not to develop new BP control rules, but investigate the impact of BP control schemes from the systems' viewpoint.

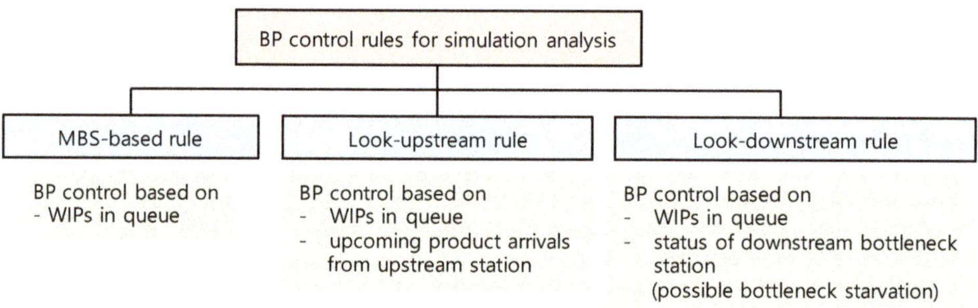

Figure 3. Three BP control rules under consideration.

(1) MBS-based rule (MBS#): Processing is initiated when the batch size of a job type is greater than or equal to a predetermined minimum batch size (MBS). The notation MBS# is used to represent the MBS rule with minimum batch size #. For example, suppose we have a batch processor whose batch capacity is six wafer lots. Then, MBS4 is an MBS rule where a batch of wafer lots is loaded when there are at least four wafer lots in the waiting queue. Two extreme cases are MBS1 and MBS6. In MBS1, whenever an idle BP finds at least one wafer lot waiting in the queue, it immediately starts loading and

processing a batch of the wafer lots. In MBS6, an idle BP starts loading only when the number of wafer lots at the queue is equal to the batch capacity (full batch loading).

(2) Look-upstream (LKUP) rule: This BP control scheme considers near-future job arrivals from the upstream stations. When a batch processor becomes idle and finds wafer lots of a product type more than the batch capacity, the wafer lots are loaded immediately. When the batch size is less than the batch capacity, the scheduler looks at the upstream stations. If a new wafer lot is expected to arrive at the BP station shortly, waiting for the new wafer lot to arrive to form a larger batch size may lead to less total waiting time at the BP station. The LKUP loading rule is similar to a look-ahead batch loading rule in References [5,13]. The arrival time of the new lot is anticipated, and the trade-off between the expected waiting time savings in the new lot and the waiting time increase in the existing wafer lots in the queue is considered. If the reduction in the waiting time is less than the increase in the waiting time, the batch is immediately loaded.

Suppose we are at the decision point, t_0, and job type i^* is selected for the next loading with batch size q. There may be jobs from the other job families waiting in the queue. When loading is delayed until the arrival time of the next job of type i^*, t_1, existing jobs waiting in the queue will experience additional delays while the new upcoming job is loaded without any waiting time. For job type i^*, the additional waiting time is $q(t_1 - t_0)$. For the other jobs in the queue, additional waiting takes place only when the jobs are loaded on the machine triggering the control decision. If there are m parallel batch processors, it is expected that the other jobs in the queue will experience delayed loading with the probability of $1/m$. Then the loading time delay (increased waiting time) can be calculated by $\frac{Q^c(t_1-t_0)}{m}$ where Q^c is the number of jobs (except job type i^*) in the BP queue. Then, the total increased waiting time, due to delayed loading is $q(t_1 - t_0) + \frac{Q^c(t_1-t_0)}{m}$. Now, we need to calculate the time saved by the delayed loading. The saved time occurs only for the incoming job. If the loading is done right away, the loading time of the next incoming job depends on the number of job types, n, and the number of batch processors. When a batch processor is available in the future, the jobs of different job types will compete with each other for the idle BP. Suppose each job type is loaded next with the same probability. Then, on average, the new job is loaded in $\frac{n+1}{2}$ loading cycles later. The interval of the machine availability (loading cycle) is $\frac{T}{m}$, where T is the BP processing time. Then, the expected waiting time is $\frac{(n+1)T}{2m}$. Now the saved wait time can be calculated by (total saved time − total increased time) as $\frac{(n+1)T}{2m} - \left[q(t_1 - t_0) + \frac{Q^c(t_1-t_0)}{m} \right]$. If the saved wait times are negative, the machine starts processing the partial batch immediately. The LKDN control rule is summarized in Figure 4.

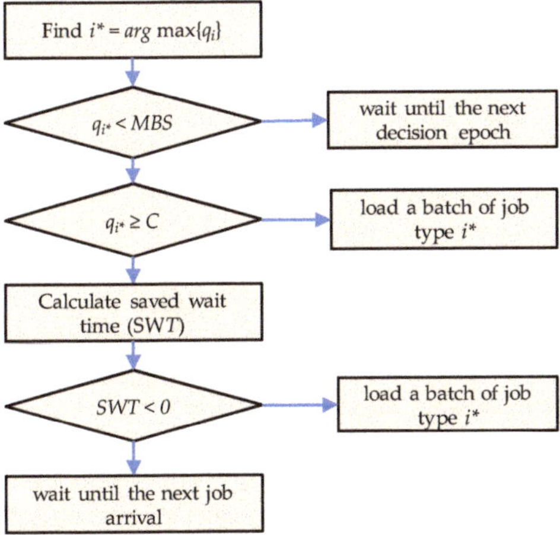

Figure 4. Decision procedure of look-upstream rule (LKUP). MBS, minimum batch size.

(3) Look-downstream (LKDN) rule: This rule considers the status of downstream machines in BP control decisions. Unlike existing control rules looking at the immediate downstream [19,22], this rule looks at a downstream bottleneck (BOT) station possibly a few steps ahead. The bottleneck station is a workstation whose facility utilization is the highest of all workstations. The detailed definition of facility utilization will be given in the subsequent section. Since the bottleneck station determines the capacity of the whole system, any idle time at the bottleneck machine undermines the system capacity. Therefore, it is important for the bottleneck machine to run without a stoppage. In this control scheme, the loading decision is made with the MBS rule in an ordinary situation. However, when the bottleneck station is expected to be idle in the near future, loading activities are initiated by forming a batch with wafer lots in the queue no matter how many wafer lots are waiting in the queue at the BP station. Here, we define the concept of virtual WIP as the work content of all jobs either in the waiting queue at the BOT station or on the way to the BOT station after visiting a BP station. Let V be the virtual WIP, p be the mean processing time per batch at the BP station and m be the number of bottleneck processors. Then, it is expected to take pV hours for the BOT station to finish all the jobs in the virtual WIP. With m BP processors in the BP station, the expected BOT starvation start time, s, is $s = pV/m$ hours when no more new jobs are additionally released from the BP station. Let t be the time required to arrive at the BOT station after being loaded on a batch processor under the condition that no waiting time is experienced on the way to the BOT station. If $s < t$, then the BOT station is expected to be idle, due to starvation. As a result, it is recommended to load the jobs immediately at the BP station to minimize the idle time of the bottleneck machines. To minimize the downtime due to starvation, some buffer WIPs are needed to absorb the flow variability. The loading decision is made based on the target BOT workload, at, where α is the buffer factor considering system variability. An appropriate value for α (>1) is selected through preliminary simulation experiments. The LKDN control rule is summarized in Figure 5.

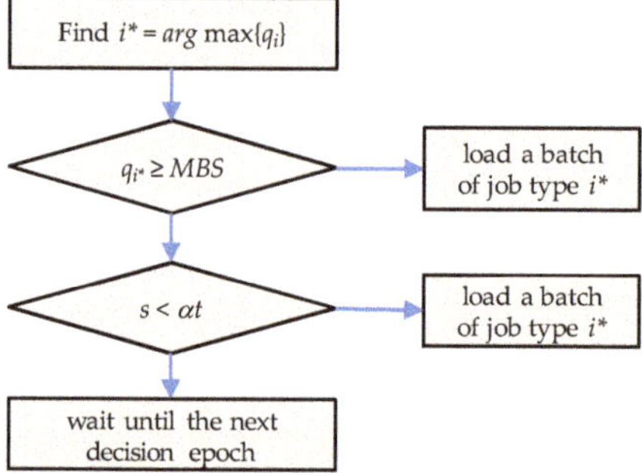

Figure 5. Decision procedure of look-downstream rule (LKDN).

4. Simulation Experiments and Results

4.1. Simulation Model Description

Simulation is a useful tool to analyze the complex and integrated system of wafer fab [29]. The simulation model in this paper is based on the wafer fab configuration from Reference [30] with a slight modification for batch processors. The wafer fab consists of 12 workstations with single or multiple processors. Even though this model is a simple version of actual wafer fabs, it contains the characteristics that a wafer fab should have, which involve reentrant product flows, batch processors, parallel machines, machine breakdown, process time variability, and different job types. The parameters describing the wafer fab are shown in Table 1, including the number of machines in each station, the number of reentrant visits, the mean process time (MPT) per wafer lot, the mean time between failure (MTBF), and the mean time to repair (MTTR). Among the 12 workstations, station 1 is a BP station while all the other stations are DP stations. The batch processors are able to process six wafer lots together as a batch. Each job follows a single product flow with 52 processing steps. The process sequence is as follows: Wafer lot release-11-12-1-9-2-3-4-5-6-7-5-1-9-2-3-4-5-10-7-5-11-12-1-9-2-3-4-5-6-7-5-1-9-2-3-4-5-10-7-5-11-12-1-9-2-3-4-5-6-7-5-8-OUT.

Table 1. Summary of The Basic Simulation Model.

Station Number	1	2	3	4	5	6	7	8	9	10	11	12
Number of Machines	2	2	3	2	3	2	2	2	2	2	3	2
Number of Reentrant Visits	5	5	5	5	10	3	5	2	5	2	3	3
MPT [1]/lot (h)	1.800	0.300	0.540	0.250	0.258	0.380	0.300	0.680	0.330	0.570	0.620	0.523
MTBF [2] (h)	38.0	38.0	38.0	38.0	38.0	38.0	38.0	38.0	38.0	38.0	38.0	38.0
MTTR [3] (h)	2.0	2.0	2.0	2.0	2.0	2.0	2.0	2.0	2.0	2.0	2.0	2.0
Scheduled Utilization	80.0%	80.0%	95.0%	67.5%	91.0%	62.0%	80.0%	73.0%	87.5%	62.0%	67.0%	83.5%

[1] MPT, mean process time; [2] MTBF, mean time between failure; [3] MTTR, mean time to repair.

Each wafer lot visits the same station more than once (reentrant flow). The total net processing time of each wafer lot is 26.106 h on average. The processing time at each station is assumed to follow an Erlang distribution with a coefficient of variation (CV) of 0.1. The small CV value is taken because processing activities on a workstation cycle is usually computer-controlled, and as a result, typically have a low variation in the processing time. The Erlang distribution with a shape parameter k is a sum of k independent exponential variables. With a small value of CV ($=1/\sqrt{k}$), Erlang distribution tends toward a Normal distribution by the central limit theorem, bounded on the lower side, which is a good alternative for processing times. The processing time at the BP station is independent of the number of lots in the batch and much greater than processing times at the DP stations. Each machine is individually subject to random downtime as wafer fab is subject to many sources of variability. The time between failure (uptime) and the time to repair (downtime) is randomly generated from an exponential distribution with given mean values.

In most wafer fabs, the photolithography, station 3 in our case, is a bottleneck station with the highest facility utilization which determines the system capacity. In this study, facility utilization (u) is defined as:

$$utilization\,(u) \quad = \frac{MTTR}{MTBF+MTTR} + \frac{(production\ rate)(number\ of\ operation\ steps\ visited)(MPT)}{(number\ of\ processors)(batch\ capacity)} \tag{1}$$

For example, let us take station 1 with two batch processors in Table 1. Suppose 24 wafer lots are produced daily, i.e., one wafer lot/hour. Every wafer lot visits this station five times before leaving the wafer fab, and the processing time for each visit is 1.8 h. Each processor in station 1 is subject to downtime of 2 h every 38 h on average. Then, the utilization of station 1 is calculated by $\frac{2}{38+2} + \frac{1 \times 5 \times 1.8}{2 \times 6} = 0.05 + 0.75 = 0.8$, i.e., 80.0%.

The simulation model is written using the ARENA modeling tool, version 14.50.00, while specialized requirements are handled by using Visual Basic for Applications (VBA) subroutines. Simulation runs are done on a laptop computer with an Intel Core i7-4500U Processor @ 1.8 GHz. Experimental runs have been replicated ten times with ten different random seeds for each manufacturing scenario to reduce the effect of randomness on the performance. Each simulation run was made for the system under a one-year 24-h-a-day operation. To obtain meaningful steady-state results from the simulation runs, statistics on the initial transient period (the first three months) of each run were excluded from the analysis. After the warm-up period, statistical data is collected through simulation runs for nine months (270 days, 6480 h).

Since this paper focuses on the batch processors, commonly used control rules are applied for lot release decisions and dispatching decisions for DP stations. For lot release rules, two input control rules, an open-loop CONTIME, and a closed-loop CONWIP are applied. The CONTIME rule is selected because the production rate is anticipated with certainty, while CONWIP is selected because it is an easy-to-use closed-loop control rule widely used in industries. The inter-arrival time of the CONTIME rule is one hour, leading to a 95% utilization of the bottleneck station as in Reference [31]. (See Table 1 for the utilization for every workstation.) The number of lots maintained in the fab for the closed-loop CONWIP rule is 70, which is chosen through preliminary experiments so that the average throughput rate is roughly the same as the average throughput rate of the CONTIME lot release cases. For dispatching in the discrete processing stations, the FCFS rule is used because it is easy to apply and widely used in industries.

4.2. Results from Simulation Experiments

Two performance measures are considered to examine the system performance: Throughput rate and lead times. For the CONTIME input rule, since the production rate is determined by the job inter-arrival time, the system lead time is used as a performance measure. On the other hand, for the CONWIP input rule, since the WIPs in the system are maintained at a constant value, the production rate is used as a performance measure.

4.2.1. Effect of BP Utilization on System Performance

Simulation experiments have been performed under the CONTIME lot release condition to see the effect of batch processors on lead time. The utilization for each station remains the same as seen in Table 1, except for the BP station that has three different utilizations. To have different utilizations, the MPT for station 1 is adjusted by using Equation (1). For example, to have a utilization of 70% at station 1, $\mathrm{MPT} = \left(0.70 - \frac{2}{38+2}\right)\left(\frac{2\times6}{1\times5}\right) = 1.56$ h. The reason why we change the MPT instead of changing the arrival rate to have different BP utilization levels is that we want to have the utilization of the other stations remain the same. Figure 6 shows the waiting time at each station. It is seen that station 1, the BP station, has the longest waiting time even with a low BP utilization, such as 60%. The waiting time at the BP station is even higher than the waiting time at the bottleneck station, station 3 with 95% utilization. The result indicates that to reduce production lead time, operation managers should pay close attention to the control of batch processors.

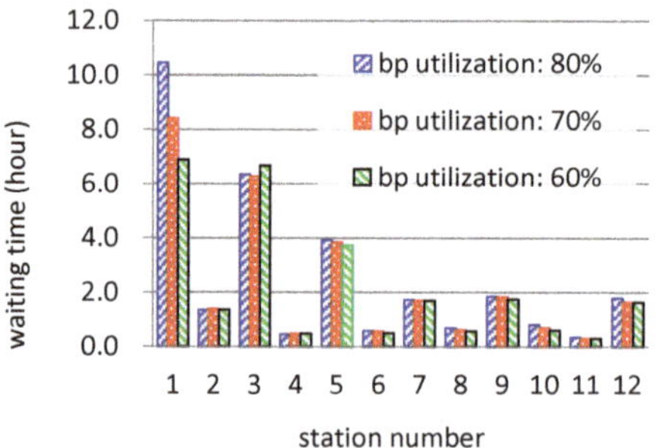

Figure 6. Waiting time at each station with different BP utilization levels.

To examine the impact of the BP utilization and DP utilization on system performance, we have performed experiments for the systems with varying utilization levels for station 1 (BP) and station 7 (DP). Here, to see the throughput change over different utilization levels, the closed-loop lot release rule, CONWIP, is used. The experimental results are given in Figure 7. When we examine the performance over varying BP utilization levels, the throughput decreases as the BP utilization increases. The decrease rate is larger in higher BP utilization levels than in lower BP utilization levels. When BP utilization is 80%, the system produces 6492 wafer lots, which is 94.9% of the capacity. (Note that the system capacity for nine months is 6840 wafer lots.) The lead time also increases slowly with the lower BP utilization, but increases rapidly with the higher BP utilization. When we compare the impact of BDs and DPs, we can find that the change of the throughput and lead time under different utilization levels are smaller in the DP case than in the BP case. The BP station is more sensitive than the DP station in terms of the effect of the utilization levels on the system performance. For example, the system with 60% BP utilization and 80% DP utilization produces 6656 wafer lots, while the system with 80% BP utilization and 60% DP utilization produces 6537 wafer lots (−119 lots). The results indicate that the batch processor should have more planned excessive capacity than the discrete processors. This is quite an important result for practitioners.

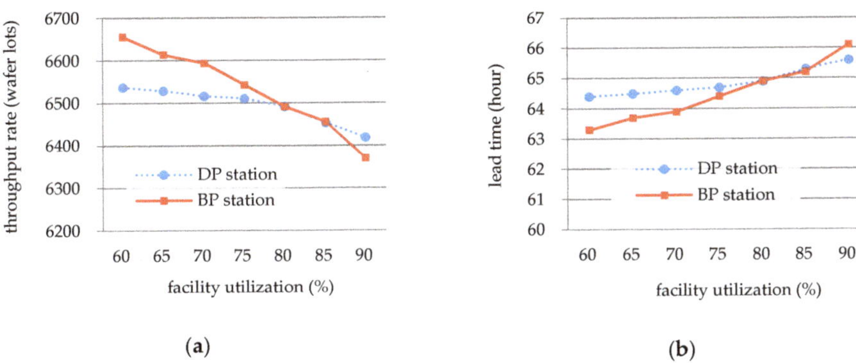

Figure 7. Performance compared over different BP and DP utilization levels in terms of (**a**) throughput rate per nine months and (**b**) system lead time.

4.2.2. Effect of the MBS Size on System Performance

In MBS-based control schemes, the determination of the MBS threshold value is a primary decision. It is known from previous research studies that the performance of the MBS threshold values depends on the BP utilization levels. To examine the combined impact of MBS threshold values and BP utilization levels, simulation experiments are carried out with six different MBS threshold levels with four different BP utilizations. Here, to see the changes in waiting time and lead time over different BP utilization levels, the open-loop lot release rule, CONTIME, is used. The experimental results are shown in Figure 8. It is seen that overall, lower MBS threshold values provide less system lead time, as well as less waiting time at batch processors. This is especially true regarding lower BP utilization levels, such as 60% or 70%. However, when the BP utilization is high, such as 80% or 90%, the MBS values have little effect on the performance. This is because the batch size is large, with a high BP utilization level regardless of the MBS threshold values.

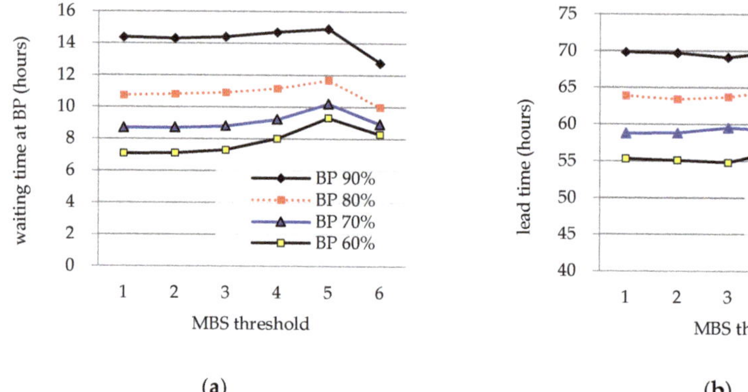

Figure 8. Performance of different MBS levels batch sizes in the MBS control in terms of (**a**) waiting time at BP station and (**b**) system lead time.

One interesting finding is that as MBS levels increase from one to five, BP waiting time tends to increase slowly. However, with an MBS threshold of six, BP waiting time drops sharply compared to an MBS threshold of five. This is arguably due to the reentrant characteristics. Note that in the system under consideration, the batch capacity of a BP is six. Hence, in MBS6, only six wafer lots are formed as a batch to have a process service and are released at the same time to a downstream workstation. They visit several discrete workstations before arriving at the next batch processor where six lots are again required to form a batch. Their arrival at the next BP station should be within a relatively small period, so less waiting time is needed to form a batch at the BP station. When the BP utilization is high, the BP waiting time drops more sharply. In the case of a BP utilization of 80% and 90%, MBS6 provides the lowest waiting time at the BP station. However, from Figure 8b, it is seen that the sharp decrease in BP waiting time under MBS6 does not lead to less system lead time. The lead time is increased with MBS6 compared to MBS5 except in the case of high BP utilization. The results indicate that less BP waiting time at the local station does not always lead to less system lead time. Industries often utilize a full load control policy in the BP station. This may be a good idea when the BP station is under a heavy workload. However, in most cases, the BP station is not critically constrained in its capacity. In this case, a lower MBS threshold is recommended for better system performance.

4.2.3. Performance of Look Upstream Control Rule (LKUP)

This rule attempts to reduce the waiting time at the batch processing station by considering jobs incoming shortly. The loading with a partial batch is delayed until a new job arrives shortly if this delay is expected to lead to less waiting time. Table 2 compares the waiting time at the BP station when MBS1 and LKUP rule are utilized for BP loading, under the CONTIME job release environment. It is seen that the LKUP rule provides consistently less waiting time at the BP station than the base case regardless of BP utilization levels.

Table 2. Comparison of the Performance of LKUP (Look-Upstream) and MBS1 (Minimum Batch Size 1).

BP Utilization	Waiting Time at BP Station (h)			System Lead Time (h)		
	MBS1	LKUP	% Difference	MBS1	LKUP	% Difference
60%	7.08	6.79	−4.1%	55.30	55.61	0.6%
65%	7.82	7.60	−2.8%	57.17	57.08	−0.2%
70%	8.69	8.40	−3.3%	58.78	58.69	−0.2%
75%	9.61	9.24	−3.9%	61.71	60.23	−2.4%
80%	10.72	10.39	−3.1%	63.87	62.58	−2.0%
85%	12.30	11.74	−4.6%	65.89	64.78	−1.7%
90%	14.37	14.16	−1.5%	69.82	68.94	−1.3%

Production operation managers are more likely interested in such global performance measures as system lead time and throughput than the local performance measures, such as waiting time at the BP station. From Table 2, it is found that the LKUP performs better than MBS1 when the BP utilization is high. However, MBS1 and LKUP rules provide almost the same lead time with lower BP utilization around 60% or 70% (Note that in these BP utilization ranges, The LKUP performs better than MBS1.) The simulation results indicate that the locally good control strategy may not lead to good system performance. This insight into the control scheme may be especially true when the product flow has a reentrant characteristic. To examine the effect of reentrant flow on the performance, we have carried out some simulation experiments with scenarios with short serial product flow. To have a serial flow system, the original product flow requiring 52 operations is divided into five independent jobs where each job visits the bottleneck machine and BP station only once, respectively. Table 3 compares the BP waiting time and lead time for the serial and reentrant product flows. Here, the lead time for the serial product flow is multiplied by five to make a fair comparison. It is seen that the local waiting time is reduced for both reentrant and serial product flow environments with the LKUP control rule. However, the system-level lead time remains almost unchanged in the reentrant flow case, whereas the lead time is reduced in the serial product flow case.

Table 3. Performance of LKUP (Look-Upstream) for the Serial and Reentrant Product Flows.

Flow Type	Waiting Time at BP Station (h)			System Lead Time (h)		
	MBS1	LKUP	% Difference	MBS1	LKUP	% Difference
serial flow	10.4	10.2	−2.4%	60.3	59.6	−1.2%
reentrant flow	8.7	8.4	−3.3%	58.8	58.7	−0.1%

4.2.4. Performance of Look Downstream Control Rule (LKDN)

The LKDN rule loads the wafer lots even with a smaller batch size than MBS level when the downstream bottleneck workload is less than the predefined target workload, 35 wafers in this case. The target workload is obtained from preliminary experiments. For the benchmark scenario, the MBS6 rule is used. Figure 9a,b show the throughput rate and lead time under the LKDN control rule and the MBS6, under the CONWIP job release environment. It is seen that the LKDN rule gives better performance than the MBS6 with more throughput rate and less lead time, especially with lower BP utilization levels. Since most wafer fabs keep the BP utilization less than 80%, the LKDN rule may be a good alternative to the MBS6. The high performance of LKDN can be realized by utilizing the bottleneck station (BOT) with less starvation downtime. Figure 9c compares the performance of LKDN and MBS6 in terms of the BOT utilization over different BP utilizations. The figure shows that the LKDN provides higher BOT capacity utilization than the MBS6, especially when the BP utilization is not too high. It should be noted that an increase in BOT capacity utilization directly leads to in greater throughput, which means a great deal in the capital-intensive wafer fab.

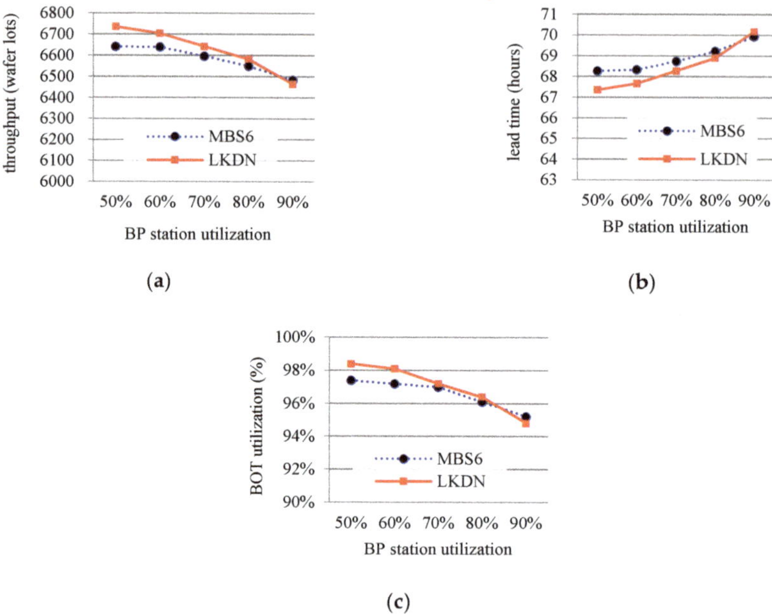

Figure 9. Performance of LKDN (Look-Downstream) and MBS6 (Minimum Batch Size 6) over different BP utilization levels in terms of (**a**) throughput, (**b**) production lead time, and (**c**) BOT (bottleneck) station utilization.

5. Conclusions

This paper attempts to provide managerial insights about the operational control policies at batch processing stations in wafer fabs. Batch processors have distinct characteristics different from discrete processors: The long batch processing time and non-smooth product flow caused by repetitive batching and splitting increase the flow variability, resulting in long lead times. Most previous studies on batch processors focus on a stand-alone BP station. Not much study has been done for batch processor control from a systems perspective. We have constructed a simulation model for a wafer fab and performed experiments with a variety of operational environments. From simulation studies, we have collected some interesting findings as follows:

(1) Batch processors are more sensitive than discrete processors in terms of capacity changes. A small change in capacity in the batch processors has more effect on the system performance than discrete processors. As a result, batch processors should have more excess capacity than discrete processors.

(2) When the BP utilization is not very high, lower MBS threshold values perform better. In the case of very high BP utilization levels, the selection of the MBS level has little effect on the performance of the system. This claim is parallel to the same results studied in References [11,32] in which only a stand-alone BP station is considered.

(3) In the wafer fab with the reentrant product flow, the batch size of a full load may decrease the waiting time at the BP stations. However, it is seen that the decrease in the BP waiting time does not guarantee a good system-level performance. Our experiments show that the full load batching policy provides longer lead times than the MBS policy with lower threshold values,

especially in the cases of moderate and lower BP utilization. The result is contradictory to the common belief about batch size determination in industries where BP operation with a full load is widely used. It is believed that the result is one of the major contributions of this paper.

(4)   When a look-upstream control policy is used, it leads to the lower waiting time at the BP station. However, simulation experiments show that the look-upstream control policy does not lead to less system lead time. When the manufacturing system is characterized as a reentrant flow, the control schemes considering future job arrivals may not provide better system performance as expected. When the benefit of the look-upstream strategy is minimal from the systems perspective, it may be a good idea to apply simple rules like MBS without considering upcoming job arrivals.

(5)   The look-downstream control policy performs better for the system, especially with low or moderate BP utilization levels than the full load batching policy. The result is contradictory to the argument from Reference [19] insisting the control decision considering the state of downstream DP station has little effect on the system performance. The contradictory result is arguably due to the manufacturing environment with which the downstream DP station is a highly utilized bottleneck with 95% utilization in our system, while the utilization of downstream DP station is not that high, i.e., 80% in Reference [19]. When the utilization of the BP station is high, LKDN and MBS provide almost the same performance.

The study in this paper may be improved on different levels. Firstly, a variety of testbeds may be modeled to examine the impact of batch processors with more scenarios. The limitation of our study in this paper is that we only consider a specific wafer fab model, and hence, the experimental analysis is done for a limited manufacturing case. Research work is invited to perform simulation experiments with a variety of wafer fab models, especially within the real-life wafer fab settings to examine the system behavior affected by batch processors. Secondly, some manufacturing settings not covered in this paper, such as sequence-dependent BP setup times, waiting time constraints, BP stations with different batch capacity, multiple product types with different product flows, and orders with due dates, are interesting issues to be addressed. When due dates are involved for each job (order), performance measures, such as tardiness and lateness may be more important than the lead time, the performance measure in this paper. In this case, new BP control rules need to be devised to improve upon the presented rules. Finally, it would be interesting to examine how lot release, DP dispatching (for bottleneck and non-bottleneck stations) and BP control policies interact with each other and how the optimal combination of the control policies is selected under different manufacturing settings.

Author Contributions: Conceptualization, P.-H.K.; methodology, P.-H.K. and R.R.; software, P.-H.K.; validation, P.-H.K. and R.R.; writing—original draft preparation, P.-H.K.; writing—review and editing, R.R. All authors have read and agreed to the published version of the manuscript.

Funding: This work was supported by the Pukyong National University Research Abroad Fund (C-D-2016-0843).

Conflicts of Interest: The authors declare no conflict of interest.

References

1.   Wang, L.C.; Chu, P.-C.; Lin, S.-Y. Impact of capacity fluctuation on throughput performance for semiconductor wafer fabrication. *Robot. Comput. Integr. Manuf.* **2019**, *55*, 208–216. [CrossRef]

2.   Ham, M. Integer programming-based real-time dispatching (i-RTD) heuristic for wet-etch station at wafer fabrication. *Int. J. Prod. Res.* **2012**, *50*, 2809–2822. [CrossRef]

3.   Mathirajan, M.; Sivakumar, A. A literature review, classification and simple meta-analysis on scheduling of batch processors in semiconductor. *Int. J. Adv. Manuf. Technol.* **2006**, *26*, 990–1001. [CrossRef]

4.   Koo, P.; Mansoer, P. A look-ahead control strategy at parallel batch processors with multiple product types. *ICIC Express Lett. Part B Appl.* **2015**, *6*, 3197–3203.

5.   Fowler, J.W.; Hogg, G.L.; Phillips, D.T. Control of multiproduct bulk service diffusion/oxidation processes. Part 2: Multiple servers. *IIE Trans.* **2000**, *32*, 167–176. [CrossRef]

6. van der Zee, D. Adaptive scheduling of batch servers in flow shops. *Int. J. Prod. Res.* **2002**, *40*, 2811–2833. [CrossRef]

7. Wang, J.; Zheng, P.; Zhang, J. Big data analytics for cycle time related feature selection in the semiconductor wafer fabrication system. *Comput. Ind. Eng.* **2020**, *143*, 106362. [CrossRef]

8. Neuts, M.F. A general class of bulk queues with Poisson input. *Ann. Math. Stat.* **1967**, *38*, 759–770. [CrossRef]

9. Deb, R.K.; Serfozo, R.F. Optimal control of batch service queues. *Adv. Appl. Probab.* **1973**, *5*, 340–361. [CrossRef]

10. Gurnani, H.; Anupindi, R.; Akella, R. Control of batch processing systems in semiconductor wafer fabrication facilities. *IEEE Trans. Semicond. Manuf.* **1992**, *5*, 319–328. [CrossRef]

11. Avramidis, A.N.; Healy, K.J.; Uzsoy, R. Control of a batch-processing machine: A computational approach. *Int. J. Prod. Res.* **1998**, *36*, 3167–3181. [CrossRef]

12. Fowler, J.W.; Phojanamongkolkij, N.; Cochran, J.K.; Montgomery, D.C. Optimal batching in a wafer fabrication facility using a multiproduct G/G/c model with batch processing. *Int. J. Prod. Res.* **2002**, *40*, 275–292. [CrossRef]

13. Glassey, C.; Weng, W. Dynamic batching heuristic for simultaneous processing. *IEEE Trans. Semicond. Manuf.* **1991**, *4*, 77–82. [CrossRef]

14. Fowler, J.W.; Phillips, D.T.; Hogg, G.L. Real-time control of multiproduct bulk-service semiconductor manufacturing processes. *IEEE Trans. Semicond. Manuf.* **1992**, *5*, 158–163. [CrossRef]

15. Sarin, S.C.; Varadarajan, A.; Wang, L. A survey of dispatching rules for operational control in wafer fabrication. *Prod. Plan. Control* **2011**, *22*, 4–24. [CrossRef]

16. Koo, P.; Moon, D. A review on control strategies of batch processing machines in semiconductor manufacturing. *IFAC Pap. Ser. Title Manuf. Model. Manag. Control* **2013**, *7*, 1690–1695. [CrossRef]

17. Leachman, R.C.; Kang, J.; Lin, V. SLIM: Short cycle time and low inventory in manufacturing at Samsung Electronics. *Interfaces* **2002**, *32*, 61–77. [CrossRef]

18. Robinson, J.K.; Fowler, J.W.; Bard, J.F. The use of upstream and downstream information in scheduling semiconductor batch operations. *Int. J. Prod. Res.* **1995**, *33*, 1849–1869. [CrossRef]

19. Neale, J.; Duenyas, I. Control of manufacturing networks which contain a batch processing machine. *IIE Trans.* **2000**, *32*, 1027–1041. [CrossRef]

20. Solomon, L.; Fowler, J.W.; Pfund, M.; Jensen, P.H. The inclusion of future arrivals and downstream setups into wafer fabrication batch processing decisions. *J. Electron. Manuf.* **2002**, *11*, 149–159. [CrossRef]

21. Cerekci, A.; Banerjee, A. Effect of upstream re-sequencing in controlling cycle time performance of batch processors. *Comput. Ind. Eng.* **2015**, *88*, 206–216. [CrossRef]

22. Kim, Y.; Lee, D.H.; Kim, J.U. A simulation study on lot release control, mask scheduling, and batch scheduling in semiconductor wafer fabrication facilities. *J. Manuf. Syst.* **1998**, *17*, 107–117.

23. Bahaji, N.; Kuhl, M.E. A simulation study of new multi-objective composite dispatching rules, CONWIP, and push lot release in semiconductor fabrication. *Int. J. Prod. Res.* **2008**, *46*, 3801–3824. [CrossRef]

24. Li, Y.; Jiang, Z.; Jia, W. An integrated release and dispatch policy for semiconductor wafer fabrication. *Int. J. Prod. Res.* **2014**, *52*, 2275–2292. [CrossRef]

25. Spearman, M.L.; Woodruff, D.L.; Hopp, W.J. CONWIP: A pull alternative to Kanban. *Int. J. Prod. Res.* **1990**, *28*, 879–894. [CrossRef]

26. Wein, L.W. Scheduling semiconductor wafer fabrication. *IEEE Trans. Semicond. Manuf.* **1998**, *1*, 115–130. [CrossRef]

27. Glassey, C.R.; Resende, M.G.C. Closed-loop job release control for VLSI circuit manufacturing. *IEEE Trans. Semicond. Manuf.* **1998**, *1*, 36–46. [CrossRef]

28. Qi, C.; Sivakumar, A.I.; Gershwin, S.B. An effective new job release control methodology. *Int. J. Prod. Res.* **2009**, *47*, 703–731. [CrossRef]

29. Fowler, J.W.; Monch, L. Discreet-event simulation for semiconductor wafer fabrication facilities: A tutorial. *Int. J. Ind. Eng. Theoryappl. Pr.* **2015**, *22*, 661–682.

30. El-Khouly, I.; El-Kilany, K.S.; Young, P. A simulation study of lot flow control in wafer fabrication facilities. In Proceedings of the 2012 International Conference on Industrial Engineering and Operations Management, Istanbul, Turkey, 3–6 July 2012; pp. 2240–2249.

31. Kim, Y.; Kim, G.; Choi, B.; Kim, H. Production scheduling in a semiconductor wafer fabrication facility producing multiple product types with distinct due dates. *IEEE Trans. Robot. Autom.* **2001**, *17*, 589–598.
32. Akcali, E.; Uzsoy, R.; Hiscock, D.G.; Moser, A.L.; Teyner, T.J. Alternative loading and dispatching policies for furnace operations in semiconductor manufacturing: A comparison by simulation. In Proceedings of the 2000 Winter Simulation Conference, Orlando, FL, USA, 10–13 December 2000; pp. 1428–1435.

Article

Automatic Supervisory Controller for Deadlock Control in Reconfigurable Manufacturing Systems with Dynamic Changes

Husam Kaid [1,*], Abdulrahman Al-Ahmari [1], Zhiwu Li [2,3] and Reggie Davidrajuh [4]

[1] Department of Industrial Engineering, College of Engineering, King Saud University, Riyadh 11421, Saudi Arabia; alahmari@ksu.edu.sa

[2] Institute of Systems Engineering, Macau University of Science and Technology, Taipa, Macau 999078, China; systemscontrol@gmail.com

[3] School of Electro-Mechanical Engineering, Xidian University, Xi'an 710071, China

[4] Department of Electrical Engineering and Computer Science, Faculty of Science and Technology, University of Stavanger, 4036 Stavanger, Norway; reggie.davidrajuh@uis.no

*   Correspondence: yemenhussam@yahoo.com

Received: 10 July 2020; Accepted: 27 July 2020; Published: 30 July 2020

Abstract: In reconfigurable manufacturing systems (RMSs), the architecture of a system can be modified during its operation. This reconfiguration can be caused by many motivations: processing rework and failures, adding new products, adding new machines, etc. In RMSs, sharing of resources may lead to deadlocks, and some operations can therefore remain incomplete. The objective of this article is to develop a novel two-step solution for quick and accurate reconfiguration of supervisory controllers for deadlock control in RMSs with dynamic changes. In the first step, the net rewriting system (NRS) is used to design a reconfigurable Petri net model under dynamic configurations. The obtained model guarantees boundedness behavioral property but may lose the other properties of a Petri net model (i.e., liveness and reversibility). The second step develops an automatic deadlock prevention policy for the reconfigurable Petri net using the siphon control method based on a place invariant to solve the deadlock problem with dynamic structure changes in RMSs and achieve liveness and reversibility behavioral properties for the system. The proposed approach is tested using examples in the literature and the results highlight the ability of the automatic deadlock prevention policy to adapt to RMSs configuration changes.

Keywords: reconfigurable manufacturing system; Petri net; deadlock; siphon; supervisory controller

1. Introduction

A typical example of discrete event systems is an automated manufacturing system (AMS) [1,2]. It enables various product types to be entered at discrete times by sharing resources like machines, automatic controlled vehicles, automated tools, robots, and buffers at asynchronous or simultaneous operations. AMSs have to cope with unexpected and rapid market changes on a competitive global market. They must make rapid modifications to their software and hardware to meet these dynamic changes. This requirement cannot, however, be satisfied successfully with traditional automated manufacturing systems, which require large capital investments. Reconfigurable manufacturing systems have now been developed to deal with those drawbacks in traditional automated manufacturing systems [3–5]. Reconfigurable manufacturing systems are a new kind of production systems that are randomly and dynamically configured in real time. Such configurations involve processing rework and failures, adding new products and machines, and adding new handling device. In RMSs, a set of system resources can be used to process each component according to a specific process sequence.

This sharing of resources, however, may lead to deadlocks, and some operations can therefore remain incomplete. Therefore, dealing with deadlock problem is critical for RMSs.

Petri nets (PNs) are widely used for the scheduling, deadlock analysis and control in AMSs as graphical and mathematical modelling tools [6–14]. They can be used to describe characteristics and behaviors of AMSs such as synchronization, concurrency, conflict, causal dependence, and sequencing. Petri nets can be used for behavioral features, for example boundedness and liveness [15,16]. From a technical point of view, several policies based on Petri nets have been proposed. These policies are based on three strategies: (i) deadlock detection and recovery, (ii) deadlock avoidance, and (iii) deadlock prevention [15,17]. Most of these policies have proposed deadlock control in Petri nets through structural analysis [6,18] and reachability graph analysis [19–21]. In addition, three criteria to evaluate and construct an AMS supervisor have been proposed, namely behavioral permissiveness, computational complexity, and structural complexity [15,22].

Recently, several approaches have been adapted to deal with dynamic changes in manufacturing systems [7,23–36]. They primarily concentrate in two directions: direct and indirect. Direct approaches provide modification mechanisms or particular rules for system structure configurations, while indirect approaches typically import additional mechanisms for system reconfiguration specifications. The event–condition–action (ECA) paradigm is developed by Almeida et al. [30] for the design of reconfigurable logic controllers. Their research has demonstrated that the reconfiguration process is highly dependent on the modularity level of the logical control system and that not all "modular" structures can be reconfigured. For a class of discrete event systems (DESs), Sampath et al. [26] presented a reconfiguration approach for their control specifications, subject to linear constraint. This approach is suited to systems such as hospital management systems and can be reconfigured in non-real time. In order to evaluate and improve the performance of the control architecture, Dumitrache et al. [27] developed a real-time reconfigurable supervised control architecture for large manufacturing systems. A model-based control design for reconfigurable manufacturing systems is developed by Ohashi and Shin [28] through state transition diagrams and general graph representation taking into account configuration and reuse of design data. Kalita and Khargonekar [29] introduced a hierarchical structure and a framework for modeling, analysis, specification, and design of logic controllers for RMSs, which allows rapid reconfigurability and reusability of the controller during reconfiguration. In [23], reconfigurable manufacturing systems were used to replace the existing manufacturing systems to offer higher convertibility and flexibility such as dedicated production systems. Serial and parallel configurations, a rules-based matrix approach has been developed and implemented. In addition, a higher-level deadlock control method is presented for the serial and parallel configurations.

Net Rewriting Systems (NRS) are another graph-based reconfiguration mechanism [34]. In terms of pattern matching and dynamic structure replacements, the reconfiguration occurs. By the implementation of a Turing machine the expressive power was shown to be Turing equivalent. A subset of net rewriting systems, called reconfigurable nets, have also been provided with an algorithm to flatten a Petri net to standard. This subset only restricts NRS to those transformations that remain unchanged in the number of places and transitions, that is, only the flow relation can be changed. Flattening significantly increases the size of transitions by multiplying the number of reconfigurations by the amount of transitions. The NRS is used in logic controllers with improved net rewriting systems [35]. The improved NRS version restricts the rewriting rules to ensure important structural characteristics such as boundedness, liveness, and reversibility are not invalidated. In addition, in [24], an improved net rewriting system (INRS) was developed with the aim of reconfiguring an RMS supervisory controller based on PNs. Changes to an RMS modification were made to rewrite rules that were then applied in the initial PN controller. The INRS is first proposed as a reconfiguration basis. The structure of a Petri net model can be changed dynamically. Then, the study provided three representations of the RMS modification and suggested an INRS-based method to the design of the Petri net controller of an RMS. In this approach, the properties of behavioral, i.e., the boundedness, reversibility, and liveness of a modified system, were not verified or validated.

In [31], colored timed PNs (CTPN) were used in the modelling of RMSs and a mechanism to describe reconfigurability in the CTPN architecture was introduced that leads to a new architecture supporting the reconfiguration. This mechanism includes reconfigurable transitions, specific places, and inhibitor arcs. Wu and Zhou introduced intelligent token Petri net (ITPN) [25]. In their model, tokens representing job instances carry real-time knowledge about system states and changes, just like intelligent cards in practice such that dynamical changes of a system can be easily modeled. These formalisms can describe the reconfiguration behavior of the system. However, some of dynamic changes do not clearly define the modularity, which brings confusion to engineers in designing, understanding, and future redevelopment. Correctness of the system such as coherence of states before and after system reconfigurations is not considered. In addition, temporal constraints, which are of great significance in real-time systems are not mentioned. In [32], reconfigurable object nets (RONs) are used to model, simulate, and analyze RMSs. A formal method was proposed for fulfilling a new production requirement. The configuration consists of new extrusion and cutting machines. The reconfiguration is represented as graph transformations, RON tool was used to simulate the reconfigured systems and TINA [37] and PIPE [38] software tools were used to carry out the analysis.

The work of Silva et al. [36] explored the principles of the different approaches and takes from them the best practices. Configuration mechanisms were proposed using Holonic and multiagent system methods to allow a reconfigurable distributed production control system to systematically detect faults. To describe communication interfaces, the principle of service-oriented architecture was used. Hybrid top-down and bottom-up approaches were presented using Petri net models. In [33], object-oriented Petri nets (ORPNs) and π-calculus were used as two complementary formalisms. Initial RMSs structure and system behavior were modeled by ORPN while the π-calculus was used to describe RMSs' reconfiguration. To evaluate, check, and validate RMSs, Petri nets and π-calculus supporting tools were used. The reconfigurability mechanism and consistency of RMSs could be analyzed by π-calculus. In [7], a new model is proposed, namely the intelligent colored token Petri net (ICTPN), which simulated dynamic configurations of systems such as adding new machines, processing failures and rework, machine failures, processing routes changes, removing old machines, and adding new products. The primary idea is that smart colored tokens were part types which represented real-time knowledge of system status and configurations. This allowed for the effective modeling of dynamic system configurations. The proposed ICTPN could modularly model dynamic system changes to generate a very compact model. Moreover, when configurations appear, only the colored token of the part type, which is changed from the current model was changed. The resulting ICTPN model ensures that the behavioral properties such as deadlock-free, conservative, and reversible were guaranteed.

All of the above methods with PNs attempted to deal with dynamic configuration issues in manufacturing systems. However, most of them do not include an algorithm or mechanism for reconfiguration, could not guarantee the properties of behavioral Petri net (i.e., boundedness (or safeness), liveness, and reversibility), or could not ensure that the results of the reconfiguration are correct, accurate or valid. In addition, few techniques for rapid and valid reconfiguration of literature deadlock control supervisors were presented.

The objective of this article is to develop a novel two-step solution for quick and accurate reconfiguration of supervisory controllers for deadlock control in RMSs with dynamic changes. In the first step, the net rewriting system used in [34,39] was adapted to design a reconfigurable Petri net model under dynamic configurations. The obtained model guarantees boundedness behavioral property but may lose the other properties of a Petri net model (i.e., liveness and reversibility). This means that the reconfigured Petri net model has finite states, deadlocks, and does not behave cyclically. For this issue, the second step develops an automatic deadlock prevention policy for reconfigurable Petri net using the siphon control method based on place invariant to solve the deadlock problem with dynamic structure changes in RMSs and achieve liveness and reversibility behavioral properties for the system. Thus, the developed approach has the ability of adapting to RMS configuration changes.

The major applications of the developed approach are as follows:

1. Mass customization manufacturing can use the proposed approach to address its difficulties. For example, by trying to make products available rapidly to consumers, a high quality production of a wide variety of products can be maintained and achieve low costs in line with standard products.
2. Lean productivity concept can also use the proposed approach to enable a company to implement an RMS in order to improve the exploitation of the part of the resources for various family products and to minimize waste from the idle resource of an RMS.
3. Agile manufacturing can use the proposed approach to facilitate rapid products changeovers, rapid introduction of new products and unattended operation.
4. Flexible manufacturing systems can use the proposed approach to increase response to a variety of customers and markets. Moreover, scalability to the desired volume of products and convertibility to current systems, machines, robots, and controls are increased in accordance with the new production requirements.

This paper is organized as follows. Section 2 describes basic concepts of Petri nets, reconfigurable Petri nets. Section 3 presents the deadlock prevention policy for reconfigurable Petri net based on the concept of minimal siphons and place invariants. The behavioral and quantitative analysis of the proposed reconfigurable Petri net are presented in Section 4. A real-world case study is presented in Section 5 to demonstrate the application of the proposed approach. Conclusions and future research are presented in Section 6.

2. Preliminaries

2.1. S^3PR NET

Definition 1. *A simple sequential process (S^2P) is a Petri net model with $N = (\{p^0\} \cup P_A, T, F)$ if (1) N is a strongly connected state machine and (2) each circuit N contains place p^0, where p^0 is a process idle place, $P_A = \{p_1, p_2, \dots, p_m\}$ is a set of operation places, $T = \{t_1, t_2, \dots, t_n\}$ is a set of transitions, $P_B = P_A \cup \{p^0\}$, $P_B \cap T = \emptyset$, $P_B \cup T \neq \emptyset$, and $F: (P_B \times T) \cup (T \times P_B) \rightarrow \mathbf{IN}$ is a set of weighted arcs called flow relations, where $\mathbf{IN} = \{0, 1, 2, \dots \}$.*

Definition 2. *A simple sequential process with resources (S^2PR) is a Petri net model with $N = (\{p^0\} \cup P_A \cup P_R, T, F)$ if*

1. *the subnet created by $Y = P_A \cup \{p^0\} \cup T$ is an S^2P;*
2. *$P_R \neq \emptyset$ and $(P_A \cup \{p^0\}) \cap P_R = \emptyset$, where P_R is called a set of resource places;*
3. *$P_C = P_A \cup \{p^0\} \cup P_R$, $F \subseteq (P_C \times T) \cup (T \times P_C)$ is flow relations;*
4. *$^{\bullet\bullet}(p^0) \cap P_R = (p^0)^{\bullet\bullet} \cap P_R \neq \emptyset$;*
5. *$\forall p \in P_A, \forall t \in {}^{\bullet}p, \forall t' \in p^{\bullet}, \exists r_p \in P_R, {}^{\bullet}t \cap P_R = t'^{\bullet} \cap P_R = \{r_p\}$;*
6. *$\forall r \in P_R, {}^{\bullet\bullet}r \cap P_A = r^{\bullet\bullet} \cap P_A \neq \emptyset$ and ${}^{\bullet}r \cap r^{\bullet} \neq \emptyset$;*

Definition 3. *Let $N = (\{p^0\} \cup P_A \cup P_R, T, F)$ be an S^2PR with M_0 being an initial marking of net N. An S^2PR is called acceptably marked if (1) $M_0(p^0) \geq 1$, (2) $M_0(p) = 0, \forall p \in P_A$, and (3) $M_0(r) \geq 1, \forall r \in P_R$. Recursively, a system of S^2PR is called an S^3PR.*

Definition 4. *A system of S^2PR, S^3PR, is defined recursively as follows:*

1. *An S^2PR is an S^3PR;*
1. *Let $N_i = (\{p^0{}_i\} \cup P_{Ai} \cup P_{Ri}, T_i, F_i), i = \{1, 2\}$, be two S^3PRs such that $(\{p^0{}_1\} \cup P_{A1}) \cap (\{p^0{}_2\} \cup P_{A2}) = \emptyset$, $P_{R1} \cap P_{R2} = P_D, P_{A1} \cap P_{A2} \neq P_D$, and $T_1 \cap T_2 \neq \emptyset$; then, the net $N = (\{p^0\} \cup P_A \cup P_R, T, F)$ is an S^3PR resulting from the integration of N_1 and N_2 by the set of common P_D (denoted as $N_1 \circ N_2$) and expressed as: (1) $p^0 = \{p^0{}_1\} \cup \{p^0{}_2\}$, (2) $P_A = P_{A1} \cup P_{A2}$, (3) $P_R = P_{R1} \cup P_{R2}$, (4) $T = T_1 \cup T_2$, and (5) $F = F_1 \cup F_2$.*

The integration of n S^2PR N_1-N_n via P_D is expressed by $\otimes_{i=1}^{n} N_i$. $\overline{N_i}$ is used to indicate the S^2P from which the S^2PR N_i is built.

Definition 5. *Let* $N_i = (\{p^0_i\} \cup P_{Ai} \cup P_{Ri}, T_i, F_i)$, $i = \{1, 2\}$, *be two* S^3*PRs.* M_0 *is an initial marking of N.* (N, M_0) *is called acceptably marked if* (1) (N, M_0) *is an acceptably marked* S^2*PR, and* (2) $N = N_1 \circ N_2$, *where* (N_i, M_{io}) *is called an acceptably marked* S^3*PR and*

1. $\forall i \in \{1,2\}, \forall p \in P_{Ai} \cup \{p^0_i\}, M_0(p) = M_{io}(p)$.
2. $\forall i \in \{1,2\}, \forall r \in P_{Ri} \backslash P_D, M_0(r) = M_{io}(r)$.
3. $\forall i \in \{1,2\}, \forall r \in P_D, M_0(p) = max\{M_{1o}(r), M_{2o}(r)\}$.

Definition 6. *Let* $N = (\{p^0\} \cup P_A \cup P_R, T, F, W, M_0)$ *be an* S^3*PR, where* $W: (P_C \times T) \cup (T \times P_C) \rightarrow$ **IN** *is a mapping that assigns a weight to an arc and* $M_0: P_C \rightarrow$ **IN** *is the initial marking.*

Definition 7. *Let* $N = (\{p^0\} \cup P_A \cup P_R, T, F, W, M_0)$ *be an* S^3*PR.* N *is said to be an ordinary net if* $p \in P_C$, $t \in T$, $\forall(p, t) \in F$, *and* $W(p, t) = 1$.

Definition 8. *Let* $N = (\{p^0\} \cup P_A \cup P_R, T, F, W, M_0)$ *be an* S^3*PR.* N *is said to be a weighted net if* $\exists p \in P_C$, $\exists t \in T$, $(p, t) \in F$, *and* $W(p, t) > 1$.

Definition 9. *Let* $N = (\{p^0\} \cup P_A \cup P_R, T, F, W, M_0)$ *be an* S^3*PR, where p and t are a place and a transition in N, respectively. The preset (postset) of p is the set of all input (output) transitions of p, i.e.,* $^\bullet p = \{t \in T \mid (t, p) \in F\}(p^\bullet = \{t \in T \mid (p, t) \in F\})$. *The preset (postset) of t is the set of all input (output) places of t, i.e.,* $^\bullet t = \{p \in P_C \mid (p, t) \in F\}(t^\bullet = \{p \in P_C \mid (t, p) \in F\})$.

Definition 10. *Let* $N = (\{p^0\} \cup P_A \cup P_R, T, F, W, M_0)$ *be an* S^3*PR.* N *is self-loop free if for all* $p, t \in P_C \cup T$; $W(p, t) > 0$ *implies* $W(t, p) = 0$ *and has a self-loop if* $W(t, p) > 0$.

Definition 11. *Let* $N = (\{p^0\} \cup P_A \cup P_R, T, F, W, M_0)$ *be an* S^3*PR and M be a marking of N, where M is a mapping* $M: P_C \rightarrow$ **IN** *and the pth element of M, expressed by* $M(p)$, *is the number of tokens in place p.*

Definition 12. *Let* $N = (\{p^0\} \cup P_A \cup P_R, T, F, W, M_0)$ *be an* S^3*PR. A transition* $t \in T$ *is enabled if* $\forall p \in {}^\bullet t$, $M(p) \geq W(p, t)$.

Definition 13. *Let* $N = (\{p^0\} \cup P_A \cup P_R, T, F, W, M_0)$ *be an* S^3*PR. The marking* M' *resulting from the firing of an enabled transition* $t \in T$ *at marking M is denoted by* $M[t\rangle M'$ *and expressed as follows:*

$$M'(p) = \begin{bmatrix} M(p) + W(p, t) & \text{if } p \in {}^\bullet t \backslash t^\bullet \\ M(p) - W(t, p) & \text{if } p \in t^\bullet \backslash {}^\bullet t \\ M(p) + W(t, p) - W(p, t) & \text{if } p \in t^\bullet \cap {}^\bullet t \\ M(p) & \text{otherwise} \end{bmatrix} \tag{1}$$

Definition 14. *Let* $N = (\{p^0\} \cup P_A \cup P_R, T, F, W, M_0)$ *be an* S^3*PR.* $R(N, M)$ *is a set of reachable markings from M in N, which is expressed by nodes and arcs; nodes represent markings that are labeled with* M_i *and arcs represent transition firings that are labeled with t. If t fires, then there is an arc from marking* M_i *to marking* M_j *and* M_j *is reached.*

Definition 15. *Let* $N = (\{p^0\} \cup P_A \cup P_R, T, F, W, M_0)$ *be an* S^3*PR. A transition* $t \in T$ *is live at* M_0 *if* $\forall M \in R(N, M_0)$, $\exists M' \in R(N, M)$ *such that* $M'[t\rangle$ *holds.* (N, M_0) *is dead at* M_0 *if there does not exist* $t \in T$ *such that* $M_0[t\rangle$ *holds.* (N, M_0) *is weakly live or live-locked if* $\forall M \in R(N, M_0)$, $\exists t \in T, M[t\rangle$ *holds.* (N, M_0) *is quasi-live if* $\forall t \in T, \exists M \in R(N, M_0)$ *such that* $M[t\rangle$ *holds.*

Definition 16. *Let $N = (\{p^0\} \cup P_A \cup P_R, T, F, W, M_0)$ be an S^3PR. [N] is said to be the incidence matrix of net N, where [N] is a $|P| \times |T|$ integer matrix with $[N](p, t) = W(t, p) - W(p, t)$. For a place p (transition t), its incidence vector, a row (column) in [N], is expressed as $[N](p, .)$ ($[N](., t)$).*

Definition 17. *Let $N = (\{p^0\} \cup P_A \cup P_R, T, F, W, M_0)$ be an S^3PR. A marking M' is called reachable from M if there exists a sequence of transitions $\delta = t_1 \, t_2 \, t_3 \ldots t_n$ that can be fired, and markings $M_1, M_2, M_3, \ldots$, and M_{n-1} are such that $M[t_0\rangle M_1[t_1\rangle M_2[t_2\rangle M_3 \ldots M_n \, [t_n\rangle M'$ holds, expressed as $M[\delta\rangle M'$, satisfies the state equation $M' = M + [N] \, \vec{\delta}$. $\vec{\delta} : T \to IN$ is called a firing count vector or a Parikh vector that maps t in T to the number of occurrences of t in δ.*

Definition 18. *Let $N = (\{p^0\} \cup P_A \cup P_R, T, F, W, M_0)$ be an S^3PR. N is said to be bounded if there exists $q \in IN$, $\forall M \in R(N, M_0)$, $\forall p \in P_C$, $M(p) \leq q$. (N, M_0) is structurally bounded if it is bounded for any M_0.*

Definition 19. *LLet $N = (\{p^0\} \cup P_A \cup P_R, T, F, W, M_0)$ be an S^3PR. N is called safe if $\forall M \in R(N, M_0)$, $\forall p \in P_C$, $M(p) \leq 1$. (N, M_0) is q -safe if it is q-bounded.L*

Consider the example of AMS illustrated in Figure 1a. The system has one robot R1 and one machine M1. Machine M1 processes one part at a time and robot R1 holds one part at a time. There are buffers for loading/unloading. Furthermore, one part type is considered to be processed in the system. The part operation sequence is illustrated in Figure 1b. Figure 2 shows the S^3PR net of the AMS example. It has six places and four transitions. The following sets of places can be used: $P^0 = \{p_1\}$, $P_R = \{p_5, p_6\}$, and $P_A = \{p_2, p_3, p_4\}$. There are five reachable markings on the Petri model. The initial marking is $M_o = (5, 0, 0, 0, 1, 1)^T$, which represents the different raw parts that are to be processed synchronously within the system, including preconditions, input signals, buffers and resource status, such as machines and robot. Places are generally used to represent the resource status, operations, and activities. The transitions are used to express control changes from one state to another. Directed arcs correspond to the material, resource, information flow, and control flow direction between states. Material, information, and resources are represented by tokens.

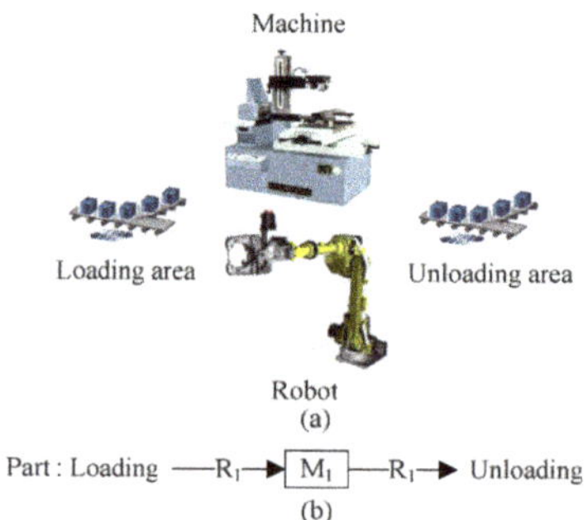

Figure 1. (**a**) Automated manufacturing system (AMS) example and (**b**) operation sequence.

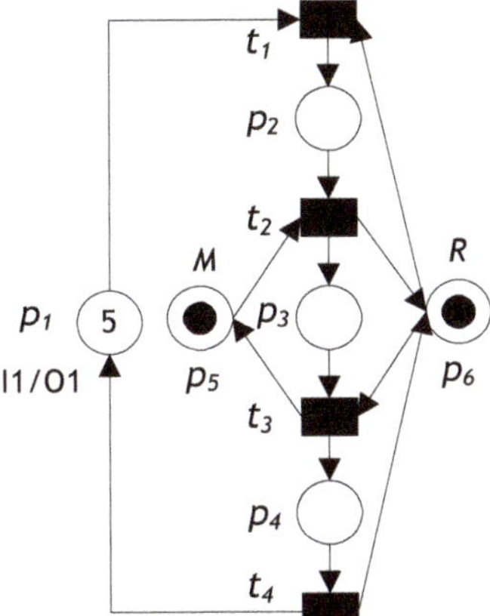

Figure 2. A system of (S^2P) (simple sequential process) (S^3PR) net of the AMS.

2.2. Reconfigurable S^3PR Net

This section presents definitions and theorems in the reconfigurable S^3PR nets, which are originally proposed by [34,35,39].

Definition 20. *Let $N = (\{p^0\} \cup P_A \cup P_R, T, F, W, M_0, K)$ be a finite-capacity S^3PR, where p^0, P_A, P_R, T, F, W, and M_0 are defined in Definitions 1–6. $K: P_C \rightarrow IN$ is the function of capacity that assigns to each place p the maximal number of tokens $K(p)$.*

Definition 21. *Let (N_i, M_i) be two S^3PR nets with $N_i = (P_{Ci}, T_i, F_i, W_i, M_i, K_i)$, $i = 1, 2$. N_1 and N_2 are called morphism nets if there exists a bijection $\Psi: N_1 \rightarrow N_2$, $\Psi = (\Psi_{PC}: P_{C1} \rightarrow P_{C2}, \Psi_T: T_1 \rightarrow T_2)$ such that for all a, $b \in P_{C1} \cup T_1$, $F_1(a, b) \in N_1 = F_2(\Psi(a), \Psi(b)) \in N_2$, and for all $p \in P_{C1}$, $M_1(p) \leq M_2(\Psi_{PC}(p))$.*

Definition 22. *Let (N_i, M_i) be two S^3PR nets with $N_i = (P_{Ci}, T_i, F_i, W_i, M_i, K_i)$, $i = 1, 2$. N_1 is called the full subnet of N_2 if there exists an injection function that maps places to places and transitions to transitions, denoted by $\xi: N_1 \rightarrow N_2$, $\xi (P_{C1}) \subseteq P_{C2}$, and $\xi (T_1) \subseteq T_2$ such that for all a, $b \in P_{C1} \cup T_1$, $F_1(a, b) = F_2(\xi (a), \xi (b))$.*
In the algebraic, a rewriting rule is a transformation approach that can change and combine the Petri nets dynamically. The main idea is to define and change the system configurations as a graph rewriting rule.

Definition 23. *Let N_R be a reconfigurable S^3PR with $N_R = ((N, M_0), R)$, where(N, M_0) is an S^3PR net with $N = (P_C, T, F, W, M_0, K)$ and $R = \{rr_1, rr_2, rr_3, \dots, rr_m\}$ is called a set of rewriting rules or dynamic configurations if*

1. *For all $rr \in R$, $rr = \{L, R, \varphi, {}^\bullet\varphi, \varphi^\bullet\}$;*
2. *$L = (P_{CL}, T_L, F_L, W_L, M_{0L}, K_L)$ is called the left-hand side;*
3. *$R = (P_{CR}, T_R, F_R, W_R, M_{0R}, K_R)$ is called the right-hand side;*
4. *$\varphi \subseteq (P_{CL} \times P_{CR}) \cup (T_L \times T_R)$ is said to be an interface transfer relation of r that relates places of L to places of R and transitions of L to transitions of R, $P_{CL}\, \varphi \subseteq P_{CR}$, $\varphi P_{CR} \subseteq P_{CL}$, $T_L \varphi \subseteq T_R$, and $\varphi T_R \subseteq P_L$;*
5. *${}^\bullet\varphi \subseteq \varphi$ is said to be an input interface transfer relation, expressed as ${}^\bullet\varphi = \{((L.p_i\}, \{R.p_i\})\}$ or $\{((L.t_i\}, \{R.t_i\})\}$, and $L.^*$ or $R.^*$ means to input nodes "*" in L or R;*
6. *$\varphi^\bullet \subseteq \varphi$ is named output interface transfer relation, $\varphi^\bullet = \{((L.p_j\}, \{R.p_j\})\}$ or $\{((L.t_j\}, \{R.t_j\})\}$, and $L.^*$ or $R.^*$ means to output nodes "*" in L or R;*
7. *for all rr_i, $rr_j \in R$ $(i \neq j)$, $\xi (L_i) \cap \xi (L_j) \neq \emptyset$, a rewriting must be guaranteed without overlap; moreover, the order of rr_i, rr_j does not impact the result of the rewriting.*

Definition 24. *Let N_R be a reconfigurable S^3PR with $N_R = ((N, M_o), \mathcal{R})$. A new rewriting reconfigurable net N_R is an S^3PR net (N_R, M_R) with $N_R = (P_C, T, F, W, M_R, K)$, and a net (N, M_o) is called the initial state of the rewriting net model.*

Definition 25. *Let N_R be a reconfigurable S^3PR with $N_R = ((N, M_o), \mathcal{R})$. A state graph in N_R is a labeled directed graph whose nodes are the marking of N_R, expressed as:*

1. *Transition firing: If Arcs labeled with t can fire in the net (N_1, M_1), leading to (N_2, M_2): $(N_1, M_1) \xrightarrow{t} (N_2, M_2) \Leftrightarrow (N_1 = N_2$ and $M_1[t_2\rangle M_2$ in $N_1)$.*
2. *Configuration changing: Arcs labeled with $r = \{L, R, \varphi, {}^\bullet\varphi, \varphi^\bullet\}$ from state (N_1, M_1) to state (N_2, M_2) if there is $\xi: L \to N_1$ so that, $\forall a \notin \xi(L)$ and $b \in L$ if*

   2.1. *$a \in {}^\bullet\xi(b) \Rightarrow b \in {}^\bullet\varphi$ and $a \in \xi(b)^\bullet \Rightarrow b \in \varphi^\bullet$.*
   2.2. *$N_1 = (P_{C1}, T_1, F_1, W_1, M_1, K_1)$ and $N_2 = (P_{C2}, T_2, F_2, W_2, M_2, K_2)$ holds the following: $P_{C2} = P_{C1} - \xi(P_{C1L}) + P_{C1R}$ and $T_2 = T_1 - \xi(T_{1L}) + T_{1R}$. Note that $-(+)$ means deleting(inserting) places or transitions from (to) N_1 and the places name of P_{C1R} and T_{1R} inserted into N_1 must be different to prevent clashes.*

Definition 26. *Let N_R be a reconfigurable S^3PR with $N_R = ((N, M_o), \mathcal{R})$. Let N_1 and N_2 be two states in N_R with $N_1 = (P_{C1}, T_1, F_1, W_1, M_{1o}, K_1)$ and $N_2 = (P_{C2}, T_2, F_2, W_2, M_{2o}, K_2)$. A net N_1 is the restriction of a net N_2 if $P_{C1} \subseteq P_{C2}, T_1 \subseteq T_2$, and $F_1 = F_2 \cap ((P_{C1} \times T_1) \cup (T_1 \times P_{C1}))$ and expressed by $N_1 \subseteq N_2$.*

Definition 27. *Let N_R be a reconfigurable S^3PR with $N_R = ((N, M_o), \mathcal{R})$. Let N_1 and N_2 be two states in N_R with $N_1 = (P_{C1}, T_1, F_1, W_1, M_{1o}, K_1)$ and $N_2 = (P_{C2}, T_2, F_2, W_2, M_{2o}, K_2)$. The set of weighted arcs (flow relation) F_2 is expressed as:*

$$F_2(a, b) = \begin{bmatrix} F_1(a, b) & \text{if } a \notin R \wedge b \notin R \\ F_R(a, b) & \text{if } a \in R \wedge b \in R \\ \sum_{b_i \in \bullet\varphi\, b} F_1(a, \xi(y_i)) & \text{if } a \notin R \wedge b \in R \\ \sum_{a_i \in \varphi\, \bullet a} F_1(\xi(a_i), b) & \text{if } a \in R \wedge b \notin R \end{bmatrix} \tag{2}$$

Definition 28. *Let N_R be a reconfigurable S^3PR with $N_R = ((N, M_o), \mathcal{R})$. Let N_1 and N_2 be two states in N_R with $N_1 = (P_{C1}, T_1, F_1, W_1, M_{1o}, K_1)$ and $N_2 = (P_{C2}, T_2, F_2, W_2, M_{2o}, K_2)$. The marking of $M'(p), p \in P_{C2}$, is expressed as:*

$$M'(p) = \begin{bmatrix} M(p) & \text{if } p \notin R \\ \sum_{p' \in \varphi p} M(\xi(p')) & \text{if } p \in R \end{bmatrix} \tag{3}$$

Theorem 1. *Let N_R be a reconfigurable S^3PR with $N_R = ((N, M_o), \mathcal{R})$. Let N_1 and N_2 be two states in N_R with $N_1 = (P_{C1}, T_1, F_1, W_1, M_{1o}, K_1)$ and $N_2 = (P_{C2}, T_2, F_2, W_2, M_{2o}, K_2)$, $P_{C1}, T_1 \neq \emptyset$ and $\mathcal{R} = \{rr\}$, $rr = \{L, R, \varphi, {}^\bullet\varphi, \varphi^\bullet\}$. If L and R are a single place or single transition, then the obtained N_2 by rr is equal to N_1.*

Proof. Straightforward. □

Theorem 2. *Let N_R be a reconfigurable S^3PR with $N_R = ((N, M_o), \mathcal{R})$. Let N_1 and N_2 be two states in N_R with $N_1 = (P_{C1}, T_1, F_1, W_1, M_{1o}, K_1)$ and $N_2 = (P_{C2}, T_2, F_2, W_2, M_{2o}, K_2)$, $P_{C1}, T_1 \neq \emptyset$ and $\mathcal{R} = \{rr\}$, $rr = \{L, R, \varphi, {}^\bullet\varphi, \varphi^\bullet\}$. If (N_1, M_1) is bounded, L is a single place or single transition and R is an S^3PR net, then the resulting (N_2, M_{2o}) net by rr is bounded.*

Proof. The rewriting of N_2 using rr is similar to replacing a place/transition by the S^3PR net. Therefore, the boundedness can be established by checking if the S^3PR net is well constructed and behaved. The resulting net (N_2, M_{2o}) maintains the boundedness because the S^3PR net is well constructed and behaved. $\square$

Corollary 1. *Let N_R be a reconfigurable S^3PR with $N_R = ((N, M_o), \mathcal{R})$. Let N_1 and N_2 be two states in N_R with $N_1 = (P_{C1}, T_1, F_1, W_1, M_{1o}, K_1)$ and $N_2 = (P_{C2}, T_2, F_2, W_2, M_{2o}, K_2)$, $P_{C1}, T_1 \neq \emptyset$ and $\mathcal{R} = \{rr\}$, $rr = \{L, R, \varphi, {}^\bullet\varphi, \varphi^\bullet\}$. If (N_1, M_{1o}) is bounded, L is an S^3PR Petri net and R is a single place or single transition, then the resulting net (N_2, M_{2o}) by rr is bounded.*

Corollary 2. *An S^3PR net (N_2, M_{2o}) can be a bounded net and a full subnet of (N_1, M_{1o}).*

Theorem 3. *Let N_R be a reconfigurable S^3PR with $N_R = ((N, M_o), \mathcal{R})$. Let N_1 and N_2 be two states in N_R with $N_1 = (P_{C1}, T_1, F_1, W_1, M_{1o}, K_1)$ and $N_2 = (P_{C2}, T_2, F_2, W_2, M_{2o}, K_2)$, $P_{C1}, T_1 \neq \emptyset$ and $\mathcal{R} = \{rr\}$, $rr = \{L, R, \varphi, {}^\bullet\varphi, \varphi^\bullet\}$. If (N_1, M_{1o}) is bounded, L is an S^3PR net and R is an S^3PR net, then the resulting net (N_2, M_{2o}) by rr is bounded.*

Proof. The rewriting of N_2 using rr is similar to replacing an S^3PR net by another S^3PR net. Therefore, the boundedness can be established by checking if the S^3PR net is well constructed and behaved. The resulting net (N_2, M_{2o}) maintains the boundedness because the S^3PR net is well constructed and behaved. $\square$

Based on Definitions 20–28 and Theorems 1–3, the developed reconfiguration procedures for S^3PR net algorithm are constructed as follows:

Algorithm 1: *Reconfiguration procedures for S^3PR net*

Input: *An S^3PR net (N_o, M_o)*
Output: *A reconfigurable S^3PR net (N_R, M_{Ro})*
Initialization: *Generate dynamic configurations $\mathcal{R} = \{rr_1, rr_2, rr_3, \dots, rr_m\}$ $k=0$.*
Step 1: *while $\mathcal{R} \neq \emptyset$ do*
      $k = k+1$
1.1. *Build $rr_k = \{L_k, R_k, \varphi_k, {}^\bullet\varphi_k, \varphi_k{}^\bullet\}$.*
1.2. *Build $L_k = (P_{CLk}, T_{Lk}, F_{Lk}, W_{Lk}, M_{Lko}, K_{Lk})$.*
1.3. *Build $R_k = (P_{CRk}, T_{Rk}, F_{Rk}, W_{Rk}, M_{Rko}, K_{Rk})$.*
1.4. *Build ${}^\bullet\varphi_k$ and $\varphi_k{}^\bullet$.*
1.5. *Build $\xi_k: N_{k-1} \rightarrow N_k$.*
1.6. *Apply rewriting rule $rr_k: N_k \xrightarrow{rr_k} N_{k-1}$.*
1.7. *Update the flow relation F_k as follows:*

$$F_k(a,b) = \begin{bmatrix} F_{k-1}(a, b) & \text{if } a \notin R_k \wedge b \notin R_k \\ F_{(k-1)R}(a,b) & \text{if } a \in R_k \wedge b \in R_k \\ \sum\limits_{b_i \in {}^\bullet\varphi b} F_{k-1}(a, \xi(y_i)) & \text{if } a \notin R_k \wedge b \in R_k \\ \sum\limits_{a_i \in \varphi^\bullet a} F_{k-1}(\xi(a_i), b) & \text{if } a \in R_k \wedge b \notin R_k \end{bmatrix}$$

1.8. *Calculate the initial marking of N_k*

$$M_{ko}(p) = \begin{bmatrix} M_{(k-1)o}(p) & \text{if } p \in P_R, P_R \in R_k \\ 0 & \text{if } p \in P_A, P_A \in R_k \end{bmatrix}$$

1.9. *$\mathcal{R} = \mathcal{R} \backslash CR$. /* CR is covered rr_k*/*
      end while
Step 2: *Output a reconfigurable S^3PR net (N_R, M_{Ro})*
Step 3: *End*

To illustrate the proposed Algorithm 1, reconsider the initial S^{3}PR net (N_0, M_0) illustrated in Figure 2. Suppose that the first system configuration includes adding new machine. In this scenario, a new machine M2 is assigned to the system (N_0, M_0) to process a part after M, a robot is needed to load/unload a part to/from M2. To model the addition of new machine by using the synthesis procedure of Algorithm 1, we construct a configuration as a rewriting rule $\mathcal{R} = \{rr_1\}$ with $rr_1 = \{L_1, R_1, \varphi_1, {}^\bullet\varphi_1, \varphi_1{}^\bullet\}$, where L_1 and R_1 are illustrated in Figures 3a and 3b, respectively. We have ξ_1: $N_1 \rightarrow N_0$, $\varphi_1 = (\{p_1, p_6, p_7, p_8, p_9\}, \{t_4, t_5, t_6\})$, ${}^\bullet\varphi_1 = (\{L_1.t_4\}, \{R_1.t_4\})$, and $\varphi_1{}^\bullet = (\{L_1.p_1, L_1.p_6\}, \{R_1.p_1\})$. Then the obtained reconfigurable S^{3}PR net (N_1, M_{10}) is illustrated in Figure 3c.

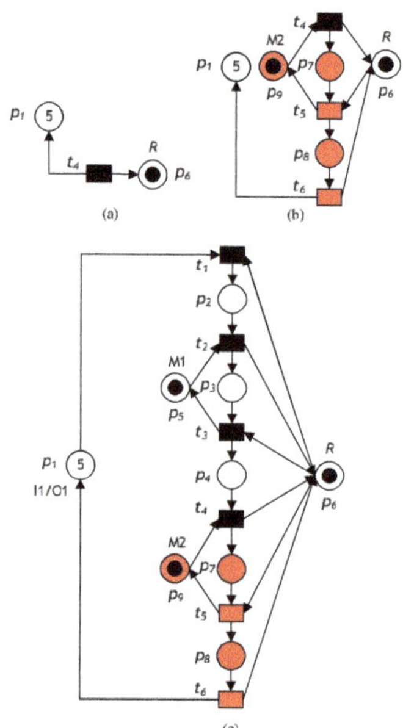

Figure 3. A reconfigured S^{3}PR net by addition of new machine. (**a**) Left hand side net L. (**b**) Right hand side net R. (**c**) A reconfigurable S^{3}PR net (N_1, M_{10}).

The second configuration includes adding a new product. In this scenario, a new product (part B) is assigned to a system, which indicates that a new operation sequence is assigned and the system requires an adjustment to its Petri net model structure. To model the addition of new product by using the synthesis procedure of Algorithm 1, we constructed a configuration as a rewriting rule $\mathcal{R} = \{rr_2\}$ with $rr_2 = \{L_2, R_1, \varphi_2, {}^\bullet\varphi_2, \varphi_2{}^\bullet\}$, where L_2 and R_2 are illustrated in Figures 4a and 4b, respectively. We have ξ_2: $N_2 \rightarrow N_1$, $\varphi_2 = (\{p_5, p_6, p_{10}, p_{11}, p_{12}, p_{13}\}, \{t_7, t_8, t_9, t_{10}\})$, ${}^\bullet\varphi_2 = (\{L_2.p_5, L_2.p_6\}, \{R_2.t_7\})$, and $\varphi_2{}^\bullet = (\{L_2.p_5, L_2.p_6\}, \{R_2.t_{10}\})$. Then the obtained reconfigurable S^{3}PR net (N_2, M_{20}) is illustrated in Figure 4c.

The third system configuration involves rework. In this scenario, a part can be inspected after all operations have been completed. The system can proceed on the basis of the original sequence of operation if the configuration is carried out properly. Otherwise, rework is needed. By using Algorithm 1, the production operations of the reworked part can be exactly and easily modeled by considering rework operations as alternative sequences. Reconsider the reconfigurable S^{3}PR net (N_2, M_2) illustrated in Figure 4c. Suppose that an inspection machine M3 is added to a system and that part B is processed in M1. Then, part B is moved to M3 by Robot 1 to check if there are defects in part B. If part B performed properly, then it will leave the system by Robot 1. Otherwise, if part B has defects, rework is needed, and part B is moved to M1 by Robot 1. To model the rework operation by using the synthesis procedure of Algorithm 1, we construct a configuration as a rewriting rule $\mathcal{R} = \{rr_3\}$ with $rr_3 = \{L_3, R_3, \varphi_3, {}^\bullet\varphi_3, \varphi_3{}^\bullet\}$, where L_3 and R_3 are illustrated in Figure 5a,b, respectively. We have

$\xi_3\colon N_3 \to N_2$, $\varphi_3 = (\{p_5, p_6, p_{10}, p_{11}, p_{12}, p_{13}, p_{14}, p_{15}, p_{16}, p_{17}\}, \{t_7, t_8, t_9, t_{10}, t_{11}, t_{12}, t_{13}, t_{14}, t_{15}\})$, ${}^{\bullet}\varphi_3 = (\{L_3.t_7\}, \{R_3.t_7\})$, and $\varphi_3{}^{\bullet} = (\{L_3.t_{10}\}, \{R_3.t_{14}\})$. Then the obtained reconfigurable S^3PR net (N_3, M_{3_0}) is illustrated in Figure 5c.

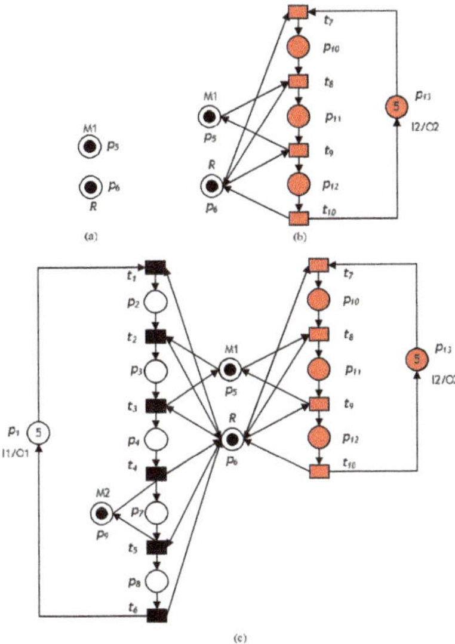

Figure 4. A reconfigured S^3PR net by addition of new product. (**a**) Left hand side net L. (**b**) Right hand side net R. (**c**) A reconfigurable S^3PR net (N_2, M_{2_0}).

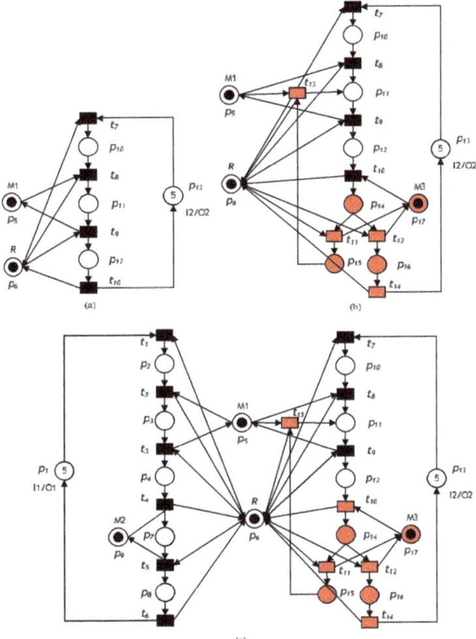

Figure 5. A reconfigured S^3PR net by rework. (**a**) Left hand side net L. (**b**) Right hand side net R. (**c**) A reconfigurable S^3PR net (N_3, M_{3_0}).

Finally, a configuration includes adding a new robot. In this scenario, a new robot R2 is assigned to the system (N_3, M_{3_0}) to load/unload a part A to/from M1 and M2. To model the addition of the new robot by using the synthesis procedure of Algorithm 1, we construct a configuration as a rewriting rule $\mathcal{R} = \{rr_4\}$ with $rr_4 = \{L_4, R_4, \varphi_4, {}^{\bullet}\varphi_4, \varphi_4{}^{\bullet}\}$, where L_4 and R_4 are illustrated in Figure 6a,b, respectively. We have $\xi_4: N_4 \rightarrow N_3$, $\varphi_4 = (\{p_1, p_2, p_3, p_4, p_{6\_1}, p_{6\_2}, p_7, p_8, p_{10}, p_{11}, p_{12}, p_{14}, p_{15}, p_{16}\}, \{t_1, t_2, t_3, t_4, t_5, t_6, t_7, t_8, t_9, t_{10}, t_{11}, t_{12}, t_{13}, t_{14}, t_{15}\})$, ${}^{\bullet}\varphi_4 = (\{L_4.t_1, L_4.t_7\}, \{R_4. t_1, R_4. t_7\})$, and $\varphi_4{}^{\bullet} = (\{L_4. t_6, L_4. t_{14}\}, \{R_4.t_6, R_4.t_{14}\})$. Then the obtained reconfigurable S³PR net (N_4, M_{4_0}) is illustrated in Figure 6c.

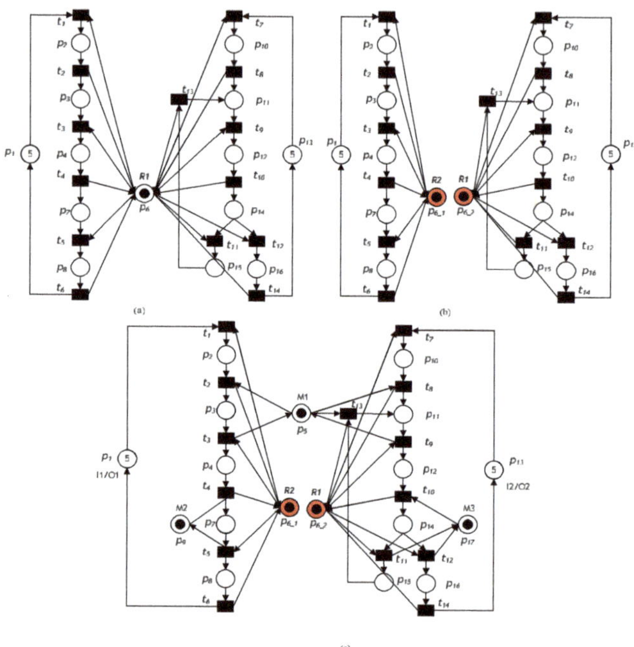

Figure 6. A reconfigured S³PR net by addition of a new robot. (**a**) Left hand side net *L*. (**b**) Right hand side net *R*. (**c**) A reconfigurable S³PR net (N_4, M_{4_0}).

3. Deadlock Prevention Policy for Reconfigurable S³PR Net Based on Siphons

This section presents definitions on siphons in reconfigurable S³PR nets. Next, the siphon control method based on place invariants is introduced. Finally, a deadlock prevention algorithm is proposed to solve the deadlock problems in reconfigurable S³PR nets.

Definition 29. *Let N_R be a reconfigurable S³PR net with $N_R = ((N, M_0), \mathcal{R})$. Let N_1 be a state of N_R with $N_1 = (P_{C1}, T_1, F_1, W_1, M_{1_0}, K_1)$. A place vector of N_1 is expressed as a column vector I: $P_{C1} \rightarrow \mathbf{Z}$ indexed by P_{C1}, and a transition vector of N_1 is defined as a column vector J: $T_1 \rightarrow \mathbf{Z}$ indexed by T_1, where $\mathbf{Z} = \{ \ldots, -2, -1, 0, 1, 2, \ldots \}$.*

Definition 30. *Let N_R be a reconfigurable S³PR net with $N_R = ((N, M_0), \mathcal{R})$. Let N_1 be a state of N_R with $N_1 = (P_{C1}, T_1, F_1, W_1, M_{1_0}, K_1)$. A place vector I of N_1 is expressed as a place invariant (PI) if $I^T. [N_1] = \mathbf{0}^T$ and I $\neq \mathbf{0}$, and a transition vector of N_1 is defined as a transition invariant (TI) if $[N_1]. J = \mathbf{0}$ and J $\neq \mathbf{0}$.*

Definition 31. *Let N_R be a reconfigurable S^3PR net with $N_R = ((N, M_o), \mathcal{R})$. Let N_1 be a state of N_R with $N_1 = (P_{C1}, T_1, F_1, W_1, M_{1o}, K_1)$. A place invariant I of N_1 is expressed as a place semi-flow if each element of I is non-negative. $\|I\| = \{p \mid I(p) \neq 0\}$ is said to be the support of place invariant of I. $\|I\|^+ = \{p \mid I(p) > 0\}$ is said to be the positive support of place invariant I. $\|I\|^- = \{p \mid I(p) < 0\}$ is said to be the negative support of place invariant I. I is a minimal place invariant if $\|I\|$ is not a superset of the support of any other one and its components are mutually prime.*

Definition 32. *Let N_R be a reconfigurable S^3PR net with $N_R = ((N, M_o), \mathcal{R})$. Let N_1 be a state of N_R with $N_1 = (P_{C1}, T_1, F_1, W_1, M_{1o}, K_1)$. A transition invariant J of N_1 is expressed as a transition semi-flow if each element of J is non-negative. $\|J\| = \{t \mid J(t) \neq 0\}$ is said to be the support of transition invariant of J. $\|J\|^+ = \{t \mid J(t) > 0\}$ is said to be the positive support of transition invariant J. $\|J\|^- = \{t \mid J(t) < 0\}$ is said to be the negative support of transition invariant J. J is a minimal transition invariant, if $\|J\|$ is not a superset of the support of any other one, and its components are mutually prime.*

Definition 33. *Let N_R be a reconfigurable S^3PR net with $N_R = ((N, M_o), \mathcal{R})$. Let N_1 be a state of N_R with $N_1 = (P_{C1}, T_1, F_1, W_1, M_{1o}, K_1)$. l_i is said to be the coefficients of place invariant I if for all $p_i \in P_{C1}$, $l_i = I(p_i)$.*

Definition 34. *Let N_R be a reconfigurable S^3PR net with $N_R = ((N, M_o), \mathcal{R})$. Let N_1 be a state of N_R with $N_1 = (P_{C1}, T_1, F_1, W_1, M_{1o}, K_1)$. A non-empty set $S \subseteq P_{C1}$ is called a siphon in N_1 if ${}^\bullet S \subseteq S^\bullet$. $S \subseteq P_{C1}$ is called a trap in N_1 if $S^\bullet \subseteq {}^\bullet S$. $S \subseteq P_{C1}$ is called a minimal siphon (trap) if a siphon (trap) contains no other siphons. A minimal siphon S is called a strict minimal siphon if $S^\bullet \subsetneq {}^\bullet S$. Let $\Pi = \{S_1, S_2, S_3, \ldots, S_k\}$ be a set of strict minimal siphons of N_1. We have $S = S_A \cup S_R$, $S_R = S \cap P_R$, and $S_A = S \backslash S_R$, where S_A and S_R are sets of operations and resources places, respectively.*

Definition 35. *Let N_R be a reconfigurable S^3PR net with $N_R = ((N, M_o), \mathcal{R})$. Let N_1 be a state of N_R with $N_1 = (P_{C1}, T_1, F_1, W_1, M_{1o}, K_1)$. A siphon S in N_1 is called marked at marking M if $\sum_{p \in S} M(p) \geq 1$, and otherwise is called unmarked at marking M.*

Definition 36. *Let N_R be a reconfigurable S^3PR net with $N_R = ((N, M_o), \mathcal{R})$. Let N_1 be a state of N_R with $N_1 = (P_{C1}, T_1, F_1, W_1, M_{1o}, K_1)$. A siphon S in N_1 is called an emptiable siphon if there exists $M \in R(N_1, M_{1o})$ such that $\sum_{p \in S} M(p) = 0$, and otherwise is called non-emptiable siphon.*

Theorem 4. *Let N_R be a reconfigurable S^3PR net with $N_R = ((N, M_o), \mathcal{R})$. Let N_1 be a state of N_R with $N_1 = (P_{C1}, T_1, F_1, W_1, M_{1o}, K_1)$ and Π the set of N_1 siphons. The net N_1 is deadlock-free if for all $S \in \Pi$, for all $M \in R(N_1, M_{1o})$, $\sum_{p \in S} M(p) \geq 1$.*

Proof. Let S be a siphon in N_1 and $p \in S$. p is marked at marking M and satisfies $\sum_{p \in S} M(p) \geq 1$. The net N_1 has at least one transition t enabled at any marking reachable from M and S is never be an unmarked, and it is therefore deadlock-free. $\square$

Theorem 5. *Let N_R be a reconfigurable S^3PR net with $N_R = ((N, M_o), \mathcal{R})$. Let N_1 be a state of N_R with $N_1 = (P_{C1}, T_1, F_1, W_1, M_{1o}, K_1)$ and Π the set of N_1 siphons. The net (N_1, M_{1o}) is in a deadlock state, i.e., M is a dead marking of N_1. Then, $\{p \in P_{C1} \mid M(p) = 0\}$ is a siphon S.*

Proof. Since M is a dead marking, each t has an empty input place p at M, $\forall p \in {}^\bullet t$, $M(p) < W(p, t)$, and thus $S^\bullet$ includes each transition of N_1. In fact, we have ${}^\bullet S \subseteq S^\bullet$. Therefore, S is a siphon. Since the net has at least one transition $t \in T_1$, S is not an empty set. $\square$

Corollary 3. *Let N_R be a reconfigurable S^3PR net with $N_R = ((N, M_o), \mathcal{R})$. Let N_1 be a state of N_R with $N_1 = (P_{C1}, T_1, F_1, W_1, M_{1o}, K_1)$, a deadlocked N_1 net includes at least one unmarked siphon S.*

Corollary 4. *Let N_R be a reconfigurable S^3PR net with $N_R = ((N, M_0), \mathcal{R})$. Let N_1 be a state of N_R with $N_1 = (P_{C1}, T_1, F_1, W_1, M_{1o}, K_1)$, N_1 is a deadlocked net at marking M. Then, N_1 has at least one unmarked siphon S such that for all $p \in S$, there exists $t \in p^\bullet$ such that $W_1(p, t) > M(p)$.*

To develop a deadlock prevention policy for reconfigurable S^3PR net, we reviewed the approach of designing a control place (monitor) for a place invariant developed by Yamalidou et al. [40]. Then we develop a deadlock prevention policy for reconfigurable S^3PR net to achieve an optimal place invariant. Yamalidou et al. propose a computationally efficient method based on place invariants that enforces algebraic constraints on the elements of a marking of a net system by constructing control places. The control purpose is to ensure a siphon to be a marked siphon, i.e., ensure a siphon be non-emptiable at all elements of a marking.

Assume that a reconfigurable S^3PR net with $N_R = ((N, M_0), \mathcal{R})$ and N_k (state of N_R) with $N_k = (P_{Ck}, T_k, F_k, W_k, M_{ko}, K_k)$, $k = 1, 2, \ldots, |\mathcal{R}|$ is a net to be controlled, which includes n places and m transitions. Let $[N_k]$ be the incidence matrix of a plant reconfigurable S^3PR net. The control places can be represented by $[N_c]$ a matrix that shows the connection relationship between control places to transitions of the net N_k. The controlled net with incidence matrix $[N]$ comprises both the original reconfigurable S^3PR net and the monitors, i.e.,

$$[N] = \begin{bmatrix} N_k \\ N_c \end{bmatrix} \tag{4}$$

The control purpose is to impose a set of linear constraints to prevent unwanted markings being reached. The constraints are formulated in a matrix form:

$$\mathcal{L}.M \geq \mathcal{B} \tag{5}$$

where M denotes the marking vector of net N_k, $\mathcal{L}$ is an integer $n_c \times n$ matrix (n_c - the number of constraints), and $\mathcal{B}$ is an integer column vector. After the introduction of a non-negative slack variable that corresponds to the initial marking M_{ko} of N_k, constraint (5) can be reformulated as:

$$M_{co} = \mathcal{B} - \mathcal{L}.M_{ko}. \tag{6}$$

where M_{co} represents the initial marking of monitor c.

If $[N_k]$ is the incidence matrix, we have: $M_k = M_{ko} + [N_k].\vec{\delta}$. Therefore, $M_c = \mathcal{B} - \mathcal{L}.(M_{ko} + [N_k].\vec{\delta})$, which also can be reformulated as:

$$M_c = M_{co} + (-\mathcal{L}.[N_k].\vec{\delta}) \tag{7}$$

The place invariant computed by (5) must meet the place invariant equation $I^T[N] = 0^T$. Therefore, the monitor $[N_c]$ can be formulated as:

$$[N_c] = -\mathcal{L}.[N_k] \tag{8}$$

Consequently, M_c may be considered as a marking of some additional monitors, where the supervised reconfigurable S^3PR net has an incidence matrix $[N] = \begin{bmatrix} N_k \\ N_c \end{bmatrix}$, and a marking vector $M = \begin{bmatrix} M_k \\ M_c \end{bmatrix}$.

Theorem 6. *Let N_R be a reconfigurable S^3PR net with $N_R = ((N, M_o), \mathcal{R})$. Let N_k be a state of N_R with $N_k = (P_{Ck}, T_k, F_k, W_k, M_{ko}, K_k)$, incidence matrix $[N_k]$ and initial marking M_{ko} be given. A set of n_c linear constraints $\mathcal{L}.M_k \geq \mathcal{B}$ are to be imposed. If $\mathcal{B} - \mathcal{L}.M_k \geq 0$ then a Petri net controller with incidence matrix $[N_c] = -L.[N_k]$ and initial marking $M_{co} = \mathcal{B} - \mathcal{L}.M_{ko}$ enforces the constraint $\mathcal{L}.M_k \geq \mathcal{B}$ when included in the closed loop system $[N] = \begin{bmatrix} N_k \\ N_c \end{bmatrix}$. In addition, the controller is maximally permissive.*

Proof. See [40,41]. $\square$

Now, we consider the place invariant approach to control the siphon. Let S be an unmarked siphon. The control purpose is to ensure that S is never unmarked through the system evolution (N, M_o) and eliminate markings that break the linear constraint (5) from the reachable markings.

Let $V_S \backslash S \in \Pi$ be the monitor resulting from controlling the siphon S. There are siphons S such that if $\sum_{p \in S} M_o(p) \geq 1$ for the initial marking M_o, then $\sum_{p \in S} M(p) \geq 1$ for all reachable markings M. Therefore, a siphon S does not require control. In order to reduce the supervisor's complexity, these siphons are identified and no monitors are added. Thus, we have two sets of constraints: $\mathcal{L}.M \geq \mathcal{B}$ and $\mathcal{L}_o.M \geq \mathcal{B}_o$ rather than a single set of constraints $\mathcal{L}.M \geq \mathcal{B}$. The deadlock prevention supervision of the original net needs enforcing $\mathcal{L}.M \geq \mathcal{B}$ and selecting an initial marking M_o such that $\mathcal{L}_o.M_o \geq \mathcal{B}_o$ and $\mathcal{L}.M_o \geq \mathcal{B}$. The constraints $\mathcal{L}_o.M \geq \mathcal{B}_o$ are the constraints that all reachable markings satisfy when the initial markings satisfy them. Therefore, there are two cases to control a siphon:

If $V_S^\bullet \subseteq {}^\bullet S$, then S does not require monitor and V_S is not assigned to a net N. Furthermore, $V_S^\bullet \subseteq {}^\bullet S$ if and only if S is a trap. Thus, when S is also a siphon, it is (trap) controlled for all initial markings M_o that satisfy $\sum_{p \in S} M_o(p) \geq 1$. Therefore, a siphon S is assigned to $(\mathcal{L}_o; \mathcal{B}_o)$.

A.   If $V_S^\bullet \subsetneq {}^\bullet S$, then S needs a monitor and V_S is assigned to N. Therefore, the S is assigned to $(\mathcal{L}; \mathcal{B})$.

Definition 37. *Let N_R be a reconfigurable S^3PR net with $N_R = ((N, M_o), \mathcal{R})$. Let N_k be a state in N_R with $N_k = (P_{Ck}, T_k, F_k, W_k, M_{ko}, K_k)$. A siphon S in N_k is called controlled if for all $M \in R(N_k, M_{ko})$, $\sum_{p \in S} M(p) \geq 1$ and satisfy $\mathcal{L}.M \geq \mathcal{B}$ and $\mathcal{L}_o.M \geq \mathcal{B}_o$.*

Definition 38. *Let N_R be a reconfigurable S^3PR net with $N_R = ((N, M_o), \mathcal{R})$. Let N_k be a state in N_R with $N_k = (P_{Ck}, T_k, F_k, W_k, M_{ko}, K_k)$. The deadlock controller for (N_k, M_{ok}) is expressed as $(V, M_{Vo}) = (P_V, T_V, F_V, M_{Vo})$, where (1) $P_V = \{V_S \backslash S \in \Pi\}$ is set of monitors. (2) $T_V = \{t \backslash t \in {}^\bullet V_S \cup V_S^\bullet\}$. (3) $F_V \subseteq (P_V \times T_V) \cup (T_V \times P_V)$ is called a flow relation of V. (4) for all $V_S \in P_V$, $M_{Vo}(V_S) = \mathcal{B} - \mathcal{L}.M_{ko}(V_s)$, where $M_{Vo}(V_S)$ is called an initial marking of a monitor. (N_{RC}, M_{RCo}) is said to be a controlled reconfigurable S^3PR net resulting from the integration of (N_k, M_{ko}) and (V, M_{Vo}), expressed as $(N_k, M_{ko}) \parallel (V, M_{Vo})$, where $N_{RC} = (P_{RC}, T_{RC}, F_{RC}, W_{RC}, M_{RCo}, K_{RC})$, $P_{RC} = P_{Ck} \cup P_V$, $T_{RC} = T_k \cup T_V$, $F_{RC}: (P_{RC} \times T_{RC}) \cup (T_{RC} \times P_{RC}) \rightarrow \mathbf{IN}$ is called flow relations, $W_{RC}: (P_{RC} \times T_{RC}) \cup (T_{RC} \times P_{RC}) \rightarrow \mathbf{IN}$ is a mapping that assigns a weight to an arc, $M_{RCo}: P_{RC} \rightarrow \mathbf{IN}$ is the initial marking, and $K_{RC}: P_{RC} \rightarrow \mathbf{IN}$ is the function of capacity that assigns to each place p the maximal number of tokens $K_{RC}(p)$.*

Based on the concept of place invariant and siphon control, the deadlock prevention algorithm for reconfigurable S^3PR net is developed as follows:

Algorithm 2: *Deadlock prevention algorithm for reconfigurable S^3PR net based on siphon control*

Input: *An S^3PR net (N_o, M_o)*

Output: *A controlled reconfigurable S^3PR net (N_{RC}, M_{RCo}).*

Initialization: *Generate dynamic configurations $\mathcal{R} = \{rr_1, rr_2, rr_3, \ldots, rr_m\}$ $k=0$, $P_V = \emptyset$, $T_V = \emptyset$, $F_V = \emptyset$, $(N_{RC}, M_{RCo}) = \emptyset$.*

Step 1: *while $\mathcal{R} \neq \emptyset$ do*

$k=k+1$

1.1. Build (N_k, M_{ko}) by using Algorithm 1.

1.2. Compute minimal siphons Π for (N_k, M_{ko}).

1.3. for each $S \in \Pi$ do

if $V_S{}^\bullet \subsetneq {}^\bullet S$, then

a.   *Add S to $(\mathcal{L}; \mathcal{B})$.*

b.   *$\left[N_{V_S} \right] = -\mathcal{L}.[N_k]$*

c.   *$M_{Vo}(V_S) = \mathcal{B} - \mathcal{L}.M_{ko}$.*

d.   *$P_V := P_V \cup \{Vs\}$*

e.   *$T_V := T_V \cup \{t \backslash t \in {}^\bullet V_S \cup V_S{}^\bullet\}$.*

f.   *$F_V := F_V \cup ((P_V \times T_V) \cup (T_V \times P_V))$*

elseIf $V_S{}^\bullet \subseteq {}^\bullet S$ and $\sum_{p \in S} M_o(p) \geq 1$, then

Add S to $(\mathcal{L}_o; \mathcal{B}_o)$.

end if

end for

1.4. $(N_{RC}, M_{RCo}) := (N_k, M_{ko}) \parallel (V_k, M_{Vko})$

1.5. $\mathcal{R} = \mathcal{R} \backslash CR.$ / CR is covered rr_k*/*

end while

Step 2: *Output a controlled reconfigurable S^3PR net (N_{RC}, M_{RCo}).*

Step 3: *End*

To illustrate the proposed Algorithm 2, reconsider the initial S^3PR net (N_o, M_o) illustrated in Figure 2. The initial net has four minimal siphons $S_1 = \{p_1, p_2, p_3, p_4\}$, $S_2 = \{p_3, p_5\}$, $S_3 = \{p_2, p_4, p_6\}$, and $S_4 = \{p_4, p_5, p_6\}$. The N_o incidence matrix is

$$[N_o] = \begin{bmatrix} -1 & 0 & 0 & 1 \\ 1 & -1 & 0 & 0 \\ 0 & 1 & -1 & 0 \\ 0 & 0 & 1 & -1 \\ 0 & -1 & 1 & 0 \\ -1 & 1 & -1 & 1 \end{bmatrix} \tag{9}$$

while its initial marking is:

$$M_o = \begin{bmatrix} 5 & 0 & 0 & 0 & 1 & 1 \end{bmatrix}^T \tag{10}$$

S_4 creates monitor V_{S1}, therefore one monitor V_{S1} is added, which enforces:

$$M(p_4) + M(p_5) + M(p_6) \geq 1 \tag{11}$$

The following place invariant is generated:

$$M(V_{S1}) = M(p_4) + M(p_5) + M(p_6) - 1 \tag{12}$$

The current matrices $\mathcal{L}$ and $\mathcal{B}$ represent the Equation (12).

$$\mathcal{L} = \begin{bmatrix} 0 & 0 & 0 & 1 & 1 & 1 \end{bmatrix}, \mathcal{B} = [1] \tag{13}$$

while the others minimal siphons create constraints in $(\mathcal{L}_o; \mathcal{B}_o)$.

$$\mathcal{L}_o = \begin{bmatrix} 1 & 1 & 1 & 1 & 0 & 0 \\ 0 & 0 & 1 & 0 & 1 & 0 \\ 0 & 1 & 0 & 0 & 0 & 1 \end{bmatrix}, \; \mathcal{B}_o = \begin{bmatrix} 1 \\ 1 \\ 1 \end{bmatrix} \tag{14}$$

The controller net incidence matrix is calculated by Equation (8):

$$\left[N_{V_S} \right] = -\mathcal{L}.[N_o] = \begin{bmatrix} -1 & 0 & 1 & 0 \end{bmatrix} \tag{15}$$

The controller's initial place marking is calculated as:

$$M_0(V_{S1}) = M_0(p_4) + M_0(p_5) + M_0(p_6) - 1 = 1$$

The controlled net of (N_0, M_0) is illustrated in Figure 7. The place and arcs of the controller are shown with blue lines.

Now, reconsider the reconfigured S^{3}PR net by addition of new machine (N_1, M_{10}) illustrated in Figure 3c. The reconfigured net has seven minimal siphons $S_1 = \{p_3, p_5\}$, $S_2 = \{p_7, p_9\}$, $S_3 = \{p_2, p_4, p_6, p_8\}$, $S_4 = \{p_4, p_5, p_6, p_8\}$, $S_5 = \{p_2, p_6, p_8, p_9\}$, $S_6 = \{p_5, p_6, p_8, p_9\}$, and $S_7 = \{p_1, p_2, p_3, p_4, p_7, p_8\}$.

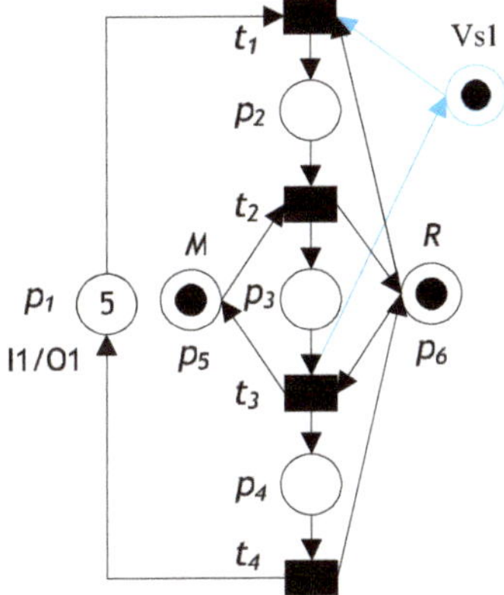

Figure 7. Controlled S^{3}PR net by Algorithm 2.

The N_1 incidence matrix is:

$$[N_1] = \begin{bmatrix} -1 & 0 & 0 & 0 & 0 & 1 \\ 1 & -1 & 0 & 0 & 0 & 0 \\ 0 & 1 & -1 & 0 & 0 & 0 \\ 0 & 0 & 1 & -1 & 0 & 0 \\ 0 & -1 & 1 & 0 & 0 & 0 \\ -1 & 1 & -1 & 1 & -1 & 1 \\ 0 & 0 & 0 & 1 & -1 & 0 \\ 0 & 0 & 0 & 0 & 1 & -1 \\ 0 & 0 & 0 & -1 & 1 & 0 \end{bmatrix} \tag{16}$$

while its initial marking is:

$$M_{10} = \begin{bmatrix} 5 & 0 & 0 & 0 & 1 & 1 & 0 & 0 & 1 \end{bmatrix}^T \tag{17}$$

S_4, S_5, and S_6 create monitor V_{S1}, V_{S2}, and V_{S3}, respectively. Thus, three monitors are added, V_{S1}, V_{S2}, and V_{S3}, which enforce:

$$M(p_4) + M(p_5) + M(p_6) + M(p_8) \geq 1 \tag{18}$$

$$M(p_2) + M(p_6) + M(p_8) + M(p_9) \geq 1 \tag{19}$$

$$M(p_5) + M(p_6) + M(p_8) + M(p_9) \geq 1 \tag{20}$$

The following place invariants are accordingly generated:

$$M(V_{S1}) = M(p_4) + M(p_5) + M(p_6) + M(p_8) - 1 \tag{21}$$

$$M(V_{S2}) = M(p_2) + M(p_6) + M(p_8) + M(p_9) - 1 \tag{22}$$

$$M(V_{S3}) = M(p_5) + M(p_6) + M(p_8) + M(p_9) - 1 \tag{23}$$

The current matrices $\mathcal{L}$ and $\mathcal{B}$ represent the Equations (18)–(20).

$$\mathcal{L} = \begin{bmatrix} 0 & 0 & 0 & 1 & 1 & 1 & 0 & 1 & 0 \\ 0 & 1 & 0 & 0 & 0 & 1 & 0 & 1 & 1 \\ 0 & 0 & 0 & 0 & 1 & 1 & 0 & 1 & 1 \end{bmatrix}, \; \mathcal{B} = \begin{bmatrix} 1 \\ 1 \\ 1 \end{bmatrix} \tag{24}$$

while the other minimal siphons create constraints in $(\mathcal{L}_0; \mathcal{B}_0)$.

$$\mathcal{L}_0 = \begin{bmatrix} 0 & 0 & 1 & 0 & 1 & 0 & 0 & 0 & 0 \\ 0 & 0 & 0 & 0 & 0 & 0 & 1 & 0 & 1 \\ 0 & 1 & 0 & 1 & 0 & 1 & 0 & 1 & 0 \\ 1 & 1 & 1 & 1 & 0 & 0 & 1 & 1 & 0 \end{bmatrix}, \; \mathcal{B}_0 = \begin{bmatrix} 1 \\ 1 \\ 1 \\ 1 \end{bmatrix} \tag{25}$$

The controller's net incidence matrix is calculated by Equation (12);

$$\left[N_{V_S} \right] = -\mathcal{L}.[N_1] = \begin{bmatrix} -1 & 0 & 1 & 0 & 0 & 0 \\ 0 & 0 & -1 & 0 & 1 & 0 \\ -1 & 0 & 0 & 0 & 1 & 0 \end{bmatrix} \tag{26}$$

The initial marking controllers are calculated as:

$$M_0(V_{S1}) = M_0(p_4) + M_0(p_5) + M_0(p_6) + M_0(p_8) - 1 = 1$$

$$M_0(V_{S2}) = M_0(p_2) + M_0(p_6) + M_0(p_8) + M_0(p_9) - 1 = 1$$

$$M_0(V_{S3}) = M_0(p_5) + M_0(p_6) + M_0(p_8) + M_0(p_9) - 1 = 2$$

The controlled reconfigurable net of (N_1, M_{10}) is illustrated in Figure 8. The place and arcs of the controllers are shown with blue lines.

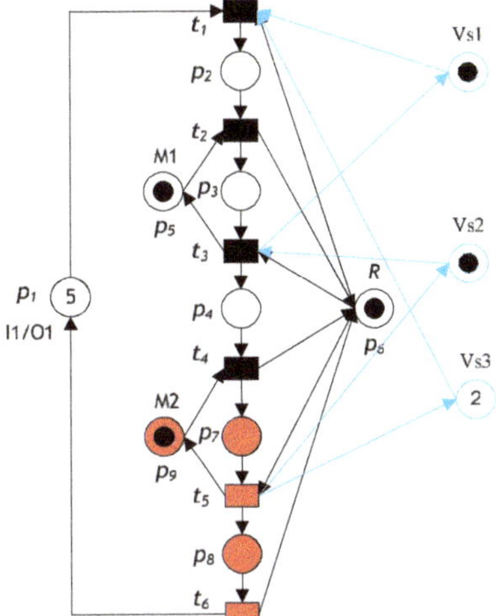

Figure 8. Controlled reconfigurable S³PR net by addition of new machine.

Then, reconsider the reconfigured S³PR net by addition of new product (N_2, M_{2O}) illustrated in Figure 4c. The reconfigured net has 11 minimal siphons $S_1 = \{p_7, p_9\}$, $S_2 = \{p_3, p_5, p_{11}\}$, $S_3 = \{p_{10}, p_{11}, p_{12}, p_{13}\}$, $S_4 = \{p_4, p_5, p_6, p_8, p_{12}\}$, $S_5 = \{p_5, p_6, p_8, p_9, p_{12}\}$, $S_6 = \{p_1, p_2, p_3, p_4, p_7, p_8\}$, $S_7 = \{p_2, p_4, p_6, p_8, p_{10}, p_{12}\}$, $S_8 = \{p_2, p_6, p_8, p_9, p_{10}, p_{12}\}$, $S_9 = \{p_4, p_5, p_6, p_8, p_{12}\}$, $S_{10} = \{p_5, p_6, p_8, p_9, p_{12}\}$, and $S_{11} = \{p_2, p_6, p_8, p_9, p_{10}, p_{12}\}$. The N_2 incidence matrix is

$$
[N_2] = \begin{bmatrix}
-1 & 0 & 0 & 0 & 0 & 1 & 0 & 0 & 0 & 0 \\
1 & -1 & 0 & 0 & 0 & 0 & 0 & 0 & 0 & 0 \\
0 & 1 & -1 & 0 & 0 & 0 & 0 & 0 & 0 & 0 \\
0 & 0 & 1 & -1 & 0 & 0 & 0 & 0 & 0 & 0 \\
0 & -1 & 1 & 0 & 0 & 0 & 0 & -1 & 1 & 0 \\
-1 & 1 & -1 & 1 & -1 & 1 & -1 & 1 & -1 & 1 \\
0 & 0 & 0 & 1 & -1 & 0 & 0 & 0 & 0 & 0 \\
0 & 0 & 0 & 0 & 1 & -1 & 0 & 0 & 0 & 0 \\
0 & 0 & 0 & -1 & 1 & 0 & 0 & 0 & 0 & 0 \\
0 & 0 & 0 & 0 & 0 & 0 & 1 & -1 & 0 & 0 \\
0 & 0 & 0 & 0 & 0 & 0 & 0 & 1 & -1 & 0 \\
0 & 0 & 0 & 0 & 0 & 0 & 0 & 0 & 1 & -1 \\
0 & 0 & 0 & 0 & 0 & 0 & -1 & 0 & 0 & 1
\end{bmatrix}
\tag{27}
$$

while its initial marking is:

$$
M_{2O} = \begin{bmatrix} 5 & 0 & 0 & 0 & 1 & 1 & 0 & 0 & 1 & 0 & 0 & 0 & 5 \end{bmatrix}^{\mathrm{T}}
\tag{28}
$$

S_4, S_5, and S_8 create monitors V_{S1}, V_{S2}, and V_{S3}, respectively. Thus, three monitors are added, V_{S1}, V_{S2}, and V_{S3}, which enforce:

$$
M(p_4) + M(p_5) + M(p_6) + M(p_8) + M(p_{12}) \geq 1
\tag{29}
$$

$$
M(p_5) + M(p_6) + M(p_8) + M(p_9) + M(p_{12}) \geq 1
\tag{30}
$$

$$
M(p_2) + M(p_6) + M(p_8) + M(p_9) + M(p_{10}) + M(p_{12}) \geq 1
\tag{31}
$$

The following place invariants are accordingly generated:

$$M(V_{S1}) = M(p_4) + M(p_5) + M(p_6) + M(p_8) + M(p_{12}) - 1 \tag{32}$$

$$M(V_{S2}) = M(p_5) + M(p_6) + M(p_8) + M(p_9) + M(p_{12}) - 1 \tag{33}$$

$$M(V_{S3}) = M(p_2) + M(p_6) + M(p_8) + M(p_9) + M(p_{10}) + M(p_{12}) - 1 \tag{34}$$

The current matrices $\mathcal{L}$ and $\mathcal{B}$ represent the Equations (29)–(31).

$$\mathcal{L} = \begin{bmatrix} 0 & 0 & 0 & 1 & 1 & 1 & 0 & 1 & 0 & 0 & 0 & 1 & 0 \\ 0 & 0 & 0 & 0 & 1 & 1 & 0 & 1 & 1 & 0 & 0 & 1 & 0 \\ 0 & 1 & 0 & 0 & 0 & 1 & 0 & 1 & 1 & 1 & 0 & 1 & 0 \end{bmatrix}, \ \mathcal{B} = \begin{bmatrix} 1 \\ 1 \\ 1 \end{bmatrix} \tag{35}$$

while the other minimal siphons create constraints in $(\mathcal{L}_o; \mathcal{B}_o)$.

$$\mathcal{L}_o = \begin{bmatrix} 0 & 0 & 0 & 0 & 0 & 0 & 1 & 0 & 1 & 0 & 0 & 0 & 0 \\ 0 & 0 & 1 & 0 & 1 & 0 & 0 & 0 & 0 & 0 & 1 & 0 & 0 \\ 0 & 0 & 0 & 0 & 0 & 0 & 0 & 0 & 0 & 1 & 1 & 1 & 1 \\ 1 & 1 & 1 & 1 & 0 & 0 & 1 & 1 & 0 & 0 & 0 & 0 & 0 \\ 0 & 1 & 0 & 1 & 0 & 1 & 0 & 1 & 0 & 1 & 0 & 1 & 0 \\ 0 & 0 & 0 & 1 & 1 & 1 & 0 & 1 & 0 & 0 & 0 & 1 & 0 \\ 0 & 0 & 0 & 0 & 1 & 1 & 0 & 1 & 1 & 0 & 0 & 1 & 0 \\ 0 & 1 & 0 & 0 & 0 & 1 & 0 & 1 & 1 & 1 & 0 & 1 & 0 \end{bmatrix}, \ \mathcal{B}_o = \begin{bmatrix} 1 \\ 1 \\ 1 \\ 1 \\ 1 \\ 1 \\ 1 \\ 1 \end{bmatrix} \tag{36}$$

The controller's net incidence matrix is calculated by Equation (12);

$$\left[N_{V_S} \right] = -\mathcal{L} \cdot [N_{2o}] = \begin{bmatrix} -1 & 0 & 1 & 0 & 0 & 0 & -1 & 0 & 1 & 0 \\ -1 & 0 & 0 & 0 & 1 & 0 & -1 & 0 & 1 & 0 \\ 0 & 0 & -1 & 0 & 1 & 0 & 0 & 0 & 0 & 0 \end{bmatrix} \tag{37}$$

The initial marking controllers are calculated as:

$$M_o(V_{S1}) = M_o(p_4) + M_o(p_5) + M_o(p_6) + M_o(p_8) + M_o(p_{12}) - 1 = 1$$

$$M_o(V_{S2}) = M_o(p_5) + M_o(p_6) + M_o(p_8) + M_o(p_9) + M_o(p_{12}) - 1 = 2$$

$$M_o(V_{S3}) = M_o(p_2) + M_o(p_6) + M_o(p_8) + M_o(p_9) + M_o(p_{10}) + M_o(p_{12}) - 1 = 1$$

The controlled reconfigurable net of (N_2, M_{2o}) is illustrated in Figure 9. The place and arcs of the controllers are shown with blue lines.

Then, reconsider the reconfigured S^{3}PR net by rework (N_3, M_{3o}) illustrated in Figure 5c. The reconfigured net has 13 minimal siphons $S_1 = \{p_4, p_5, p_6, p_8, p_{12}, p_{16}\}$, $S_2 = \{p_5, p_6, p_8, p_9, p_{12}, p_{16}\}$, $S_3 = \{p_4, p_5, p_6, p_8, p_{16}, p_{17}\}$, $S_4 = \{p_5, p_6, p_8, p_9, p_{16}, p_{17}\}$, $S_5 = \{p_2, p_6, p_8, p_9, p_{10}, p_{12}, p_{15}, p_{16}\}$, $S_6 = \{p_2, p_4, p_6, p_8 \, p_{10}, p_{15}, p_{16}, p_{17}\}$, $S_7 = \{p_2, p_6, p_8, p_9, p_{10}, p_{15}, p_{16}, p_{17}\}$, $S_8 = \{p_2, p_4, p_6, p_8, p_{10}, p_{12}, p_{15}, p_{16}\}$, $S_9 = \{p_7, p_9\}$, $S_{10} = \{p_1, p_2, p_3, p_4, p_7, p_8\}$, $S_{11} = \{p_3, p_5, p_{11}\}$, $S_{12} = \{p_{14}, p_{17}\}$, and $S_{13} = \{p_{10}, p_{11}, p_{12}, p_{13}, p_{14}, p_{15}, p_{16}\}$. Siphons S_1–S_7, create monitors V_{S1}- V_{S7}, respectively. Thus, seven monitors are added, V_{S1}- V_{S7}, which enforce:

$$M(p_4) + M(p_5) + M(p_6) + M(p_8) + M(p_{12}) + M(p_{16}) \geq 1 \tag{38}$$

$$M(p_5) + M(p_6) + M(p_8) + M(p_9) + M(p_{12}) + M(p_{16}) \geq 1 \tag{39}$$

$$M(p_4) + M(p_5) + M(p_6) + M(p_8) + M(p_{16}) + M(p_{17}) \geq 1 \tag{40}$$

$$M(p_5) + M(p_6) + M(p_8) + M(p_9) + M(p_{16}) + M(p_{17}) \geq 1 \tag{41}$$

$$M(p_2) + M(p_6) + M(p_8) + M(p_9) + M(p_{10}) + M(p_{12}) + M(p_{15}) + M(p_{16}) \geq 1 \tag{42}$$

$$M(p_2) + M(p_4) + M(p_6) + M(p_8) + M(p_{10}) + M(p_{15}) + M(p_{16}) + M(p_{17}) \geq 1 \tag{43}$$

$$M(p_2) + M(p_6) + M(p_8) + M(p_9) + M(p_{10}) + M(p_{15}) + M(p_{16}) + M(p_{17}) \geq 1 \tag{44}$$

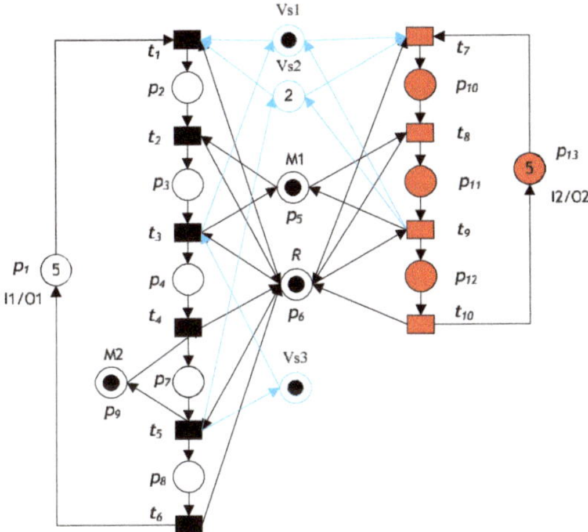

Figure 9. Controlled reconfigurable S^{3}PR net by addition of new product.

The following place invariants are accordingly generated:

$$M(V_{S1}) = M(p_4) + M(p_5) + M(p_6) + M(p_8) + M(p_{12}) + M(p_{16}) - 1 \tag{45}$$

$$M(V_{S2}) = M(p_5) + M(p_6) + M(p_8) + M(p_9) + M(p_{12}) + M(p_{16}) - 1 \tag{46}$$

$$M(V_{S3}) = M(p_4) + M(p_5) + M(p_6) + M(p_8) + M(p_{16}) + M(p_{17}) - 1 \tag{47}$$

$$M(V_{S4}) = M(p_5) + M(p_6) + M(p_8) + M(p_9) + M(p_{16}) + M(p_{17}) - 1 \tag{48}$$

$$M(V_{S5}) = M(p_2) + M(p_6) + M(p_8) + M(p_9) + M(p_{10}) + M(p_{12}) + M(p_{15}) + M(p_{16}) - 1 \tag{49}$$

$$M(V_{S6}) = M(p_2) + M(p_4) + M(p_6) + M(p_8) + M(p_{10}) + M(p_{15}) + M(p_{16}) + M(p_{17}) - 1 \tag{50}$$

$$M(V_{S7}) = M(p_2) + M(p_6) + M(p_8) + M(p_9) + M(p_{10}) + M(p_{15}) + M(p_{16}) + M(p_{17}) - 1 \tag{51}$$

The current matrices $\mathcal{L}$ and $\mathcal{B}$ represent the Equations (38)–(44).

$$\mathcal{L} = \begin{bmatrix}
0 & 0 & 0 & 1 & 1 & 1 & 0 & 1 & 0 & 0 & 0 & 1 & 0 & 0 & 0 & 1 & 0 \\
0 & 0 & 0 & 0 & 1 & 1 & 0 & 1 & 1 & 0 & 0 & 1 & 0 & 0 & 0 & 1 & 0 \\
0 & 0 & 0 & 1 & 1 & 1 & 0 & 1 & 0 & 0 & 0 & 0 & 0 & 0 & 0 & 1 & 1 \\
0 & 0 & 0 & 0 & 1 & 1 & 0 & 1 & 1 & 0 & 0 & 0 & 0 & 0 & 0 & 1 & 1 \\
0 & 1 & 0 & 0 & 0 & 1 & 0 & 1 & 1 & 1 & 0 & 1 & 0 & 0 & 1 & 1 & 0 \\
0 & 1 & 0 & 1 & 0 & 1 & 0 & 1 & 0 & 1 & 0 & 0 & 0 & 0 & 1 & 1 & 1 \\
0 & 1 & 0 & 0 & 0 & 1 & 0 & 1 & 1 & 1 & 0 & 0 & 0 & 0 & 1 & 1 & 1
\end{bmatrix}, \mathcal{B} = \begin{bmatrix} 1 \\ 1 \\ 1 \\ 1 \\ 1 \\ 1 \\ 1 \end{bmatrix} \tag{52}$$

The controller's net incidence matrix is calculated by Equation (12);

$$[N_{V_S}] = \begin{bmatrix} -1 & 0 & 1 & 0 & 0 & 0 & -1 & 0 & 1 & 0 & -1 & 0 & 0 & 0 \\ -1 & 0 & 0 & 0 & 1 & 0 & -1 & 0 & 1 & 0 & -1 & 0 & 0 & 0 \\ -1 & 0 & 1 & 0 & 0 & 0 & -1 & 0 & 0 & 0 & 0 & 1 & 0 & 0 \\ -1 & 0 & 0 & 0 & 1 & 0 & -1 & 0 & 0 & 0 & 0 & 1 & 0 & 0 \\ 0 & 0 & -1 & 0 & 1 & 0 & 0 & 0 & 0 & 0 & 0 & 0 & 0 & 0 \\ 0 & 0 & 0 & 0 & 0 & 0 & 0 & 0 & -1 & 0 & 1 & 1 & 0 & 0 \\ 0 & 0 & -1 & 0 & 1 & 0 & 0 & 0 & -1 & 0 & 1 & 1 & 0 & 0 \end{bmatrix} \tag{53}$$

The initial marking controllers are calculated as:

$M_o(V_{S1}) = 1$, $M_o(V_{S2}) = 2$, $M_o(V_{S3}) = 2$, $M_o(V_{S4}) = 3$, $M_o(V_{S5}) = 1$, $M_o(V_{S6}) = 1$, and $M_o(V_{S7}) = 2$.

The controlled reconfigurable net of (N_3, M_{30}) is illustrated in Figure 10. The place and arcs of the controller are shown with blue lines.

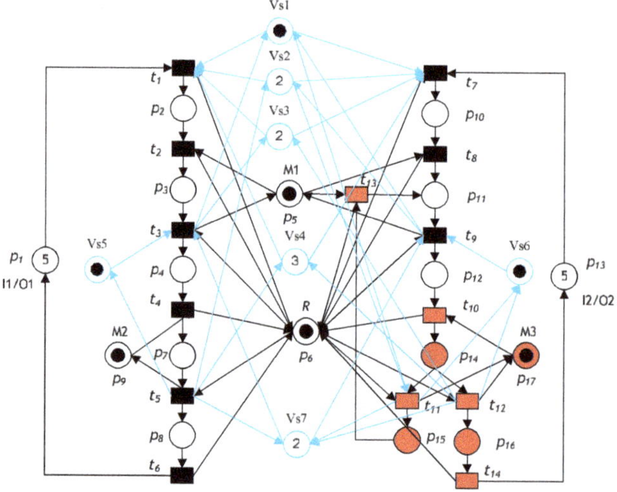

Figure 10. Controlled reconfigurable S³PR net by rework.

Finally, reconsider the reconfigured S³PR net by addition of a new robot (N_4, M_{40}) illustrated in Figure 6c. The reconfigured net has 17 minimal siphons, ten of which S_1–S_{10} that create monitors V_{S1}–V_{S10}, respectively, which are siphons $S_1 = \{p_2, p_6, p_{6\text{-}1}, p_8, p_9\}$, $S_2 = \{p_4, p_5, p_{6\text{-}1}, p_8, p_{11}\}$, $S_3 = \{p_5, p_{6\text{-}1}, p_8, p_9, p_{11}\}$, $S_4 = \{p_4, p_5, p_{6\text{-}1}, p_{6\text{-}2}, p_8, p_{12}, p_{16}\}$, $S_5 = \{p_4, p_5, p_{6\text{-}1}, p_{6\text{-}2}, p_8, p_{16}, p_{17}\}$, $S_6 = \{p_5, p_{6\text{-}1}, p_{6\text{-}2}, p_8, p_9, p_{12}, p_{16}\}$, $S_7 = \{p_5, p_{6\text{-}1}, p_{6\text{-}2}, p_8, p_9, p_{16}, p_{17}\}$, $S_8 = \{p_3, p_5, p_{6\text{-}2}, p_{12}, p_{16}\}$, $S_9 = \{p_3, p_5, p_{6\text{-}2}, p_{16}, p_{17},\}$, and $S_{10} = \{p_{6\text{-}2}, p_{10}, p_{15}, p_{16}, p_{17}\}$. The current matrices $\mathcal{L}$ and $\mathcal{B}$ are expressed as:

$$\mathcal{L} = \begin{bmatrix} 0 & 1 & 0 & 0 & 0 & 1 & 0 & 0 & 1 & 1 & 0 & 0 & 0 & 0 & 0 & 0 & 0 & 0 \\ 0 & 0 & 0 & 1 & 1 & 1 & 0 & 0 & 1 & 0 & 0 & 1 & 0 & 0 & 0 & 0 & 0 & 0 \\ 0 & 0 & 0 & 0 & 1 & 1 & 0 & 0 & 1 & 1 & 0 & 1 & 0 & 0 & 0 & 0 & 0 & 0 \\ 0 & 0 & 0 & 1 & 1 & 1 & 1 & 0 & 1 & 0 & 0 & 0 & 1 & 0 & 0 & 0 & 1 & 0 \\ 0 & 0 & 0 & 1 & 1 & 1 & 1 & 0 & 1 & 0 & 0 & 0 & 0 & 0 & 0 & 0 & 1 & 1 \\ 0 & 0 & 0 & 0 & 1 & 1 & 1 & 0 & 1 & 1 & 0 & 0 & 1 & 0 & 0 & 0 & 1 & 0 \\ 0 & 0 & 0 & 0 & 1 & 1 & 1 & 0 & 1 & 1 & 0 & 0 & 0 & 0 & 0 & 0 & 1 & 1 \\ 0 & 0 & 1 & 0 & 1 & 0 & 1 & 0 & 0 & 0 & 0 & 0 & 1 & 0 & 0 & 0 & 1 & 0 \\ 0 & 0 & 1 & 0 & 1 & 0 & 1 & 0 & 0 & 0 & 0 & 0 & 0 & 0 & 0 & 0 & 1 & 1 \\ 0 & 0 & 0 & 0 & 0 & 0 & 0 & 1 & 0 & 0 & 0 & 1 & 0 & 0 & 0 & 0 & 1 & 1 \end{bmatrix}, \mathcal{B} = \begin{bmatrix} 1 \\ 1 \\ 1 \\ 1 \\ 1 \\ 1 \\ 1 \\ 1 \\ 1 \\ 1 \end{bmatrix} \tag{54}$$

The controller's net incidence matrix is calculated by Equation (12);

$$
[N_{V_S}] =
\begin{bmatrix}
0 & 0 & -1 & 0 & 1 & 0 & 0 & 0 & 0 & 0 & 0 & 0 & 0 & 0 & 0 \\
-1 & 0 & 1 & 0 & 0 & 0 & 0 & 0 & 0 & 0 & 0 & 0 & 0 & 0 & 0 \\
-1 & 0 & 0 & 0 & 1 & 0 & 0 & 0 & 0 & 0 & 0 & 0 & 0 & 0 & 0 \\
-1 & 0 & 1 & 0 & 0 & 0 & -1 & 0 & 1 & 0 & -1 & 0 & 0 & 0 & 0 \\
-1 & 0 & 1 & 0 & 0 & 0 & -1 & 0 & 0 & 0 & 0 & 1 & 1 & 0 & 0 \\
-1 & 0 & 0 & 0 & 1 & 0 & -1 & 0 & 1 & 0 & -1 & 0 & 0 & 0 & 0 \\
-1 & 0 & 0 & 0 & 1 & 0 & -1 & 0 & 0 & 0 & 0 & 1 & 1 & 0 & 0 \\
0 & 0 & 0 & 0 & 0 & 0 & -1 & 0 & 1 & 0 & -1 & 0 & 0 & 0 & 0 \\
0 & 0 & 0 & 0 & 0 & 0 & -1 & 0 & 0 & 0 & 0 & 1 & 1 & 0 & 0 \\
0 & 0 & 0 & 0 & 0 & 0 & 0 & 0 & -1 & 0 & 1 & 1 & 1 & 0 & 0
\end{bmatrix}
\tag{55}
$$

The initial marking controllers are calculated as $M_0(V_{S1}) = 1$, $M_0(V_{S2}) = 1$, $M_0(V_{S3}) = 2$, $M_0(V_{S4}) = 2$, $M_0(V_{S5}) = 3$, $M_0(V_{S6}) = 3$, $M_0(V_{S7}) = 4$, $M_0(V_{S8}) = 1$, $M_0(V_{S9}) = 2$, and $M_0(V_{S10}) = 1$.

The controlled reconfigurable net of (N_4, M_{40}) is illustrated in Figure 11. The place and arcs of the controller are shown with blue lines.

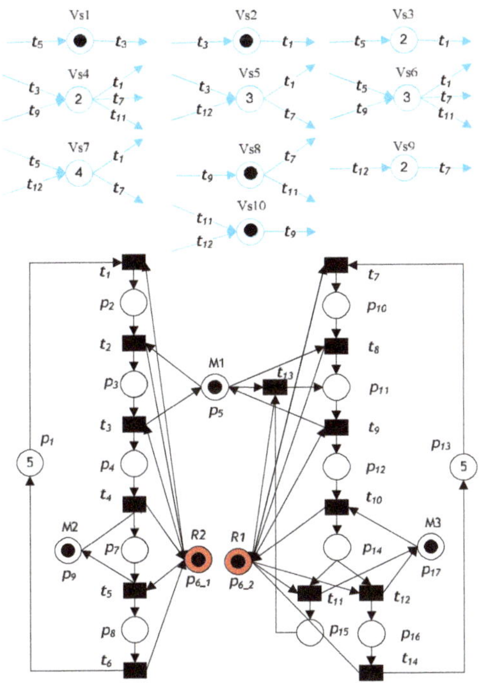

Figure 11. Controlled reconfigurable S^{3}PR net by rework.

4. Behavioral and Quantitative Analysis of Reconfigurable S^{3}PR Net

4.1. Liveness

Liveness is one of the most important issues in reconfigurable manufacturing systems with dynamic changes. Conversely, in these systems, deadlock is usually unwanted. When a system is not live, tasks could never be performed because of local or global deadlocks. Liveness of a transition means that, irrespective of the current state of the net, it can always eventually fire.

Theorem 7. *The controlled reconfigurable S^3PR net (N_{RC}, M_{RC0}) with $N_{RC} = (P_{RC}, T_{RC}, F_{RC}, W_{RC}, M_{RC0}, K_{RC})$ is live.*

Proof. All transitions T_{RC} in (N_{RC}, M_{RC0}) must be proven to be live. There is no unmarked siphon, $p \in S$. p is marked at marking M and satisfies $\sum_{p \in S} M(p) \geq 1$, since all $t \in T_{RC}$ are live. For all $t \in T_{RC}$, if for all $p \in {}^\bullet t$, $M_{RC0}(p) > 0$, then t can fire in any case. Therefore, the controlled reconfigurable S^3PR net (N_{RC}, M_{RC0}) is live. $\square$

To demonstrate the liveness of a reconfigurable S^3PR net, consider the model illustrated in Figure 9. Its reachability graph with all model markings is illustrated in Figure 12 and it is apparent that all transitions are live, which means that the system is live.

4.2. Boundedness

The boundedness is associated with a place, indicating that the number of tokens in a place never exceeds a certain number. This means that there is no overflow in a place.

Theorem 8. *Let a reconfigurable S^3PR net (N_{RC}, M_{RC0}) with $N_{RC} = (P_{RC}, T_{RC}, F_{RC}, W_{RC}, M_{RC0}, K_{RC})$ be a controlled net. Then (N_{RC}, M_{RC0}) is bounded.*

Proof. Theorem 7 proves that the net (N_{RC}, M_{RC0}) is live. Therefore, the boundedness can be established by checking if the net (N_{RC}, M_{RC0}) is well constructed, behaved, and controlled. The resulting net (N_{RC}, M_{RC0}) maintains the boundedness as the net is well constructed, behaved and has a finite reachability set. $\square$

To demonstrate the boundedness of a controlled reconfigurable S^3PR net, consider the net illustrated in Figure 9. Its reachability graph is illustrated in Figure 12. It is obvious that markings reachable from initial marking are five-bounded, which indicates that the system is bounded.

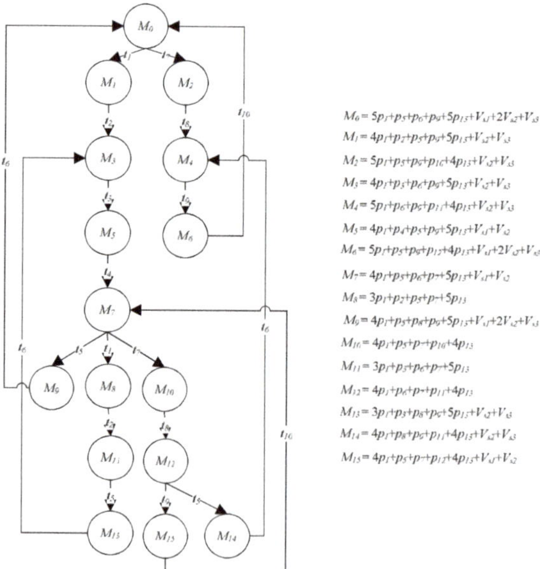

$$M_0 = 5p_1+p_5+p_6+p_8+5p_{15}+V_{s1}+2V_{s2}+V_{s3}$$
$$M_1 = 4p_1+p_2+p_5+p_8+5p_{15}+V_{s2}+V_{s3}$$
$$M_2 = 5p_1+p_5+p_8+p_{16}+4p_{13}+V_{s2}+V_{s3}$$
$$M_3 = 4p_1+p_5+p_6+p_8+5p_{13}+V_{s2}+V_{s3}$$
$$M_4 = 5p_1+p_6+p_8+p_{11}+4p_{13}+V_{s2}+V_{s3}$$
$$M_5 = 4p_1+p_4+p_5+p_8+5p_{13}+V_{s1}+V_{s2}$$
$$M_6 = 5p_1+p_5+p_6+p_{12}+4p_{13}+V_{s1}+2V_{s2}+V_{s3}$$
$$M_7 = 4p_1+p_5+p_6+p_7+5p_{13}+V_{s1}+V_{s2}$$
$$M_8 = 3p_1+p_2+p_5+p_7+5p_{13}$$
$$M_9 = 4p_1+p_5+p_8+p_9+5p_{15}+V_{s1}+2V_{s2}+V_{s3}$$
$$M_{10} = 4p_1+p_5+p_7+p_{16}+4p_{13}$$
$$M_{11} = 3p_1+p_5+p_6+p_7+5p_{13}$$
$$M_{12} = 4p_1+p_6+p_7+p_{11}+4p_{13}$$
$$M_{13} = 3p_1+p_5+p_8+p_6+5p_{15}+V_{s2}+V_{s3}$$
$$M_{14} = 4p_1+p_8+p_6+p_{11}+4p_{15}+V_{s2}+V_{s3}$$
$$M_{15} = 4p_1+p_5+p_7+p_{12}+4p_{15}+V_{s1}+V_{s2}$$

Figure 12. Reachable markings of a controlled reconfigurable S^3PR net, as illustrated in Figure 8.

4.3. Reversibility

Reversibility means that a system can always return to its initial marking. A controlled reconfigurable S^{3}PR Petri net model (N_{RC}, M_{RC_0}) is reversible if for each marking $M \in R(N_{RC}, M_{RC_0})$, initial marking M_{RC_0} is reachable from M.

Theorem 9. *Let a reconfigurable S^{3}PR net* (N_{RC}, M_{RC_0}) *with* N_{RC} = (P_{RC}, T_{RC}, F_{RC}, W_{RC}, M_{RC_0}, K_{RC}) *be a live and controlled net. N_{RC} is reversible if for each marking* $M \in R(N_{RC}, M_{RC_0})$, *initial marking M_{RC_0} is reachable from M, M and M_{RC_0} satisfying all place invariants and M marks each trap of N_{RC}.*

Proof. Suppose that M is reachable. Then there exists a finite transition sequence $\delta = t_1\, t_2\, t_3 \cdots t_n$ that can be fired, and markings M_1, M_2, M_3, ... , and M_{n-1} are such that $M_{RC_0}[t_1\rangle M_1[t_2\rangle M_2[t_3\rangle M_3 \cdots M_{n-1}[t_n\rangle M$, expressed as $M_{RC_0}[\delta\rangle M$, agrees with the state equation $M = M_{RC_0} + [N_{RC}]\, \vec{\delta}$. In addition, M and M_{RC_0} satisfy all place invariants, $I^T.M = I^T.M_{RC_0}$. Therefore, we can say that M_{RC_0} is the home marking of the net (N_{RC}, M_{RC_0}), M is reachable from M_{RC_0}, and we get $M_{RC_0} \xrightarrow{\vec{\delta}} M$. Thus, the reconfigurable S^{3}PR net (N_{RC}, M_{RC_0}) is reversible. $\square$

To demonstrate the reversibility of a controlled reconfigurable S^{3}PR net, consider the model illustrated in Figure 8. Its reachability graph is illustrated in Figure 13. In the net shown in Figure 8, there are seven minimal place invariants: $I_1 = p_3 + p_5$, $I_2 = p_2 + p_3 + p_{10}$, $I_3 = p_7 + p_9$, $I_4 = p_4 + p_7 + p_{11}$, $I_5 = p_2 + p_3 + p_4 + p_7 + p_{12}$, $I_6 = p_2 + p_4 + p_6 + p_8$, $I_7 = p_1 + p_2 + p_3 + p_4 + p_7 + p_8$, since $\forall i \in \{1,2,3,4,5,6,7\}$, $I_i^T.[N_{RC}] = 0^T$. $M_6 \in R(N_{RC}, M_{RC_0})$, $I_1^T.M_6 = I_1^T.M_{RC_0} = M_6(p_3) + M_6(p_5) = M_{RC_0}(p_3) + M_{RC_0}(p_5) = 1$. The net has a unique T-invariant $J = t_1 + t_2 + t_3 + t_4 + t_5 + t_6$ and the transition sequence $\delta = t_1 t_2 t_3 t_4 t_5 t_6$ is firable. As a result, $M_{RC_0}[t_1\rangle M_1[t_2\rangle M_2[t_3\rangle M_3[t_4\rangle M4[t_5\rangle M6[t_6\rangle M_{RC_0}$. Therefore, the reconfigurable S^{3}PR net (N_{RC}, M_{RC_0}) is live, bounded, and reversible.

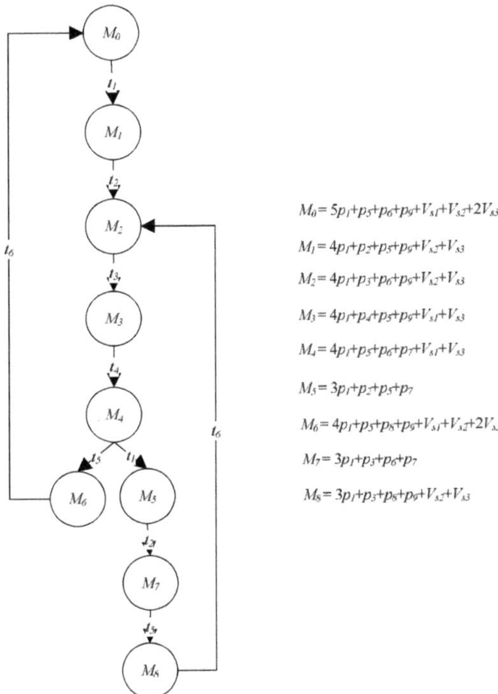

Figure 13. Reachable markings of a controlled reconfigurable S^{3}PR net, as illustrated in Figure 7.

4.4. Computational Complexity

Algorithm 1 is used to design a reconfigurable S^3PR net with $N_R = ((N, M_o), \mathcal{R})$. In addition, Algorithm 2 computes the control places to a reconfigurable S^3PR net with $N_R = ((N, M_o), \mathcal{R})$.

Theorem 10. *Given a reconfigurable S^3PR net with $N_R = ((N, M_o), \mathcal{R})$, where N_R with $N_k = (P_{Ck}, T_k, F_k, W_k, M_{ko}, K_k)$, the time complexity of Algorithm 1 is polynomial.*

Proof. Let N_k be states in N_R with $N_k = (P_{Ck}, T_k, F_k, W_k, M_{ko}, K_k)$, $\mathcal{R} = \{rr_1, rr_2, rr_3, \ldots, rr_k\}$, $rr_k = \{L_k, R_k, \varphi_k, {}^\bullet\varphi_k, \varphi_k{}^\bullet\}$, and net (N_R, M_{Ro}) be the obtained reconfigurable S^3PR net. Let x be the cardinality of $\mathcal{R}$, i.e., $|\mathcal{R}| = x$. The "While" loop is executed x times to design state N_k in a reconfigurable S^3PR net (N_R, M_{Ro}). Therefore, in the worst case, the computational complexity of algorithm 1 is $O(x)$. Thus, the computational complexity of the Algorithm 1 has polynomial time complexity. $\square$

Theorem 11. *Given a reconfigurable S^3PR net with $N_R = ((N, M_o), \mathcal{R})$, where N_R with $N_k = (P_{Ck}, T_k, F_k, W_k, M_{ko}, K_k)$, the time complexity of Algorithm 2 is polynomial.*

Proof. Algorithm 2 is used to design a control place V_S to each minimal siphon S, $V_S{}^\bullet \subsetneq {}^\bullet S$ in each state N_k in a reconfigurable S^3PR net (N_R, M_{Ro}) to achieve the liveness of net (N_R, M_{Ro}). Obviously, each V_S is associated with the minimal siphon S in net (N_k, M_{ko}). Let x be the cardinality of $\mathcal{R}$, i.e., $|\mathcal{R}| = x$. Let y be the number of minimal siphons S (denoted as S') that requires V_S i.e., $|S'| = y$. The "While" loop is executed x times to design state N_k in reconfigurable S^3PR net (N_R, M_{Ro}). The "FOR loop" loop is executed y times to design V_S for the S' in (N_k, M_{ko}). Therefore, the computational complexity of Algorithm 2 is $O(xy)$. Thus, the computational complexity of the Algorithm 2 has polynomial time complexity. $\square$

4.5. GPENSIM Code and Validation

We coded the developed approach using the GPenSIM tool [6,42] to verify and validate it and compared the developed code with the studies by Ezpeleta et al. [43], Li and Zhou [44], and Kaid et al. [6]. There were three files generated: (1) the Petri net definition file (PDF) that represents the static model by stating the sets of places, transitions, and arcs, (2) the common processor file (COMMON_PRE file) that represents the conditions for activation of the enabling fire transitions, and (3) the main simulation file (MSF) that calculates the results of the simulation. The developed approach was implemented on MATLAB R2015a. A PC with Windows 10, 64-bit and Intel(R) Core (TM) i7-4702MQ CPU @ 2.20 GHz, 16 GB RAM.

Simulation leads to a better time performance in the designed model including total throughput time (total time in system), total throughput, and utilization of the robots and machines. Consider the model illustrated in Figure 8. The simulation was undertaken for 480 min. The results summarized in Table 1 were obtained after simulation in MATLAB. Table 1 shows the results for the time performance criteria mentioned above. All methods achieve approximately the same values for the utilization of resources as illustrated in Figure 14. In addition, the proposed method, as illustrated in Figure 14, can achieve approximately the same values with other techniques for throughput. In term of throughput time of Part A, the proposed method can achieve approximately the same values with other techniques as illustrated in Figure 14. Therefore, the proposed method is valid, sufficiently accurate results can be obtained and other cases can be applied.

Table 1. Time performance comparison with the existing methods.

Performance	Ezpeleta et al. [43]	Li and Zhou [44]	Kaid et al. [6]	The Proposed Method
M1 utilization (%)	29.05	29.05	29.60	29.05
M2 utilization (%)	29.61	29.61	30.50	29.61
R1 utilization (%)	48.04	48.04	47.56	48.04
Throughput (parts)	34	34	34	34
Throughput time (min/part)	14.12	14.12	14.12	14.12

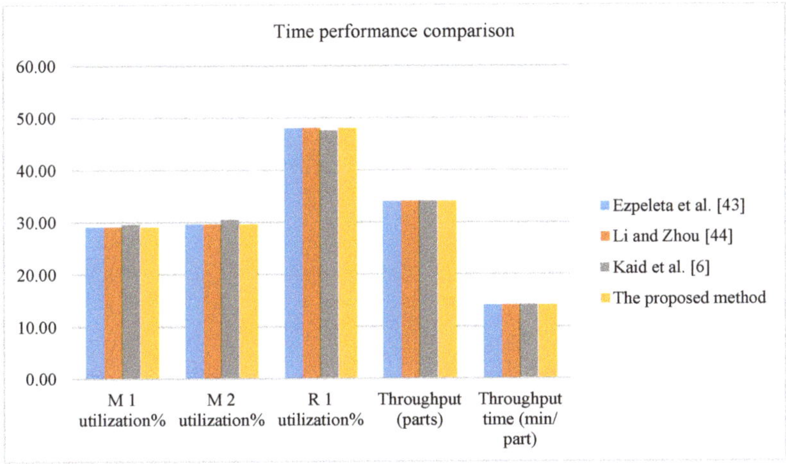

Figure 14. Comparison of the proposed method with the existing methods.

5. Numerical Example

In this section, an example is used to present the application of the proposed approach. Consider an AMS example illustrated in Figure 15a. Its Petri net model is given in [6,7,15,22,45,46]. The system consists of four machines M1–M4 for processing parts; two robots R1 and R2 for loading and unloading parts. Each machine (robot) can process (hold) one part at a time. There are two input buffers I1 and I2 and two output buffers O1 and O2. Two raw part types, A and B, are considered to be processed in the system. Figure 15b shows the operation sequences of the two raw part types. The S^3PR net of this AMS example is illustrated in Figure 16. It comprises 19 places and 14 transitions. The places can be defined as the following set partitions: $P_A = \{p_2, p_3, \ldots, p_{12}\}$, $P_R = \{p_{13}, p_{14}, \ldots, p_{18}\}$, and $P^0 = \{p_1, p_{19}\}$. The S^3PR net contains 282 reachable markings.

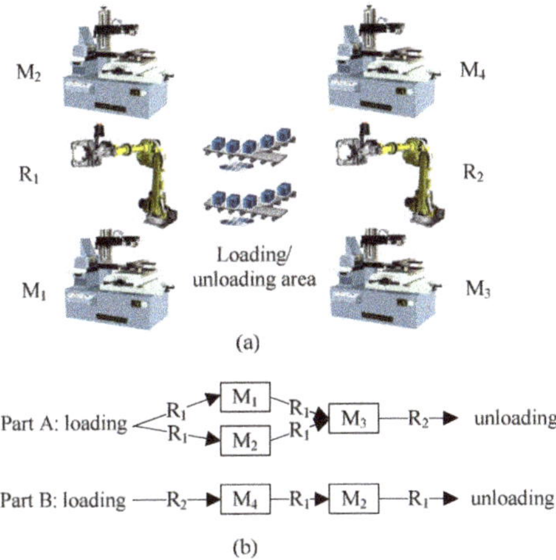

Figure 15. (**a**) An AMS example and (**b**) production sequence.

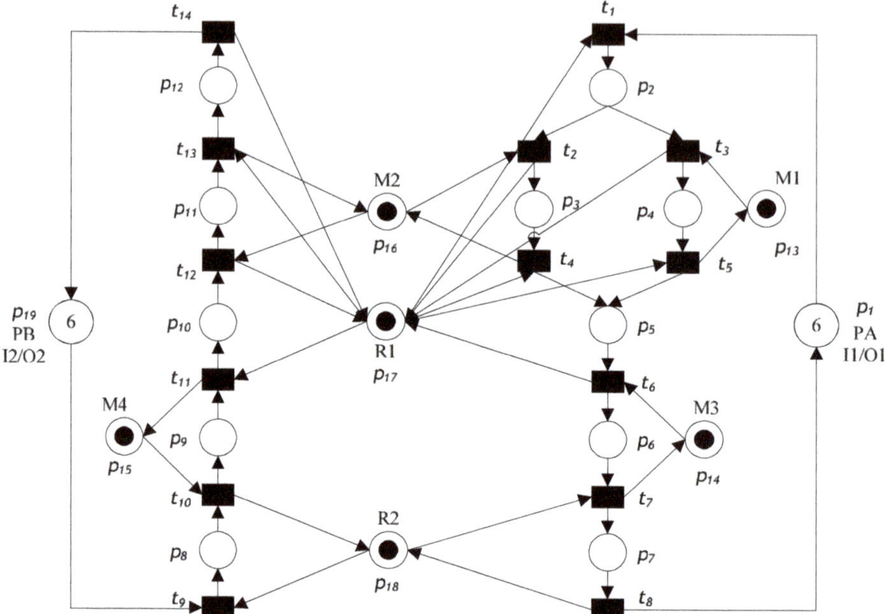

Figure 16. S^3PR net (N_o, M_o) of the AMS illustrated in Figure 13a.

Suppose that the first configuration of the system involves removing old machine. In this case, an old machine M1 is removed from the system (N_o, M_o). To model the removed machine by using the synthesis procedure of Algorithm 1, we construct a configuration as a rewriting rule $\mathcal{R} = \{rr_1\}$ with $rr_1 = \{L_1, R_1, \varphi_1, {}^{\bullet}\varphi_1, \varphi_1{}^{\bullet}\}$, where L_1 and R_1 are illustrated in Figures 17a and 17b, respectively. In addition, we have $\xi_1: N_1 \rightarrow N_o$, $\varphi_1 = (\{p_2, p_3, p_4, p_5, p_{13}, p_{16}, p_{17}\}, \{t_1, t_2, t_3, t_4, t_5, t_6\})$, ${}^{\bullet}\varphi_1 = (\{L_1.t_1\}, \{R_1.t_1\})$, and $\varphi_1{}^{\bullet} = (\{L_1.t_6\}, \{R_1.t_6\})$. The second configuration includes adding new product. If a new product (part C) is assigned to a system, which indicates that a new operation sequence is assigned and the system requires an adjustment to its Petri net model structure. To model the addition of new product by using the synthesis procedure of Algorithm 1, we construct a configuration as a rewriting rule $\mathcal{R} = \{rr_2\}$ with $rr_2 = \{L_2, R_1, \varphi_2, {}^{\bullet}\varphi_2, \varphi_2{}^{\bullet}\}$, where L_2 and R_2 are illustrated in Figures 18a and 18b, respectively. Moreover, we have $\xi_2: N_2 \rightarrow N_1$, $\varphi_2 = (\{p_{15}, p_{17}, p_{20}, p_{21}, p_{22}, p_{23}\}, \{t_{15}, t_{16}, t_{17}, t_{18}\})$, ${}^{\bullet}\varphi_2 = (\{L_2.p_{15}, L_2.p_{17}\}, \{R_2.t_{15}\})$, and $\varphi_2{}^{\bullet} = (\{L_2.p_{15}, L_2.p_{17}\}, \{R_2.t_{18}\})$.

The third system configuration involves rework. In this scenario, a part can be inspected after all operations have been completed. By using the proposed Algorithm 1, the production operations of the reworked part can be exactly and easily modeled by considering rework operations as alternative sequences. Suppose that an inspection machine M5 is added to a system and that part A is processed in M1 and M3. Then, part A is moved to an M5 by Robot 2 to check if there are defects in part A. If part A performs properly, then it will leave the system by Robot 2. Otherwise, if part A has defects, rework is needed, and part A is moved to M3 by Robot 2. To model the rework operation by using the synthesis procedure of Algorithm 1, we construct a configuration as a rewriting rule $\mathcal{R} = \{rr_3\}$ with $rr_3 = \{L_3, R_3, \varphi_3, {}^{\bullet}\varphi_3, \varphi_3{}^{\bullet}\}$, where L_3 and R_3 are illustrated in Figures 19a and 19b, respectively, $\xi_3: N_3 \rightarrow N_2$, $\varphi_3 = (\{p_6, p_7, p_{14}, p_{18}, p_{24}, p_{25}, p_{26}, p_{27}\}, \{t_6, t_7, t_8, t_{19}, t_{20}, t_{21}, t_{22}\})$, ${}^{\bullet}\varphi_3 = (\{L_3.t_6\}, \{R_3.t_6\})$, and $\varphi_3{}^{\bullet} = (\{L_3.t_8\}, \{R_3.t_{21}\})$. The Specifications of S^3PR net illustrated in Figure 16 under changeable control specifications are shown in Table 2. In addition, the required monitors using Algorithm 2 of the system illustrated in Figure 16 under changeable control specifications are shown in Table 3.

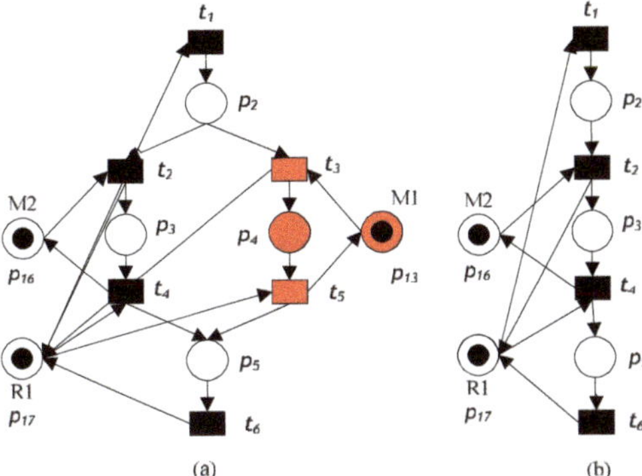

Figure 17. A reconfigured S^{3}PR net by removing a machine. (**a**) Left hand side net *L*. (**b**) Right hand side net *R*.

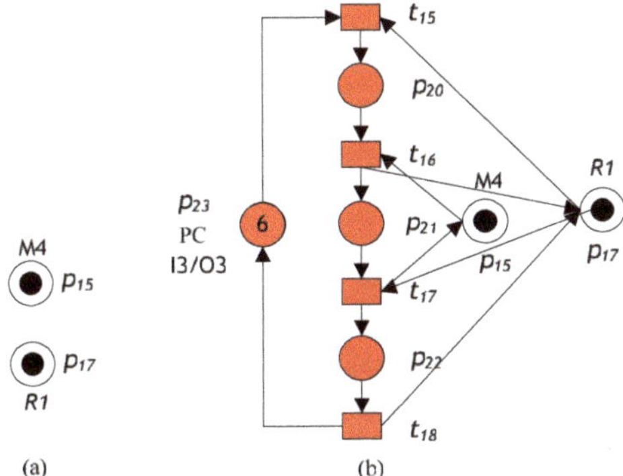

Figure 18. A reconfigured S^{3}PR net by adding a product. (**a**) Left hand side net *L*. (**b**) Right hand side net *R*.

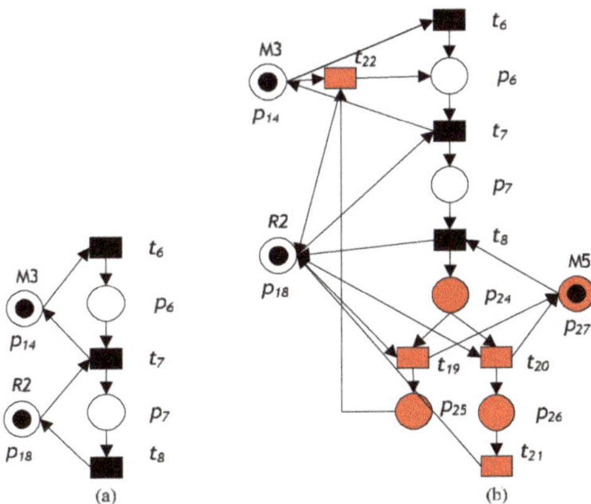

Figure 19. A reconfigured S^{3}PR net by rework. (**a**) Left hand side net *L*. (**b**) Right hand side net *R*.

Table 2. The Specifications of S^{3}PR net illustrated in Figure 16 under configurations.

Parameter	Configuration			
	An Initial S^{3}PR Net	**Removal of an Old Machine**	**Addition of a New Product**	**Rework**
No. of monitors	5	3	5	10
No. of arcs	21	12	26	48
Liveness	Live	Live	Live	Live
Boundedness	Bounded	Bounded	Bounded	Bounded
Reversibility	Reversible	Reversible	Reversible	Reversible

Table 3. Required monitors using Algorithm 2 of the system illustrated in Figure 16 under configurations.

Configuration	i	Siphon	$^\bullet V_{Si}$	$V_{Si}^\bullet$	$M_{RCo}(V_{si})$
An initial S^{3}PR net	1	S_1	t_7,t_{13}	t_1,t_9	5
	2	S_2	t_4,t_5,t_{13}	t_1,t_{11}	2
	3	S_3	t_7,t_{13}	t_1,t_9	4
	4	S_4	t_7,t_{11}	t_1,t_9	3
	5	S_5	t_4,t_{13}	t_2,t_{11}	1
Removal of an old machine	1	S_1	t_4,t_{13}	t_1,t_9	1
	2	S_2	t_7,t_{11}	t_4,t_9	3
	3	S_3	t_7,t_{13}	t_1,t_9	4
Addition of a new product	1	S_1	t_{11},t_{17}	t_{10},t_{15}	1
	2	S_2	t_4,t_{13},t_{17}	t_1,t_{10},t_{15}	2
	3	S_3	t_4,t_{13}	t_1,t_9	1
	4	S_4	t_7,t_{11},t_{17}	t_4,t_9,t_{15}	3
	5	S_5	t_7,t_{13},t_{17}	t_1,t_9,t_{15}	4
Rework	1	S_1	t_7	t_6,t_{19}	1
	2	S_2	t_{20}	t_6	2
	3	S_3	t_6,t_{20}	t_7	1
	4	S_4	t_7,t_{11},t_{17}	t_4,t_9,t_{15},t_{19}	3
	5	S_5	t_7,t_{13},t_{17}	t_1,t_9,t_{15},t_{19}	4
	6	S_6	t_{11},t_{17},t_{20}	t_4,t_9,t_{15}	4
	7	S_7	t_{13},t_{17},t_{20}	t_1,t_9,t_{15}	5
	8	S_8	t_{11},t_{17}	t_8,t_{15}	1
	9	S_9	t_4,t_{13},t_{17}	t_1,t_{10},t_{15}	2
	10	S_{10}	t_4,t_{13}	t_1,t_9	1

The controlled net after adding above changeable control specifications is illustrated in Figure 20. The place and arcs of the controller are illustrated with blue lines.

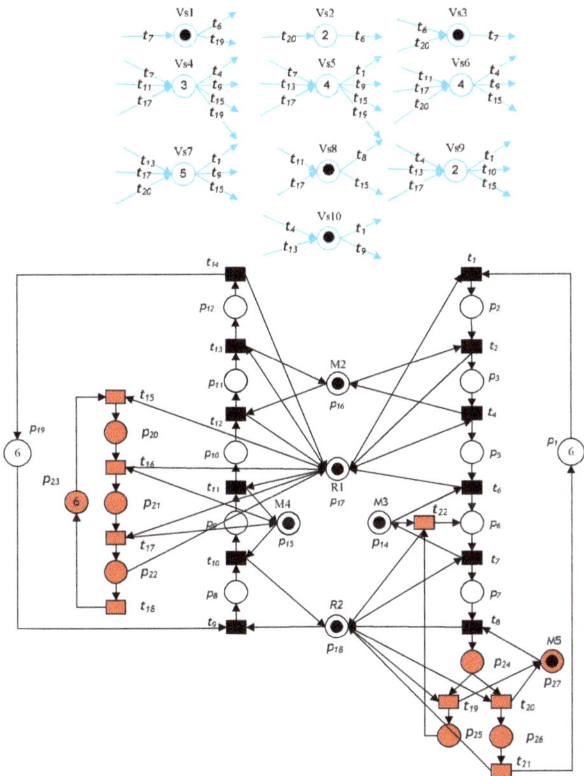

Figure 20. Controlled reconfigurable S^{3}PR net after adding changeable control specifications.

6. Conclusions

This paper develops a novel two-step solution for quick and accurate reconfiguration of supervisory controllers for deadlock control in RMSs with dynamic changes. In the first step, the net rewriting system is used to design a reconfigurable PN model under dynamic configurations. The obtained model guarantees boundedness behavioral property but may not guarantee the other properties of a Petri net model (i.e., liveness and reversibility). The second step proposes an automatic deadlock prevention policy for reconfigurable Petri net using the siphon control method based on a place invariant to solve the deadlock problem with dynamic structure changes in RMSs and guarantee the liveness and reversibility properties for the system. The proposed method is validated using the GPenSIM tool and compared with existing methods in the literature to highlight its ability of adapting to RMS configuration changes.

The major advantages of the developed approach are as follows: (1) It does not need to compute reachability graphs as illustrated in Algorithm 2, Section 3, and has low-overhead computation as proved in Theorems 10 and 11, Section 4.4. (2) It can automatically and dynamically modify the structure of a Petri net model without affecting its behavioral properties, i.e., liveness, boundedness, and reversibility as illustrated in Algorithm 2, Section 3. (3) It allows rapid reconfigurability and reusability of the controller during reconfiguration as shown in Algorithm 2, Section 3. (4) It can easily handle any dynamical changes in RMSs compared with the studies in Badouel et al. [39], Llorens and Oliver [34], Wu and Zhou [25], and Kaid et al. [7] as shown in Algorithm 2, Section 3. (5) The GPenSIM code is developed for designing, simulation, validation, and performance analysis of deadlock problems with dynamic structure changes in RMSs and the correctness of the proposed approach is proven and compared with the studies in Ezpeleta et al. [43], Li and Zhou [44], and Kaid et al. [6] as shown in Section 4.5. (6) Based on Theorems 10 and 11, the computational complexity of the proposed approach has polynomial time complexity. Therefore, it has low computational complexity and can be applicable to other types of complex systems such as mass customization manufacturing, lean productivity, agile manufacturing, and flexible manufacturing systems. (7) It can consider systems with sequential and complex resource requirements, meaning that a set of system resources can be used and shared to process each component according to sequential processes that depend on the step-by-step discrete execution and multiple processes that depend on the execution at the same time as shown in numerical example.

The limitation of the developed approach is that the obtained models lack an appropriate conversion approach from the PN model into control languages for application. Thus, our future research will examine the developed approach to have an automatic method to examine the applicability of the obtained models for real world manufacturing systems.

Author Contributions: Conceptualization, H.K. and A.A.-A.; software, H.K. and R.D.; resources, H.K., A.A.-A. and R.D.; formal analysis, H.K. and A.A.-A.; investigation, H.K., A.A.-A. and Z.L.; validation, H.K., A.A.-A., Z.L. and R.D.; writing—original draft preparation, H.K., A.A.-A. and Z.L.; writing—review and editing, H.K., A.A.-A., Z.L. and R.D.; visualization, H.K., A.A.-A. and Z.L.; supervision, A.A.-A., and Z.L. All authors have read and agreed to the published version of the manuscript.

Funding: This research was funded by King Saud University through Researchers Supporting Project Number (RSP-2020/62).

Acknowledgments: The authors would like to thank King Saud University for funding and supporting this research through Researchers Supporting Project Number (RSP-2020/62).

Conflicts of Interest: The authors declare no conflict of interest.

References

1. Hu, Y.; Ma, Z.; Li, Z. Design of Supervisors for Active Diagnosis in Discrete Event Systems. *IEEE Trans. Autom. Control* **2020**. [CrossRef]
2. Wang, D.; Wang, X.; Li, Z. Nonblocking Supervisory Control of State-Tree Structures with Conditional-Preemption Matrices. *IEEE Trans. Ind. Inform.* **2019**, *16*, 3744–3766. [CrossRef]
3. Mehrabi, M.G.; Ulsoy, A.G.; Koren, Y. Reconfigurable manufacturing systems: Key to future manufacturing. *J. Intell. Manuf.* **2000**, *11*, 403–419. [CrossRef]
4. Katz, R. Design principles of reconfigurable machines. *Int. J. Adv. Manuf. Technol.* **2007**, *34*, 430–439. [CrossRef]
5. Patel, R.; Gojiya, A.; Deb, D. Failure Reconfiguration of Pumps in Two Reservoirs Connected to Overhead Tank. In *Innovations in Infrastructure*; Springer: Berlin/Heidelberg, Germany, 2019; pp. 81–92.
6. Kaid, H.; Al-Ahmari, A.; Li, Z.; Davidrajuh, R. Single controller-based colored Petri nets for deadlock control in automated manufacturing systems. *Processes* **2020**, *8*, 21. [CrossRef]
7. Kaid, H.; Al-Ahmari, A.; Li, Z.; Davidrajuh, R. Intelligent colored token Petri nets for modeling, control, and validation of dynamic changes in reconfigurable manufacturing systems. *Processes* **2020**, *8*, 358. [CrossRef]
8. Li, Z.; Zhou, M. *Deadlock Resolution in Automated Manufacturing Systems: A Novel Petri Net Approach*; Springer Science & Business Media: Berlin/Heidelberg, Germany, 2009.
9. Zan, X.; Wu, Z.; Guo, C.; Yu, Z. A Pareto-based genetic algorithm for multi-objective scheduling of automated manufacturing systems. *Adv. Mech. Eng.* **2020**, *12*, 1687814019885294. [CrossRef]
10. Li, L.; Basile, F.; Li, Z. An approach to improve permissiveness of supervisors for GMECs in time Petri net systems. *IEEE Trans. Autom. Control* **2019**, *65*, 237–251. [CrossRef]
11. Liu, Y.; Cai, K.; Li, Z. On scalable supervisory control of multi-agent discrete-event systems. *Automatica* **2019**, *108*, 108460. [CrossRef]
12. Chen, Q.; Yin, L.; Wu, N.; El-Meligy, M.A.; Sharaf, M.A.F.; Li, Z. Diagnosability of vector discrete-event systems using predicates. *IEEE Access* **2019**, *7*, 147143–147155. [CrossRef]

13. Kaid, H.; Al-Ahmari, A.; Nasr, E.A.; Al-Shayea, A.; Kamrani, A.K.; Noman, M.A.; Mahmoud, H.A. Petri Net Model Based on Neural Network for Deadlock Control and Fault Detection and Treatment in Automated Manufacturing Systems. *IEEE Access* **2020**, *8*, 103219–103235. [CrossRef]
14. Zhao, M. An integrated control method for designing non-blocking supervisors using Petri nets. *Adv. Mech. Eng.* **2017**, *9*, 1687814017700829. [CrossRef]
15. Chen, Y.; Li, Z.; Khalgui, M.; Mosbahi, O. Design of a maximally permissive liveness-enforcing Petri net supervisor for flexible manufacturing systems. *Autom. Sci. Eng. IEEE Trans.* **2011**, *8*, 374–393. [CrossRef]
16. Al-Ahmari, A.; Kaid, H.; Li, Z.; Davidrajuh, R. Strict Minimal Siphon-Based Colored Petri Net Supervisor Synthesis for Automated Manufacturing Systems with Unreliable Resources. *IEEE Access* **2020**. [CrossRef]
17. Kaid, H.; Al-Ahmari, A.; El-Tamimi, A.M.; Abouel Nasr, E.; Li, Z. Design and implementation of deadlock control for automated manufacturing systems. *South Afr. J. Ind. Eng.* **2019**, *30*, 1–23. [CrossRef]
18. Chao, D.Y. Improvement of suboptimal siphon-and FBM-based control model of a well-known S^{3}PR. *IEEE Trans. Autom. Sci. Eng.* **2011**, *8*, 404–411. [CrossRef]
19. Ghaffari, A.; Rezg, N.; Xie, X. Design of a live and maximally permissive Petri net controller using the theory of regions. *IEEE Trans. Robot. Autom.* **2003**, *19*, 137–141. [CrossRef]
20. Uzam, M. The use of the Petri net reduction approach for an optimal deadlock prevention policy for flexible manufacturing systems. *Int. J. Adv. Manuf. Technol.* **2004**, *23*, 204–219. [CrossRef]
21. Sun, D.; Chen, Y.; El-Meligy, M.A.; Sharaf, M.A.F.; Wu, N.; Li, Z. On algebraic identification of critical states for deadlock control in automated manufacturing systems modeled with Petri nets. *IEEE Access* **2019**, *7*, 121332–121349. [CrossRef]
22. Nasr, E.A.; El-Tamimi, A.M.; Al-Ahmari, A.; Kaid, H. Comparison and Evaluation of Deadlock Prevention Methods for Different Size Automated Manufacturing Systems. *Math. Probl. Eng.* **2015**, *501*, 1–19. [CrossRef]
23. Lee, S.; Tilbury, D.M. Deadlock-free resource allocation control for a reconfigurable manufacturing system with serial and parallel configuration. *IEEE Trans. Syst. ManCybern. Part C (Appl. Rev.)* **2007**, *37*, 1373–1381. [CrossRef]
24. Li, J.; Dai, X.; Meng, Z. Automatic reconfiguration of petri net controllers for reconfigurable manufacturing systems with an improved net rewriting system-based approach. *IEEE Trans. Autom. Sci. Eng.* **2009**, *6*, 156–167. [CrossRef]
25. Wu, N.; Zhou, M. Intelligent token Petri nets for modelling and control of reconfigurable automated manufacturing systems with dynamical changes. *Trans. Inst. Meas. Control* **2011**, *33*, 9–29.
26. Sampath, R.; Darabi, H.; Buy, U.; Liu, J. Control reconfiguration of discrete event systems with dynamic control specifications. *IEEE Trans. Autom. Sci. Eng.* **2008**, *5*, 84–100. [CrossRef]
27. Dumitrache, I.; Caramihai, S.; Stanescu, A. Intelligent agent-based control systems in manufacturing. In Proceedings of the 2000 IEEE International Symposium on Intelligent Control. Held jointly with the 8th IEEE Mediterranean Conference on Control and Automation (Cat. No. 00CH37147), Rio Patras, Greece, 19–19 July 2000; pp. 369–374.
28. Ohashi, K.; Shin, K.G. Model-based control for reconfigurable manufacturing systems. In Proceedings of the 2001 ICRA. IEEE International Conference on Robotics and Automation (Cat. No. 01CH37164), Seoul, Korea, 21–26 May 2001; pp. 553–558.
29. Kalita, D.; Khargonekar, P.P. Formal verification for analysis and design of logic controllers for reconfigurable machining systems. *IEEE Trans. Robot. Autom.* **2002**, *18*, 463–474. [CrossRef]
30. Almeida, E.E.; Luntz, J.E.; Tilbury, D.M. Event-condition-action systems for reconfigurable logic control. *IEEE Trans. Autom. Sci. Eng.* **2007**, *4*, 167–181. [CrossRef]
31. Zhang, L.; Rodrigues, B. Modelling reconfigurable manufacturing systems with coloured timed Petri nets. *Int. J. Prod. Res.* **2009**, *47*, 4569–4591. [CrossRef]
32. Kahloul, L.; Bourekkache, S.; Djouani, K.; Chaoui, A.; Kazar, O. Using high level Petri nets in the modelling, simulation and verification of reconfigurable manufacturing systems. *Int. J. Softw. Eng. Knowl. Eng.* **2014**, *24*, 419–443. [CrossRef]
33. Yu, Z.; Guo, F.; Ouyang, J.; Zhou, L. Object-oriented Petri nets and π-calculus-based modeling and analysis of reconfigurable manufacturing systems. *Adv. Mech. Eng.* **2016**, *8*, 1687814016677698. [CrossRef]
34. Llorens, M.; Oliver, J. Structural and dynamic changes in concurrent systems: Reconfigurable Petri nets. *IEEE Trans. Comput.* **2004**, *53*, 1147–1158. [CrossRef]

35. Li, J.; Dai, X.; Meng, Z. Improved net rewriting systems-based rapid reconfiguration of Petri net logic controllers. In Proceedings of the 31st Annual Conference of IEEE Industrial Electronics Society, Raleigh, NC, USA, 6–10 November 2005.

36. da Silva, R.M.; Benítez-Pina, I.F.; Blos, M.F.; Santos Filho, D.J.; Miyagi, P.E. Modeling of reconfigurable distributed manufacturing control systems. *IFAC-Pap.* **2015**, *48*, 1284–1289. [CrossRef]

37. Berthomieu, B.; Ribet, P.-O.; Vernadat, F. The tool TINA–construction of abstract state spaces for Petri nets and time Petri nets. *Int. J. Prod. Res.* **2004**, *42*, 2741–2756. [CrossRef]

38. Bonet, P.; Catalina, M.L.; Puigjaner, R. A Petri Net tool for performance modeling. In Proceedings of the 23rd Latin American Conference on Informatics (CLEI 2007), San Jose, Costa Rica, October 2007.

39. Badouel, E.; Llorens, M.; Oliver, J. Modeling concurrent systems: Reconfigurable nets. In Proceedings of the PDPTA, Las Vegas, NV, USA, 23–26 June 2003; pp. 1568–1574.

40. Yamalidou, K.; Moody, J.; Lemmon, M.; Antsaklis, P. Feedback control of Petri nets based on place invariants. *Automatica* **1996**, *32*, 15–28. [CrossRef]

41. Moody, J.O.; Antsaklis, P.J. *Supervisory Control of Discrete Event Systems Using Petri Nets*; Springer Science & Business Media: Berlin/Heidelberg, Germany, 2012; Volume 8.

42. Davidrajuh, R. *Modeling Discrete-Event Systems with GPenSIM: An Introduction*; Springer: Berlin/Heidelberg, Germany, 2018.

43. Ezpeleta, J.; Colom, J.M.; Martinez, J. A Petri net based deadlock prevention policy for flexible manufacturing systems. *IEEE Trans. Robot. Autom.* **1995**, *11*, 173–184. [CrossRef]

44. Li, Z.; Zhou, M. Elementary siphons of Petri nets and their application to deadlock prevention in flexible manufacturing systems. *IEEE Trans. Syst. Man Cybern. Part A Syst. Hum.* **2004**, *34*, 38–51. [CrossRef]

45. Piroddi, L.; Cordone, R.; Fumagalli, I. Selective siphon control for deadlock prevention in Petri nets. *IEEE Trans. Syst. Man Cybern. Part A: Syst. Hum.* **2008**, *38*, 1337–1348. [CrossRef]

46. Chen, Y.; Li, Z.; Zhou, M. Behaviorally optimal and structurally simple liveness-enforcing supervisors of flexible manufacturing systems. *IEEE Trans. Syst. Man Cybern. Part A Syst. Hum.* **2012**, *42*, 615–629. [CrossRef]

Article

The Level of the Additive Manufacturing Technology Use in Polish Metal and Automotive Manufacturing Enterprises

Justyna Patalas-Maliszewska *, Marcin Topczak and Sławomir Kłos

Institute of Mechanical Engineering, University of Zielona Gora, 65-516 Zielona Góra, Poland;
p.m.topczak@gmail.com (M.T.); s.klos@iizp.uz.zgora.pl (S.K.)
* Correspondence: j.patalas@iizp.uz.zgora.pl

Received: 31 December 2019; Accepted: 19 January 2020; Published: 21 January 2020

Abstract: (1) Background: Products, manufactured using additive manufacturing technologies (AM) are increasingly present on the market. The research was undertaken to determine the possibilities of increasing the use of AM technology in Polish manufacturing companies. The aim of the paper is to determinate the level of the AM technology use of Polish Metal and Automotive Manufacturing and the influence of AM technology use on the increase of manufacturing company's competitiveness–in the context of Polish Manufacturing Companies. (2) Methods: This paper uses literature studies to determinate the AM technology used within the production processes in the automotive and metal industry companies (so called dimensions) and a questionnaire survey, which was carried out on a sample of 250 Polish Metal and Automotive Manufacturing Enterprises. (3) Results: The results were verified by a statistical analysis, using correlation coefficients. Based on the data obtained, it was determined that both metal and automotive Polish companies use, or have in their investment plans, the implementation of AM technology, due to the need to reduce production costs and increase speed and flexibility when responding to customer needs. Moreover, the relationship between applied additive manufacturing technologies and the effects of their use, in enterprises, was analysed. The novelty of our work is defining the dimensions of the AM technology use for our empirical research and determining the influence of AM technology use on the increase Polish manufacturing company's competitiveness. (4) Conclusions: The possibilities of using the results of research in economic practice were demonstrated. We also highlighted the impracticality for managers to support the selection and implementation of AM technology in the context of obtaining possible benefits for a manufacturing company.

Keywords: polish manufacturing company; additive manufacturing technology; questionnaire survey; empirical research

---

1. Introduction

Additive manufacturing (AM) technologies are, currently, being increasingly used in manufacturing companies, particularly in view of the need to reduce the time required to launch a new product onto the market. AM is treated as a combination of materials with the aim of obtaining a real object, based on 3D CAD data [1], which is applicable in the production area of, among other things, prototyping processes, the production of products in small series and the production of tools [2].

AM technologies are particularly applicable in the field of design and construction, in manufacturing companies in the automotive, aviation, military and metal industries [3]. However, according to data from the Central Statistical Office of Poland [4], only 2.4% of Polish enterprises uses AM technology, with only 1.4% of these enterprises having their own 3D printers. Other entities commission 3D printing from external entities. In addition, large enterprises (11.2%) and industrial processing units (5.0%) use AM technology, taking into account their type of business. Information and communications industry companies (3.2%) use their own 3D printers and outsource printing to entities from the industrial processing industry (2.9%). Bearing in mind the research results presented, regarding the application of AM technology in Polish production enterprises, research was undertaken to determine the possibilities of increasing the use of AM technology in Polish production companies.

The second chapter of this article indicates a research gap, based on an analysis of the current state of research in the field of the application of AM technology in Polish manufacturing companies. Based on the analysis carried out, the need was pointed out, on the one hand, to conduct advanced, empirical research into Polish enterprises, regarding the use of AM technology; on the other hand, restrictions on their use was pointed out. In the first stage of the research, the dimensions of the AM technology use in Polish manufacturing companies, were defined. Research was narrowed to the automotive and metal industries. A detailed description of the AM technologies used was then made for each individual dimension. The fourth part of the article describes the research methodology and gives details of the research samples. The following sections analyse the data acquired and indicate that both metal and automotive companies use—or actually have investment plans to implement—additive manufacturing technologies, due to the need to reduce production costs, optimise processes and increase speed and flexibility in responding to customer needs. The summary presents the possibilities of using the results of research in economic practice and formulates the direction of further work.

2. The State of Research in the Application of AM Technology in Polish Manufacturing Companies

Polish manufacturing enterprises make little use of additive technologies (2.4% according to GUS data). In 2017, 3D printing technology was most often used by enterprises in the Dolnośląskie (3.5%), Podkarpackie (3.2%) and Świętokrzyskie (3.1%) voivodships, yet least frequently so in the Lubelskie (1.0%) and Lubuskie voivodships (1.1%). In 2017, 3D printing technology was mainly used to create prototypes or models for personal use (1.8%), of which 10% were mainly for large enterprises. The least frequently used 3D printing was to create goods for sale, excluding prototypes or models (0.4%) and to create goods for use in production processes, excluding prototypes or models (0.5%). Spatial printing was mainly used by large enterprises employing 250 people or more.

Based on the analysis of the literature on the subject [5–29], the characteristics of AM technologies used have been broken down into separate dimensions:

- In automotive industry companies:

  (1)   Production process of machinery parts and equipment made of plastic and metal, including high precision mechanical parts and sub-assemblies,
  (2)   Production process for functional prototypes and co-operating mechanisms,
  (3)   Production process in models used in strength tests, tests and modelling.

- In metal industry companies:

  (1)   Prototype and model production process,
  (2)   The production process for injection moulds, foundry moulds, highly precise metal structures and other elements with a complex geometry or requiring high mechanical properties, the production of which would often not be possible using foundry technologies,
  (3)   The process of the repair and regeneration of complex damaged metal parts.

The following AM technologies according to the defined dimensions were identified for further research (Table 1).

Table 1. Characteristics of additive manufacturing (AM) technology.

Automotive Industry Companies				
Dimensions	**Method**	**Additive Manufacturing Technologies**	**Characteristics**	**Raw Material**
Production process for machinery and equipment parts made of plastic and metal, including high-precision mechanical parts and components, the production process of models used in strength tests, tests and modelling	Layered extrusion of molten thermoplastic	FDM (Fused Deposition Modelling) MEM (Moulded and Extruded Manufacturing), FFF (Fused Filament Fabrication)	The method is based on the layered joining of a plastic polymer material, extruded through a nozzle, with the material from which the element is made being a line made of thermoplastic material	ABS (acrylonitrile butadiene styrene), ABSi (acrylonitrile-butadiene -styrene-biocompatible), PC (polycarbonate), Ultem 9085, PLA (Polylactide), TPU (Thermoplastic Polyurethane), Nylon
Production process for machinery and equipment parts made of plastic and metal, including high-precision mechanical parts and sub-assemblies, production process for functional prototypes and co-operating mechanisms	Material application and UV curing	PolyJet, ProJet, MJM (Multi Jet Modelling)	Liquid resin in the form of droplets is applied, in layers, by means of a piezo-crystalline nozzle head, on the working platform and fully cured by a UV lamp integrated with the head. The support material, in the form of a gel, is removed with water.	Polymers with the trade names: Polimer PA 12, VeroWhitePlus, VeroCyan, VeroClear Object Agilus, Agilus Black Composites with specific mechanical properties: DM_Shore A40, DM_Shore A50, DM_Shore A60, DM_Shore A70, DM_Shore A85, DM_Shore A95
Production process for functional prototypes and co-operating mechanisms	Production of laminated objects (lamination of sheets)	LOM (Laminated Object Manufacturing), PSL (Plastic Sheet Lamination), SDL (Selective Deposition Lamination)	Sheets of material are laminated and then, with the help of, for example, a laser, an object of the desired shape is cut out	Paper, polymers, metals, ceramics, cellulose, poly-carbonate composites
Production process for functional prototypes and co-operating mechanisms, production process for models used in strength tests, tests and modelling.	Curing photosensitive resins	SLA (Stereo-lithography), DLP (Digital Light Processing)	The method involves the localised UV cure of the applied liquid resin layer in which photo-incised polymerisation takes place.	Polymers, resins and polymer composites with the trade names: Poly1500, Tusk Somos SolidGrey3000, Taurus, NeXt, ProtoGen, PerFORM.
Metal Industry Companies				
Production Process	**Method**	**Additive Manufacturing Technologies**	**Characteristics**	**Raw Material**
Production process for injection moulds, foundry moulds, highly precise metal structures and other elements with a complex geometry or requiring high mechanical properties, the production of which would often not be possible by means of foundry technologies.	Powder sintering	SLS (Selective Laser Sintering), SLM (Selective Laser Melting), DMLS (Direct Metal Laser Sintering), EBM (Electron Beam Melting) *	The method is based on the relationship in which the contacting powder grains bind each other together by melting their surfaces as a result of having been heated with a laser beam.	Polymer (polyamides, thermoplastic polyurethane, alumide), composite (Glass-filled Polyamide) and metallic powders (various metal powder mixtures, e.g., stainless steels, titanium, aluminium)
Production process for injection moulds, foundry moulds, highly precise metal structures and other elements with a complex geometry or requiring high mechanical properties, the production of which would often not be possible by means of foundry technologies; process of repair and regeneration of complex, damaged metal parts.	Hardening of material by direct energy supply	LENS (Laser Engineering Net Shape), EBAM (Electron Beam Additive Manufacturing)	In the method, the material stream is supplied through a nozzle and cured, by means of a laser or electron stream, directly onto the surface on which it falls; the LENS method, which is a specific type of laser surfacing, is used for laser performance of the final shape and dimension of the element.	Raw material deposited by means of wire or powdered metals (titanium, tantalum and nickel), alloys, ceramics or composites

* Technologies are also used in the automotive industry (spare parts).

In the processes carried out in AM technologies, the following technologies can be distinguished: technologies using a powder-filled platform, the powder bed technology, so-called, occurring in such technologies as Selective Laser Melting (SLM), Direct Metal Laser Sintering (DMLS), Electron Beam Melting (EBM) and also direct deposition technology, in such technologies as Electron Beam Additive Manufacturing (EBAM), Laser Engineering Net Shape (LENS) [30].

By evenly placing the powder on the working platform, the material feeding system has the advantage because the model does not need to have brackets generated. Moreover, the powder bed gives the opportunity to make models in the higher parts of the working chamber [27]. Selective powder micro-metallurgy (SLM) has many advantages related to accuracy of performance, versatility and the possibility of the complete elimination of the assembly phase, by sintering ready-made constructions. In the selective laser melting of powders of various metals, stainless steel powders are most often used. Unlike Selective Laser Sintering (SLS), the powder is re-melted through a laser beam, which significantly affects the tightness of the model and the lack of pores [27]. Metal additive manufacturing (MAM) is becoming a popular industrial production solution. Metal AM is characterised by two leading methods using metal powders, namely SLM/DLMS which is used mostly in the automotive industry and electron beam melting (EBM) [31,32]. The benefits of using additive manufacturing techniques are observed in the production of engine parts, fuel systems or turbines with a complex geometry and with defined aerodynamic properties. In addition, the technology of laser sintered metal powders enables the development of advanced and lightweight structures that combine high strength with a weight reduction of up to 60%. Even very complex elements, made of high-strength material, can easily be manufactured by means of additive manufacturing techniques, where the use of traditional processes, for the production of such parts, is impossible or very expensive [33]. Powder sintering methods are used to make injection moulds for mass production as also for the production of pressure casting moulds and other elements requiring high mechanical properties, the production of highly precise metal structures, making surgical instruments, passive and active implants, as well as for medical parts/devices, manufacturing elements complex shapes and structures that would not have been possible with casting technologies (DMLS).

FDM (fused deposition modelling) technology is used for rapid prototyping. In FDM technology, one of the nozzles of the printing device feeds the material for modelling, while the other feeds the material to produce supports, when necessary. Depending on the printing machine and the material used, the supports, produced in the final phase, can be broken off (BST series machines) or dissolved in water (SST series machines). Manufacturing supports is sometimes necessary due to the presence of hanging elements that affect the stability of the manufactured item. Molten thermoplastic layer extrusion methods are used for the production of plastic machinery parts and equipment, the production of prototypes of co-operating mechanisms and the production of models, used in strength tests, etc. [19].

Material application and UV hardening methods are used for the production of high-precision parts, mechanical components, the production of flexible elements, such as gaskets and washers and the production of high-quality prototypes over a short time.

- Methods based on the technology of curing photosensitive resins are used in the production of prototypes, among other things, for empirical research, in tests and in modelling processes, the production of components used in automotive sectors, electronic housings, latch assemblies, car bodies, the production of machine covers and the production of functional prototypes.
- Methods for producing laminated objects are used to produce models, prototypes and machine parts made from plastic, cellulose or composite materials.

2.1. Technologies Based on the Photo-Initiated Polymerisation Method (SLA, DPL)

Liquid photo-polymer is placed in a tank (ladle) and then selectively cured, layer by layer at source, emitting heat using a digital screen, LCD screen, UV radiation or laser beam, until the element is completely manufactured. In the case of Stereo-lithography (SLA), a laser beam is mainly used which is focussed on the surface of the tank, creating each layer of the desired 3D object by cross-linking or polymer degradation. Digital Light Processing (DLP) uses a digital projector screen, through which the entire layer of the element is cured simultaneously over the entire surface. This reduces the time it takes to manufacture a part [3,6,25].

2.2. Powder Sintering Technologies (SLS, SLM, DMLS, EBM)

Basically, the SLS process uses a laser or electron beam to sinter or coalesce powdered material, layer by layer, in order to create a solid structure. The powder is placed in a container where the sintering takes place. The molten powder is supplemented with further powder supplies and levelled with a roller. The final product, covered with loose powder, is then cleaned with brushes and compressed air. SLM relies on selective laser melting to form the element, layer by layer, while DMLS, operating on the same principle, consists in the direct sintering of the metal, by laser and is used for the sintering of metal powders only [15–18,20,22,25–28,30,33].

2.3. Technologies Based on the Method of Layered Extrusion of Molten Thermoplastics (FDM, MEM, FFF)

3D printing technology uses continuous filament from thermoplastic material as the basic material while additional material is used to make the support. The fibre from the material spool is fed from the coil through a movable, heated extruder head. The molten material is ejected from the extruder nozzle and deposited on a 3D printing platform that can be heated, for additional grip. The movement of the extruder head is controlled by computer. The next layer is applied onto the previous layer until the manufacturing process of the object is completed [19,21,22,24,25,27].

2.4. Technologies Based on the Method of Curing Material by Direct Energy Supply (LENS, EBAM)

This technology allows elements to be produced by directly melting materials and depositing them on the workpiece, layer by layer. The LENS process must take place in a hermetically sealed, argon-filled chamber so that the oxygen and moisture levels are kept very low. This keeps the manufactured item clean and prevents oxidation. The metal powder is delivered directly to the head which deposits the material. After depositing a single layer, the head which deposits the material advances to the next layer. The whole section is constructed by building successive layers. Once completed, the element is removed and can be processed or finished in any way. EBAM is an additive manufacturing technology that produces large-scale metal structures. The EB gun deposits the metal with a wire, layer by layer, until the element reaches a grid-like shape and is ready for its final treatment [11–13,22,24,25,27].

This article offers an explanation of the influence the AM technology use on increase the manufacturing company's competitiveness. On the basis of the current analysis of the literature on the subject and of the defined dimensions of the AM technology use a questionnaire was developed, based on which, pilot studies of the first stage were conducted, followed by the main research into the Polish automotive and metal manufacturing companies, regarding the use of additive manufacturing technology. This study assumes also that those workers who were involved in the survey realize at least 80% of the defined dimensions associated with the dimensions of the of the AM technology use. The surveys applied for testing the research model (Figure 1) were developed by defining scales to fit the knowledge codification meaning.

Factors of dimensions of the AM technology use were based on feedback surveys and their sources are listed here: The dimensions of the of the Additive Manufacturing Technology Use: I know that in my company the use AM technologies during the business processes, namely in automotive industry: DA11-production processing of machinery and equipment parts made of plastic and metal, including high-precision mechanical parts and components, DA12-production process of functional prototypes and co-operating mechanisms, DA13-production process of models used in strength tests, tests and modelling; in metal industry: DM11-prototype and model production process, DM12-production process of injection moulds, foundry moulds, highly precise metal constructions and other elements with a complex geometry or requiring high mechanical properties, the production of which would often not be possible by means of foundry technologies, DM13-process of repair and regeneration of complex damaged metal parts is: factor0: not very important/ factor1: very important.

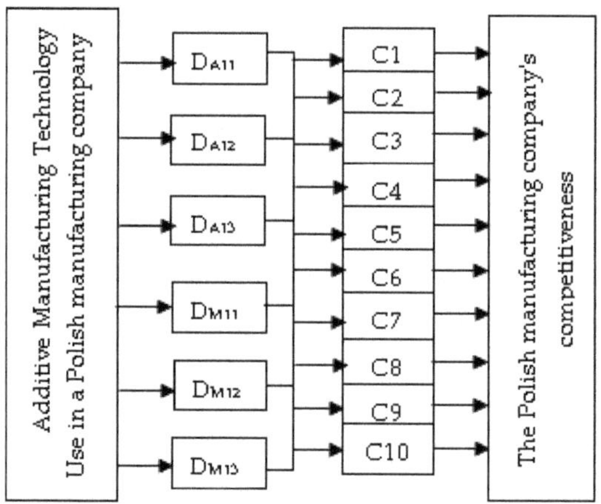

Figure 1. Research model.

The Polish manufacturing company's competitiveness: I know that in my company the realization of the work activities using Additive Manufacturing Technology, is: factor0: not very important/factor1: very important for C1-reduction of production costs, C2-effective use of material, C3-freedom in product design, C4-no assembly stage, C5-product personalisation to meet specific customer requirements, C6-quick response to market needs, C7-optimisation of product functions, C8-development, C9-elimination of human labour, C10-waste reduction/lower energy consumption.

The data for this study were collected from 250 Polish manufacturing enterprises between October to November, 2019.

3. Research Method

The studies were carried out in two stages: (1) pilot studies and (2) main studies. The tests were carried out using the survey method. The survey questionnaire included multi-alternative closed questions.

The pilot study was intended, initially, to identify the needs of the industry's representatives in carrying out work in the field of material testing and additive manufacturing technologies. The pilot study involved 10 companies from the automotive industry and 10 companies from the metal industry from the Lubuskie Voivodeship, employing between 30 and 1400 employees. The survey respondents from the automotive industry were mostly representatives of the management staff of operational, technological and R&D areas (managers (2), directors (2), supervisors (3), principal (1)) as well as the cost leader and technical and the commercial adviser. In the metal industry, half of the respondents were representatives of the production management staff (director (1), manager (1) managers (3) and representatives of specialist technological, quality and sales staff, (5)). Among the 20 companies surveyed, 7 declared that AM technologies were used in their enterprises, of which 3 companies were from the automotive industry and 4 were from the metal industry.

The main research was then carried out in 250 small, medium and large manufacturing companies from western Poland, including 125 representatives from the metal and automotive industries. The research group accounts for 1% of the total number of manufacturing companies from the automotive industry and metal industry group in Western Poland. Among the automotive companies there were 92 small enterprises, 29 medium enterprises and 4 large enterprises, while representatives of the metal industry comprised 120 small enterprises, 27 medium enterprises and 3 large enterprises. The respondents in the automotive industry were mostly representatives of the management (69) and company owners (32), specialist employees, including technologists, logisticians, designers, marketing employees (16) and company support employees, including assistants, accountants, administrative staff, etc. (8). In the metal industry, the respondents to the survey were mainly representatives of the management (72), owners and shareholders (36) and specialist employees, among others, in the fields of quality, production, purchasing, design and construction (13) and company support employees, including accountants and sales/sales department staff (4).

A total of 109 of the 250 enterprises surveyed declared themselves to be users of additive manufacturing technology, of which 57 were from the metal industry and 52 were from the automotive industry.

4. Research Results

4.1. Pilot Studies

In the enterprises in which AM technologies were used, none of the proposed methods were indicated, i.e., FDM, Laminated Object Manufacturing (LOM), Plastic Sheet Lamination (PSL), Selective Deposition Lamination (SDL), DLP, PolyJet, DMLS, SLS, SLA, EBM, MEM, Fused Filament Fabrication (FFF), ProJet, Multi Jet Modelling (MJM), LENS, EBAM, SLM (Table 1). However, it was emphasised that plasma-burning, hot-dip galvanising technologies and anti-corrosion coating were used. Unfortunately, pilot studies have shown that only traditional, additive manufacturing methods, such as welding, were used. Therefore, the main research is looking for possible answers in order to increase the use of modern, additive technologies in Polish enterprises; subsequently, it was then determined which of the defined dimensions of the Additive Manufacturing Technology Use (Table 1):

- Automotive industry companies:

    (1) Production process of functional prototypes and co-operating mechanisms—in two companies,
    (2) (Production process of models used in strength tests, tests and modelling—in one company.

- Metal industry companies:

    (1) Production process of injection moulds, foundry moulds and other elements requiring high mechanical properties—in four companies.

Only one of the companies surveyed indicated their willingness to adopt AM technology. The company identified the following factors that could potentially influence their decision:

(a) Reduction of production costs
(b) Effective use of material
(c) Freedom in product design
(d) No assembly stage
(e) Product personalisation to meet specific customer requirements
(f) Quick response to market needs
(g) Optimisation of product functions

In addition, it was emphasised that any technology implemented should be associated with the welding and bonding processes, indicating at the same time the limitations existing in the company, namely the outdated machine park and the need to update workstations.

Pilot studies showed that in the manufacturing companies surveyed—which use AM technologies—traditional and specific methods were used, such as welding, plasma-burning and the production of an anti-corrosion layer. The use of traditional additive methods probably has something to do with the lack of appropriate technological facilities, along with a low recognition/perception of modern methods of additive manufacturing.

4.2. Main Research

Among those manufacturing enterprises to declare themselves users of additive technologies, 29 enterprises from the automotive industry indicated the use of the FDM and EBM methods, while 1 enterprise indicated the use of FDM with 1 company, from the metal industry, indicating the use of EBM technology (Table 1).

A total of 78 companies indicated "other" technologies, of which 23 were from the automotive industry and 55 were from the metal industry. Welding was indicated by 39 enterprises, while for those companies using welding technology, 29 companies indicated using Metal Inert Gas (MIG), Metal Active Gas (MAG), Tungsten Inert Gas (TIG)technology. Material processing was indicated by 18 companies, of which 4 indicated machining with CNC machines. Other companies indicated methods such as aluminium casting, metallisation, plasma-burning, laser cutting, joining sheets, carburising, etc.

Basic research has shown that AM technologies are used in 44% of manufacturing companies in the metal and automotive industries. It is interesting that the number of those respondents, classifying the welding method as the AM method, came to about 15%. It is worth noting that over 11% of respondents, in the automotive industry, declared the use of two AM technologies, namely: FDM and EBM.

Subsequently, the dimensions of the Additive Manufacturing Technology Use were then determined; these are indicated in Table 1:

- Automotive industry (52 companies in total):

  (1)   Production processing of machinery and equipment parts made of plastic and metal, including high-precision mechanical parts and components—27 marked answers,
  (2)   Production process of functional prototypes and co-operating mechanisms—29 marked answers,
  (3)   Production process of models used in strength tests, tests and modelling—no company marked an answer.

Four companies simultaneously indicated the production of machinery and equipment parts and the production of prototypes. One company gave the answer "other" indicating that it uses AM technology for the production of machinery and equipment parts and for the provision of services.

- Metal industry (57 companies in total):

  (1)   Prototype and model production process—4 marked answers,
  (2)   Production process of injection moulds, foundry moulds, highly precise metal constructions and other elements with a complex geometry or requiring high mechanical properties, the production of which would often not be possible by means of foundry technologies—54 marked answers,
  (3)   Process of repair and regeneration of complex damaged metal parts—3 marked answers.

Two enterprises, marking their answers "other", indicated that they used AM for the provision of services (1 company) and for machine construction (1 company).

Three companies marking their answer "the injection mould production process" also marked "prototyping", whereas three companies that indicated the process of manufacturing injection moulds also indicated protection and regeneration (2 companies) and the production of semi-finished products (1 company).

A total of 72 enterprises declared their willingness to adopt additive technologies, including 3 companies were interested in FDM technology, while 2 interested companies also marked DMLS; 2 companies in LOM technology, with 1 company also indicating DMLS; 5 companies in DLP technology, including 5 companies also marked DMLS; 1 company in PolyJet technology, with DMLS technology also indicated; 20 companies in DMLS technology, 3 companies also indicated SLS, 2 companies also indicated SLA, 2 companies also indicated EBM; 3 companies in SLS technology; 2 companies in SLA technology; 2 companies in EBM technology and 52 companies marked the answer "other" (Figure 2).

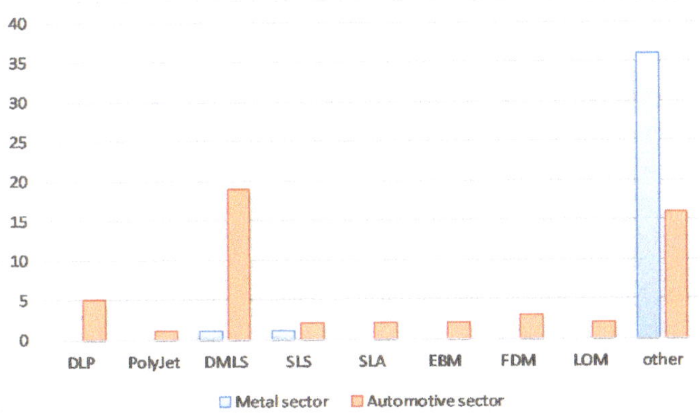

Figure 2. Interest in implementing AM in the manufacturing companies surveyed in the metaland automotive industries.

Respondents who ticked "other" indicated, among others, laser technologies (12), robotic technologies (4), powder technologies (1), hardening technologies (4), welding (1), production/application of layers of material (3) bonding of materials (2), regeneration of machine elements (1) (Figure 3). As many as 22 companies answering "other" claimed that, currently, they could not, or were not able to determine what technology they would be interested in, while 2 did not provide any justification for using additive technologies. As results from the analysis of the data obtained, companies are interested in implementing laser technology. This is indicated by declarations of intent to implement the laser sintering of metals-DMLS (20 companies) with justification of the respondents who indicated the answer "other", pointing to technologies using welding, cutting and laser processing (12 companies).

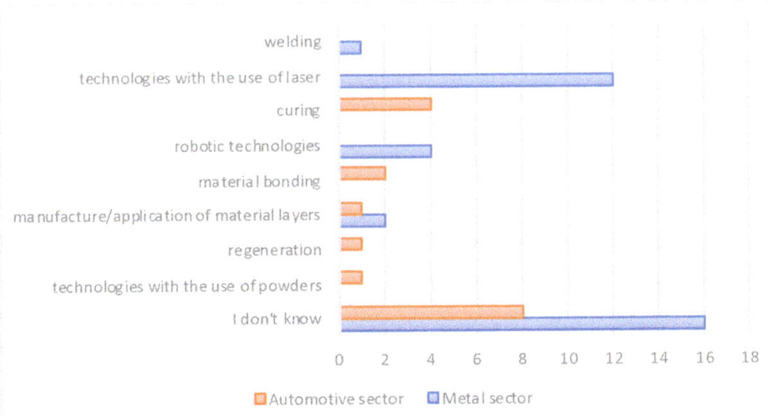

Figure 3. Interest in implementing AM. Answers declared by respondents from the automotive and metal industries.

Companies declaring an interest in implementing additive technologies indicated the following factors that could potentially influence their decision (Figure 4):

(a)   Reduction of production costs—37 marked answers
(b)   Effective use of material—26 marked answers
(c)   Freedom in product design—26 marked answers
(d)   No assembly stage—25 marked answers
(e)   Product personalisation to meet specific customer requirements—25 marked answers
(f)   Quick response to market needs—32 marked answers
(g)   optimisation of product functions—29 marked answers
(h)   Other—39 marked answers, of which 21 focussed on the development of the company, 10 on increasing work efficiency and processes, 4 on reducing employment, 2 on reducing exhaust emissions and waste; 2 companies did not provide justification or indicated insufficient knowledge of this area.

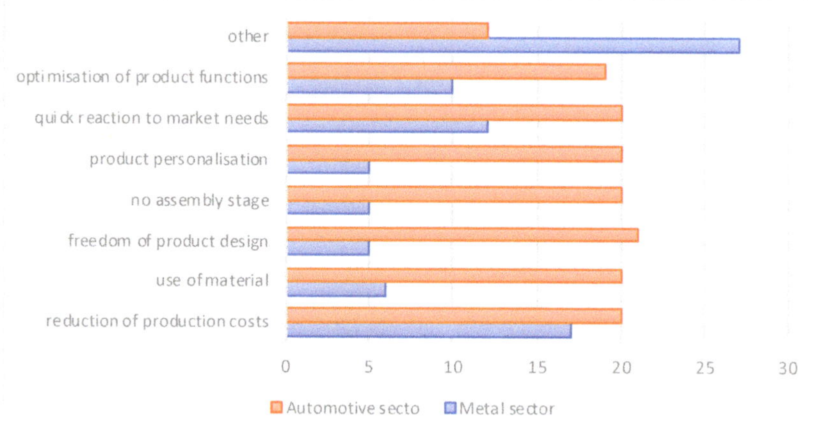

Figure 4. Factors affecting interest in implementing AM in the manufacturing companies surveyed in the metal and automotive industries.

Analysis of the responses received, with reference to the restrictions (Figure 5) determining the implementation of AM technology, showed that 21 companies indicated the below par performance of devices which had a lengthy printing time, with 13 indicating the poor quality of the products manufactured, 14 indicating the low strength poor clarity of the printed details, while 51 marked the answer "other".

The respondents marked the answer "other" (Figure 6) and indicated their justifications of the answer, with 23 indicating lack of funds, 9 indicating lack of space, 7 did not see the need, 3 indicated the unstable market situation, 2 companies indicated that they were under-staffed, while 3 were unable to give any reasons. Moreover, 3 companies said, in their self-justification, that they were at the 'searching for the right technology' stage or were those companies who intimated that, 'so far, no method had been developed that would meet our expectations'.

Analysing the results obtained, it was found that the main determinants influencing the decision to implement AM technology, in automotive and metal industry manufacturing companies, was the need to reduce costs and the ability to respond quickly to market needs. The research results are in the line with the current development trend of production companies, whose managers are constantly looking for a production system that will allow the company to adapt it to current market needs [34].

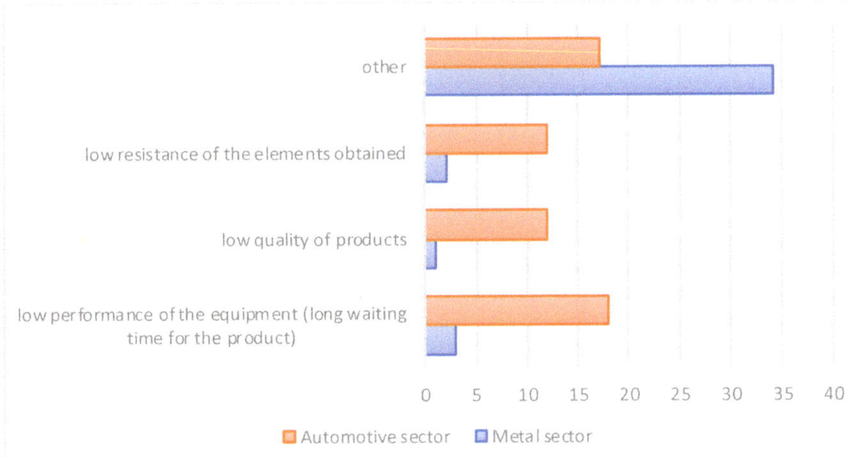

Figure 5. Restrictions affecting the decision to implement AM in the metal and automotive industry manufacturing companies surveyed.

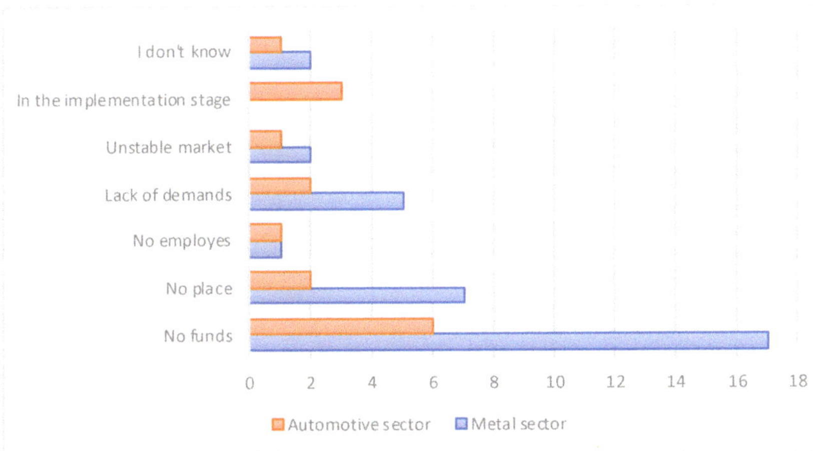

Figure 6. Restrictions affecting the decision to implement AM, in the metal and automotive industry manufacturing companies surveyed. Answers given by the respondents.

4.3. Analysis of the Research Results

Based on the research results of 250 Polish metal and automotive industry manufacturing companies, the research model (Figure 1) was studied using a correlation and regression approach in order to estimate the impact of using AM technologies in Polish manufacturing enterprises. An analysis of the research results was carried out using Statistica ver.13.3. The data were carefully examined with respect to linearity, equality of variance and normality. No significant deviations were detected. Table 2 presents the correlation analysis, where only the variables, for which the relationships were significant, were summarised, that is, where the technologies used were: FDM, EBM, welding, heat bonding, laser cutting.

Table 2. Correlation analysis.

Relations	Correlation	r2	t	p
AM used: FDM/production cost reduction	0.1927	0.0371	3.0931	0.0022
AM used: FDM/efficient use of material	0.2774	0.0770	4.5474	0.0000
AM used: FDM/freedom in product design	0.3177	0.1010	5.2774	0.0000
AM used: FDM/no assembly stage	0.2462	0.0606	4.0000	0.0001
AM used: FDM/product personalisation	0.2872	0.0825	4.7220	0.0000
AM used: FDM/quick response to market needs	0.1901	0.0361	3.0495	0.0025
AM used: FDM/optimisation of product functions	0.2506	0.0628	4.0769	0.0001
AM used: FDM/development	−0.1046	0.0109	−1.6569	0.0988
AM used: FDM/elimination of human labour	0.0510	0.0026	0.8044	0.4219
AM used: FDM/waste reduction/lower energy consumption	−0.0332	0.0011	−0.5225	0.6018
AM used: EBM/production cost reduction	0.0299	0.0009	0.4709	0.6381
AM used: EBM/efficient use of material	−0.0027	0.0000	−0.0421	0.9664
AM used: EBM/freedom in product design	−0.0027	0.0000	−0.0421	0.9664
AM used: EBM/no assembly stage	0.0000	0.0000	0.0000	1.0000
AM used: EBM/product personalisation	0.0000	0.0000	0.0000	1.0000
AM used: EBM/quick response to market needs	0.0440	0.0019	0.6934	0.4887
AM used: EBM/optimisation of product functions	−0.0102	0.0001	−0.1606	0.8725
AM used: EBM/development	−0.0171	0.0003	−0.2695	0.7878
AM used: EBM/elimination of human labour	0.1367	0.0187	2.1724	0.0308
AM used: EBM/waste reduction/energy consumption	−0.0183	0.0003	−0.2887	0.7730
AM used: welding/production cost reduction	0.0483	0.0023	0.7622	0.4467
AM used: welding/efficient use of material	−0.0313	0.0010	−0.4930	0.6224
AM used: welding/freedom in product design	−0.0682	0.0047	−1.0764	0.2828
AM used: welding/no assembly stage	−0.0638	0.0041	−1.0073	0.3148
AM used: welding/product personalisation	−0.0638	0.0041	−1.0073	0.3148
AM used: welding/quick response to market needs	0.0089	0.0001	0.1402	0.8886
AM used: welding/optimisation of product functions	−0.0103	0.0001	−0.1618	0.8716
AM used: welding/development	0.0426	0.0018	0.6718	0.5024
AM used: welding/elimination of human labour	0.1264	0.0160	2.0067	0.0459
AM used: welding/waste reduction/lower energy consumption	−0.0374	0.0014	−0.5898	0.5558
AM used: heat bonding/production cost reduction	0.0366	0.0013	0.5772	0.5643
AM used: EBM/efficient use of material	−0.0434	0.0019	−0.6848	0.4941
AM used: heat bonding/freedom in product design	−0.0434	0.0019	−0.6848	0.4941
AM used: heat bonding/no assembly stage	−0.0425	0.0018	−0.6700	0.5035
AM used: heat bonding/personalisation of the product	−0.0425	0.0018	−0.6700	0.5035
AM used: heat bonding/quick response to market needs	−0.0489	0.0024	−0.7703	0.4419
AM used: heat bonding/optimisation of product functions	−0.0462	0.0021	−0.7282	0.4672
AM used: heat bonding/development	0.0466	0.0022	0.7341	0.4672
AM used: heat bonding/elimination of human labour	−0.0163	0.0003	−0.2561	0.7981
AM used: heat bonding/waste reduction/lower energy consumption	0.3464	0.1200	5.8151	0.0000
AM used: laser cutting/production cost reduction	0.1162	0.0135	1.8429	0.0665
AM used: laser cutting/efficient use of material	0.0870	0.0076	1.3747	0.1705
AM used: laser cutting/freedom in product design	0.0870	0.0076	1.3747	0.1705
AM used: laser cutting/no assembly stage	0.0909	0.0083	1.4376	0.1518
AM used: laser cutting/product personalisation	0.0909	0.0083	1.4376	0.1518
AM used: laser cutting/quick response to market needs	0.0664	0.0044	1.0479	0.2957
AM used: laser cutting/optimisation of product functions	0.0761	0.0058	1.2016	0.2307
AM used: laser cutting/development	0.3385	0.1146	5.6652	0.0000
AM used: laser cutting/elimination of human labour	−0.0232	0.0005	−0.3652	0.7153
AM used: laser cutting/waste reduction/lower energy consumption	−0.0163	0.0003	−0.2572	0.7973

where: r2—coefficient of determination; t—the value of t statistics examining the significance of the correlation coefficient; p—probability value.

The analysis shows significant relationships between:

(a) The FDM technology used and the impact of its use vis-à-vis the increased freedom gained when it comes to product design.

(b) Heat bonding and reducing waste/lower energy consumption as a result of using this technology.

(c) Between the introduction and application of laser cutting technology and the company's development.

The importance of a company's dependence on such an impact should be emphasised since such results may encourage further production companies to implement AM technology. Other relationships basically confirm the assumed benefits of using AM technology. FDM technology consists in the layered joining of plastic polymer material extruded through a nozzle, allowing the production of objects of any shape, with the only restriction being the dimensions of the designed element. The relationship between heat bonding the material and reducing the amount of waste can be justified by the maximum use of raw materials. To define the nature of significant interactions of the influence of AM technology use on the increase manufacturing company's competitiveness—in the context of Polish Manufacturing Companies, the study tests the hypotheses using regression analyses which estimate this impact (Equations (1)–(3)).

Based on the analysis, it was noticed that as the use of FDM technology increases, the possibilities for product design (Equation (1), Figure 7).

$$\text{Increasing the possibilities for product design} = 0.0682 + 0.2985 \times \text{FDM} \tag{1}$$

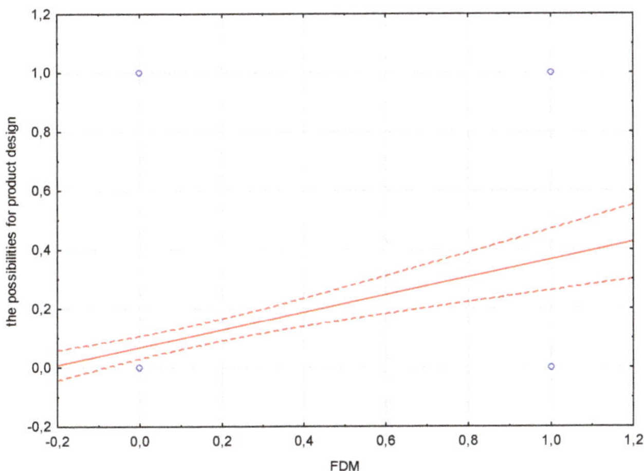

Figure 7. Interactions between using Fused Deposition Modelling (FDM) and increasing the possibilities for product design.

The results show that as the use of heat bonding increases the level of reduction in post-production waste increases (Equation (2), Figure 8).

$$\text{waste reduction} = 0.0041 + 0.2459 \times \text{heat bonding} \tag{2}$$

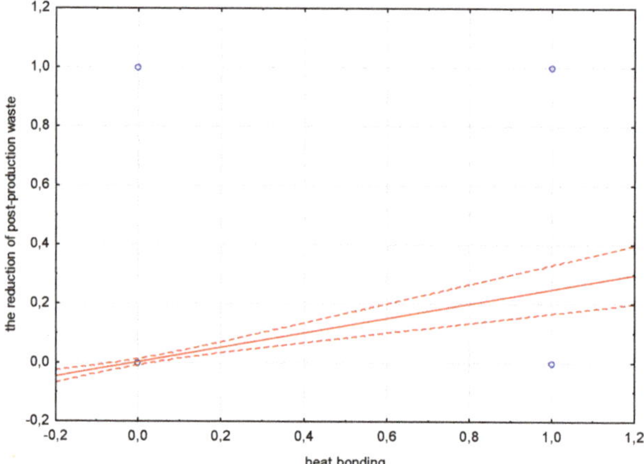

Figure 8. Interactions between the use of heat bonding and the reduction of post-production waste.

The results obtained showed that with the increase in the use of laser cutting technology, the level of company development increases significantly (Equation (3), Figure 9).

$$\text{development} = 0.1074 + 0.6426 \times \text{laser cutting} \tag{3}$$

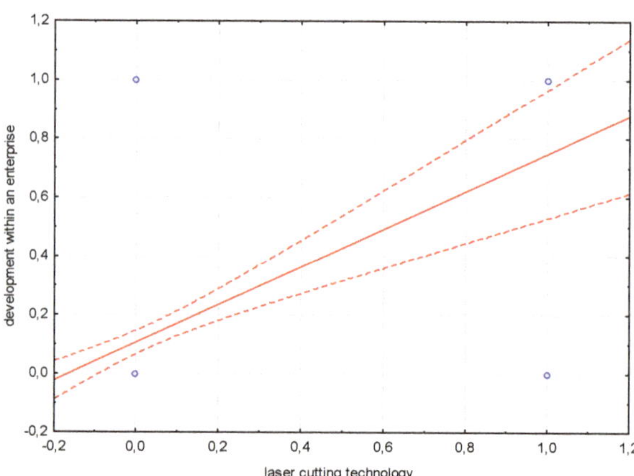

Figure 9. Interactions between the use of laser cutting technology and the future possibilities for increased development within an enterprise.

The statistical results showed that the three AM technologies examined are held to have the strongest positive effect on the company's competitiveness. Figure 10 presents the structural model.

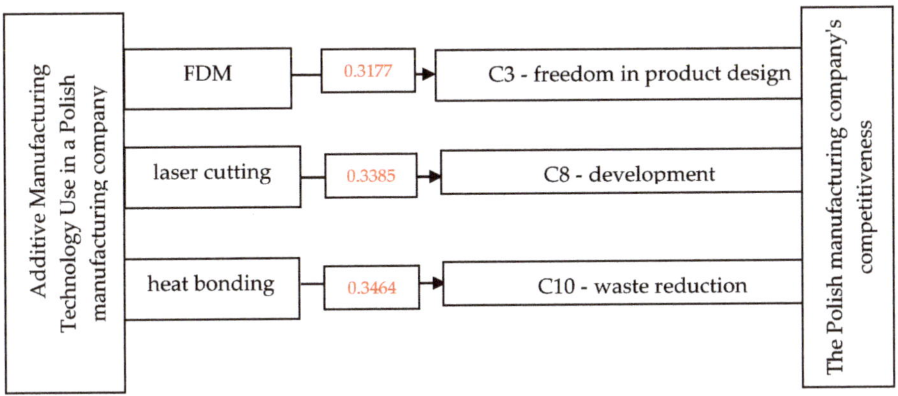

Figure 10. The structural model.

This study explores the influence of AM Technology use on the increase manufacturing company's competitiveness—in the context of Polish Manufacturing Companies. The statistical results showed that the three AM technologies examined have the strongest positive effect on the company's competitiveness.

In addition, an analysis of dependencies was then made to assess the impact of the factors defined which determine management decisions to change business processes within an enterprise. The research was carried out using Statistica ver.13.3. The data was thoroughly examined for linearity, equality of variance and normality. No significant deviations were found. Tables 3–5 present the correlation analysis, with only those variables for which the relationships were significant—that is, the production process, the prototyping process and the services—being listed.

Table 3. Correlation analysis: production process.

Relations	Correlation	r2	*t*	*p*
production process in the metal industry/reduction of production costs	0.2665	0.0710	3.0662	0.0027
production process in the automotive industry/reduction of production costs	0.1887	0.0356	2.1315	0.0350
production process in the metal industry/efficient use of material	0.1064	0.0113	0.1865	0.2377
production process in the automotive industry/efficient use of material	0.1550	0.0240	1.7397	0.0844
production process in the metal industry/freedom in product design	0.0692	0.0048	0.7696	0.4430
production process in the automotive industry/freedom in product design	0.0913	0.0083	1.0165	0.3114
production process in the metal industry/no assembly stage	0.0692	0.0048	0.7696	0.4430
production process in the automotive industry/no assembly stage	0.0913	0.0083	1.0165	0.3114
production process in the metal industry/ product personalisation	0.0692	0.0048	0.7696	0.4430
production process in the automotive industry/product personalisation	0.0913	0.0083	1.0165	0.3114
production process in the metal industry/quick response to market needs	0.1544	0.0238	1.7329	0.0856
production process in the automotive industry/quick response to market needs	0.0929	0.0086	1.0350	0.3027
production process in the metal industry/optimisation of product functions	0.1000	0.0100	1.1148	0.2671
production process in the automotive industry/optimisation of product functions	0.1319	0.0174	1.4752	0.1427
production process in the metal industry/development	0.1483	0.0220	1.6627	0.0989
production process in the automotive industry/development	0.1148	0.0132	0.2812	0.2025
production process in the metal industry/elimination of human labour	0.1167	0.0136	1.3034	0.1949
production process in the automotive industry/elimination of human labour	0.1255	0.0157	1.4027	0.1632
production process in the metal industry/waste reduction/lowering energy consumption	0.0175	0.0003	0.1942	0.8464
production process in the automotive industry/ reduction of waste/lowering energy consumption	−0.0669	0.0045	−0.7440	0.4583

Table 4. Correlation analysis: prototyping process.

Relations	Correlation	r2	*t*	*p*
prototyping in the metal industry/reduction of production costs	0.0605	0.0037	0.6718	0.5029
prototyping in the automotive industry/reduction of production costs	0.1690	0.0285	1.9012	0.0596
prototyping in the metal industry/efficient use of material	0.1718	0.0295	1.9342	0.0554
prototyping in the automotive industry/efficient use of material	0.0539	0.0029	0.5987	0.5505
prototyping in the metal industry/freedom in product design	0.1948	0.0380	2.2032	0.0294
prototyping in the automotive industry/freedom in product design	0.0812	0.0066	0.9040	0.3678
prototyping in the metal industry/no assembly stage	0.1948	0.0380	2.2032	0.0294
prototyping in the automotive industry/no assembly stage	0.0812	0.0066	0.9040	0.3678
prototyping in the metal industry/ product personalisation	0.0950	0.0090	1.0589	0.2917
prototyping in the automotive industry/product personalisation	0.1426	0.0203	1.5974	0.1127
prototyping in the metal industry/quick response to market needs	0.1139	0.0130	1.2719	0.2058
prototyping in the automotive industry/quick response to market needs	0.0475	0.0023	0.5274	0.5988
prototyping in the metal industry/optimisation of product functions	0.0812	0.0066	0.9040	0.3678
prototyping in the automotive industry/optimisation of product functions	0.1948	0.0380	2.2032	0.0294
prototyping in the metal industry/development	0.2741	0.0751	3.1604	0.0020
prototyping in the automotive industry/development	0.2437	0.0594	2.7863	0.0062
prototyping in the metal industry/elimination of human labour	0.2252	0.0507	2.5635	0.0116
prototyping in the automotive industry/elimination of human labour	0.2231	0.0498	2.5385	0.0124
prototyping in the metal industry/ waste reduction/lower energy consumption	−0.0232	0.0005	−0.2572	0.7975
prototyping in the automotive industry/waste reduction/lower energy consumption	0.0810	0.0066	0.9009	0.3694

Table 5. Correlation analysis: services.

Relations	Correlation	r2	*t*	*p*
services in the metal industry/reduction of production costs	−0.0356	0.0013	−0.3954	0.6932
services in the automotive industry/reduction of production costs	−0.0356	0.0013	−0.3954	0.6932
services in the metal industry/efficient use of material	−0.0202	0.0004	−0.2237	0.8234
services in the automotive industry/efficient use of material	−0.0202	0.0004	−0.2237	0.8234
services in the metal industry/freedom in product design	−0.0183	0.0003	−0.2033	0.8392
services in the automotive industry/freedom in product design	−0.0183	0.0003	−0.2033	0.8392
services in the metal industry/no assembly stage	−0.0183	0.0003	−0.2033	0.8392
services in the automotive industry/no assembly stage	−0.0183	0.0003	−0.2033	0.8392
services in the metal industry/product personalisation	−0.0183	0.0003	−0.2033	0.8392
services in the automotive industry/product personalisation	−0.0183	0.0003	−0.2033	0.8392
services in the metal industry/quick response to market needs	−0.0293	0.0009	−0.3247	0.7460
services in the automotive industry/quick response to market needs	−0.0293	0.0009	−0.3247	0.7460
services in the metal industry/optimisation of product functions	−0.0265	0.0007	−0.2938	0.7694
services in the automotive industry/optimisation of product functions	−0.0265	0.0007	−0.2938	0.7694
prototyping in the metal industry/development	−0.0415	0.0017	−0.4607	0.6458
services in the automotive industry/development	0.1943	0.0378	2.1969	0.0299
services in the metal industry/elimination of human labour	−0.0163	0.0003	−0.1811	0.8566
services in the automotive industry/elimination of human labour	0.4939	0.2440	6.2998	0.0000
services in the metal industry/ waste reduction/lower energy consumption	−0.0115	0.0001	−0.1270	0.8991
services in the automotive industry/waste reduction/lower energy consumption	−0.0115	0.0001	−0.1270	0.8991

In both the metal and automotive industries, in analysing the relationship between the use of AM technology in a given process and the effects of their implementation, it was shown that there are significant relationships between the production process and the impact on reducing production costs (Table 3).

Based on Table 4, significant relationships between the prototyping process and the company's development were indicated. Making prototypes of elements using AM technology along with further research in this area will contribute to making more effective construction and production decisions.

Based on the data in Table 5, it was found that enterprises in the automotive industry strive to search for modern solutions, the use of which will reduce the involvement of employees to perform certain services for business partners (Figure 11).

This is understandable, since the boards of production companies indicate that in the west of Poland there is still a shortage of suitably qualified employees. This is an area of research that requires further work, in order to define the possible actions that should be taken by research and development units in the west of Poland, to contribute to the strengthening and intensification of their co-operation with manufacturing companies. In this area, further in-depth research is planned regarding the possibilities of intensifying the co-operation of Polish manufacturing companies with scientific and research units.

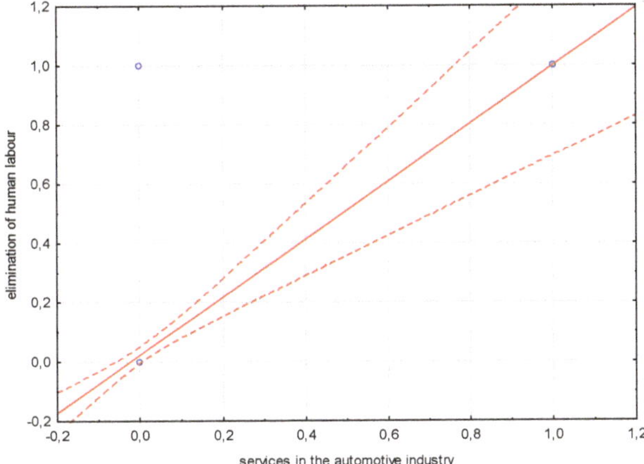

Figure 11. Interactions between the process of providing services in the automotive industry. vis-à-vis the elimination of human labour.

5. Discussion

According to the research results within Polish Metal and Automotive Manufacturing Enterprises, it can be concluded that mangers of both metal and automotive companies are aware of the need to implement additive manufacturing technologies, due to the need to be more flexible in responding to customer needs.

Firstly, we discuss the dimensions of the AM technology use. The level of the AM technology use of Polish Automotive Manufacturing Enterprise was determined for two dimensions: (1) production processing of machinery and equipment parts made of plastic and metal and (2) production process of functional prototypes and co-operating mechanisms and of Polish Metal Manufacturing Enterprise for 1 dimension: (1) production process of injection moulds, foundry moulds, highly precise metal constructions and other elements with a complex geometry or requiring high mechanical properties. Research has shown that almost half of the respondents use AM technologies, while in the automotive industry AM are used in a similar ratio in the production of machinery and equipment and for the production of functional prototypes. In the case of the metal industry, AM technologies are almost entirely used in the production of high-precision components with complex geometry, which would be impossible or inefficient to produce using traditional methods.

Secondly, the obtained data allowed to indicate further directions and areas of research, experiments and improvement activities. The research has shown that the automotive industry, due to the continuous development of technology and the dynamic product market, is constantly looking for new solutions that will allow for more free design and construction of elements, reduction of time, stages and costs of production, maximum use of material and quick response to customer needs. This is important information due to the level of awareness of manufacturing companies and understanding of market processes. Managers of manufacturing companies from the automotive industry know the potential and possibilities to achieve the desired effects by applying additive manufacturing technology in the company.

The added value of our research to the recent state of the research field is the definition of knowledge about AM technology in Polish manufacturing enterprises. Based on empirical research conducted in 250 production companies of western Poland, the projected benefits of AM implementation (reduction of costs, personnel, waste) and limitations in the implementation of AM (funds, operational site, staff) were determined. This study allowed to identify the state of knowledge about AM technology and to identify implementation needs both in the area of machine and device production, production of functional prototypes, but also in the area of production of high-precision components with complex geometry. Managers of production companies have indicated that they are mainly interested in technologies that use laser. The data obtained allow the research to be targeted at a specific group of methods using laser technologies. The practical significance of our work is determined in the form of a recommendation for managers to support the selection and implementation of AM technology in the context of obtaining possible benefits for a manufacturing company.

Based on the obtained research, which clearly identify the general limitations that may affect the decision to implement AM in the enterprise, research initiations should be undertaken in the following directions:

- designing a system to help enterprises decide on the implementation of AM into production, analysing the current parameters of the production process and implementation possibilities,
- designing the matching system for the most optimal method of additive manufacturing
- improving the methods of additive manufacturing used in the area of production efficiency and reducing the costs of implementing and operating devices.

Based on the obtained data, another area requiring research was identified in order to increase the efficiency of devices and reduce production time. The second limitation indicated is the lack of funds and hence the inability to finance the implementation and operation of additive manufacturing devices. The metal industry uses AM technologies mainly in the production of high-precision components with complex geometry. The main limitation to implementing AM in enterprises from the metal industry is the lack of funds. This area requires further research to reduce the costs of application and operation of AM devices so that the profitability of AM implementation is achievable in a shorter period of time. Managers see the application potential for AM technology precisely in the area of production of high-precision elements whose production would be impossible or inefficient using traditional production methods. Decisions on the implementation of AM representatives of the automotive industry argue factors influencing both the development of the organization, processes and products themselves, balancing this with a reduction in production costs. In the metal industry, production cost reduction comes first.

The results of the empirical research, here presented, are the response to the need in-depth research, into Industry 4.0. A manufacturing company can be competitive in the market due to the high-quality of the products and services it offers and by implementing new solutions and technologies, e.g., intelligent new material handling systems, multi-agent system of autonomous automated guided vehicle [35,36], in the context of the Industry 4.0 concept [37,38]. Managers are also looking for solutions that will be helpful when deciding on the purchase of new technologies in order to adapt the enterprise to the Industry 4.0. concept.

Like all studies, this one has certain limitations that further research should aim to overcome. Firstly, because the intention is to analyse the influence of AM technology use on the increase manufacturing company's competitiveness, this study focuses on Polish manufacturing industries. It would be unwise to generalize the findings too broadly to other enterprises. Furthermore, all the variables were measured at the same moment in time. So, it would be useful to provide such research over a longer time period. These conclusions and limitations suggest proposals for future research directions.

6. Conclusions

Research conducted in 250 Polish manufacturing companies has shown that a significant number of companies in the west of Poland use, or are interested in implementing, AM technology. In the pilot studies, the desire to implement AM technology was indicated by some 5% of respondents, while in the main research, this was indicated by some 28.8%, who gave as the deciding factor, the need to reduce production costs and increase flexibility in responding to customer needs. Further work will require research into areas affecting the decision of the management boards of Polish manufacturing companies in order to implement AM technology and develop possible scenarios for co-operation between R&D units and production companies in order to increase the use of AM technology in the west of Poland. Moreover, the results of the empirical research, here presented, have become the motivation for conducting research into the construction of a decision-making system to support the selection and implementation of AM technology in the context of obtaining possible benefits for a production company.

Author Contributions: Conceptualization, J.P.-M.; data curation, M.T.; formal analysis, J.P.-M. and M.T.; funding acquisition, J.P.-M.; methodology, J.P.-M., M.T. and S.K.; resources, J.P.-M., M.T. and S.K.; software, S.K.; validation, J.P.-M., M.T. and S.K.; visualization, J.P.-M. and M.T.; writing—original draft, J.P.-M., M.T. and S.K.; writing—review and editing, J.P.-M. and M.T. All authors have read and agreed to the published version of the manuscript.

Funding: This research was funded by the programme of the MINISTER OF SCIENCE AND HIGHER EDUCATION under the name: "Regional Initiative of Excellence" in 2019–2022 project number 003/RID/2018/19; funding amount 11.936.596.10 PLN.

Conflicts of Interest: The authors declare no conflicts of interest.

References

1. The Economist. Special Report: A Third Industrial Revolution, 2012. Available online: http://www.economist.com/node/21552901 (accessed on 21 August 2019).
2. Goole, J.; Amighi, K. 3D printing in pharmaceutics: A new tool for designing customized drug delivery systems. *Int. J. Pharm.* **2016**, *499*, 376–394. [CrossRef] [PubMed]
3. Cichoń, K.; Brykalski, A. The use of 3D printers in industry. *Electrotech. Rev.* **2017**, *3*, 156–158.
4. Central Statistical Office of Poland; Information Society in Poland. *Results of Statistical Surveys in the Years 2014–2018*; Central Statistical Office of Poland: Warsaw, Poland, 2018.
5. Boyer, R.R. An overview on the use of titanium in the aerospace industry. *Mater. Sci. Eng.* **1996**, *A213*, 103–114. [CrossRef]
6. Cader, M. 3D printing in the industry. The latest trends and uses. *Automation* **2018**, *11*, 26–33.
7. Cader, M.; Trojnacki, M. Analysis of the possibility of using incremental technologies to manufacture mobile robot construction elements. *Meas. Autom. Robot.* **2013**, *2*, 200–207.
8. Deckard, C.R. Method and Apparatus for Producing Parts by Selective Sintering. U.S. Patent 4,863,538, 9 September 1989.
9. Flynn, J.M.; Shokran, A.; Newman, S.T.; Dhokia, V. Hybrid additive and subtractive machine tools—Research and industrial developments. *Int. J. Mach. Tools Manuf.* **2016**, *101*, 79–101. [CrossRef]
10. Zhao, G.; Ma, G.; Feng, J.; Xiao, W. Nonplanar slicing and path generation methods for robotic additive manufacturing. *Int. J. Adv. Manuf. Technol.* **2018**, *96*, 3149–3159. [CrossRef]
11. Gill, D.; Smugeresky, J.E.; Atwood, C.J. *Laser Engineered Net Shaping TM (LENS®) for the Repair and Modification of NWC Metal Components*; Sandia National Laboratories: Albuquerque, NM, USA, 2006.
12. Grobelny, P.; Furmanński, Ł.; Legutko, S. Investigations of Surface Topography of Hot Working Tool Steel Manufactured with the Use of 3D Print. In Proceedings of the 13th International Conference Modern Technologies in Manufacturing MTeM 2017—AMATUC, Cluj-Napoca, Romania, 12–13 October 2017; MATEC Web of Conferences. Volume 137 (02004).
13. Grobelny, P.; Furmanński, Ł.; Legutko, S. Selected Parameters of Surface Topography of Hot Working Tool Steel (1.2709) Manufactured with the Use of 3D Print. In Proceedings of the 12th International Multidisciplinary Conference, Johor Bahru, Malaysia, 1–2 May 2017.
14. Hull, C.W. Apparatus for Production of Three-Dimensional Objects by Stereolithography. U.S. Patent 4,575,330, 11 March 1986.
15. Karoluk, M.; Pawlak, A.; Chlebus, E. The use of SLM incremental technology in the processing of Ti-6Al-7Nb titanium alloy for biomedical applications. *Curr. Probl. Biomech.* **2014**, *8*, 81–86.
16. Kozłowski, E.; Mazurkiewicz, D.; Żabiński, T.; Prucnal, S.; Sęp, J. Assessment model of cutting tool condition for real-time supervision system. *Maint. Reliab.* **2019**, *21*, 679–685. [CrossRef]
17. Królikowski, M.A.; Krawczyk, M.B. Machining and incremental techniques as integral stages of the hybrid production process from metals in Industry 4.0. *Mechanic* **2018**, *129*, 8–9.

18. Liu, S.; Zhu, H.; Peng, G.; Yin, J.; Zeng, X. Microstructure prediction of selective laser melting AlSi10Mg using finite element analysis. *Mater. Des.* **2018**, *142*, 319–328. [CrossRef]
19. Liu, Y.J.; Liu, Z.; Jiang, Y.; Wang, G.W.; Yang, Y.; Zhang, L.C. Gradient in microstructure and mechanical property of selective laser melted AlSi10Mg. *J. Alloys Compd.* **2018**, *735*, 1414–1421. [CrossRef]
20. Mazurkiewicz, A. Analysis of the print quality of an element made of ABS thermoplastic made in FDM technology, Buses—Technology. *Oper. Transp. Syst.* **2017**, *6*, 956–960.
21. Mehesh, M.; Brandon, M.; Lane, M.; Donmez, A.; Feng, S.C.; Moylan, S.P. A review on measurement science needs for real-time control of additive manufacturing metal powder bed fusion processes. *Int. J. Prod. Res.* **2017**, *55*, 1–19.
22. Michopoulos, J.G.; Lambrakos, S.; Iliopoulos, A. Multiphysics challenges for controlling layered manufacturing. In Proceedings of the 34th Computers and Information in Engineering Conference, Buffalo, NY, USA, 17–20 August 2014.
23. Oczoś, K.E. The growing importance of Rapid Manufacturing in the incremental development of products. *Mechanic* **2008**, *4*, 241–257.
24. Rao, P. The final step toward quality in additive manufacturing. Analyzing and fixing problems on the fly in AM can smooth out defects, variances. *ISE Mag.* **2019**, *51*, 4–39.
25. Rumman, R.; Lewis, D.A.; Hascoet, J.Y.; Quinton, J.S. Laser metal deposition and wire arc additive manufacturing of materials: An overview. *Arch. Metall. Mater.* **2019**, *64*, 467–473.
26. Siemiński, P.; Budzik, G. *Additive Manufacturing Technologies. 3D Printing*; Warsaw University of Technology: Warsaw, Poland, 2015.
27. Srivastava, M.; Rathee, S.; Maheshwari, S.; Kundra, T. *Additive Manufacturing Fundamentals and Advancements*; CRC Press: Boca Raton, FL, USA, 2019. [CrossRef]
28. Takata, N.; Kodaira, H.; Suzuki, A.; Kobashi, M. Size dependence of microstructure of AlSi10Mg alloy fabricated by selective laser melting. *Mater. Charact.* **2017**, *143*, 18–26. [CrossRef]
29. Yang, Y.; Wang, G.; Liang, H.; Gao, C.; Peng, S.; Shen, L.; Shuai, C. Additive manufacturing of bone scaffolds. *Int. J. Bioprint* **2019**, *5*, 148. [CrossRef]
30. Herzog, D.; Seyda, V.; Wycisk, E.; Emmelmann, C. Additive manufacturing of metals. *Acta Mater.* **2016**, *117*, 371–392. [CrossRef]
31. Dvorak, F.; Micali, M.; Mathieu, M. Planning and Scheduling in Additive Manufacturing. *Intel. Artif.* **2018**, *21*, 40–52. [CrossRef]
32. Feldshtein, E.; Patalas-Maliszewska, J.; Kłos, S.; Kałasznikow, A.; Andrzejewski, K. The use of Plackett-Burman plans and the analysis of expert opinions, in order to assess the significance of controllable parameters of the plasma cutting process. *Maint. Reliab.* **2018**, *20*, 443–449. [CrossRef]
33. Ciurana, J.; Hernandez, L.; Delgado, J. Energy density analysis on single tracks formed by selective laser melting with CoCrMo powder material. *Int. J. Adv. Manuf. Technol.* **2013**, *680*, 1103–1110. [CrossRef]
34. Gola, A. Reliability analysis of reconfigurable manufacturing system structures using computer simulation methods. *Maint. Reliab.* **2019**, *21*, 90–102. [CrossRef]
35. Gola, A.; Kłosowski, G. Development of computer-controlled material handling model by means of fuzzy logic and genetic algorithms. *Neurocomputing* **2019**, *338*, 381–392. [CrossRef]
36. Bocewicz, G.; Wojcik, R.; Banaszak, Z. AGVs Distributed Control Subject to Imprecise Operation Times. In Proceedings of the KES International Symposium on Agent and Multi-Agent Systems: Technologies and Applications, Incheon, Korea, 26–28 March 2008; Volume 4953, pp. 421–430.
37. Li, L. China's manufacturing locus in 2025: With a comparison of "Made-in-China 2025" and "Industry 4.0". *Technol. Forecast. Soc. Chang.* **2018**, *135*, 66–74. [CrossRef]
38. Schwab, K. *The Fourth Industrial Revolution 2017*; Crown Publishing Group: New York, NY, USA, 2017.

 *applied sciences*

Article

Eco-Design of Energy Production Systems: The Problem of Renewable Energy Capacity Recycling

Svetlana Ratner [1,2], Konstantin Gomonov [1,*], Svetlana Revinova [1] and Inna Lazanyuk [1]

[1]   Department of Economic and Mathematical Modelling, Peoples' Friendship University of Russia (RUDN University), 6 Miklukho-Maklaya Street, 117198 Moscow, Russia; ratner-sv@rudn.ru (S.R.); revinova-syu@rudn.ru (S.R.); lazanyuk-iv@rudn.ru (I.L.)

[2]   Economic Dynamics and Innovation Management Laboratory, V.A. Trapeznikov Institute of Control Sciences, Russian Academy of Sciences, 65 Profsoyuznaya Street, 117997 Moscow, Russia

*   Correspondence: gomonov-kg@rudn.ru; Tel.: +7-495-433-4065

Received: 23 May 2020; Accepted: 22 June 2020; Published: 24 June 2020

Abstract: Due to the rapid development of recycling technologies in recent years, more data have appeared in the literature on the environmental impact of the final stages of the life cycle of wind and solar energy. The use of these data in the eco-design of modern power generation systems can help eliminate the mistakes and shortcomings when planning wind and solar power plants and make them more eco-efficient. The aim of this study is to extend current knowledge of the environmental impacts of most common renewables throughout the entire life cycle. It examines recent literature data on life cycle assessments of various technologies for recycling of wind turbines and photovoltaic (PV) panels and develops the recommendations for the eco-design of energy systems based on solar and wind power. The study draws several general conclusions. (i) The contribution of further improvements in PV's recycling technologies to environmental impacts throughout the entire life cycle is insignificant. Therefore, it is more beneficial to focus further efforts on economic parameters, in particular, on achieving the economic feasibility of recycling small volumes of PV-waste. (ii) For wind power, the issue of transporting bulky components of wind turbines to and from the installation location is critical for improving the eco-design of the entire life cycle.

Keywords: eco-design; end-of-life treatment; recycling; solar power plant; wind power plant; life cycle analysis

---

1. Introduction

The transition to a circular economy can significantly contribute to the achievement of the Sustainable Development Goals (SDGs), in particular, Goal 12 ("ensuring sustainable consumption and production patterns"). The foundations for the successful circulation of resources are laid long before the manufacturing of the products, namely, at the design stage of the production systems. A better design can make a product more durable, more suited for repair, modernization, or restoration. Thus, a better design of manufacturing processes will reduce the environmental burden throughout the entire life cycle of the product.

In the last decade, there has been a growing interest in the issues of optimal choice of energy technologies in academic literature. For instance, Turconi and co-authors [1] compared the main technologies of electricity generation based on hard coal, lignite, natural gas, oil, nuclear power, and several renewable sources by the amount of emissions. Poinssot and colleagues [2] weigh environmental footprints of a closed cycle of nuclear energy against the environmental footprint of an open cycle. Paper [3] examined carbon emissions and water consumption of electricity generation from pulverized coal, wind power, and solar PV, while [4] did the same kind of comparative analysis for a wider mix of technologies, including oil, natural gas, hydropower, biomass, and nuclear energy. Chang and coauthors [5] tried to choose a more ecologically friendly technology from coal and shale-gas-fired power generation, using the amount of greenhouse gas emissions and water consumption as the criteria. Paper [6] compared the environmental impacts of different PV-technologies, while [7] extended the analysis to include wind power and biogas electricity generation for consideration. The interest in this research topic has been determined by the rapid development of renewable energy (RE), which has a lower negative impact on the environment than traditional hydrocarbon energy throughout the entire life cycle [1,8–10]. However, the issues of decommissioning and end-of-life (EoL) treatment of renewable energy capacities, such as solar and wind plants, have not been thoroughly studied [11]. Disposal at landfills or incineration at waste processing plants remains the most common EoL treatment technologies for used photovoltaic modules and blades of wind turbines. Various mechanical, thermal, and chemical technologies of photovoltaic (PV) panel recycling, as well as recycling technologies for composite materials used in blades of wind turbines, are just at the beginning of their industrial applications. Therefore, the estimates of their environmental impact are quite different and not widely understood. There is no sufficient evidence for the comparison of available renewable energy technology by environmental impact throughout the entire life cycle, including the EoL stage. These knowledge constraints make the problem of a better eco-design of energy systems difficult to solve.

The significant growth of economic feasibility of solar and wind energy, achieved in recent years, opens up new prospects for even more rapid development of RE around the world, including the countries previously quite skeptical about renewables. This fully applies to Russia, which in recent years has been trying to reduce its lag from leading countries in the development of RE [12]. The hasty introduction of reactive innovations in the field of RE as a necessary response to the lag in technological development cannot be considered as an advantage from an economic point of view. However, it allows taking the experience and knowledge of other countries into account and making better decisions in the design of energy systems from an environmental point of view. In other words, the existing lag in technological development can be turned into a competitive advantage in the eco-efficiency of energy systems.

The environmental impact of the EoL stage is still a challenging area in the field of life cycle assessment. Slowly growing expertise in large-scale decommissioning of RE plants built 20–25 years ago as demonstration projects [13,14] brings new data and allows managers and policy-makers to formulate and solve the problems of optimal eco-design of energy systems. Monitoring of new technologies of EoL treatment of wind and solar plants is critical for effective regulation of the scale and type of RE installations, the method of recycling, and the location of recycling plants. The solution to such problems will contribute to the development of a circular economy.

The aim of this study is to extend current knowledge of the environmental impacts of most common RE technologies throughout the entire life cycle. It examines recent literature data on the life cycle assessment (LCA) of various technologies for the recycling of wind turbines and PV panels and develops recommendations for the eco-design of energy systems based on renewables. The rest of the paper is organized as follows: Section 2 presents the results of the systematic literature review of state-of-art recycling technologies for PV panels and wind turbines. Section 3 describes the features of inventory and life cycle assessments for the case of RE's recycling. Section 4 investigates the results of life cycle inventory and assessment of RE's recycling obtained under different assumptions and within different system boundaries. This section also presents general conclusions regarding the possibilities for improvements in the eco-design of the energy technologies under consideration. Section 5 discusses the practical applications of the study for the eco-design of energy systems in Russia. Section 6 debates the main contribution of the study to the existing pool of scientific literature, as well as its limitations and possibilities for future research.

2. Modern Recycling Technologies for PV Panels and Wind Turbines: A Systematic Literature Review

2.1. Ecology Issues of Renewable Energy Sources (RES)-Capacity Recycling

At the end of 2018, the global installed PV capacity reached 486 GW (REN21, 2019; Figure 1). Given that the average life of modern solar panels is 25 years [15], and the average weight of the most common 200 Wp multi-Si PV model is 16.8 kg [16], the world may face the need to dispose of more than 24 million tons of expired multi-Si PV models by 2035. In the future, these volumes will only increase due to the rapid development of solar energy in all countries [17,18]. According to expert forecasts, PV panels will reach 4500 GW of cumulative installed capacity worldwide by the middle of the century [15].

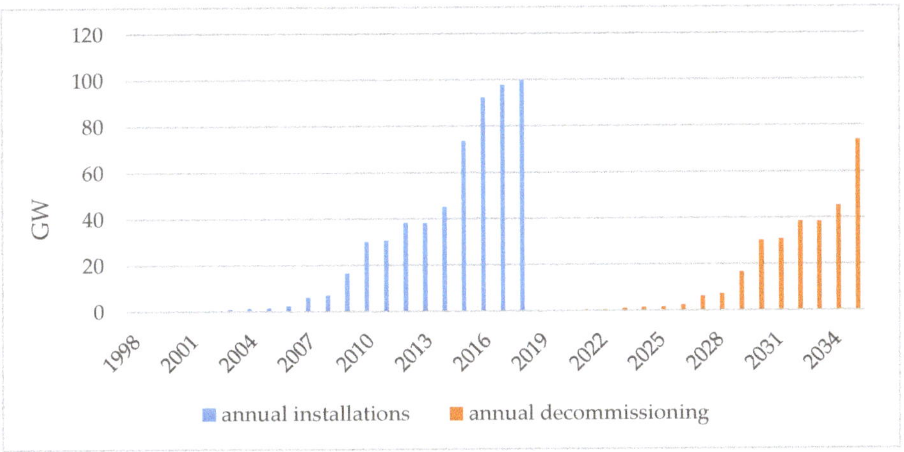

Figure 1. Historical development of annual new grid-connected photovoltaic (PV) installations (in GW) and anticipated decommissioning until 2035. Source: based on [19].

The PV solar panels contain lead (Pb), cadmium (Cd), and many other harmful chemicals. So far, the most common EoL treatment technology for PV models remains their disposal at landfills. [15]. It can be quite dangerous since harmful chemicals can leak into the ground, causing drinking water contamination [20]. Incineration is also used for PV models, as with regular municipal waste. It is crucial to understand that incinerating all kinds of electronic waste, including PV modules, can release toxic heavy metals into the atmosphere [21]. It is also known that some of the materials in PV modules are persistent and accumulative when released. This can cause long-term adverse ecology effects. Incineration also abolishes the opportunity of recovering raw materials. The only advantage of this method is that PV modules do not need to be separated from other commercial or industrial waste [22].

Industrial recycling capacities of used or defective PV models are currently limited and represented only by the Czech company Retina, several factories of First Solar in the United States, Germany, and Malaysia, Toshiba Environmental Solutions [15], and Veolia's new factory in France. At the legislative level, only the EU has established the rules for the collection and processing of solar panels in Waste of Electrical and Electronic Equipment (WEEE) Directive (Directive 2012/19/EU). Some countries with developed solar energy, such as South Korea, Japan, and the USA, are actively working on the problem of organizing the recycling of solar waste, while others, including China, are only exploring ways to solve this problem [23].

As has been previously reported in the literature, existing industrial technologies for recycling solar panels make it possible to achieve 95–97% recovery of cadmium and tellurium for thin-film solar cells [24], 90% recovery of glass [15,24], and 80% recovery of silicon for multi-Si PV models [25]. They also achieve 67–81% recovery of aluminum and 80% recovery of ethylene-vinyl acetate (EVA), which is used to fix individual PV models into a single panel. The use of recovered materials in the production cycle reduces resource consumption. It can achieve a drastic reduction in waste; however, it is associated with enormous expenditures of energy and/or chemical materials, as well as the emission of significant amounts of harmful substances into the atmosphere. Therefore, according to the data [16], the recycling of all the components of 1-kW multi-Si PV panels by modern industrial thermal and chemical methods emits 6.32 kg SO_2, 23.4 kg NO_2, 13.8 kg CO_2, as well as 0.97 kg of ammonia (NH_3), 2.59 kg of ethylene, and 4.26 kg of methane.

At the end of 2019, the worldwide installed capacity of wind turbines reached 621 GW and 29 GW for onshore and offshore wind energy, respectively [26] (Figure 2). Considering the historical pace of development of wind energy in the world and an average 20-year lifetime of wind turbines, one can expect that in the coming years, up until 2030, the volumes of decommissioning of wind turbines will grow exponentially.

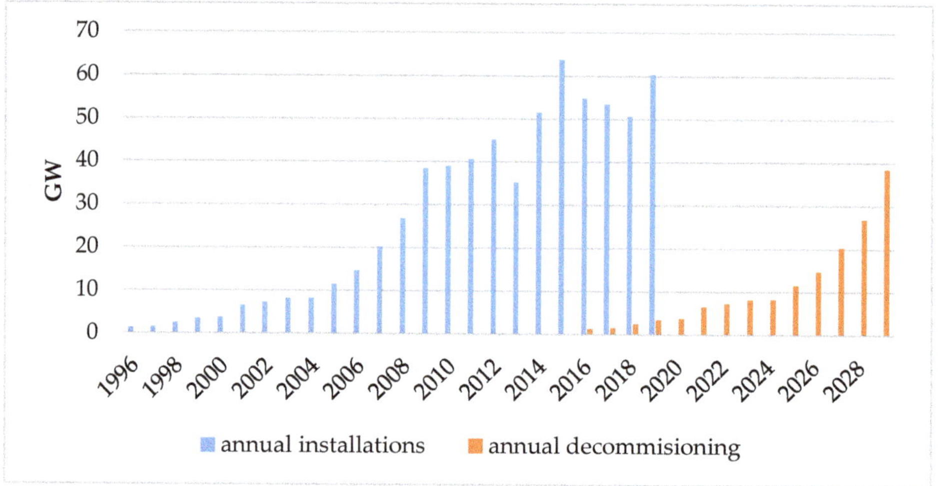

Figure 2. Historical development of annual new wind turbines installations (in GW) and anticipated decommissioning until 2035. Source: based on [26].

A closer look at the literature on the EoL of wind turbines reveals that the most difficult components for recycling are the blades and the concrete foundation [27,28]. The blades are currently simply landfilled or, in some cases, incinerated at municipal waste plants. As they are made from composite materials, their recycling is very complicated. In cases where it is technologically possible, the recovered materials have much lower quality than virgin materials [29,30]. It does not allow us to reuse them for the same purposes. In addition, the large sizes of the blades, which can reach 80–90 m in length on modern turbines, make it difficult to transport them to the place of recycling.

The landfill of large volumes of wind turbine blades is a severe environmental and economic problem [31]. The mass of the blades of modern wind turbines reaches 12.37–13.41 tons per 1 MW, and for turbines ending their life in the coming years, it is about 8.34 tons per 1 MW [32]. It means that if in 2020, it is necessary to landfill more than 31,000 tons of blades, in 2025, this amount will be already more than 142,000 tons, while in 2030, this number will be more than half a million tons. Landfilling of such large objects will require their preliminary energy-intensive shredding or cutting.

The EoL treatment of offshore wind turbines is an even more significant problem. Although their share in the total volume of wind turbine installations is much lower (Figure 3), they are more susceptible to adverse environmental effects and have a shorter lifetime. Thus, for example, in 2015, the first 10 MW Yttre Stengrund Swedish wind farm was decommissioned after only 15 years of operation [33]. In 2018, Utgrunden I Marine Wind Park (10.5 MW, Sweden) was decommissioned, which had operated for 18 years. In 2013, the first marine park built in the UK by Blyth (4 MW) ceased to operate after 13 years. Due to the lack of a clear demolition program, it is still abandoned in its original location. The Beatrice Demonstration Project (10 MW, UK) met the same fate.

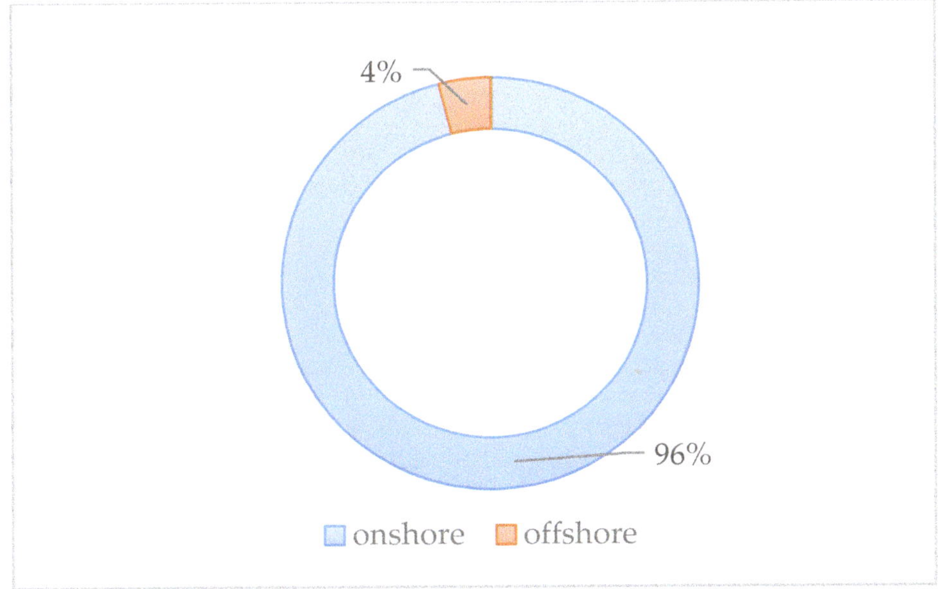

Figure 3. Onshore and offshore wind turbine installations in global wind capacity [26].

From a logistical point of view, the process of dismantling a marine wind park is much more complicated. Besides, there is currently no clear understanding of whether it is necessary to completely dismantle the foundation of wind turbines or they must be left in the same place [34]. Several methods are being developed in the literature to extend the lifetime of marine parks, which are based on the common idea of their more or less large-scale re-equipment. Some academicians have suggested replacing the engine, gearbox, and turbine blades and reusing the old tower, foundation, and power grid [33]. Others have suggested replacing all wind farm equipment except the foundation, which can last up to 100 years [35]. The implementation of such re-equipment is expected to reduce costs compared with the construction of new offshore wind farms [33]. Still, it does not entirely solve the disposal problem.

From an economic point of view, the extraction of valuable materials from used blades is of more interest. For example, Vestas' blades are made up of glass fiber, carbon fiber (CF), epoxy resin, and polyurethane [36]. Carbon fiber (CF), in addition to wind energy, is also actively used in aircraft, automotive, shipbuilding, and other industries. Carbon fiber is produced from light fractions of crude oil, propane, and propylene through a multistage chemical treatment that requires significant energy costs. The production of one kg of carbon fiber requires 165 kWh, while the recovery is only 8.8 kWh. According to [37], the energy intensity of the glass fiber recycling is 10 times less than that of its production.

Some European countries, especially the ones the most committed to the concept of developing a circular economy, prohibit the disposal of unprocessed waste with a high organic component at the legislative level. Due to the presence of carbon fiber in the composition of blades material, the organic component in them reaches 30%. It obliges energy companies to look for other ways to treat old blades from decommissioned wind turbines.

Thus, the issues of optimizing the entire life cycle of the most common renewable energy technologies are far from their final solution. The last stages of the life cycle of solar and wind power plants represent the most prospective area for improvements and development of new technologies.

2.2. Modern Technologies of PV Panel Recycling

The modern industrial algorithm for recycling silicon-based photovoltaic panels begins with their disassembly, during which aluminum (frame) and glass parts (coating) are separated. Almost 95% of the glass is recyclable, and all external metal parts can be reused to form new frames for solar modules [38]. The remaining materials are heat-treated at a temperature of 500 °C, during which the encapsulating plastic evaporates, leaving the silicon elements ready for further processing. In modern recycling plants, evaporating plastic is not released into the environment but used as a heat source for further heat treatment (energy recovery technology), which partially reduces adverse environmental effects [15,16,39].

After heat treatment, 80% of the PV modules (by mass) can be reused, while the rest is subject to additional cleaning. Silicon particles contained in cracked and scratched wafers are etched with acid and then melted for reuse to produce new silicon modules, resulting in an 85% level of recovery of the raw silicon material. The reuse of silicon can significantly reduce the adverse environmental effects of silicon photovoltaic panel manufacturing by saving, in addition to silicon sand itself, a significant amount of energy and water used in the production stage of metallurgical silicon. Based on the data of Huang and co-authors [16], it is possible to estimate the energy savings in the manufacturing of 1 kW of a photovoltaic panel due to the reuse of silicon as 76.2 kWh and water savings as 76.2 L. It avoids an average emission of 123.34 kg of CO_2, 3.36 kg of SO_2, and 0.73 kg of NO_x.

The recycling of thin-film photovoltaic panels is more complicated [21,40]. In the first step, the panel is crushed to a particle size of not more than 4–5 mm, which allows the removal of the lamination that holds the internal materials. Unlike silicon panels, the remaining substance consists of both solid and liquid materials, which need to be separated. Liquids pass through a precipitation and dehydration process to allow the recovery of semiconductor materials. Separation of semiconductor metals is carried out in various ways, depending on the technology used in the manufacture of panels; however, an average recovery rate of 95% is achieved. Solids contaminated with so-called interlayer materials are cleaned by vibration. Furthermore, the material undergoes washing, as a result of which there remains clean glass that must be melted and reused [39].

There are some new technologies for recycling photovoltaic silicon modules at the stage of laboratory research: technology for dissolving a laminating film in an organic solvent (tetrahydrofuran, o-dichlorobenzene or toluene), technology for dissolving with additional ultrasonic treatment, hot cutting, pyrolysis, technology for dissolving in nitric acid, and some others [41,42]. For thin-film modules, in addition to organic solvent dissolution and hot cutting technologies, laser exposure, vacuum processing, and flotation technologies are also being developed [43,44].

Previous studies have shown that the recycling of silicon modules becomes feasible at a scale of at least 19,000 tons per year [45]. Only when such volumes are achieved, the cost of processing decreases due to the economies of scale. The economic feasibility of recycling an entire PV panel is much higher than the expediency of recycling PV modules only, since the metal frame and electrical cables are easier to recycle, and the recovered materials (aluminum, copper) are valued higher [46,47]. The economic feasibility of recycling silicon panels substantially depends on the price of new modules, which has fallen significantly in recent years. For example, for Europe, the commercial attractiveness of silicon recycling is also determined by the lack of its mining (Figure 4). Of all the European countries, only Norway has a share of more than 4% in the global silicon production market.

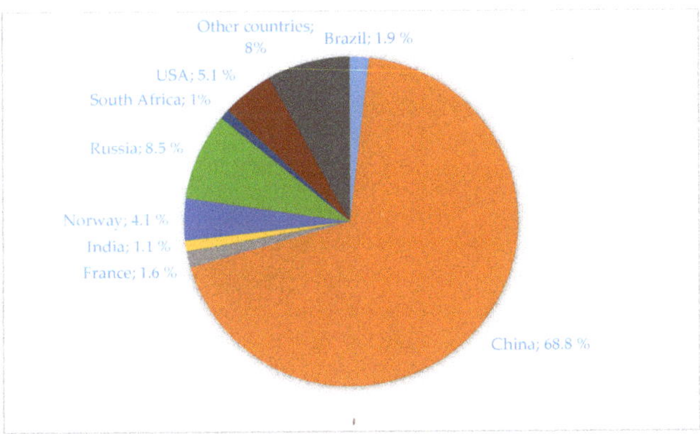

Figure 4. The share of leading countries in the production of the world's silicon supply. Source: based on data [48].

2.3. Modern Technologies of Blade Recycling

The simplest method of recycling a composite material is mechanical (cutting, crushing, crumpling). However, it damages individual fibers, which leads to a decrease in mechanical characteristics. According to the data [49], the tensile strength of recycled fiberglass compared to new is not more than 78%; for carbon fiber, it is less than 50%, with the fiber recycling ratio (recyclate yield rate) of 55–58% of the initial mass. The material recycled in this way has a low cost and cannot be a substitute for the virgin material. Typically, such a recycled composite material is used for less demanding mechanical applications, for instance, aggregates for artificial wood or asphalt and concrete in the construction industry [50,51]. Recycled carbon fiber (CF) is also suitable for short-fiber nonwoven composites used in aircraft and vehicle interiors [52]. Manufacturers can use recycled material as a filler or strength enhancer in new products, reducing the cost and environmental impact of waste disposal. In addition, recycled CF can be mixed with a polymer to produce thermally and electrically conductive materials used to improve the durability of paint, cement, and other construction materials [36,53].

Other actively developed methods for processing composite materials are pyrolysis, oxidation in a fluidized bed, and chemical decomposition of polymer resins (binders) [54]. During pyrolysis, the composite is heated to 450–700 °C in the absence of oxygen. The polymer resin is evaporated, and the fibers remain intact and can be restored entirely [36]. The recyclate yield rate of both carbon and glass fibers during pyrolysis is 67–70% of the original mass. The residual tensile strength is 52% for fiberglass and 78% for carbon fiber [49].

During oxidation in a fluidized bed, the polymer component (resin) is burned in a stream of hot (450–550 °C) air enriched with oxygen. The carbon fiber recyclate yield rate is higher and reaches 86–90% of the initial mass, while for fiberglass, it is 42–44% [49]. The residual tensile strength is 50% for fiberglass and 75% for carbon fiber. The chemical influence on the binder polymer element, selected directly for a specific combination of fiber and polymer matrix, helps to get the maximum possible amount of fiber suitable for reuse with minimal time and resources. The recyclate yield rate of both carbon and fiberglass is almost 100% [49], the residual tensile strength for fiberglass is 58%, and for carbon fiber, it is 95%.

To the best of our knowledge, the economic characteristics of the various processes of recycling of composite materials are absent in the literature because all of the methods mentioned above are still at the stage of laboratory research. Meanwhile, the data on the current energy intensity of these methods are known, which in combination with data on recycling efficiency, allow us to judge the competitiveness of methods from an economic point of view indirectly (Table 1). However, economic feasibility depends not only on the cost of the production process but on the availability and size of the market for recycled materials. Therefore, the prospects of the above methods for recycling blades of wind turbines are also determined by the fact that they are suitable for the recycling of composite wastes from other sectors of the economy (aircraft, automotive). Therefore, the cost of their development and practical application can be reduced due to economies of scale.

Table 1. Comparison of end-of-life (EoL) treatment methods for wind turbine blades. Source: based on data of [49].

Method	Fibrous Recyclate Yield Rate, %	Retained Tensile Strength of Recycled Fibre Compared to Virgin Fibre, %	Energy Demand, MJ/kg
Landfill	0	0	0.26
Mechanical	55–58%	GF—78% CF < 50%	0.27
Pyrolysis	67–70%	GF—52% CF—78%	21.2
Fluidised-bed process	GF—42–44% CF—86–90%	GF—50% CF—75%	GF—22.2 CF—9.0
Chemical	100%	GF—58% CF—95%	19.2

The possibility of incinerating composite materials in municipal waste plants for energy recovery depends on the ratio of polymer to carbon fiber in its composition [50]. A significant disadvantage of incinerating composite material is that approximately 60% of the initial mass remains in the form of ash. Ashes must also be disposed of in some way. At the moment, it is either subject to disposal at landfills (which also contradicts the legislation in the field of waste management in some countries) or to reuse in construction materials [55].

Another possibility for recycling wind turbine blades is to reuse them as load-bearing structures in the construction of buildings, technical infrastructure (e.g., bridges), or to create artificial reefs [55,56]. However, cases of such reuse of the blades are more likely experimental than industrial methods of disposal.

As can be seen from the analysis of the currently existing technologies for processing used capacities for wind and solar energy, the final stage of the life cycle of renewable energy facilities can impact the environment significantly. Therefore, the choice of electricity generation technology in planning the development of the energy system in a specific territory (region, municipality) should take into account the total environmental effects throughout the life cycle.

3. Methodology and Data

In the last decades, life cycle assessment (LCA) methodology is commonly used in the literature for evaluating the environmental performance of the entire life cycle of the products, processes, or production systems. The LCA algorithm is described in the series of ISO 14040–14043 standards and consists of the following four main stages:

(1) Goal and scope definition. At this stage, a researcher has to determine the research objective, define the functional unit, and set the system's boundaries. When determining the system's boundaries, most researchers use the "cradle-to-grave" or "cradle-to-gate" approaches. In a "cradle-to-gate" approach, life cycle stages such as extraction of raw materials, transportation, processing of materials, and manufacturing of the product are taken into account. When using the "cradle-to-grave" approach, steps such as the usage and final disposal of the product are additionally taken into account.

(2) Life cycle inventory (LCI). At this stage, a complete map of the studied life cycle is constructed as a sequence of production and transportation processes. The inputs of each production/transportation process (such as raw materials, water, and energy consumption), intermediate processes, and outputs (such as main product, by-product, wastes, and emissions to the air, water, and soil) are determined. As a rule, a specific production process that occurs at a particular enterprise, including its supply chain, is investigated. If some stage of the life cycle exists so far only as a laboratory (experimental) process and has various scenarios, then averaged data from various sources can be used for calculations. If the

processes are not yet profoundly understood, their inventories are incomplete. In such cases, the results of inventory analysis often complement sensitivity analysis.

(3) Life cycle impact assessment (LCIA). At this stage, the identified potential environmental impacts of the production system are translated into such categories as global warming, acidification, ecotoxicity (human, marine, or terrestrial), ozone depletion, abiotic depletion, and eutrophication. Indicators for measuring these environmental impact categories are determined by one of the following mature methods of LCIA: CML 2001, Cumulative Energy Demand (CED), eco-indicator 99, EDIP, ILCD, ReCiPe, IPCC, or IMPACT 2002+. Some methods (for example, CML 2001) allow the translation of all the adverse environmental effects of the product life cycle into physical units, such as kg CO_2-eq, kg NO_x-eq, kg PO_4-eq, and kg 1.4-DCB-eq. Others (for example, ReCiPe endpoint) allow the aggregation of all the negative effects in all categories into a single dimensionless quantity). The choice of methodology mainly depends on the preferences of the researcher and the objectives of the study.

(4) Interpretation. At this stage, a researcher summarizes the LCI and LCIA results, identifies critical points of the life cycle with the most considerable adverse effects, and makes recommendations for possible improvements.

If the life cycle of the product also includes the recycling stage, the modeling and calculation of the total adverse environmental effects are complicated by the fact that the recycled product can enter the production cycle again. In this case, it reduces the need for some raw materials. In modeling the processes of recycling, two main approaches are used: the "cut-off approach" and the "end-of-life approach". When using the first approach, the recycling efforts are economically allocated among the treatment process and all the recovered materials with a positive economic value. In the second approach, the potential benefits from the usage of recycled materials are calculated by awarding credits for the avoided environmental impacts caused by the primary production of replaced products. At the same time, some researchers use a simplified approach called open-loop recycling (OLR), in which further flows of secondary materials are not taken into account [57]. More details about differences in allocation approaches can be found in [38].

When choosing the technology for generating electricity from several possible options by environmental parameters, we are faced with the need to take into account the full life cycle of the electricity plant, including the recycling of massive generating equipment. The first stages of the life cycle of wind and solar energy are relatively well studied using LCA. For example, [6,39,58] compare a broad spectrum of environmental categories of photovoltaic technologies, [59] studies a life cycle of organic photovoltaics, and [60–62] compare GHG emissions of several photovoltaic technologies. Other studies [63–65] give life cycle assessments of solar panels manufactured in specific locations. Ecological aspects of the disposal of used equipment have also been studied (e.g., [66]). In contrast, accounting for recycling as a primary route for end-of-life treatment is a complex problem [38].

First, as a rule, when conducting an LCA for an electricity plant, 1 kWh of energy produced over the entire life of the equipment is used as a functional unit. Such a choice of a functional unit is convenient because it allows one to compare the results of LCA for different technologies and make a choice in favor of the most environmentally friendly. In LCA of the recycling of used generating equipment, the mass of the equipment itself (PV panel or wind turbine) usually is considered as a functional unit. Therefore, it is impossible to directly aggregate the LCA results of the first and last stages of the life cycle.

The second difficulty is the lack of a universal approach to determining the system's boundaries for recycling processes. Some researchers include the collection and transportation of used PV panels and wind turbines, while others suggest that recycling is done directly at RES plant location. The third difficulty is that, as noted above, most of the recycling processes of RES capacities are at the stage of laboratory research, so the data on the inputs and outputs of these processes are minimal and have great uncertainty.

This study systematizes the LCA results of various methods of EoL treatment of used PV panels and wind turbines, brings them to single units of measurement, and converts them to functional units used to evaluate the earlier stages of the life cycle of electricity plants. The conversion to the functional unit of 1 kWh of all estimations from the literature was carried out under the assumption that the capacity factor of PV plant is 14% [12], solar irradiation in the location of PV plant is 1200 kWh/m^2/year [12], the weight of the 200 Wp c-Si module is 16.8 kg [16], the weight of 1 m^2 of the module is 13.2 kg [67], the weight of 1 m^2 of the CdTe module is 14.63 kg [67], and the module efficiency is 10.5% [67].

The primary difference of the present research in comparison with similar ones is that we did not restrict ourselves with a certain LCIA method and considered a maximally wide spectrum of impact categories. The purpose of our comparative analysis of LCA results is to identify common conclusions that remain valid even with different approaches, assumptions, and system boundaries and can serve as guidelines for the eco-design of energy systems.

4. Results and Discussion

4.1. PV Panels LCA

To operate with LCA results performed for different system boundaries and functional units, we used data only from the sources where the results were given in physical units (not in dimensionless points). For this reason, the analysis did not include the results of studies where environmental impact assessments were presented in normalized values (e.g., [16,68–71]). Additionally, because recycling technologies are rapidly progressing, we selected for comparison only the results of the decade. Table 2 presents the results of a comparative analysis of the photovoltaic plant's life cycle.

The results of LCA in [41] were calculated according to the ReCiPe endpoint method with GaBi software. The functional unit was 1 kg of silicon-based PV waste modules. The study considered two types of PV modules (multi- and monocrystalline silicon modules) and several end-of-life treatment scenarios: landfill, incineration, and thermal, chemical, and mechanical recycling. In the case of recycling, the system boundaries included manufacturing, installation, operation, recycling, and reuse of recycled materials in a new cycle of manufacturing. One of the notable features of the study was a separate analysis of the transportation phase. In this case, the impacts do not depend on the recycling technology but the waste collection system and distance from it.

The environmental effects of transporting used modules to the place of recycling or disposal were taken into account in two ways: (1) assuming a transportation distance of 50 km; (2) assuming a transportation distance of 100 km (for the case of recycling only). Modeling results showed that transportation for 50 km increases impacts on human health up to 2.1–2.3 $\times$ 10^{-4} DALY in case of landfill or incineration, and up to 1–1.2 $\times$ 10^{-4} in case of recycling. Transportation for 100 km for recycling increases the impact on human health up to 1.6–1.7 $\times$ 10^{-4} DALY, which is comparable to the landfill or incineration indicators on site (without transportation). Similar results were obtained in two other categories of environmental impact. Transportation for 50 km increased ecosystem effects to 5.8–6.2 $\times$ 10^{-11} species·year in case of landfill or incineration and 3–3.5 $\times$ 10^{-11} species·year in case of recycling. Transportation per 100 km for processing increased the adverse effects up to 4.5–5.5 $\times$ 10^{-11} species·year, which offset all the positive effects of recycling compared to landfill or incineration.

Thus, the results of this analysis demonstrate that the recycling plant should be at most 80 km away from a PV plant. Otherwise, landfill and incineration scenarios have lower impacts when considering human health and ecosystems. This is a critical practical conclusion for the eco-design of the energy system regarding the territorial location of PV-plants. Solar plants must be placed compactly in a relatively small radius from the processing plant, or they should be some kind of mobile recycling plant. In addition, it is essential to consider not only the distance but also the method of transportation.

In [67], an LCIA is made for the recycling of c-Si and CdTe PV modules using both the cut-off and end-of-life approaches. Data were collected from several recycling companies in central Europe. The functional unit was 1 kg of used framed c-Si and unframed CdTe PV modules, but the final results were presented for 3 kW modules, mounted on a slanted roof. The life cycle impact assessment competed with the ILCD Midpoint 2011 method, but only six of the most relevant impact categories were taken into consideration.

Table 2. Life cycle assessment (LCA) results for PV panels.

Source	Functional Unit	System's Boundaries	Life Cycle Impact Assessment Method, Software	Results
Lunardi et al., 2018 [41]	1 kg of c-Si PV waste modules	Manufacturing, installation, operation, recycling (thermal/chemical/mechanical) and reuse of recycled materials in a new circle of manufacturing. Transportation is not included.	ReCiPe endpoint, GaBi	**Landfill and incineration:** Human health 1.6–1.7×10^{-4} DALYs Ecosystems 4.0–4.6×10^{-11} species·year Resources 8–10 \$ US **Recycling:** Human health 0.5–0.6×10^{-4} DALYs Ecosystems 1.1–$1.9*10^{-11}$ species·year Resources 2.1–4.1 \$ US
Latunussa et al., 2016 [72]	1000 kg of c-Si PV waste panels, including internal cables	Delivery of the waste to the recycling plant, recycling (pilot method) PV modules into metallurgical grade silicon scrap, sorting of the recyclable metal and glass fractions, disposal of residues. Transportation within a radius of 500 km by lorry	ILCD midpoint, SimaPro	Abiotic resource depletion (mineral) 4.32×10^{-3} kg Sb eq. CED 2780 MJ Freshwater ecotoxicity 1.31×10^{3} CTUe Marine eutrophication 1.05 kg N eq. Freshwater eutrophication 0.0456 kg P eq. Terrestrial eutrophication 11.7 molc N eq. Acidification 2.41 molc H+ eq. Photochemical ozone formation 2.86 kg NMVOC eq Ionizing radiation ecosystem 6.96×10^{-5} CTUe Ionizing radiation human health 22.9 kg U235 eq Particulate matter 0.0819 kg PM2.5 eq Human toxicity (non-cancer) 1.84×10^{-5} CTUh Human toxicity (cancer) 2.83×10^{-5} CTUh Ozone depletion 2.35×10^{-5} kg CFC-11 eq. Climate change 370 kg CO_2 eq.
Ardente et al., 2019 [11]	1000 kg of c-Si PV waste panel with TPT (Tedlar-polyethylene terephthalate-Tedlar) back sheet	**For innovative FRELP recycling process**: Delivery of the waste to the recycling plant, recycling (pilot method) PV modules into metallurgical grade silicon scrap, sorting of the recyclable metal and glass fractions, disposal of residues. Transportation within a radius of 300/500 km by lorry **For new pyrolysis recycling process:** fluorine-free PV waste is treated in a fixed-bed pyrolysis plant in place of incineration	EDIP2003 and ReCiPe, SimaPro	**FRELP/new pyrolysis recycling process** Climate change 461.0/361.3 kg CO_2 eq Ozone depletion 0.00003/0.00004 kg CFC-11 eq Human toxicity, cancer effects 0.00001/0.00001 CTUh Human toxicity, non-cancer effects 0.00002/0.00001 CTUh Human toxicity (ReCiPe) 349.9/290.7 kg 1,4-DB eq Particulate matter 0.1/0.1 kg PM2.5 eq Ionizing radiation HH 32.0/30.4 kg U235 eq Photochemical ozone formation 3.0/3.8 kg NMVOC eq Acidification 2.7/2.7 molc H+ eq Acidification (EDIP) 52.1/32.5 m2 Terrestrial eutrophication 12.2/12.0 molc N eq Freshwater eutrophication 0.054/0.1 kg P eq Marine eutrophication 1.095/1.1 kg N eq Freshwater ecotoxicity 255.4/246.3 CTUe Mineral depletion 0.004/0.004 kg Sb eq CED Non-renewable 3074.8/2899.4 MJ

Table 2. *Cont.*

Source	Functional Unit	System's Boundaries	Life Cycle Impact Assessment Method, Software	Results
Stolz & Frischknecht, 2016 [67]	1 kg of used 3kw c-Si framed module, mounted on a slanted roof	Manufacturing, installation, operation, recycling (mechanical) and reuse of recycled materials in a new circle of manufacturing. Transportation on 500 km by lorry	ILCD Midpoint 2011	Particulate matter 0.0375 kg PM2.5 eq Freshwater ecotoxicity 391 CTUe Human ecotoxicity (non-cancer) 4.55×10^{-6} CTUh Human ecotoxicity (cancer) 5.0×10^{-5} CTUh Abiotic resource depletion 6.18×10^{-3} kg Sb eq. Climate change 27.4 kg CO_2 eq.
	1 kg of used 3 kW CdTe unframed module, mounted on a slanted roof	Manufacturing, installation, operation, recycling (mechanical, dissolving, precipitation, and dewatering) and reuse of recycled materials in a new circle of manufacturing. Transportation on 678 km by lorry	ILCD Midpoint 2011	Particulate matter 0.0327 kg PM2.5 eq Freshwater ecotoxicity 197 CTUe Human ecotoxicity (non-cancer) 1.32×10^{-6} CTUh Human ecotoxicity (cancer) 2.05×10^{-7} CTUh Abiotic resource depletion 1.74×10^{-3} kg Sb eq. Climate change 6.05 kg CO_2 eq.
Held and Ilg, 2011 [73]	1 m^2 of CdTe module	Manufacturing, installation, operation, recycling (mechanical, dissolving, precipitation, and dewatering). Transportation of produced and used modules is not included	CML2001	**Module/module with a ground-mounted balance of systems** Primary Energy 831/1270 MJ Acidification potential 0.223/0.358 kg SO_2-eq Eutrophication potential $1.91 \times 10^{-2}/2.90 \times 10^{-2}$ kg PO_4-eq Global warming potential 513/861 kg CO_2-eq Photochemical Ozone Creation Potential $1.68 \times 10^{-2}/3.18 \times 10^{-2}$ kg Ethene-eq
	1 kWh of electricity	Manufacturing, ground-mounted installation, operation, recycling (mechanical, dissolving, precipitation, and dewatering). Transportation of produced and used modules is not included	CML2001	**Solar irradiation in place of installation 1200/1700/1900 kWh/m2/yr** Primary Energy (fossil) 0.43/0.31/0.28 MJ Acidification potential $1.23 \times 10^{-4}/8.68 \times 10^{-5}/7.77 \times 10^{-5}$ kg SO_2-eq Eutrophication potential $9.95 \times 10^{-6}/7.02 \times 10^{-6}/6.29 \times 10^{-6}$ kg PO4-eq Global warming potential $2.95 \times 10^{-2}/2.09 \times 10^{-2}/1.87 \times 10^{-2}$ kg CO_2-eq Photochemical Ozone Creation Potential $1.09 \times 10^{-5}/7.71 \times 10^{-6}/6.90 \times 10^{-6}$ kg Ethene-eq

Table 2. *Cont.*

Source	Functional Unit	System's Boundaries	Life Cycle Impact Assessment Method, Software	Results
Ilias et al., 2018 [74]	1 kg of reusable Si	From the collection, transportation (600 km) and recycling of PV waste (thermal treatment of the PV modules and the parallel removal of the organic EVA in a noble gas environment) to final production of the recovered Si	Gabi 6.0	GWP (100 years) 679.84 kg CO_2-eq. ODP 5.50 kg R11/CFC-eq. Human Toxic (Cancer) 2.70 CTUh Human Toxic (Non-Cancer) 3.90 CTUh Particulate Matter 0.54 kg PM2.5-eq. Ionising Radiation 0.26 kg U235-eq. Photochemical ozone formation 1.35kg NMVOC–eq. Acidification 5.95 moles of H+eq. Eco Tox 35.65 CTUe Resource depletion, water 406.11 kg Resource depletion, mineral, fossil 1.549 kg Sb eq.
Corcelli et al., 2018 [75]	1 m^2 of EoL c-Si PV panel	The thermal treatment of the decommissioned PV panel and the subsequent recycling of the recoverable fractions. High-rate (HR) recovery scenario assumed: the heat produced by the plastics thermal treatment is recovered and then exploited for hot water generation or for heating purposes within the plant where the process takes place. Transportation is not included.	ReCiPe Midpoint (H) v.1.10	Climate change 2.64×10^1 kg CO_2 eq Acidification potential 1.11×10^{-1} kg SO_2 eq Freshwater eutrophication 5.51×10^{-3} kg P eq Human toxicity 7.74E+00 kg 1,4-DB eq Photochemical ozone formation 6.30×10^{-2} kg NMVOC TEP 2.30×10^{-3} kg 1,4-DB eq WDP 1.88×10^0 m^3 MDP 2.17×10^0 kg Fe eq FDP 8.11×10^0 kg oil eq
		High-rate (HR) and low-rate (LR) recovery scenarios with environmental credits/LR scenarios assume that only the aluminum frame and glass are recycled, and the not-recovered fraction of copper, silicon, and ashes are disposed of in a sanitary landfill. Transportation is not included.	ReCiPe Midpoint (H) v.1.10	**HR/LR recovery scenario** GWP $-8.69 \times 10^1/-1.92 \times 10^1$ kg CO_2 eq TAP $-5.79 \times 10^{-1}/-2.29 \times 10^{-1}$ kg SO_2 eq FEP $-4.39 \times 10^{-2}/-1.31 \times 10^{-2}$ kg P eq HTP $-4.18 \times 10^1/-1.73 \times 10^1$ kg 1,4-DB eq POFP $-2.98 \times 10^{-1}/-1.10 \times 10^{-1}$ kg NMVOC TEP $-7.61 \times 10^{-3}/-2.32 \times 10^{-3}$ kg 1,4-DB eq WDP $-2.14 \times 10^0/-5.06 \times 10^{-1}$ m^3 MDP $-2.04 \times 10^0/-2.39 \times 10^{-1}$ kg Fe eq FDP $-2.06 \times 10^1/-2.64 \times 10^0$ kg oil eq

For the case of c-Si PV modules, the weighted average of multi- and monocrystalline Si PV modules was considered. It was estimated that the used modules would be transported by truck over a total distance of 500 km. The technology of recycling is mechanical. The desirable outputs of the recycling process are the bulk materials of glass cullets, aluminum scraps, and copper scraps. For the case of CdTe PV modules, the average transport distance from the place of installation to the recycling plant was considered as 678 km (data of First Solar's recycling facility located in Germany). The recycling process includes shredding and milling the used CdTe PV modules in the first step, then removing and dissolving the semiconductor film as the second step. The LCIA results for both types of panels are presented in Table 2. As one can see, CdTe PV modules are superior in environmental performance to c-Si PV modules in all categories of environmental impact. This conclusion can also be used in the eco-design of energy systems.

Note that even though both LCAs of [41,67] were performed throughout the entire life cycle, they still do not provide a complete picture of all the adverse effects of a PV plant. They do not take into account the productivity of a PV plant at the operation stage. More efficient PV plants (both by capacity factor and by energy conversion coefficient) can produce more useful products (electricity) for their life cycle and thereby reduce the need for the production and installation of additional PV capacity. Therefore, it is more appropriate to use 1 kWh of generated electricity as a functional unit for the LCAs of all energy facilities, including solar panels. Recalculation of the results of these two studies into other functional units (1 kWh of electricity), unfortunately, is impossible, since the location of the installation of solar panels is a source of uncertainty in this study. In locations with a high level of solar radiation, the total adverse environmental effects over the entire life cycle can be lower due to the greater volume of produced useful products (electricity).

Latunussa and co-authors [72] apply the LCA methodology to a pilot process of recycling of crystalline-silicon (c-Si) PV panels on the Italian "SASIL S.p.A" company. The functional unit was 1000 kg of PV waste panels, including internal cables. The system boundaries of the LCA included "gate-to-gate" recycling processes, starting from the delivery of the waste to the recycling plant and ending with the sorting of the different recyclable material fractions and the disposal of residues. The transportation of PV waste to the recycling plant was considered, while the decommissioning of the PV plant was not. Transportation was considered under the assumption that the distance from the PV plant to the nearest collection point of electronic waste is no more than 100 km, and the distance between the collection point and the recycling site is 400 km.

The novel process of recycling has a sequence of physical (mechanical and thermal) treatments, followed by acid leaching and electrolysis. The amounts of energy produced by the incineration process (for example, the incineration of the sandwich layer and plastics from cables) are considered as coproducts and their positive impact calculated as a credit (avoided environmental impact). By contrast, the environmental credits derived for potentially substituted primary materials are not included.

The modeling and calculation were implemented with SimaPro software version 8.0. The ILCD midpoint method was used for the life cycle impact assessment (Table 2). The "mineral, fossil, and renewable resource depletion" impact category is replaced with "abiotic depletion, fossil" and "cumulative energy demand (CED)" in order to distinguish the contributions of energy sources from those of nonenergy materials. The "land use" and "water resource depletion" impact categories were not taken into consideration due to their high uncertainty.

The results of this study do not have direct applications for the eco-design of the power system. However, they can be used to calculate the entire life cycle of c-Si PV modules for any method of installing panels and for any location.

Ardente [11] extended the results obtained in [72] by introducing options for the recycling process that depend on the material of the back-sheet. This study also introduced several additional impact categories. The results indicated that there is little potential to reduce the environmental impact of the recycling process in some categories due to the variation of materials for the back-sheet and recycling technologies for the auxiliary parts of the PV panel.

Held and Ilg [73] considered both individual stages and the entire life cycle of thin-film CdTe modules. They used two different functional units: 1 m^2 of the module and 1 kWh of energy produced. An interesting feature of the study is the comparison of impacts of the module with and without the balance of systems (BoS). It was revealed that the relative contribution of the BoS on the total impact of the PV power plant is around 35% to 45%. Estimates of environmental impacts for 1 kWh of electricity as a functional unit are made under three different assumptions about the amount of solar radiation in the location of the solar power plant. The disadvantage of the study is the lack of accounting for the effects of transporting new modules to the installation site and used modules to the recycling site. As shown above, transportation can have a significant influence on all environmental impact assessments. In addition, in this paper, the lifetime of the PV modules is assumed to be 30 years, which is an overestimation.

The authors of [74] presented a rather unusual approach to life cycle analysis, which the authors defined as "grave-to-cradle". They suggested that recycled silicon should replace a virgin material in the production of PV panels and considered the process in terms of industrial symbiosis. Recycling is carried out using thermal and chemical methods.

Corcelli and coauthors [75] investigated two c-Si PV panel recycling scenarios: one with a high level of material recovery and another with a low level. Environmental impacts were calculated in two versions: including credits and excluding them. The disadvantage of this study is the lack of accounting for transportation.

Comparing the environmental impacts of recycling with the impacts of all previous stages of the life cycle presented in EcoInvent (Tables 3 and 4), one can conclude that any technology for recycling PV waste is more ecologically friendly than landfilling. However, this is true only if the recycling plant is located in the same region where the panels are manufactured and used. In this case, the transportation of heavy modules over long distances was not required. This is consistent with what has been found in [46,47] for economical and in [76] for environmental parameters of recycling. By comparing the results from Tables 3 and 4, we can conclude that CdTe panels are preferable over silicon panels for the full life cycle (with EoL stage) in most categories of environmental impacts. These results go beyond previous reports [77], showing that current techniques used in the recycling of PVs produce higher impacts in the case of c-Si than in the case of CdTe. A further novel finding is the following: despite the fact that current technologies for recycling of PVs can be significantly improved from an environmental point of view [78], the contribution of these improvements to the negative impacts throughout the life cycle is insignificant. Therefore, it is advisable to focus on further improvement in economic parameters, in particular, on achieving the economic feasibility of recycling small volumes of PV waste.

4.2. LCA of Wind Turbines or Their Components

Aggregating the LCA results for wind turbines, as in the previous case, we also did not consider the results presented in dimensionless units (e.g., [68,79,80]), or received more than ten years ago (e.g., [81]). For a more detailed understanding of the development possibilities of recycling, we also separately examined the studies analyzing the LCA of the composite materials of blades (Table 5).

Garrett and Rønde [82] performed an LCA for a 50-MW wind park. The functional unit was 1 kWh of electricity produced. GaBi DfX software and primary data from Vestas were used. The analysis included all stages for the manufacturing and transportation of raw materials, turbine and wind plant components, as well as maintenance and end-of-life disposal. The study used data on primary fuel consumption collected by Vestas for truck and sea vessel transportation of turbine components. The transportation distance corresponded to the one that is part of Vestas' supply chain. Turbine recyclability (in percent turbine mass) was estimated to be between 81% and 85% (mainly metals), depending on the class of the turbine. In modeling, an avoided impact approach was used.

Table 3. Environmental impacts of different life cycle stages of c-Si PV panels (for 1 kWh [1]).

Impact	P+O+R [2]	R	P+O
	[41,67,74]	[11,72,75]	EcoInvent
Human heath, DALYs	1.35×10^{-7} [41]–4.59×10^{-7} [41]		1.66×10^{-10}
Abiotic resource depletion (mineral), kg Sb eq	1.67×10^{-5} [67]–4.18×10^{-3} [74]	1.08×10^{-8} [11]–1.17×10^{-8} [72]	9.61×10^{-6}
CED, MJ		7.51×10^{-3} [72]–8.30×10^{-3} [11]	1.26×10^{0}
Freshwater ecotoxicity, CTUe	1.06×10^{0} [67]	6.65×10^{-4} [11]–5.53×10^{-3} [72]	1.15×10^{-1}
Marine eutrophication, kg N eq.		2.84×10^{-6} [72]–2.97×10^{-6} [11]	1.06×10^{-4}
Freshwater eutrophication, kg P eq		1.23×10^{-7} [72]–1.13×10^{-6} [75]	6.33×10^{-5}
Terrestrial eutrophication, molc N eq.		3.16×10^{-5} [72]–3.29×10^{-5} [11]	1.02×10^{-3}
Acidification, molc H+ eq	1.61×10^{-2} [74]	6.51×10^{-6} [72]–7.29×10^{-6} [11]	6.46×10^{-4}
Photochemical ozone formation, kg NMVOC eq	3.65×10^{-3} [74]	7.72×10^{-6} [72]–1.29×10^{-5} [75]	3.33×10^{-4}
Ionizing radiation human health, kg U235 eq	7.02×10^{-4} [74]	6.18×10^{-5} [72]–8.64×10^{-5} [11]	7.88×10^{-3}
Particulate matter, kg PM2.5 eq	1.01×10^{-4} [67]–1.46×10^{-3} [74]	2.21×10^{-7} [72]–2.70×10^{-7} [11]	8.78×10^{-5}
Human toxicity (non-cancer), CTUh	1.23×10^{-8} [67]–1.07×10^{-2} [74]	2.70×10^{-11} [11]–5.40×10^{-11} [11]	3.42×10^{-8}
Human toxicity (cancer), CTUh	1.35×10^{-9} [67]–7.29×10^{-3} [74]	2.70×10^{-11} [11]–7.64×10^{-11} [72]	2.35×10^{-9}
Ozone depletion, kg CFC-11 eq		6.35×10^{-11} [72]–1.08×10^{-10} [11]	7.62×10^{-9}
Climate change, kg CO$_2$ eq	7.13×10^{-6} [74]–7.40×10^{-6} [67]	9.76×10^{-4} [11]–5.41×10^{-3} [75]	8.19×10^{-2}
Human toxicity, kg 1,4-DB eq	2.08×10^{-2} [74]	7.85×10^{-4} [11]–9.45×10^{-4} [11]	1.72×10^{-1}
Acidification potential, kg SO$_2$-eq		2.28×10^{-5} [75]	5.44×10^{-4}
Eutrophication potential, kg PO4-eq			2.35×10^{-4}
Photochemical Ozone Creation Potential, kg Ethene-eq			3.26×10^{-4}
Resource depletion, water, m^3	1.10×10^{-3} [74]	3.85×10^{-4} [75]	1.04×10^{-3}

[1]—The conversion to the functional unit of 1 kWh was carried out on the assumption that the capacity factor is 14%, the weight of the 200 Wp module is 16.8 kg, and the weight of 1 m^2 of the module is 13.2 kg. [2]—P, manufacturing (production) of PV panel with upstream activities; O, operation; R, recycling.

Table 4. Environmental impacts of different life cycle stages of CdTe PV panels (for 1 kWh [3]).

Impact	P+O+R	P+O (CdTe)
	[67,73]	EcoInvent
Abiotic resource depletion (mineral), kg Sb eq	8.75×10^{-6} [67]	6.12×10^{-6}
CED, MJ	2.85×10^{-1} [73]–4.36×10^{-1} [73]	5.99×10^{-1}
Freshwater ecotoxicity, CTUe	9.90×10^{-1} [67]	1.50×10^{1}
Particulate matter, kg PM2.5 eq	1.64×10^{-4} [67]	5.04×10^{-5}
Human toxicity (non-cancer), CTUh	6.63×10^{-9} [67]	8.84×10^{-8}
Human toxicity (cancer), CTUh	1.03×10^{-9} [67]	7.58×10^{-9}
Climate change, kg CO$_2$ eq	3.04×10^{-2} [73]–2.96×10^{0} [73]	3.95×10^{-2}
Acidification potential, kg SO$_2$-eq	7.66×10^{-5} [73]–1.23×10^{-4} [73]	4.28×10^{-4}
Eutrophication potential, kg PO4-eq	6.56×10^{-6} [73]–9.96×10^{-6} [73]	1.90×10^{-4}
Photochemical Ozone Creation Potential, kg Ethene-eq	5.77×10^{-6} [73]–1.09×10^{-5} [73]	1.55×10^{-5}

[3]—Recalculation into a functional unit of 1 kWh was carried out by assuming that the mass of 1 m^2 of the module is 14.63 kg (according to First Solar), module efficiency 10.5%, solar irradiation 1200 kWh/m^2/year.

Table 5. LCA data for wind turbines and components.

Source	Object/Functional Unit	System's Boundaries	Life Cycle Impact Assessment Method, Software	Results
Garrett and Rønde, 2012 [82]	Wind park/1 kWh of electricity	Extraction raw materials, manufacturing turbine components and wind plant components, installation, maintenance and end-of-life disposal. Blades are not recyclable.	CML, GaBi DfX	Abiotic depletion (elements) 0.44 mg Sb eq Abiotic depletion (fossils) 0.10 MJ Acidification 37 mg SO_2-eq Eutrophication 3.7 mg PO_4-eq Freshwater aquatic ecotoxicity 100 mg DCB-e Global warming 7.7 g CO_2-eq Human toxicity 1150 mg DCB-eq Marine aquatic ecotoxicity 1100 g DCB-eq Photochemical oxidation 4.1 mg Ethene Terrestrial ecotoxicity 19 mg DCB-eq
Bonou et al., 2016 [83]	Wind turbine/1 kWh of electricity	Extraction raw materials, manufacturing turbine components and wind plant components, installation, maintenance and end-of-life disposal. Blades are 100% recyclable (shredding, incineration), foundation is 50% recyclable (crushing).	ILCD midpoint and ReCiPe endpoint	Climate change (onshore average) 6.5 g CO_2-eq Climate change (offshore average) 9.35 g CO_2-eq
Poujol et al., 2020 [84] and Boulay et al., 2017 [85]	Floating wind park/1 kWh of electricity	Extraction raw materials, manufacturing turbine components and wind plant components, installation, maintenance and end-of-life disposal. Blades are not recyclable.	ILCD, CED AWARE ReCiPe 2016 midpoint (H)	Climate Change 22.3 g CO_2 eq Resource Depletion 14.6 mg Sb eq Water Use 6.7 L Marine Ecotoxicity 6. 91 g 1,4-DBC eq Air Quality 25.2 mg PM2.5 eq CED Renewable 22.4 kJ CED non-Renewable 429 kJ
Vestas, 2015 [86]	Wind park/1 kWh of electricity	Extraction raw materials, manufacturing turbine components and wind plant components, installation, maintenance and end-of-life disposal. Blades at EoL treatment: 50% incineration, 50% landfill	CML 2013, GaBi DfX	Abiotic resource depletion (elements) 0.06 mg Sb-e Abiotic resource depletion (fossils) 0.10 MJ Acidification potential 32 mg SO_2-e Eutrophication potential 3.7 mg PO_4-e Freshwater aquatic ecotoxicity potential 62 mg DCB-e Global warming potential 7.2 g CO_2-e Human toxicity potential 1427 mg DCB-e Marine aquatic ecotoxicity potential 793 g DCB-e Photochemical oxidant potential 3.9 mg Ethene Terrestrial ecotoxicity potential 4.1 mg DCB-e

Table 5. *Cont.*

Source	Object/Functional Unit	System's Boundaries	Life Cycle Impact Assessment Method, Software	Results
Al-Behadili et al., 2015 [87]	Wind turbine/1 kWh of electricity	Extraction raw materials, manufacturing turbine components and wind plant components, installation, maintenance and end-of-life disposal. Blades are not recyclable.	N/a	SO_2 0.0271291 g NO_X 0.0382147 g CO_2 10.420169 g N_2O 0.0001474 g CH_4 0.0001065 g NMVOC 0.0003469 g CO 0.0112237 g
Chipindula et al., 2018 [88]	Wind turbine/1 kWh of electricity	Extraction raw materials, manufacturing turbine components and wind plant components, installation, maintenance and end-of-life disposal. Blades and foundation are not recyclable.	Impact 2002+ midpoint and CED, SimaPro	**Onshore/Shallow-water offshore/Deep-water offshore:** Carcinogens 29.64/89.93/79.54 kg C_2H_3Cl-eq Non-carcinogens 23.11/59.05/62.51 kg C_2H_3Cl-eq Respiratory inorganics 0.90/2.65/1.96 kg $PM_{2.5}$-eq Ionizing radiation 4663/10,197/8838 Bq C14-eq Ozone layer depletion $4.64 \times 10^{-5}/9.59 \times 10^{-5}/9.76 \times 10^{-5}$ kg CFC11-eq Respiratory organics 0.217/0.777/0.275 kg C_2H_4-eq Aquatic ecotoxicity 65,200/178,164/181,041 kg TEG Terrestrial ecotoxicity 2627/67,364/66,740 kg TEG Terrestrial acid/nutria 8.4/22.91/19.09 kg SO_2-eq Land occupation 6.86/21.32/15.44 m2 Aquatic acidification 2.65/7.72/6.47 kg SO_2eq Aquatic eutrophication 0.40/1.01/1.29 kg PO_4 P-lim Global warming 440/1144/648 kg CO_2eq Non-renewable energy 6578/16,115/10,930 MJ Mineral extraction 223/590/702 MJ
Alsaleh and Sattler, 2019 [57]	Wind turbine/1 kWh of electricity	Extraction raw materials, manufacturing turbine components and wind plant components, installation, maintenance and end-of-life disposal. Blades and foundation are not recyclable.	TRACI, SimaPro	Ozone depletion 2.69×10^{-9} kg CFC-11-eq Global warming 1.80×10^{-2} kg CO_2-eq Smog 1.71×10^{-3} kg O_3-eq Acidification 1.04×10^{-4} kg SO_2 –eq Eutrophication 9.78×10^{-5} kg N eq Carcinogenic 1.78×10^{-9} CTUh Non-carcinogenic 5.88×10^{-9} CTUh Respiratory effects 1.46E−05 kg $PM_{2.5}$eq Ecotoxicity 1.83×10^{-1} CTUe Fossil fuel depletion 2.02×10^{-2} MJ Cumulative energy demand 1.07×10^{-1} kWh Water depletion Index 9.24×10^{-7} m^3

Table 5. *Cont.*

Source	Object/Functional Unit	System's Boundaries	Life Cycle Impact Assessment Method, Software	Results
Guezuraga et al., 2012 [89]	Wind turbine/1 kWh of electricity	Extraction raw materials, manufacturing turbine components and wind plant components, installation, maintenance and end-of-life disposal. Blades are incinerated; the foundation is landfilled.	CED, GEMIS	**2.0 MW-geared/1.8 MW-gearless** CO_2-eq 9.73/8.82 g Cumulative energy demand 0.654/0.645 kWh
Stavridou et al., 2019 [90]	Wind turbine/1 kWh of electricity	Extraction raw materials, manufacturing turbine components and wind plant components, installation, maintenance and end-of-life disposal. Blades are incinerated, the foundation is landfilled.	CED, GEMIS	**Steel tower/lattice tower** Cumulative energy demand 0.48/0.33 kWh
Lefeuvre et al., 2016 [91]	Blades, 1 kg of carbon fiber reinforced polymers (CFRPs)	The process of pyrolysis only	ReCIPe Midpoint (H), SimaPro	Global warming potential −0.08 kg CO_2 eq
Li et al., 2016 [92]	Blades, 1 t of CFRPs	CFRP treatment routes: landfilling, incineration, recycling and landfilling, recycling, and incineration	EcoInvent data	**Landfilling/incineration/recycling and landfilling/recycling and incineration** Global warming potential 24/2011/-378/750 kg CO_2-eq.

Bonou [83] used the ILCD method for the assessment of the environmental impact of two types of wind power plants (onshore and offshore) from the extraction of raw materials to the EoL. The primary data were collected from four representative European power plants with state-of-art technology provided by Siemens Wind Power. The functional unit was 1 kWh. The EoL of the power plants consisted of the management of construction and demolition wastes. The recycling of turbine blades and foundation was included. The recycling process for composite blades was mechanical shredding and incineration in cement production. The recycling process for the cement foundation was crushing with the positive output of crushed gravel. The results in physical units were obtained only for the impact category "climate change", and they indicated the preference of land-based wind parks.

Poujol [84] studied a floating 24-MW offshore wind farm's life cycle from "cradle-to-grave". The multicriteria approach was used for LCIA. It is based on the combined performance of ILCD, CED, and ReCiPe 2016 MidPoint method (H). An additional parameter of "water use" was estimated according to the AWARE method [85]. The functional unit was 1 kWh of electricity. The recycling technology was not specified. Presumably, this is the usual recycling of metals and landfill for composite blades. The contribution of the EoL stage to all the adverse effects was quite large due to the need to use diesel-powered marine vehicles for decommissioning a wind farm located 16 km from the coast.

In a report of Vestas [86], the environmental impacts associated with the production of electricity from a 50-MW onshore wind plant were studied using a cradle-to-grave LCA. The functional unit was 1 kWh of electricity. At the EoL stage, it was assumed that metals are recycled (85–87% of the total turbine mass), and composite blades are incinerated by 50% and landfilled by another 50%. Blades and foundation treatment did not bring avoided environmental impacts. An important distinguishing feature of this work is the high certainty of data on all processes. It included the process of transporting equipment to the installation site and the recycling site and the process of connecting to the grid. The estimated transportation of the turbine components ranged from 50 km for the base to 2200 km for the blades; the transportation distance to the recycling site was assumed to be 200 km.

Al-Behadili and El-Osta [87] calculated the emissions of a 1.65-MW wind turbine over its entire life cycle (including EoL). The functional unit was 1 kWh of electricity. An important feature was a fact that emissions related only to energy consumption were taken into account. The initial data were obtained from literature and taken in an averaged form. EoL treatment technologies were not specified, but from the context of the paper, one can conclude that it was recycling of the metal parts of the turbine and the landfilling of the blades. Transportation distance was not specified.

Chipindula [88] performed a classic LCA for several types of turbines: medium ground-level power (1, 2, and 2.5 MW), powerful offshore in shallow water (2 and 2.5 MW), and offshore in deep waters (2.5 and 5 MW). All intermediate and final calculated data were averaged over the class of turbines. The functional unit was 1 kWh of electricity. EoL treatment involved the recycling of metal components of the turbine in the range from 55% (for aluminum) to 90% (for iron, steel, and copper) and 100% disposal of composite parts and a concrete base. Transportation of turbine components to the installation site was taken into account, and the maximum distance was assumed to be 10,000 km. Transportation to the place of recycling was not taken into account.

The results of the study show that although offshore wind parks have a higher capacity factor (45–47% compared to 35% for onshore wind turbines), in most categories, they produce more significant adverse impacts. This is precisely due to the contribution of the EoL stage, in which concrete foundations are left in the ground.

Alsaleh and Sattler [57] performed a life cycle inventory and assessment of the various phases of the 2-MW Gamesa onshore wind turbine life cycle in terms of TRACI impact categories with SimaPro Software version 8.3.2. The study considered RES recycling (OLR) with 98%, 90%, and 50% recycling rates for metals, plastic, and electronic components, respectively. Fiberglass (blade material) and lubricants were considered 100% landfilled. The most notable result of the study is the conclusion regarding the environmental friendliness of transporting new turbines to the installation site: the impact of truck transport for the distance 656 km is, in most cases, comparable or even greater than that of transoceanic ship transport, which was 8325 km. The effect of transporting the used parts of the turbines to the place of recycling or landfill was not taken into account.

Guezuraga [89] used GEMIS software for modeling the entire life cycle of two types of wind turbines: a 2-MW turbine with a gearbox and a 1.8-MW turbine without a gearbox. The paper assumed that the 2-MW turbine was transported for a distance of 2700 km, and the 1.8-MW turbine was transported for a distance of 1100 km by truck. The transportation distance to the place of recycling or landfill was not indicated. Only energy-related impact categories were considered. Recycling of stainless steel, cast iron, and copper was considered, whereas epoxy, plastic, and fiberglass were incinerated. The concrete foundation was 100% landfilled. Comparing the results for the two turbines, the authors concluded that the impact of the 1.8-MW turbine without gearbox was a little less.

The authors of [90] compared energy demand for the entire life cycle of 2-MW onshore wind turbines with a tall (76.16 m) tower. The study compared the ecological impact of traditional steel and innovative lattice towers. A lattice tower needs around 35% less steel, and it has an almost 33% lighter foundation, which gives a significant advantage for transportation and construction. GEMIS software was used for modeling total energy demand. The transportation for 240 km by truck and 1020 km by ship to the location of installation was considered. EoL treatment included recycling of metals with 5–10% loss, the 100% incineration of epoxy, fiberglass, and plastic, and 100% landfill of the concrete foundation. The distance to the place of processing, incineration, or landfill was not indicated. The study concluded that the turbine with a lattice tower was 32% less impactful on the environment in terms of CO_2 emissions.

Of the studies modeling LCAs of composite materials, only a few can be distinguished. For example, in [91], recycling by pyrolysis of carbon fiber reinforced polymers (CFRPs) was investigated with an LCA according to the ReCIPe Midpoint (H) 1.10 method. The functional unit was 1 kg of CFRP waste. It was estimated that recycling can avoid −0.08 kg CO_2 eq. due to the recovery of methanol and ethyl acetate.

In [92], a limited version of LCA was carried out for several possible options for handling CFRP waste. Only CO_2 emissions were taken into account. The models in this study were based on hypothetical CFRP treatment routes because exact facility locations were not identified. "Gate-to-grave" models have been developed for CFRP waste treatment by landfilling and incineration, beginning at the point of waste collection and including waste processing (disassembly, shredding), transport, and waste treatment (landfill, incineration). For recycling, a "gate-to-gate" approach is taken, which includes the production of composite materials from recycled carbon fibers (r-CF) and the use and/or disposal of other recyclate materials. Maintenance and facility construction is also not included in the LCI boundaries.

The results demonstrated that landfilling produced minor GHG emissions (24 kg CO_2eq./t CFRP) due to the inert nature of CFRP waste; incineration resulted in the greatest net GHG emissions (2011 kg CO_2eq./t CFRP) from the combustion process. Energy outputs from incineration are assumed to displace the electricity and natural gas-fired heat generation, which gave a credit (−1041 kg CO_2eq./t CFRP). Mechanical recycling with landfilling of the coarse recyclate fraction and displacement of GF production resulted in a net global warming potential reduction of 378 kg CO_2eq./t CFRP.

The results of a comparative analysis of environmental assessments of the life cycle of wind turbines are summarized in Table 6. The following conclusion can be drawn: increasing the capacity of wind turbines by increasing its height and the span of the blades from an environmental point of view is unreasonable since transporting bulky components of the turbine to the installation and recycling site or landfill contributes too much to the overall ecology footprint. For the same reason, it is inappropriate to build offshore wind farms far in deep waters. Contrary to the findings of [93], we did not find solid evidence that the environmental impacts of onshore wind farms are higher. This situation can only change if environmentally friendly modes of transport (for example, electric trains) are used. Overall, these findings are in accordance with findings reported by Wang et al. [93]. The technologies for recycling the blades need further development. So far, it can be stated with a degree of certainty that incinerating the blades of wind turbines is an environmentally unacceptable alternative to their disposal. The development of technologies to reduce the weight of the tower for a wind turbine, for example, by improving its design, has significant prospects from the environmental point of view.

Table 6. Environmental impacts of different life cycle stages of wind plants (for 1 kWh).

Impact	P+O+R [4]	Blades Recycling Potential	P+O [5]
	[57,82–84,86–90]	[91,92]	EcoInvent
Climate Change, g CO_2 eq	6.5 [84]–10.42 [87]	−0.01088 [91]–0.102 [92]	95.7
Resource Depletion, mg Sb eq	0.06 [86]–14.6 [83]		11.692
Abiotic depletion (fossils), MJ	0.02 [57]–0.1 [82,86]		1.4744
Water Use, L	0.0009 [57]–6.7 [83]		40.842
Marine Ecotoxicity, g 1,4-DBC eq	6.91 [83]		179.95
Air Quality, mg PM2.5 eq	14.6 [57]–25.2 [83]		203.25
CED, kWh	0.107 [57]–0.654 [89]		1.476
Acidification potential, mg SO_2-e	27.1 [87]–104 [57]		540
Eutrophication potential, mg PO_4-e	3.7 [86]		231.48
The photochemical oxidant, mg Ethene	3.9 [86]		39.714
N_2O, g	0.0001474 [87]		0.003353
CH_4, g	0.0001065 [87]		0.26812
NMVOC, g	0.0003469 [87]		0.47615
CO, g	0.0112237 [87]		0.6853
Carcinogens, CTUh	1.78×10^{-9} [57]		1.5899×10^{-8}
Non-carcinogens, CTUh	5.88×10^{-9} [57]		1.0176×10^{-7}
Land occupation, m^2	6.86/21.32/15.44 [88]		0.012118
Ozone depletion, kg CFC-11-eq	2.69×10^{-9} [57]		8.1088×10^{-9}

[4]—Minimum and maximum values are given in single units of measurement from sources [57,82–84,86–89] of Table 5. Data from [88] were not taken into account since they are 3–4 orders of magnitude higher than data from other sources, which most likely indicates significant differences in the choice of primary data. The exception was the data on the land occupation category since they were not found in other sources, and it was impossible to compare them with any other results. [5]—Data from EcoInvent for Vestas 2 MW wind turbine (global).

A major limitation of our study is the uncertainty induced by parameters of technological, spatial, and temporal nature in LCA models of wind and solar plants. For example, recent research [94,95] suggests that there is great variability in results within sets of wind turbines with similar nominal power output. Nevertheless, we can still state several important common principles that can be applied to the eco-design on energy systems based on RES. This is particularly important when investigating new possibilities for the development of renewables in countries where the issues of eco-design have not been properly addressed yet.

5. Applications for Eco-Design of Energy Systems and Energy Policy: The Case of Renewable Energy in Russia

Therefore, despite the incomplete data and the presence of significant uncertainties, the available studies of the entire life cycle of solar and wind energy plants allow us to draw several general conclusions regarding the eco-design of their production systems. Firstly, recycling is much more preferable from an environmental point of view than disposal, so it must be foreseen (including in the organizational and regulatory aspects) at the earliest stages of development of the renewable energy technologies. Secondly, when choosing a location for solar and wind energy plants, it is necessary to take into account not only the climatic and infrastructural conditions but also the distance from the place of the proposed location of the recycling enterprises.

Let us examine how this conclusion can be applied to the improvement of the eco-design of solar and wind energy projects in Russia. The literature review shows that there has been little discussion on the problem of RES capacity recycling in this country [96]. A reasonable explanation of this knowledge gap is an initial stage of development of wind and solar energy. To date, the most developed type of renewable energy in Russia is solar. In 2019, the volume of electricity generated by solar stations in Russia amounted to 850.38 MWh, while wind stations contributed only 209.84 MWh. Other types of renewable sources are represented by small hydropower plants (48.9 MWh), biogas plants (28 MWh), biomass and waste power plants (43.5 MWh), and landfill gas (9.6 MWh).

The system of government incentives for renewable energy in Russia stimulates not only the construction of new renewable energy plants but also the development of national manufacturers of solar panels and wind turbines. Due to these incentives, Hevel (the largest producer of silicon heterostructured solar modules in Russia) and NovaWind JSC (the Rosatom division responsible for wind energy projects) are already successfully operating on the Russian market.

Hevel currently manages 1145.5 MW of solar projects that are scattered throughout southern Siberia, the Volga region, the North Caucasus, and the Far East (Figure 5). The production unit of the company is located in the city of Novocheboksarsk, the Republic of Chuvashia. Thus, the distance between the manufacturer and the most remote solar energy facility is more than 7500 km.

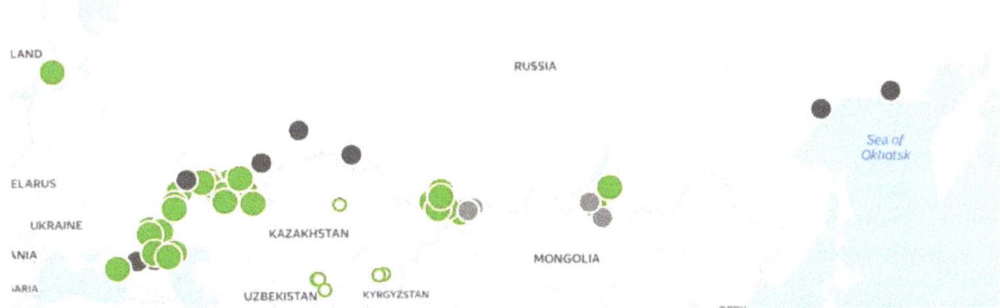

Figure 5. The locations of solar projects under Hevel management (single solar plants are marked with green color; solar plants as a part of thermal power plants are marked with grey color).

At the time it was launched in 2015, the plant's production capacity was only 90 MW per year; by 2019, it had already increased to 300 MW per year. The company plans to implement several large solar projects not only in Russia but also in Kazakhstan and further increase production to strengthen its position in the market and reduce production costs due to the economies of scale, learning-by-doing, and learning-by-researching [97,98]. Following this development strategy will likely lead to the construction of an increasing number of solar power plants, including ones at a considerable distance from the place of direct production.

Such an approach in the future can significantly complicate the development of the recycling of solar modules, making it environmentally and economically impractical due to the need to transport old modules over long distances. A more feasible solution, which allows optimizing the design of the production system even at this stage in the development of solar energy, could be the development of a network of Hevel subsidiaries in the regions with the greatest natural and infrastructural potential for solar generation. Furthermore, it is necessary to begin the development of technologies for the production of thin-film modules, the environmental friendliness of which, over the entire life cycle, exceeds the ecology parameters of the silicon modules.

NovaWind JSC was founded in 2017 as a system integrator of all wind projects launched at this point in Russia. The technology partner of the company is Dutch company Lagerwey. The production units of the company are located in the city of Volgodonsk, Rostov Region. Shortly, the company plans to reach a production capacity of 96 turbines per year. The first Adygea Wind Farm project, with a total capacity of 150 MW (60 wind turbines), was successfully launched in early March 2020. In addition, the implementation of several large-scale wind projects in the Stavropol and Krasnodar territories in the Rostov and Volgograd regions is planned for the next two years. All these regions are part of the Southern Federal District and border each other, so the logistics of wind projects can be considered optimal, from both economic and environmental points of view.

The first NovaWind JSC wind projects include the construction of wind farms using L100 direct-drive wind turbines with a capacity of 2.5 MW. The program is already under development for the start of the production of turbines with a high power of 4.5 MW. Since the negative impact of powerful turbines on the environment throughout the life cycle is greater than the impact of medium-power turbines, when developing such projects, special attention should be paid to the transportation of the large-sized parts of the turbines, as well as to ensure the possibility of their subsequent disposal.

6. Conclusions

Even though assessing the environmental impact of the life cycle of the fastest-growing types of renewable energy is a popular topic in scientific literature, it has still not been studied enough. The particular difficulty is the study of the recycling of obsolete solar and wind power equipment. The first reason for this is the complexity of the methodological approach, which involves the distribution of all the positive effects created in the recycling process to earlier stages of the life cycle. The second reason is the great uncertainty of the current estimates, both due to the lack of primary data and because of the variety of logistic solutions for supply chains.

This study systematizes the most recent and relevant LCA results and identifies, based on the comparison, the patterns that work under any assumption about the technology and organization of the recycling process. The paper concludes by arguing that the contribution of further improvements in PV recycling technologies to environmental impacts throughout the entire life cycle is insignificant. Therefore, it is more beneficial to focus further efforts on economic parameters, in particular, on achieving the economic feasibility of recycling small volumes of PV waste. In the case of wind power, broadly translated, our findings indicate that the issue of transporting bulky components of wind turbines to and from the installation location is critical for improving the eco-design of the entire life cycle.

These patterns can now be the basis for the design of production systems for generating electricity and thereby contribute to the development of a circular economy. It is important to note that those countries that are still at the beginning of the path of renewable energy development (for example, Russia) can take into account the positive and negative experiences of the leading countries and optimize their production systems at the very early stages of their development.

The main practical conclusion that can be drawn from this study is that in order to improve the environmental performance of solar and wind power plants, Russia needs to develop accurate data centers for the forecasting of waste flow. It will help to elaborate a reasonable proactive strategy and optimize the number and locations of recycling plants.

Future research should further develop and confirm these initial findings by monitoring the progress in RES environmental performance on each stage of their life cycles. In addition, the comparison of less developed RES technologies with an LCA "cradle-to-grave" approach might prove an important area for future research.

Author Contributions: Conceptualization, S.R. (Svetlana Ratner); Data curation, S.R. (Svetlana Ratner); Formal analysis, I.L.; Funding acquisition, I.L.; Investigation, S.R. (Svetlana Ratner); Methodology, S.R. (Svetlana Ratner); Project administration, S.R. (Svetlana Revinova); Resources, K.G.; Software, S.R. (Svetlana Ratner); Supervision, S.R. (Svetlana Ratner); Validation, S.R. (Svetlana Ratner), K.G. and S.R. (Svetlana Revinova); Visualization, K.G.; Writing—original draft, S.R. (Svetlana Ratner); Writing—review & editing, S.R. (Svetlana Ratner), K.G., S.R. (Svetlana Revinova) and I.L. All authors have read and agreed to the published version of the manuscript.

Funding: The publication has been prepared with the support of the «RUDN University Program 5–100».

Abbreviations

CED	cumulative energy demand
CF	carbon fiber
EoL	end-of-life
GF	glass fiber
LCA	life cycle assessment
LCI	life cycle inventory
LCIA	life cycle impact assessment
PV	photovoltaic
RE	renewable energy
RES	renewable energy sources
SDGs	Sustainable Development Goals
WEEE	waste of electrical and electronic equipment

References

1. Turconi, R.; Boldrin, A.; Astrup, T. Life cycle assessment (LCA) of electricity generation technologies: Overview, comparability and limitations. *Renew. Sustain. Energy Rev.* **2013**, *28*, 555–565. [CrossRef]
2. Poinssot, C.; Bourg, S.; Ouvrier, N.; Combernoux, N.; Rostaing, C.; Vargas-Gonzalez, M.; Bruno, J. Assessment of the environmental footprint of nuclear energy systems. Comparison between closed and open fuel cycles. *Energy* **2014**, *69*, 199–211. [CrossRef]
3. Feng, K.; Hubacek, K.; Siu, Y.L.; Li, X. The energy and water nexus in Chinese electricity production: A hybrid life cycle analysis. *Renew. Sustain. Energy Rev.* **2014**, *39*, 342–355. [CrossRef]
4. Xin, L.; Feng, K.; Siu, Y.L.; Hubacek, K. Challenges faced when energy meets water: CO_2 and water implications of power generation in inner Mongolia of China. *Renew. Sustain. Energy Rev.* **2015**, *45*, 419–430. [CrossRef]
5. Chang, Y.; Huang, R.; Ries, R.; Masanet, E. Life-cycle comparison of greenhouse gas emissions and water consumption for coal and shale gas fired power generation in China. *Energy* **2015**, *86*, 335–343. [CrossRef]
6. Ratner, S.V.; Lychev, A.V. Evaluating environmental impacts of photovoltaic technologies using Data Envelopment Analysis. *Adv. Syst. Sci. Appl.* **2019**, *19*, 12–30. [CrossRef]
7. Salnikova, A.A.; Slavjanov, A.S.; Khrustalev, E.Y.; Khrustalev, O.E. Environmental effects evaluation of innovative renewable energy projects. *J. Environ. Manag. Tour.* **2019**, *10*, 100–108. [CrossRef]
8. Padey, P.; Blanc, I.; Le Boulch, D.; Zhao, X. A simplified life cycle approach for assessing greenhouse gas emissions of wind electricity. *J. Ind. Ecol.* **2012**, *16*, S28–S38. [CrossRef]
9. Hsu, D.D.; O'Donoughue, P.; Fthenakis, V.; Heath, G.A.; Kim, H.C.; Sawyer, P.; Choi, J.-K.; Turney, D.E. Life cycle greenhouse gas emissions of crystalline silicon photovoltaic electricity generation. *J. Ind. Ecol.* **2012**, *16*, S122–S135. [CrossRef]
10. Kim, H.C.; Fthenakis, V.; Choi, J.-K.; Turney, D.E. Life Cycle Greenhouse Gas Emissions of Thin-film Photovoltaic Electricity Generation. *J. Ind. Ecol.* **2012**, *16*, S110–S121. [CrossRef]
11. Ardente, F.; Latunussa, C.E.; Blengini, G.A. Resource efficient recovery of critical and precious metals from waste silicon PV panel recycling. *Waste Manag.* **2019**, *91*, 156–167. [CrossRef] [PubMed]
12. Ratner, S.V.; Nizhegorodtsev, R.M. Analysis of renewable energy projects' implementation in Russia. *Therm. Eng.* **2017**, *64*, 429–436. [CrossRef]
13. Hocine, L.; Samira, K.M. Optimal PV panel's end-life assessment based on the supervision of their own aging evolution and waste management forecasting. *Sol. Energy* **2019**, *191*, 227–234. [CrossRef]
14. Farrell, C.; Osman, A.; Doherty, R.; Saad, M.; Zhang, X.; Murphy, A.; Harrisona, J.; Vennarda, A.S.M.; Kumaravel, V.; Al-Muhtaseb, A.H.; et al. Technical challenges and opportunities in realising a circular economy for waste photovoltaic modules. *Renew. Sustain. Energy Rev.* **2020**, *128*, 109911. [CrossRef]
15. Chowdhury, S.; Rahman, K.S.; Chowdhury, T.; Nuthammachot, N.; Techato, K.; Akhtaruzzaman, M.; Tiong, S.K.; Sopian, K.; Amin, N. An overview of solar photovoltaic panels' end-of-life material recycling. *Energy Strat. Rev.* **2020**, *27*, 100431. [CrossRef]
16. Huang, B.; Zhao, J.; Chai, J.; Xue, B.; Zhao, F.; Wang, X. Environmental influence assessment of China's multi-crystalline silicon (multi-Si) photovoltaic modules considering recycling process. *Sol. Energy* **2017**, *143*, 132–141. [CrossRef]

17. Domínguez, A.; Geyer, R. Photovoltaic waste assessment of major photovoltaic installations in the United States of America. *Renew. Energy* **2019**, *133*, 1188–1200. [CrossRef]

18. Santos, J.; Alonso-Garcia, C. Projection of the photovoltaic waste in Spain until 2050. *J. Clean. Prod.* **2018**, *196*, 1613–1628. [CrossRef]

19. REN21. Renewables 2019 Global Status Report. Paris: REN21 Secretariat. 2020. Available online: https://www.ren21.net/wp-content/uploads/2019/05/gsr_2019_full_report_en.pdf (accessed on 19 May 2020).

20. Nover, J.; Zapf-Gottwick, R.; Feifel, C.; Koch, M.; Metzger, J.W.; Werner, J.H. Reply to "Comment on 'Long-term leaching of photovoltaic modules'". *Jpn. J. Appl. Phys.* **2017**, *57*, 19102. [CrossRef]

21. Savvilotidou, V.; Antoniou, A.; Gidarakos, E. Toxicity assessment and feasible recycling process for amorphous silicon and CIS waste photovoltaic panels. *Waste Manag.* **2017**, *59*, 394–402. [CrossRef]

22. Lunardi, M.M.; Alvarez-Gaitan, J.P.; Bilbao, J.I.; Corkish, R. A review of recycling processes for photovoltaic modules. *Sol. Panels Photovolt. Mater.* **2018**. [CrossRef]

23. Xu, Y.; Li, J.; Tan, Q.; Peters, A.L.; Yang, C. Global status of recycling waste solar panels: A review. *Waste Manag.* **2018**, *75*, 450–458. [CrossRef] [PubMed]

24. First Solar Sustainability Report. 2017. Available online: http://www.firstsolar.com/-/media/First-Solar/Sustainability-Documents/FirstSolar_SustainabilityReport2017.ashx (accessed on 19 May 2020).

25. End-of-life Management Solar Photovoltaic Panels. Irena.org. 2020. Available online: https://www.irena.org/publications/2016/Jun/End-of-life-management-Solar-Photovoltaic-Panels (accessed on 14 June 2020).

26. Global Wind Report 2019. Global Wind Energy Council. 2020. Available online: https://gwec.net/global-wind-report-2019/ (accessed on 19 May 2020).

27. Mamanpush, S.H.; Li, H.; Englund, K.; Tabatabae, A.T. Recycled wind turbine blades as a feedstock for second generation composites. *Waste Manag.* **2018**, *76*, 708–714. [CrossRef] [PubMed]

28. Tazi, N.; Kim, J.; Bouzidi, Y.; Chatelet, E.; Liu, G. Waste and material flow analysis in the end-of-life wind energy system. *Resour. Conserv. Recycl.* **2019**, *145*, 199–207. [CrossRef]

29. Kalkanis, K.; Psomopoulos, C.; Kaminaris, S.; Ioannidis, G.; Pachos, P. Wind turbine blade composite materials—End of life treatment methods. *Energy Procedia* **2019**, *157*, 1136–1143. [CrossRef]

30. Rahimizadeh, A.; Tahir, M.; Fayazbakhsh, K.; Lessard, L. Tensile properties and interfacial shear strength of recycled fibers from wind turbine waste. *Compos. Part A Appl. Sci. Manuf.* **2020**, *131*, 105786. [CrossRef]

31. Lichtenegger, G.; Rentizelas, A.A.; Trivyza, N.; Siegl, S. Offshore and onshore wind turbine blade waste material forecast at a regional level in Europe until 2050. *Waste Manag.* **2020**, *106*, 120–131. [CrossRef]

32. Liu, P.; Barlow, C.Y. Wind turbine blade waste in 2050. *Waste Manag.* **2017**, *62*, 229–240. [CrossRef]

33. Topham, E.; McMillan, D.; Bradley, S.; Hart, E. Recycling offshore wind farms at decommissioning stage. *Energy Policy* **2019**, *129*, 698–709. [CrossRef]

34. Smyth, K.; Christie, N.; Burdon, D.; Atkins, J.P.; Barnes, R.; Elliott, M. Renewables-to-reefs?—Decommissioning options for the offshore wind power industry. *Mar. Pollut. Bull.* **2015**, *90*, 247–258. [CrossRef]

35. Arany, L.; Bhattacharya, S.; Macdonald, J.; Hogan, S.J. Design of monopiles for offshore wind turbines in 10 steps. *Soil Dyn. Earthq. Eng.* **2017**, *92*, 126–152. [CrossRef]

36. Cherrington, R.; Goodship, V.; Meredith, J.; Wood, B.; Coles, S.R.; Vuillaume, A.; Feito-Boirac, A.; Spee, F.; Kirwan, K. Producer responsibility: Defining the incentive for recycling composite wind turbine blades in Europe. *Energy Policy* **2012**, *47*, 13–21. [CrossRef]

37. Shuaib, N.A.; Mativenga, P. Carbon footprint analysis of fibre reinforced composite recycling processes. *Procedia Manuf.* **2017**, *7*, 183–190. [CrossRef]

38. Faircloth, C.C.; Wagner, K.H.; Woodward, K.E.; Rakkwamsuk, P.; Gheewala, S.H. The environmental and economic impacts of photovoltaic waste management in Thailand. *Resour. Conserv. Recycl.* **2019**, *143*, 260–272. [CrossRef]

39. Ludin, N.A.; Mustafa, N.I.; Hanafiah, M.M.; Ibrahim, M.A.; Teridi, M.A.M.; Sepeai, S.; Zaharim, A.; Sopian, K. Prospects of life cycle assessment of renewable energy from solar photovoltaic technologies: A review. *Renew. Sustain. Energy Rev.* **2018**, *96*, 11–28. [CrossRef]

40. Azeumo, M.F.; Germana, C.; Ippolito, N.M.; Franco, M.; Luigi, P.; Settimio, S. Photovoltaic module recycling, a physical and a chemical recovery process. *Sol. Energy Mater. Sol. Cells* **2019**, *193*, 314–319. [CrossRef]

41. Lunardi, M.M.; Alvarez-Gaitan, J.P.; Bilbao, J.I.; Corkish, R. Comparative life cycle assessment of end-of-life silicon solar photovoltaic modules. *Appl. Sci.* **2018**, *8*, 1396. [CrossRef]

42. Zhao, P.; Guo, J.; Yan, G.; Zhu, G.; Zhu, X.; Zhang, Z.; Zhang, B. A novel and efficient method for resources recycling in waste photovoltaic panels: High voltage pulse crushing. *J. Clean. Prod.* **2020**, *257*, 120442. [CrossRef]

43. Rocchetti, L.; Beolchini, F. Recovery of valuable materials from end-of-life thin-film photovoltaic panels: Environmental impact assessment of different management options. *J. Clean. Prod.* **2015**, *89*, 59–64. [CrossRef]

44. Sica, D.; Malandrino, O.; Supino, S.; Testa, M.; Lucchetti, M.C. Management of end-of-life photovoltaic panels as a step towards a circular economy. *Renew. Sustain. Energy Rev.* **2018**, *82*, 2934–2945. [CrossRef]

45. Deng, R.; Chang, N.L.; Ouyang, Z.; Chong, C.M. A techno-economic review of silicon photovoltaic module recycling. *Renew. Sustain. Energy Rev.* **2019**, *109*, 532–550. [CrossRef]

46. Li, Y.; Li, H.; Wang, G.; Liu, X.; Zhang, Q. Study on the optimal deployment for Photovoltaic components recycle in China. *Energy Procedia* **2019**, *158*, 4298–4303. [CrossRef]

47. Guo, Q.; Kluse, C. A framework of photovoltaics recycling facility location optimization. *Sustain. Prod. Consum.* **2020**, *23*, 105–110. [CrossRef]

48. Blagoeva, D.T.; Dias, P.A.; Marmier, A.; Pavel, C.C. *Assessment of Potential Bottlenecks along the Materials Supply Chain for the Future Deployment of Low-Carbon Energy and Transport Technologies in the EU: Wind Power, Photovoltaic and Electric Vehicles Technologies, Time Frame: 2015–2030*; Publications Office of the European Union: Brussels, Belgium, 2016. [CrossRef]

49. Liu, P.; Meng, F.; Barlow, C. Wind turbine blade end-of-life options: An eco-audit comparison. *J. Clean. Prod.* **2019**, *212*, 1268–1281. [CrossRef]

50. Pickering, S. Recycling technologies for thermoset composite materials—Current status. *Compos. Part A Appl. Sci. Manuf.* **2006**, *37*, 1206–1215. [CrossRef]

51. Hu, J.; Danish, M.; Lou, Z.; Zhou, P.; Zhu, N.; Yuan, H.; Qian, P. Effectiveness of wind turbine blades waste combined with the sewage sludge for enriched carbon preparation through the co-pyrolysis processes. *J. Clean. Prod.* **2018**, *174*, 780–787. [CrossRef]

52. Jiang, G.; Pickering, S.; Lester, E.; Turner, T.; Wong, K.; Warrior, N. Characterisation of carbon fibres recycled from carbon fibre/epoxy resin composites using supercritical n-propanol. *Compos. Sci. Technol.* **2009**, *69*, 192–198. [CrossRef]

53. Marsh, G. Could thermoplastics be the answer for utility-scale wind turbine blades? *Reinf. Plast.* **2010**, *54*, 31–35. [CrossRef]

54. Cousins, D.S.; Suzuki, Y.; Murray, R.E.; Samaniuk, J.R.; Stebner, A.P. Recycling glass fiber thermoplastic composites from wind turbine blades. *J. Clean. Prod.* **2019**, *209*, 1252–1263. [CrossRef]

55. Jensen, J.; Skelton, K. Wind turbine blade recycling: Experiences, challenges and possibilities in a circular economy. *Renew. Sustain. Energy Rev.* **2018**, *97*, 165–176. [CrossRef]

56. Asmatulu, E.; Twomey, J.; Overcash, M. Recycling of fiber-reinforced composites and direct structural composite recycling concept. *J. Compos. Mater.* **2013**, *48*, 593–608. [CrossRef]

57. Alsaleh, A.; Sattler, M. Comprehensive life cycle assessment of large wind turbines in the US. *Clean Technol. Environ. Policy* **2019**, *21*, 887–903. [CrossRef]

58. Soares, W.M.; Athayde, D.D.; Nunes, E.H.M. LCA study of photovoltaic systems based on different technologies. *Int. J. Green Energy* **2018**, *15*, 577–583. [CrossRef]

59. Lizin, S.; Van Passel, S.; De Schepper, E.; Maes, W.; Lutsen, L.; Manca, J.; Vanderzande, D. Life cycle analyses of organic photovoltaics: A review. *Energy Environ. Sci.* **2013**, *6*, 3136–3149. [CrossRef]

60. Peng, J.; Lu, L.; Yang, H. Review on life cycle assessment of energy payback and greenhouse gas emission of solar photovoltaic systems. *Renew. Sustain. Energy Rev.* **2013**, *19*, 255–274. [CrossRef]

61. Gerbinet, S.; Belboom, S.; Léonard, A. Life Cycle Analysis (LCA) of photovoltaic panels: A review. *Renew. Sustain. Energy Rev.* **2014**, *38*, 747–753. [CrossRef]

62. Wu, P.; Ma, X.; Ji, J.; Ma, Y. Review on life cycle assessment of greenhouse gas emission profit of solar photovoltaic systems. *Energy Procedia* **2017**, *105*, 1289–1294. [CrossRef]

63. Collins, K.; Powell, B.; Anctil, A. Life cycle assessment of silicon solar panels manufacturing in the United States. In Proceedings of the 2015 IEEE 42nd Photovoltaic Specialist Conference (PVSC), New Orleans, LA, USA, 14–19 June 2015; pp. 1–4.

64. Chen, W.; Hong, J.; Yuan, X.; Liu, J. Environmental impact assessment of monocrystalline silicon solar photovoltaic cell production: A case study in China. *J. Clean. Prod.* **2016**, *112*, 1025–1032. [CrossRef]

65. Luo, W.; Khoo, Y.S.; Kumar, A.; Low, J.S.C.; Li, Y.; Tan, Y.S.; Wang, Y.; Aberle, A.G.; Ramakrishna, S. A comparative life-cycle assessment of photovoltaic electricity generation in Singapore by multicrystalline silicon technologies. *Sol. Energy Mater. Sol. Cells* **2018**, *174*, 157–162. [CrossRef]

66. Desideri, U.; Proietti, S.; Zepparelli, F.; Sdringola, P.; Bini, S. Life Cycle Assessment of a ground-mounted 1778kWp photovoltaic plant and comparison with traditional energy production systems. *Appl. Energy* **2012**, *97*, 930–943. [CrossRef]

67. Stolz, P.; Frischknecht, R.; Wambach, K.; Sinha, P.; Heath, G. *Life Cycle Assessment of Photovoltaic Module Recycling*; treeze Ltd.: Uster, Switzerland, 2016.

68. Zhong, Z.W.; Song, B.; Loh, P. LCAs of a polycrystalline photovoltaic module and a wind turbine. *Renew. Energy* **2011**, *36*, 2227–2237. [CrossRef]

69. Mohr, N.; Meijer, A.; Huijbregts, M.A.; Reijnders, L. Environmental life cycle assessment of roof-integrated flexible amorphous silicon/nanocrystalline silicon solar cell laminate. *Prog. Photovolt. Res. Appl.* **2012**, *21*, 802–815. [CrossRef]

70. Giacchetta, G.; Leporini, M.; Marchetti, B. Evaluation of the environmental benefits of new high value process for the management of the end of life of thin film photovoltaic modules. *J. Clean. Prod.* **2013**, *51*, 214–224. [CrossRef]

71. Vellini, M.; Gambini, M.; Prattella, V. Environmental impacts of PV technology throughout the life cycle: Importance of the end-of-life management for Si-panels and CdTe-panels. *Energy* **2017**, *138*, 1099–1111. [CrossRef]

72. Latunussa, C.E.; Ardente, F.; Blengini, G.; Mancini, L. Life Cycle Assessment of an innovative recycling process for crystalline silicon photovoltaic panels. *Sol. Energy Mater. Sol. Cells* **2016**, *156*, 101–111. [CrossRef]

73. Held, M.; IIg, R. Update of environmental indicators and energy payback time of CdTe PV systems in Europe. *Prog. Photovolt. Res. Appl.* **2011**, *19*, 614–626. [CrossRef]

74. Ilias, A.V.; Meletios, R.G.; Katsigiannis, Y.A.; Nikolaos, B. Integration & assessment of recycling into c-Si photovoltaic module's life cycle. *Int. J. Sustain. Eng.* **2018**, *11*, 186–195. [CrossRef]

75. Corcelli, F.; Ripa, M.; Leccisi, E.; Cigolotti, V.; Fiandra, V.; Graditi, G.; Sannino, L.; Tammaro, M.; Ulgiati, S. Sustainable urban electricity supply chain—Indicators of material recovery and energy savings from crystalline silicon photovoltaic panels end-of-life. *Ecol. Indic.* **2018**, *94*, 37–51. [CrossRef]

76. Eskew, J.; Ratledge, M.; Wallace, M.; Gheewala, S.H.; Rakkwamsuk, P. An environmental Life Cycle Assessment of rooftop solar in Bangkok, Thailand. *Renew. Energy* **2018**, *123*, 781–792. [CrossRef]

77. Maani, T.; Celik, I.; Heben, M.; Ellingson, R.; Apul, D. Environmental impacts of recycling crystalline silicon (c-SI) and cadmium telluride (CDTE) solar panels. *Sci. Total Environ.* **2020**, *735*, 138827. [CrossRef]

78. Del Pero, F.; Delogu, M.; Berzi, L.; Escamilla, M. Innovative device for mechanical treatment of End of Life photovoltaic panels: Technical and environmental analysis. *Waste Manag.* **2019**, *95*, 535–548. [CrossRef] [PubMed]

79. Nunes, A.O.; Viana, L.R.; Guineheuc, P.-M.; Moris, V.A.S.; De Paiva, J.M.F.; Barna, R.; Soudais, Y. Life cycle assessment of a steam thermolysis process to recover carbon fibers from carbon fiber-reinforced polymer waste. *Int. J. Life Cycle Assess.* **2017**, *23*, 1825–1838. [CrossRef]

80. Tomporowski, A.; Piasecka, I.; Flizikowski, J.; Kasner, R.; Kruszelnicka, W.; Mroziński, A.; Bieliński, K. Comparison analysis of blade life cycles of land-based and offshore wind power plants. *Pol. Marit. Res.* **2018**, *25*, 225–233. [CrossRef]

81. Martínez, E.; Sanz-Adan, F.; Pellegrini, S.; Jiménez-Macías, E.; Blanco-Fernandez, J.; Cámara, E.M. Life-cycle assessment of a 2-MW rated power wind turbine: CML method. *Int. J. Life Cycle Assess.* **2008**, *14*, 52–63. [CrossRef]

82. Garrett, P.; Rønde, K. Life cycle assessment of wind power: Comprehensive results from a state-of-the-art approach. *Int. J. Life Cycle Assess.* **2012**, *18*, 37–48. [CrossRef]

83. Bonou, A.; Laurent, A.; Olsen, S.I. Life cycle assessment of onshore and offshore wind energy-from theory to application. *Appl. Energy* **2016**, *180*, 327–337. [CrossRef]

84. Poujol, B.; Prieur-Vernat, A.; Dubranna, J.; Besseau, R.; Blanc, I.; Pérez-López, P. Site-specific life cycle assessment of a pilot floating offshore wind farm based on suppliers' data and geo-located wind data. *J. Ind. Ecol.* **2020**, *24*, 248–262. [CrossRef]

85. Boulay, A.-M.; Bare, J.; Benini, L.; Berger, M.; Lathuillière, M.J.; Manzardo, A.; Margni, M.; Masaharu, M.; Núñez, M.; Pastor, A.V.; et al. The WULCA consensus characterization model for water scarcity footprints: Assessing impacts of water consumption based on available water remaining (AWARE). *Int. J. Life Cycle Assess.* **2017**, *23*, 368–378. [CrossRef]

86. Razdan, P.; Garret, P. *Life Cycle Assessment of Electricity Production from an Onshore V110-2.0 MW Wind Plant*; Vestas: Aarhus N, Denmark, 2015.

87. Al-Behadili, S.; El-Osta, W. Life Cycle Assessment of Dernah (Libya) wind farm. *Renew. Energy* **2015**, *83*, 1227–1233. [CrossRef]

88. Chipindula, J.; Botlaguduru, V.S.V.; Du, H.; Kommalapati, R.R.; Huque, Z. Life Cycle environmental impact of onshore and offshore wind farms in texas. *Sustainability* **2018**, *10*, 2022. [CrossRef]

89. Guezuraga, B.; Zauner, R.; Pölz, W. Life cycle assessment of two different 2 MW class wind turbines. *Renew. Energy* **2012**, *37*, 37–44. [CrossRef]

90. Stavridou, N.; Koltsakis, E.; Baniotopoulos, C.C. A comparative life-cycle analysis of tall onshore steel wind-turbine towers. *Clean Energy* **2019**, *4*, 48–57. [CrossRef]

91. Lefeuvre, A.; Yerro, X.; Jean-Marie, A.; Dong, P.A.V.; Azzaro-Pantel, C. Modelling pyrolysis process for CFRP recycling in a closed-loop supply chain approach. *Comput. Aided Chem. Eng.* **2017**, *40*, 2029–2034. [CrossRef]

92. Li, X.; Bai, R.; McKechnie, J. Environmental and financial performance of mechanical recycling of carbon fibre reinforced polymers and comparison with conventional disposal routes. *J. Clean. Prod.* **2016**, *127*, 451–460. [CrossRef]

93. Wang, S.; Wang, S.; Liu, J. Life-cycle green-house gas emissions of onshore and offshore wind turbines. *J. Clean. Prod.* **2019**, *210*, 804–810. [CrossRef]

94. Sacchi, R.; Besseau, R.; Pérez-López, P.; Blanc, I. Exploring technologically, temporally and geographically-sensitive life cycle inventories for wind turbines: A parameterized model for Denmark. *Renew. Energy* **2019**, *132*, 1238–1250. [CrossRef]

95. Besseau, R.; Sacchi, R.; Blanc, I.; Pérez-López, P. Past, present and future environmental footprint of the Danish wind turbine fleet with LCA_WIND_DK, an online interactive platform. *Renew. Sustain. Energy Rev.* **2019**, *108*, 274–288. [CrossRef]

96. Mahmoudi, S.; Huda, N.; Alavi, Z.; Islam, T.; Behnia, M. End-of-life photovoltaic modules: A systematic quantitative literature review. *Resour. Conserv. Recycl.* **2019**, *146*, 1–16. [CrossRef]

97. Klimek, M. Techniques of generating schedules for the problem of financial optimization of multi-stage project. *Appl. Comput. Sci.* **2019**, *15*, 18–35. [CrossRef]

98. Klimek, M. Priority algorithms for the problem of financial optimization of a multi stage project. *Appl. Comput. Sci.* **2017**, *13*, 20–34. [CrossRef]

Article

LCA as a Tool for the Environmental Management of Car Tire Manufacturing

Piotrowska Katarzyna [1,*], Piasecka Izabela [2], Bałdowska-Witos Patrycja [2], Kruszelnicka Weronika [2] and Tomporowski Andrzej [2]

[1] Faculty of Mechanical Engineering, Lublin University of Technology, 20-618 Lublin, Poland
[2] Department of Technical Systems Engineering, Faculty of Mechanical Engineering, University of Science and Technology in Bydgoszcz, 85-796 Bydgoszcz, Poland; izabela.piasecka@utp.edu.pl (P.I.); patrycja.baldowska-witos@utp.edu.pl (B.-W.P.); weronika.kruszelnicka@utp.edu.pl (K.W.); a.tomporowski@utp.edu.pl (T.A.)
* Correspondence: k.piotrowska@pollub.pl; Tel.: +48-815-384-585

Received: 14 August 2020; Accepted: 30 September 2020; Published: 9 October 2020

Abstract: Car tire manufacturing can be the cause of numerous environmental hazards. Harmful emissions from the production process are an acute danger to human health as well as the environment. To mitigate these unwanted consequences, manufacturers employ the eco-balance analysis at the product designing and development stage, when formulating general development strategies, and increasingly when investigating the entire product lifecycle management process. Since the negative effects of products are considered in a broader range of implications, it has become necessary to extend the traditional scope of analytical interest onto the production, use, and end-of-life stages. This work investigates the manufacturing of passenger car tires executed with traditional and modern manufacturing technologies. The Life Cycle Assessment (LCA) of tires reported in this study involved three LCA methods: Eco-Indicator 99, Cumulative Energy Demand (CED) and the scientific assessment methods developed by the Intergovernmental Panel on Climate Change, Global Warming Potential (IPCC). LCA as a tool for environmental analysis can be carried out for the entire life cycle or its individual phases. The implementation of the work made it possible to demonstrate that as a result of the identification of the main sources of negative impacts, it is possible to propose ways to minimize these impacts in the car tire manufacturing process. The results indicate that the most damaging impact is the depletion of natural resources, which play a key role in the production process of car tires.

Keywords: car tire production process; rubber industry; environmental impact; Life Cycle Assessment; entire product lifecycle

---

1. Introduction

The natural environment consists of many interacting phenomena and processes that must achieve a state of equilibrium. Manufacturing activity is constantly striving towards one direction—a cycle of production and consumption. In this cycle, raw materials are in the end converted to waste and dangerous emissions. Modern civilization is dependent on the automotive industry, both in terms of transport and everyday life. In order to reduce the consumption of raw materials, manufacturers must take this is into account at the design and production stage of a product.

The global production of tires currently reaches about 2 billion units, and the world production of rubbers is at nearly 30 million tonnes.T. This will not change in the near future; the number of passenger cars traveling on the road will only increase, and motor vehicles cannot be manufactured without tires.

Until recently, potential buyers of car tires paid attention primarily to the highest quality of the product, while manufacturers paid attention to the highest possible efficiency of the manufacturing process. Along with the growing ecological awareness around the world, attention began to be paid to the lowest possible harmfulness of both the production, operation, and post-use disposal of tires. Therefore, new solutions began to appear on the market, such as the green tires. Automotive companies are actively searching for solutions expected to improve the environmental soundness of their products. These efforts have contributed to reductions in fuel consumption and the emissions of harmful substances to the atmosphere, and have promoted the use of components that facilitate material recycling. Ecological tires are structurally similar to traditional tires; however, they are made of different materials. In addition, the ecological tire uses improved rubber compounds, and the tread pattern has been modified so the groove and sipes are not so deep. In numerous ways, ecological tires outperform traditional solutions, in terms of fuel efficiency, rolling resistance, and the emission of harmful substances such as carbon dioxide. Also, savings in fuel consumption are said to reach approximately 0.2 L/100 km, which compensates for the higher price of ecological tires. In addition, the novelty tire is about 20% lighter than a tire produced in a traditional manufacturing technology. Several aspects differentiate the traditional and the ecological tire, e.g., the tread pattern. In the modern tire production process, carbon black has been entirely replaced by silica. The substitution of steel and other materials contributes to the weight reduction of the green tire, which is claimed to weigh 25% less than a traditional one. Green tires are believed to reduce fuel consumption and carbon dioxide emissions by reducing rolling resistance, which in the traditional model amounts to approximately 12 kg/T, whereas the modern design is said to maintain the level of up to 8 kg/T. Another advantage of lower rolling resistance is the lesser noise generated by the tires. Therefore, green tires also have an impact on the environment, as noise is treated as a part of pollution. Furthermore, there is a marked rise in fuel economy. In urban traffic, a percentage share of fuel consumption reduction for ecological tires is close to 3%, while in rural/motorway traffic (at high speeds) it is reported to approach to 5%. Finally, the average resource used during the production of one ecological tire amounts to: 1036 MJ of electricity, 45 liters of water, 0.02 kg of solvents, and about 0.5 kg of waste is generated [manufacturer's data].

Due to changes taking place in the modern economy and the emergence of new social expectations, in the life cycle of car tires, apart from technical, design, and manufacturing aspects, environmental protection objectives must be taken into account, including reducing the depletion of raw materials and improving the quality of the environment and negative impact on human health. In order to achieve these objectives, appropriate assessment tools are required that enable the identification and reduction of emerging environmental problems. The basic tool used for this type of assessment is a product life cycle analysis known as the LCA method (Life Cycle Assessment) [1].

Environmental Life Cycle Assessment (LCA) has been developed over the last three decades. LCA developed from merely energy analysis to a comprehensive environmental burden analysis in the 1970s. Then, full-fledged life cycle impact assessment and life cycle costing models were introduced in the 1980s and 1990s, and social-LCA and particularly consequential LCA gained ground in the first decade of the 21st century [2]. Life Cycle Assessment is a cradle-to-grave approach to assessing industrial systems. Cradle-to-grave begins with collecting material from the ground to produce a product and ends when all the materials are returned to the field. LCA evaluates all stages of life from the point of view that they are interdependent, which means that one operation leads to all. This method makes it possible to use the cumulative environmental impact from all stages of the product life cycle and provides a comprehensive picture of environmental conditions [3].

The key application of the LCA method is to support the search for new solutions that will ensure a balance between economic development and environmental protection. Optimization in terms of ecology, energy, and economy can be achieved by rationalizing the use of resources and all kinds of activities aimed at reducing the amount of pollution and waste. The rationalization of resource consumption is based on reducing the material and energy consumption of the processes taking place at each stage of the life cycle. Among them, the design and production stage is of particular importance, because it is then that decisions are made that affect the course of the further part of the cycle and, consequently, have a significant impact on the overall ecological and energy characteristics of a car tire. For this reason, particular attention during this research will be focused on the differences in the potential environmental impact of the production stage of traditional versus ecological car tires [4].

The main purpose of this work was to analyze the car tire manufacturing process with the use of the LCA method. Secondly, the study aimed to show the capacity of the LCA method to determine the tire production process parameters that will mitigate adverse environmental impacts in their entire life cycle horizon.

2. Life Cycle of a Car Tires

2.1. The Production Stage

The design of a modern car tire is an intricate and complex structural process that combines a number of rubber, steel, and textile elements. Each of these materials displays different characteristics that impart a specifically selected set of properties, e.g., shape, stiffness, strength, vibration damping, heat, and electrostatic charges dissipation [5]. Each manufactured tire is expected to perform according to the design specifications; hence, the choice and quality of materials are equally important as the manufacturing technology. Tires are elements that play an absolutely critical role in ensuring the safety of the driver and passengers, which is why it is essential that their production process is performed in accordance with applicable requirements and the product undergoes specialist tests [6]. The product damage in manufacturing usually occurs at the interface of two heterogeneous materials and is categorized as inner, outer, sidewall, tread face, or shoulder damage [7]. The tire is a system combining up to 20 different basic and supplementary elements that affect car handling. The components of the tire are shown in Table 1.

Table 1. Most important components of a car tire [8,9].

Component	Material	Description	Occurrence
Inner liner	Synthetic butyl rubber	High resistance to acids, bases, and oxidizing agents. Minimizes air loss and protects inner components against water and oxygen.	Traditional and the ecological tire
Textile carcass	A textile fabric consisting of about 1.0–1.5-mm-thick cord threads	The skeleton that allows the tire to maintain the shape under internal loads. It transfers loads during acceleration, braking, and turning.	Traditional and ecological tire
Bead wires	A rubber-coated wire bundle	It ensures a proper and tight fit of the tire on the rim. Owing to friction, it transmits longitudinal forces.	Traditional and ecological tire
Filler	Synthetic rubber	Provides high sidewall stiffness, driving comfort, manageability, and durability of the tire.	Traditional and ecological tire
Bead bundle	Rubber-coated nylon or aramid cord	It guarantees tire durability and tightness, separates the wire from the rim.	Traditional and ecological tire
Sidewall	The rubber compound containing anti-ageing agents	Provides tire protection. Protects it against chafing and environmental factors (UV radiation, temperature variations, chemical compounds).	Traditional and ecological tire
Belts	Rubber-coated steel cords	Provides strength to the tire. Stabilizes the tread, prevents deep damage to the carcass, greatly reduces rolling resistance.	Traditional tire
	Rubber-coated polymer cords		Ecological tire
Tread	Main component of mixture is soot	Ensures maneuverability, optimal grip and highest wear resistance, the key to noise level reduction, and rolling resistance.	Traditional tire
	Main component of mixture is silica		Ecological tire

Depending on the class of the tire, it is reinforced with a range of supplementary elements, including insulation inserts, zero-degree belt, chafer strip, bead reinforcements, insolation bundle, sidewall reinforcement, and undertread [10,11]. After the designing, documentation, and prototyping stages, the tire components are prepared. With the top product quality in mind, all materials should be carefully selected on the basis of their respective physical and chemical properties, size, thickness, cutting angles, etc. Design specifications must be observed at every production plant, regardless of its location, thus ensuring that the same tire types provide the same performance, regardless of the place of manufacture [10].

The structure of rubber compounds is carefully adjusted to impart desired properties to the tire, so as to ensure its proper functioning and so that it will serve its intended purpose, e.g., a different compound is used in a summer versus winter tire. Moreover, rubber compounds of specially engineered properties build different parts of the tire, such as the sidewalls, the inner liner, or fillers in the bead zone, all of which are made of various types of rubber compounds. The compounds typically include the following materials: synthetic rubbers (butyl, SBR, BR), natural rubbers, oils, fillers (silica, soot), Sulphur, resins, anti-aging agents (antioxidants), vulcanization accelerators, and other agents as per specific requirements. The approximate material consumption involved in producing 1000 kg of tread rubber compound for a passenger car tire is 500 kg of rubber (mainly synthetic), 150 kg of silica, 150 kg of carbon black, 20 kg of Sulphur, 20 kg of resin, 100 kg of oils, and about 60 kg of other components [12,13]. All the ingredients are machine-mixed, their quantity and the adding sequence being crucial, until a homogeneous mass is obtained. The prepared compound is formed into plates or ribbons, whose dimensions facilitate their use in the subsequent stages of production. The compound surface is next coated with a release agent, for easier separation of rubber layers in the later stages, and the rubber compound material itself is subjected to accuracy measurements and analyses to verify its compliance with the design specifications [14]. An integral part of the tire design are the steel and textile cords. They form a carcass of the tire, which guarantees its required stiffness and shape, and which translates into a desired level of performance while driving. A single thread of a textile cord is made of twisted weaves of a large number of ultra-thin viscose, nylon, aramid, and polyester fibers (a single 195/65 R15 tire contains between 1500 and 1800 threads) [15]. The purpose of bead wires is to enable mounting a tire on a rim. There are several technologies for their production. They may come in the form of rings formed by braiding several wires with a round cross-section or by winding several layers of tape. The beads consist of separate wires of a rectangular or polygonal cross-section, coated in a rubber compound. Having all the components collected and prepared in advance, the confection stage may commence. The tire building process consists of arranging the pre-products in a strictly controlled sequence. Depending on the technology, a raw tire is obtained in a one- or two-stage process, which is followed by vulcanization in special molds, and then quality control [16,17]. The flowchart in Figure 1 presents the car tire production process.

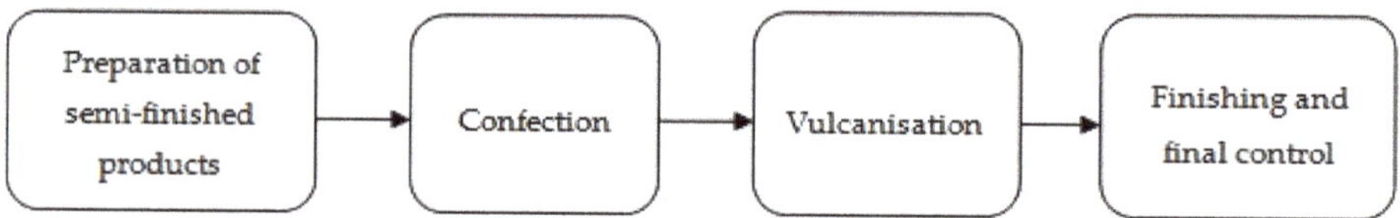

Figure 1. The tire-manufacturing process flowchart.

As mentioned, the technologies used by tire manufacturers involve one or two stages. In the former case, the entire tire is produced on a single machine; in the latter, the carcass is typically put together in the first stage, and the basic and supplementary layers and technologies (belt, tread, etc.) are fitted consecutively [17]. The layers of the tire are laid flat inside a drum, which is next filled with air, the pressure of which causes the layers to bond. In the next stage, the structure is subjected to rolling, which shapes the tire, ensures the strong bonding of components and guarantees the removal of excess air that could become trapped between the components. The process output is the raw tire, which undergoes dimensional and shape control tests for errors [15].

After successfully passing the control step, in the next stage of production, the raw tires are placed in vulcanization presses, where vulcanization molds are fitted. The molds typically have a container structure, i.e., their elements form two sides of the mold. The rings form the sidewalls and impress the branding and other inscriptions, while the side protrusions form the groove and sipes in the tread. The two halves of the rings are each responsible for the half of the tread on one side. The tire is treated with high temperatures (in excess of 150 °C), while the pneumatic load presses the tire against the sides and the protrusions of the mold. In the thermoplastic melt-freeze process, the rubber transits to fluid (thus acquiring the shape of the mold) and eventually becomes elastic. Following the treatment, the tire remains in the mold for 10 minutes and then removed to cool down [17,18].

The final quality inspection includes a variety of tests. The visual assessment of the tire condition detects any foreign bodies that might have become trapped between the layers. This is followed by an X-ray inspection of the interior structure that aims to ensure that no defects or internal damage is present, and additional analyses that involve testing the mass, shape, and rigidity parameters, heterogeneity criteria, radial force measurements, conicity, and other parameters whose uniformity ensures the safety and comfort of driving. The tires are also cut across to assess the product's compliance with the design specifications, which would otherwise impair their safety and performance properties [17].

Modern tire constructions tend to incorporate composites that combine various material properties, thus eliminating the disadvantages of particular compounds. As a result, new tire designs can conform to the challenges of constantly rising environmental and operational requirements [17,19].

2.2. The Use Stage

Each element of the tire can have a significant impact on fuel consumption and also on the reduction or increase of the negative environmental impact. On 1 November 2012, the obligation to label car tires was introduced in the European Union. The labels provide information on the most important aspects of tire performance, such as rolling resistance, wet grip, and noise emissions allowing for the quick and easy comparison of different products. They are designed to encourage manufacturers to act towards reducing rolling resistance, which leads to a reduction in carbon dioxide emissions to the atmosphere. Tires, mainly due to the rolling resistance, are responsible for 20–30% of the fuel consumption in passenger cars. Reducing fuel consumption allows for the reduction of the emission of CO_2 and other harmful compounds into the environment. The other parameters on the labels are intended to force tire manufacturers to care for all performance metrics equally. Labels are to encourage consumers to make more informed choices and mobilize manufacturers to create products of the highest quality [20].

Tire wear contributes significantly to the flow of (micro) plastics into the environment. The particles generated during their use, emitted on the roads, can be dispersed in the environment through various routes. Smaller ones are usually emitted into the air, while large particles settle on the road surface and due to rainwater runoff, get into the soil, sewage, and surface waters. It is estimated that the relative share of tire wear in the total global amount of plastics ending up in our oceans is 5–10%. In the air, 3–7% of particulate matter (PM2.5) is made up of tire wear, indicating that it can contribute to the global health burden of air pollution, which can only be effectively addressed if awareness increases in this area [21].

2.3. The End of Life Management

Car tires are the largest source of post-consumer rubber products. For years, however, the number of tires withdrawn from use has increased, causing a negative impact on the environment. However, nowadays, the application of the principles of sustainable development implies a new approach to waste using environmentally friendly concepts, and it is necessary to re-evaluate the possibilities of managing this waste in order to increase its use and reduce the amount that requires disposal or storage. To this end, a waste hierarchy should be followed, focusing on reuse, recycling, and recovery, and disposal should be interpreted as the last available option corresponding to the highest level of loss and change in resources. The disposal of used tires is a global environmental problem due to the large number of tires produced each year; according to ETRMA, around 3.4 million tons of used tires are produced in Europe annually. There are three main lines of action aimed at solving the problem of used tires: extension of service life through increased durability and retreading, material recycling, and energy recovery. A certain number of tires can also be used in full [22–24].

Due to their durability, used tires constitute nuisance waste and should be used industrially. They do not degrade in the environment within 100 years. They must not be left in landfills because their accumulation in large quantities is a fire hazard. Recycling tires is a difficult process due to their composition and construction. In addition to rubber, they contain textile and steel cords that must be properly separated. Used tires that are no longer suitable for retreading are waste. Therefore, they should be managed in an environmentally safe manner. For this purpose, the most frequently used methods are the ones involving product or material recycling. Whole, compressed, or cut worn tires have many applications, including their shape, material characteristics, and ability to absorb impacts or noise are used [22,25]. Used car tires can be also used by burning them in cement plants or other energy installations. The obtained heat is used in many technological processes, e.g., for burning clinker in cement kilns. The calorific value of rubber can be compared to the calorific value of coal. The use of used tires as fuel in the production of cement is considered a waste-free method of managing large amounts of this type of waste, because the tires burn completely without slag or ash remains. The metals contained in the tire permanently adhere to the clinker, improving its properties [26].

The adoption of three directives by the European Parliament, Landfill 1999/31/EC, End-of-Life Vehicle 2000/53/EC, and Waste Incineration 2000/73/EC have had a major impact on the management of used tires in the EU. The Landfill Directive 1999/31/EC introduced a ban on the disposal of whole used tires from July 2003, and from July 2006, on shredded tires. Moreover, it obliges the member states to create conditions enabling the implementation of this intention. The End-of-Life Vehicle Directive 2000/53/EC requires tires to be removed from vehicles prior to scrapping. The third of the Waste Incineration Directive 2000/73/EC obliges cement plants that use tires as a supplementary fuel to obtain lower limits for the pollutant content in the waste gases [27].

The issues of post-consumer management are more often being considered when analyzing all the earlier stages of the life cycle. Tire manufacturers should, as much as possible, prevent or reduce waste and harmful negative impacts on the environment, not only during the production of tires but also after their end of life [28].

3. Materials and Methods

3.1. Object and Plan of Analysis

The objects of the study were two 205/55/R16 tires designated for use in passenger cars. The first specimen was produced using a traditional manufacturing technology, the other was the modern eco-tire.

Until recently, the first three stages of the tire life cycle (design, manufacturing, and use) were of interest to people only in the economic dimension. The framework for the minimization of the environmental impact occurring in the life cycle of industrial products is rooted in the concept of Life Cycle Thinking (LCT). It is one of the concepts recommended by the European Commission that are instrumental in achieving sustainable development. Its purpose is to foster positive change in manufacturing by reducing the environmental impact of processes and manufactured products. Currently, the EU is promoting mechanisms aimed at stimulating the manufacturing of environmentally sound products with a view to improve the condition of the European natural environment and boost the attractiveness and position of EU in the world. Therefore, the notion of seeking methods to minimize the environmental impact of products and services throughout their entire life cycle is consistent with the policies of the European Commission. The basic assumption is to implement intervention measures predominantly at the life cycle stages that exhibit the highest potential for reducing the negative impact on the environment [29–32]. Therefore, the focus of this work is on establishing which stages and factors of the car tire life cycle display the biggest potential for sustainability-driven improvement. From the design, to manufacture, utilization, and to post-consumer waste management, a product life is essentially a cycle, which is why the term "life cycle" is commonly applied to all products, including car tires. The concept of life cycle management (LCM) has an important practical dimension, but at the same time it is a way to introduce life cycle thinking into business practices. Its main goal is to manage the company's operations and its products in such a way as to ensure more and more sustainable production and consumption [33,34]. A growing concern for nature leads to the development and use of increasingly complex methods to reduce the burden on the environment, hence, the development of methodologies for the assessment of the environmental impact of processes, products, and industries. One of the methods in question, the Life Cycle Assessment (LCA), applies a holistic approach to the product, which includes its impact from the acquisition of raw materials, through production and use, to final post-consumer waste utilization. It covers the environmental impact of all life cycle stages and is in line with the principles of sustainable development [35–38].

The point of reference for the analysis carried out in the subsequent sections was provided by the regulations and recommendations set out in the ISO 14040 group of International Standards. All four LCA phases were completed: the goal and scope definition, inventory analysis, impact assessment, and interpretation (Figure 2) [31].

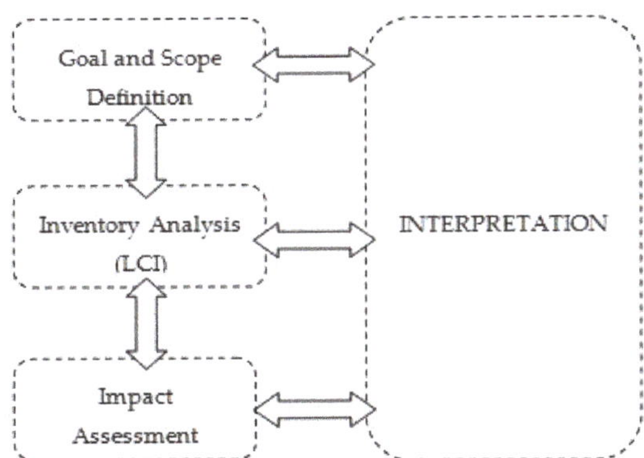

Figure 2. Four phases of the Life Cycle Assessment (LCA) study. Own work based on [31,39].

3.2. Four Steps of the Analysis

One of the possible applications of the LCA method is the assessment of production technology from the point of view of its impact on the natural environment. The aim of the analysis was to compare the environmental impacts associated with the life cycle of two different types of tires based on different manufacturing technologies. Therefore, the conducted analysis is classified as a type of comparative analysis. It has been used to determine whether there are differences in the magnitude of the environmental impact generated during the life cycles of selected car tires produced based on two different technologies. The change of the production technology of traditional tires to ecological tires has taken place in the last few years and is mainly related to the materials and materials used in their production. The key issues of the conducted analysis concern environmental effects [36,39]. The goal of the analysis was firstly to provide the description of the status quo of the industry (retrospective LCA), and secondly, to study the potential future consequences and determine recommendations for the development of more environmentally friendly solutions (prospective LCA). The procedure follows an LCA methodology whose purpose is to determine the extent of the negative environmental impact of a given object's life cycle. To that end, all four LCA phases were conducted. In the LCA analysis, two terms are particularly important: function and functional unit. In many cases, it is not possible to simply compare product A and product B as they may have different performance characteristics. The system boundaries are not always easy to define and the assumptions made during the process maybe not consistent, resulting in challenges when comparing between different results. Moreover, in most of the published cases of study, the LCA of a process, product, or service is retrospective (measured/estimated data) rather than prospective (forecasted data). Therefore, it does not typically relate to broad steps such as design, development, research, and education that can tremendously change the life cycle or even improving it by considering the prevention or reduction of manufacturing [38,39]. In a general sense, the analysis serves to estimate the potential environmental burden associated with the way that a given product performs its specific function. If possible and justified, a time frame should also be included in the functional unit. The functional unit is the quantified performance of a product system that serves as the reference unit in the life cycle analysis. The reference flow is a measure of the outputs from processes in the product system required to perform the function expressed by the functional unit. If the LCA analysis covers the entire life cycle, it is recommended that the functional unit from the product use stage be considered, as it will most comprehensively express and capture the essence of the function [40].

Due to the fact that both the analyzed products in this study are passenger vehicle tires performing in the same area of applications, it was resolved that the function would refer to this aspect. It was defined as covering the distance of 50,000 kilometers over a period of 5 years in an average style of driving.

The key issue investigated in the course of this analysis was to highlight the differences in the environmental burden between the two products that result from changes in the manufacturing process and raw materials consumed by their production processes. The geographic and time frame of the data was the same and it was the technological scope that was different. Given that the company providing the data is in Europe and has a well-established position on the market, the considered geographic range is also Europe. The time frame encompassed the same range because both traditional and ecological car tires are currently in continuous production. From the spatial dimension perspective, no relevant impact is noted since both products are produced at the same tire manufacturing site, however, with the implementation of different production methods. This is the technology of manufacturing—traditional or ecological—that produces tires of specific and different parameters. Making certain exclusions in LCA analyses is almost always necessary. The cut-off criteria are either specified by value choices or may be dictated by independent factors, such as data gaps. In any case, several parameters are distinguished for specifying the cut-off rules, the most popular of which are exemptions based on the criterion of the mass percentage share in the total physical output and the percentage share in the total economic value of the output. The cut-off criteria enable selecting and specifying the key elements of the system and simplify both the model and the analysis. However, the exclusions should be clearly defined and described [41–43]. Storage, distribution, sales, and technical testing processes were outside of the scope of this study.

3.3. Life Cycle Inventory—LCI

The life cycle inventory analysis (LCI) constituted the second phase of LCA, which is the analysis of the set of inputs and outputs. This model represents the product system structure, and its smallest elements are unit processes. A unit process is the smallest component of the product system for which data are collected. What is more, the processes are connected by material and energy flows. All individual unit processes represented in the model were subjected to a completeness check and evaluation, performed using the material and energy balancing. The data were acquired according to inventory types and collected in the inventory table. Data collection provides additional background information regarding the source of origin, geographic range, age, etc. Therefore, the acquired data help determine the spatial and temporal location of the problems. However, the key classification criterion at this stage was the fact of belonging to the inventory of the studied system. In consequence, the problems were not assigned to a time or place, which, furthermore, is the reason why LCA is considered to be a so-called general technique. During analysis, the collection of data is adapted to model requirements and the impact category indicator (Figure 3) [43–45].

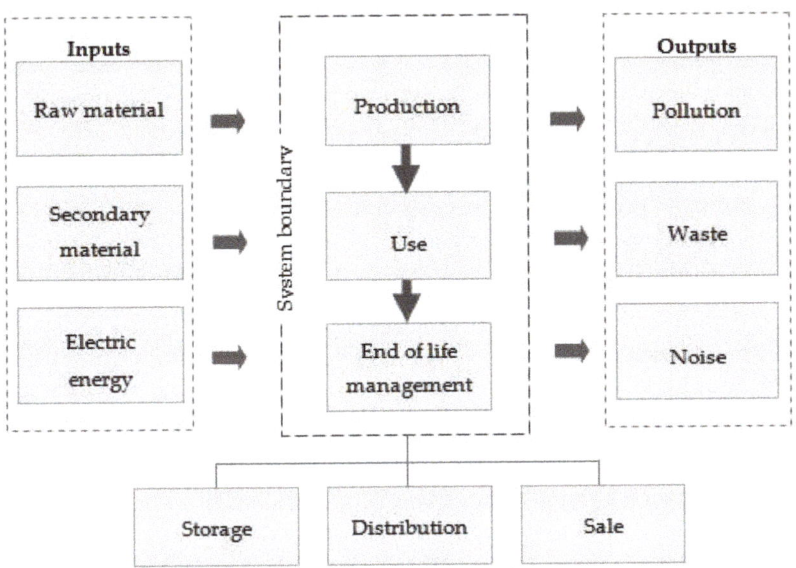

Figure 3. Inputs and outputs throughout the car tire life cycle. Own work based on [28].

The research conducted within this study was carried out with the use of SimaPro 8.4.1 Software. The data were mostly acquired from tire manufacturers, however, the data related to processes and materials, less relevant for the assessment of environmental burden, were derived from the Ecoinvent 3.4 database (a component of SimaPro product).

The aggregation of data concludes this phase and introduces the subsequent stage—LCIA [45].

3.4. Life Cycle Impact Assessment—LCIA

In essence, the primary function of the Life Cycle Impact Assessment (LCIA) stage is to determine the impact of the lifecycle of a given product on the environment. The differences in the methodologies of various approaches to LCA are principally concerned with the LCIA phase. Inventory analysis is focused on quantifying the environmental burden, and its reliability is relative to data quality and the choice of criteria and assumptions for building the product system [42,46].

LCIA consists of mandatory and optional elements. Mandatory elements include defining impact categories, category indicators, characterization models as well as classification and characterization, whereas among optional elements there are normalization, grouping, weighting, and data quality analysis. While the sequence of obligatory elements is strictly defined and should be followed as is during the analysis, with regards to the optional elements, the question is whether and which steps are to be taken. If the results from the analysis are to be publicized, normalization and weighting are typically omitted, and the interpretation is performed at the level of the LCIA environmental profile (impact categories). Although this results in the loss of transparency and increases the difficulty of the analysis, it does carry a positive outcome, specifically, the results are deprived of subjectivity [47,48].

The results from the LCIA phase are presented in Section 4.

3.4.1. Eco-Indicator 99

The Eco-indicator 99 method (EI99) is a representative of the endpoint methods for environmental impact modeling and assessment. The characterization process defines the environment through the perspective of three main categories of damage, also referred to as environmental damage endpoints: Human Health, Ecosystem Quality, and Resources, which are further described by 11 impact points. The results from the EI99 analysis of impact points are subject to further processing—normalizing, grouping and weighting to obtain a single-score indicator, expressed in Eco-indicator points (Pt). The scale is adjusted so that one point reflects one-thousandth of the annual environmental load of an average European inhabitant [49].

Human health, one of the three damage categories recognized in the EI99 method, distinguishes five effects: climate change, ozone layer depletion, carcinogenic effect, respiratory involvement, and ionizing radiation. Deciding on an indicator of the impact area from among the endpoints of the environmental mechanism enables one to define a common unit for all adverse effects to human health. Environmental damage can be responsible for a number of disorders, which is why it requires certain means for the quantification of differences between them, hence the DALY scale. Disability Adjusted Life Years (DALY) assigns binary weights to diseases, where 0 denotes ideal health and 1 death. DALY is derived from the number of year life lost due to premature death or the number of years lived disabled [49–51].

Damage to ecosystem quality is significantly more diverse and less homogeneous compared to the impact on human health. It is expressed as the loss of certain species in a given area. Currently, a temporary solution is in use that allows converting between PAF (Potentially Affected Fraction) and PDF (Potentially Disappeared Fraction) units. This method cuts off and limits the considered groups of species because, theoretically, all species could be in danger of extinction. Three impact categories are considered in damage to ecosystem quality: ecotoxicity, acidification/eutrophication, and land use [30].

Modeling within the third area of ecological burden, resources, involves resource and damage analysis. Eco-indicator 99 accounts exclusively for mineral resources and fossil fuels. The quantification of damage pertaining to this category necessitated introducing a special indicator, surplus energy, expressed in MJ, which is analogous to DALY, PAF, and PDF. The higher the surplus energy—being a derivative of the decrease in concentration—the lower the quality of the resource (Figure 4) [50,51].

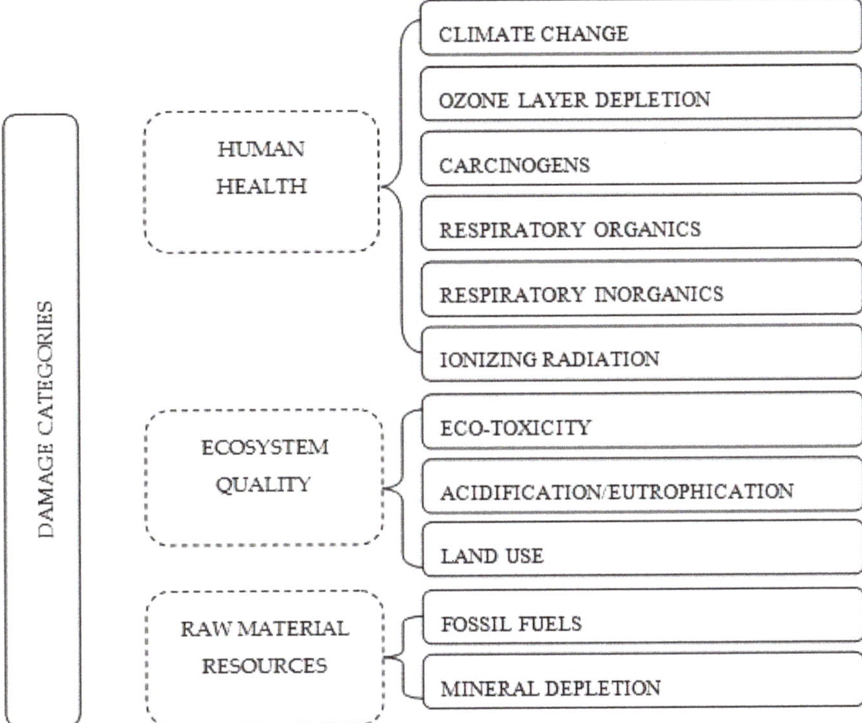

Figure 4. Grouping in Eco-indicator 99. Own work based on [30].

3.4.2. Cumulative Energy Demand (CED)

Cumulative Energy Demand is expressed in MJ-Eq and describes environmental burden through 7 damage categories: 2 non-renewable sources (nuclear energy and fossil fuels) and 5 renewable sources (solar, water, wind, geothermal, and biomass) [52].

3.4.3. IPCC

The scientific assessment method developed by the Intergovernmental Panel on Climate Change, Global Warming Potential (IPCC), serves as a quantitative assessment tool for expressing the contribution of particular greenhouse gases (GHG) to the greenhouse effect, relative to carbon dioxide. The time horizon is 20, 100, or 500 years. The total greenhouse effect damage indicator related to CO_2 is equal to 1 [49,52].

3.4.4. Interpretation

Life cycle interpretation identifies in quantitative terms the relevance of findings from LCA, specifically, the disaggregation of results obtained at the LCI stage, characterization, normalization, grouping, or weighting, to specific components [30,36]. The data for the analysis were acquired directly from the car tire manufacturer and a recycling company, whereas the remaining data was derived from the SimaPro databases (PréConsultants, Amersfoort, The Netherlands).

A detailed interpretation of the results obtained from the analysis is given in Sections 4 and 5.

4. Results and Discussion

The results are summarized in three sections, corresponding to the respective assessment methods Eco-indicator 99 (Section 4.1), CED (Section 4.2), and IPCC (Section 4.3). The results from the characterization stage are given in Table 2, and from the grouping and weighting stages in Tables 3–8 and Figures 5–10. The results describe each stage of traditional and ecological tire life cycles, accounting for their total life cycle impact and five environmental damage categories eliciting the highest levels of detrimental effects established in the study: carcinogens, respiratory inorganics, climate change, acidification/eutrophication, and fossil fuels. The outcomes of the study are the environmental factors quantified as eco-indicator points (Pt), i.e., easily comprehensible parameters facilitating the comparative analysis of dissimilar environmental impacts. One thousand points express an annual impact of an average European resident on the environment.

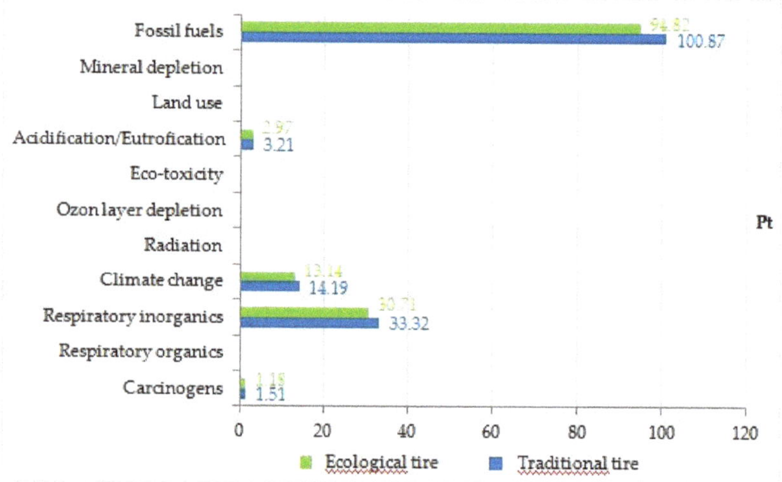

Figure 5. Results of grouping and weighting of environmental damage at individual stages of the traditional and ecological tire life cycle, including impact categories.

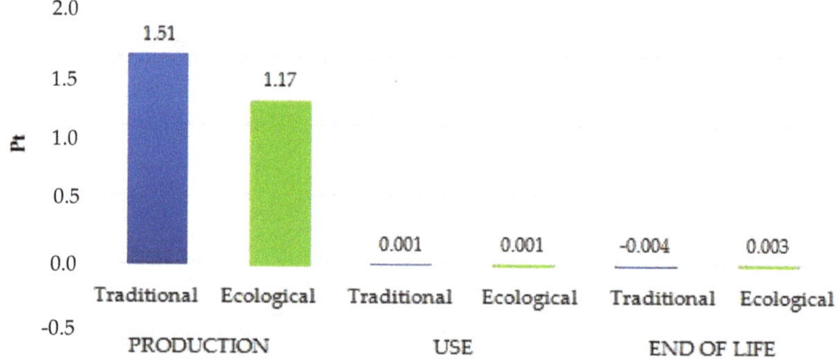

Figure 6. Results of grouping and weighting of environmental damage from carcinogens at individual stages of the traditional and ecological tire life cycle.

Table 2. Results of Eco-indicator 99 characterization of environmental damage at individual stages of the traditional and ecological tire life cycle, including impact categories.

Impact Category	Production		USE		End of Life		Unit
	Traditional Tire	**Eco Tire**	**Traditional Tire**	**Eco Tire**	**Traditional Tire**	**Eco Tire**	
Carcinogens	5.81×10^{-5}	4.53×10^{-5}	3.58×10^{-8}	3.58×10^{-8}	-1.55×10^{-7}	1.17×10^{-7}	DALY
Respiratory organics	8.97×10^{-8}	5.77×10^{-8}	6.60×10^{-7}	6.25×10^{-7}	-1.32×10^{-8}	-2.22×10^{-8}	DALY
Respiratory inorganics	1.91×10^{-4}	1.48×10^{-4}	1.09×10^{-3}	1.03×10^{-3}	-8.54×10^{-7}	-1.49×10^{-6}	DALY
Climate change	6.95×10^{-5}	5.40×10^{-5}	4.76×10^{-4}	4.51×10^{-4}	-4.41×10^{-7}	-5.31×10^{-8}	DALY
Radiation	1.04×10^{-7}	7.99×10^{-8}	6.46×10^{-12}	6.46×10^{-12}	0.00×10^{0}	0.00×10^{0}	DALY
Ozone layer	1.43×10^{-8}	2.55×10^{-9}	1.67×10^{-13}	1.67×10^{-13}	-4.22×10^{-10}	-2.12×10^{-10}	DALY
Ecotoxicity	2.90×10^{1}	1.85×10^{-1}	3.44×10^{-2}	3.44×10^{-2}	2.87×10^{-1}	4.66×10^{-1}	PAF $\times$ m^2 $\times$ year
Acidification/Eutrophica-tion	4.75×10^{0}	3.69×10^{0}	3.64×10^{1}	3.45×10^{1}	-4.45×10^{-2}	-7.83×10^{-2}	PDF $\times$ m^2 $\times$ year
Land use	2.47×10^{0}	2.30×10^{0}	5.97×10^{-5}	5.97×10^{-5}	0.00×10^{0}	0.00×10^{0}	PDF $\times$ m^2 $\times$ year
Minerals	5.16×10^{-1}	3.52×10^{-1}	2.82×10^{-4}	2.82×10^{-4}	-7.14×10^{-2}	-1.65×10^{-4}	MJ
Fossil fuels	1.34×10^{2}	9.78×10^{1}	4.11×10^{3}	3.89×10^{3}	-5.88×10^{0}	-8.32×10^{0}	MJ

Table 3. Results of grouping and weighting of environmental damage at individual stages of the traditional and ecological tire life cycle, including impact categories [unit: Pt].

Impact Category	Production		USE		End of Life	
	Traditional Tire	**Eco Tire**	**Traditional Tire**	**Eco Tire**	**Traditional Tire**	**Eco Tire**
Carcinogens	1.51×10^{0}	1.18×10^{0}	9.43×10^{-4}	9.33×10^{-4}	-4.04×10^{-3}	3.06×10^{-3}
Respiratory organics	2.34×10^{-3}	1.50×10^{-3}	1.72×10^{-2}	1.63×10^{-2}	-3.44×10^{-4}	-5.78×10^{-4}
Respiratory inorganics	4.96×10^{0}	3.87×10^{0}	2.84×10^{1}	2.69×10^{1}	-2.22×10^{-2}	-3.88×10^{-2}
Climate change	1.81×10^{0}	1.41×10^{0}	1.24×10^{1}	1.17×10^{1}	-1.15×10^{-2}	-1.38×10^{-3}
Radiation	2.72×10^{-3}	2.08×10^{-3}	1.78×10^{-7}	1.68×10^{-7}	0.00×10^{0}	0.00×10^{0}

Table 3. *Cont.*

Impact Category	Production		USE		End of Life	
	Traditional Tire	Eco Tire	Traditional Tire	Eco Tire	Traditional Tire	Eco Tire
Ozone layer	3.72×10^{-4}	6.65×10^{-5}	4.54×10^{-9}	4.34×10^{-9}	-1.10×10^{-5}	-5.51×10^{-6}
Ecotoxicity	2.26×10^{-1}	1.44×10^{-1}	2.78×10^{-4}	2.68×10^{-4}	2.24×10^{-3}	3.63×10^{-3}
Acidification/eutrophication	3.71×10^{-1}	2.88×10^{-1}	2.84×10^{0}	2.69×10^{0}	-3.47×10^{-3}	-6.10×10^{-3}
Land use	1.93×10^{-1}	1.79×10^{-1}	4.86×10^{-6}	4.66×10^{-6}	0.00×10^{0}	0.00×10^{0}
Mineral depletion	1.23×10^{-2}	8.39×10^{-3}	6.84×10^{-6}	6.71×10^{-6}	-1.70×10^{-3}	-3.93×10^{-6}
Fossil fuels	3.18×10^{0}	2.33×10^{0}	9.78×10^{1}	9.27×10^{1}	-1.40×10^{-1}	-1.98×10^{-1}
TOTAL	1.23×10^{1}	9.41×10^{0}	1.41×10^{2}	1.34×10^{2}	-1.81×10^{-1}	-2.38×10^{-1}

Table 4. Results of grouping and weighting of environmental damage from carcinogens at individual stages of the traditional and ecological tire life cycle [unit: Pt].

Substance	Production		USE		End of Life	
	Traditional Tire	Eco Tire	Traditional Tire	Eco Tire	Traditional Tire	Eco Tire
Arsenic, ions	1.35×10^{0}	1.05×10^{0}	7.92×10^{-4}	7.79×10^{-4}	-4.62×10^{-3}	1.03×10^{-3}
Cadmium, ions	1.15×10^{-1}	8.93×10^{-2}	9.28×10^{-5}	9.28×10^{-5}	-1.35×10^{-4}	2.87×10^{-5}
Fine particles < 2.5 μm	2.03×10^{-2}	1.59×10^{-2}	6.06×10^{-5}	6.06×10^{-5}	X	-
Cadmium	1.61×10^{-2}	1.24×10^{-2}	2.16×10^{-7}	2.16×10^{-7}	-3.05×10^{-4}	1.04×10^{-4}
Arsenic	1.55×10^{-2}	1.21×10^{-2}	7.59×10^{-8}	7.59×10^{-8}	-	-
Unspecified metals	3.18×10^{-7}	-	-	-	1.34×10^{-3}	2.04×10^{-3}
Dioxins	-	1.01×10^{-3}	2.71×10^{-7}	2.71×10^{-7}	-	-

Table 5. Results of grouping and weighting of environmental damage from respiratory inorganics at individual stages of the traditional and ecological tire life cycle [unit: Pt].

Substance	Production		USE		End of Life	
	Traditional Tire	**Eco Tire**	**Traditional Tire**	**Eco Tire**	**Traditional Tire**	**Eco Tire**
Sulphur oxide	1.33×10^{-2}	-	1.58×10^{1}	1.50×10^{1}	-4.07×10^{-3}	-3.60×10^{-3}
Nitrogen dioxide	9.12×10^{-3}	1.41×10^{-2}	1.00×10^{1}	9.49×10^{0}	-	-
Suspended particulate matter (SPM)	-	-	2.49×10^{0}	2.35×10^{0}	-	-
Sulphur dioxide	2.00×10^{0}	1.58×10^{0}	5.86×10^{-3}	5.83×10^{-3}	-	-
Fine particles, < 2.5 μm	1.46×10^{0}	1.14×10^{0}	4.34×10^{-3}	4.34×10^{-3}	-	-
Nitric oxide	1.31×10^{0}	1.01×10^{0}	1.16×10^{-2}	1.16×10^{-2}	-1.68×10^{-2}	-3.06×10^{-2}
Particulate matter, > 2.5 μm, and < 10 μm	1.67×10^{-1}	1.31×10^{-1}	3.99×10^{-3}	3.99×10^{-3}	-	-
Ammonia	2.56×10^{-3}	2.07×10^{-3}	1.49×10^{-6}	1.49×10^{-6}	5.99×10^{-6}	1.04×10^{-5}

Table 6. Results of grouping and weighting of environmental damage from climate change at individual stages of the traditional and ecological tire life cycle [unit: Pt].

Substance	Production		USE		End of Life	
	Traditional Tire	**Eco Tire**	**Traditional Tire**	**Eco Tire**	**Traditional Tire**	**Eco Tire**
Carbon dioxide	2.77×10^{-2}	1.07×10^{-2}	1.24×10^{1}	1.17×10^{1}	-1.03×10^{-2}	-1.24×10^{-3}
Carbon dioxide, fossil	1.69×10^{0}	1.33×10^{0}	6.62×10^{-3}	6.60×10^{-3}	-	-
Methane, fossil	8.76×10^{-2}	6.86×10^{-2}	4.92×10^{-4}	4.92×10^{-4}	-	-
Carbon dioxide, biogenic	1.63×10^{-2}	1.27×10^{-2}	3.35×10^{-5}	3.35×10^{-5}	-	-
Nitrous oxide	4.54×10^{-3}	3.55×10^{-3}	6.46×10^{-7}	6.46×10^{-7}	8.53×10^{-6}	2.60×10^{-5}
Carbon monoxide	-	-	3.08×10^{-3}	2.92×10^{-3}	-	-
Carbon dioxide in air	-2.26×10^{-2}	-1.76×10^{-2}	-1.82×10^{-7}	-1.75×10^{-7}	-	-

Table 7. Results of grouping and weighting of environmental damage from acidification and eutrophication at individual stages of the traditional and ecological tire life cycle [unit: Pt].

Substance	Production		USE		End of Life	
	Traditional Tire	Eco Tire	Traditional Tire	Eco Tire	Traditional Tire	Eco Tire
Nitrogen dioxide	1.76×10^{-3}	2.72×10^{-3}	1.93×10^{0}	1.83×10^{0}	-	-
Sulphur oxide	7.61×10^{-4}	-	9.05×10^{-1}	8.57×10^{-1}	-2.32×10^{-4}	-2.06×10^{-4}
Nitric oxide	2.52×10^{-1}	1.94×10^{-1}	2.26×10^{-3}	2.23×10^{-3}	-3.24×10^{-3}	-5.90×10^{-3}
Sulphur dioxide	1.14×10^{-1}	9.00×10^{-2}	3.43×10^{-4}	3.33×10^{-4}	-	-
Ammonia	1.41×10^{-3}	1.14×10^{-3}	8.23×10^{-7}	8.19×10^{-7}	3.29×10^{-6}	5.71×10^{-6}

Table 8. Results of grouping and weighting of environmental damage involved in the extraction of fossils at individual stages of the traditional and ecological tire life cycle [unit: Pt].

Substance	Production		USE		End of Life	
	Traditional Tire	Eco Tire	Traditional Tire	Eco Tire	Traditional Tire	Eco Tire
Natural gas, 42.7 MJ/kg, fossil	2.65×10^{-3}	-	9.77×10^{1}	9.26×10^{1}	-	-
Natural gas, fossil	1.12×10^{0}	9.08×10^{-1}	7.09×10^{-2}	7.08×10^{-2}	-	-
Crude oil, fossil	1.11×10^{0}	9.01×10^{-1}	6.45×10^{-2}	6.44×10^{-2}	-	-
Hard coal, fossil	4.23×10^{-1}	3.30×10^{-1}	1.98×10^{-4}	1.96×10^{-4}	-	-
Crude oil, 42.6 MJ/kg, fossil	2.96×10^{-1}	-	-	-	-1.69×10^{-2}	-4.77×10^{-3}
Natural gas from waste gases generated in coal mining processes/m^3	1.34×10^{-1}	1.04×10^{-1}	4.43×10^{-7}	4.35×10^{-7}	-	-
Natural gas, 30.3 MJ/kg, fossil	6.69×10^{-2}	8.40×10^{-2}	-	-	-	-
Natural gas, 35 MJ/m^3, fossil	3.21×10^{-2}	-	-	-	7.44×10^{-3}	1.42×10^{-2}
Natural gas, raw material, 35 MJ/m^3, fossil	-	-	-	-	-4.28×10^{-2}	-6.37×10^{-2}
Natural gas, 36.6 MJ/m^3, fossil	-	-	-	-	-4.12×10^{-2}	-7.21×10^{-2}
Crude oil, raw material, 41 MJ/kg, fossil	-	-	-	-	-4.35×10^{-2}	-7.26×10^{-2}

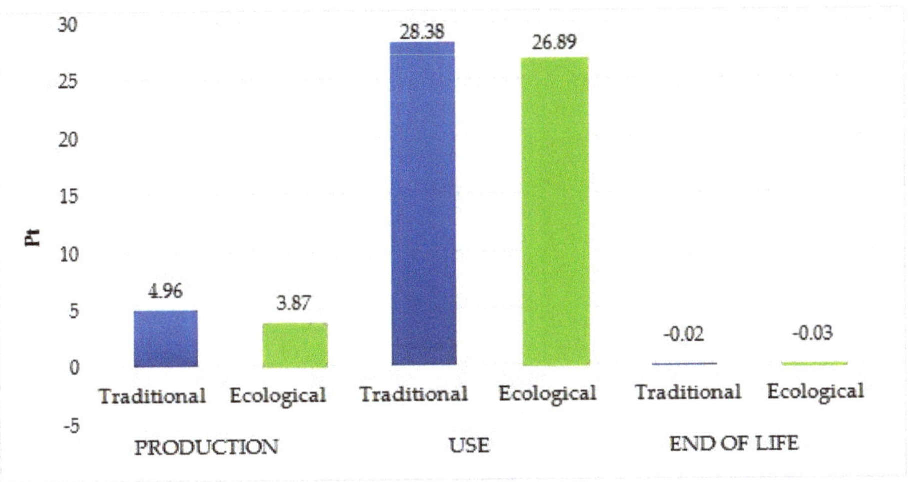

Figure 7. Results of grouping and weighting of environmental damage from respiratory inorganics at individual stages of the traditional and ecological tire life cycle.

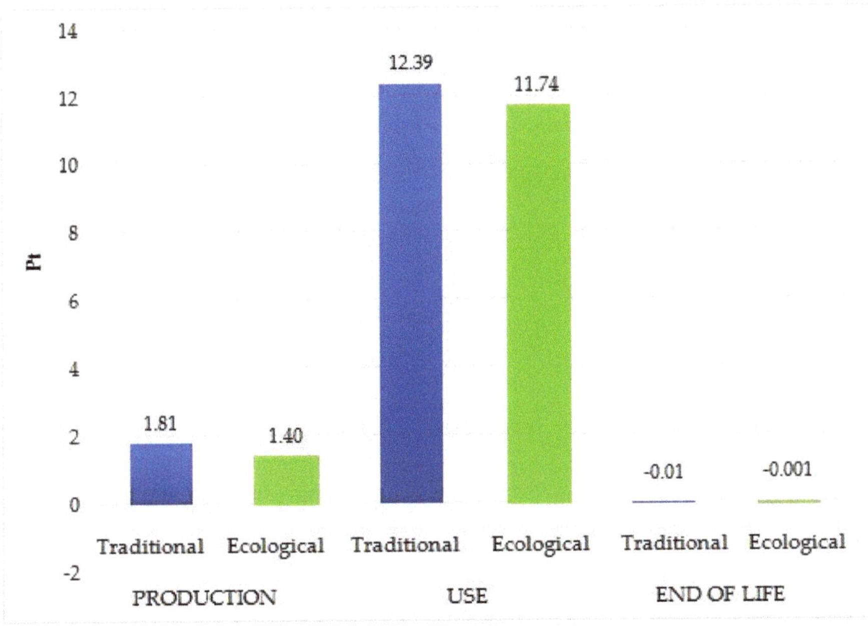

Figure 8. Results of grouping and weighting of environmental damage from climate change at individual stages of the traditional and ecological tire life cycle.

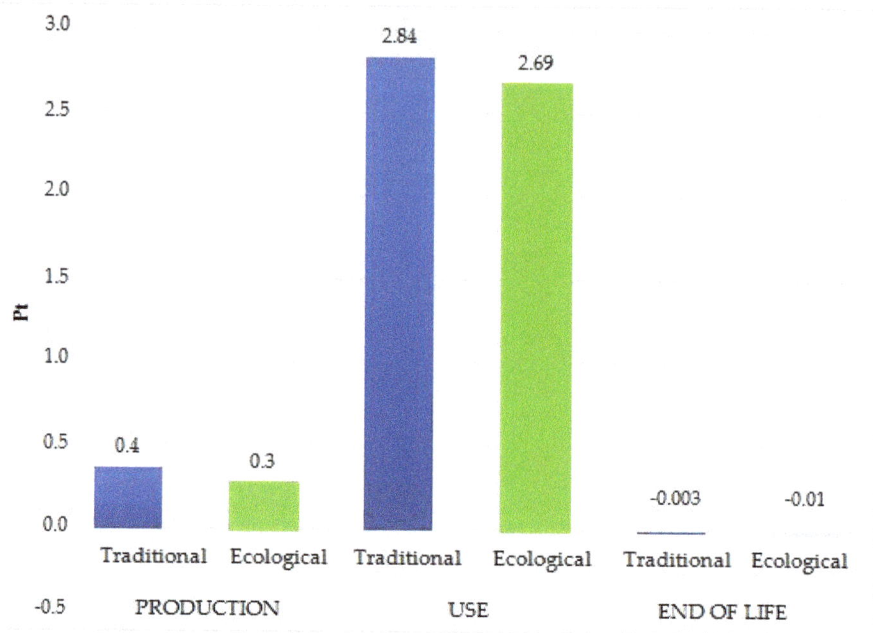

Figure 9. Results of grouping and weighting of environmental damage from acidification/eutrophication at individual stages of the traditional and ecological tire life cycle.

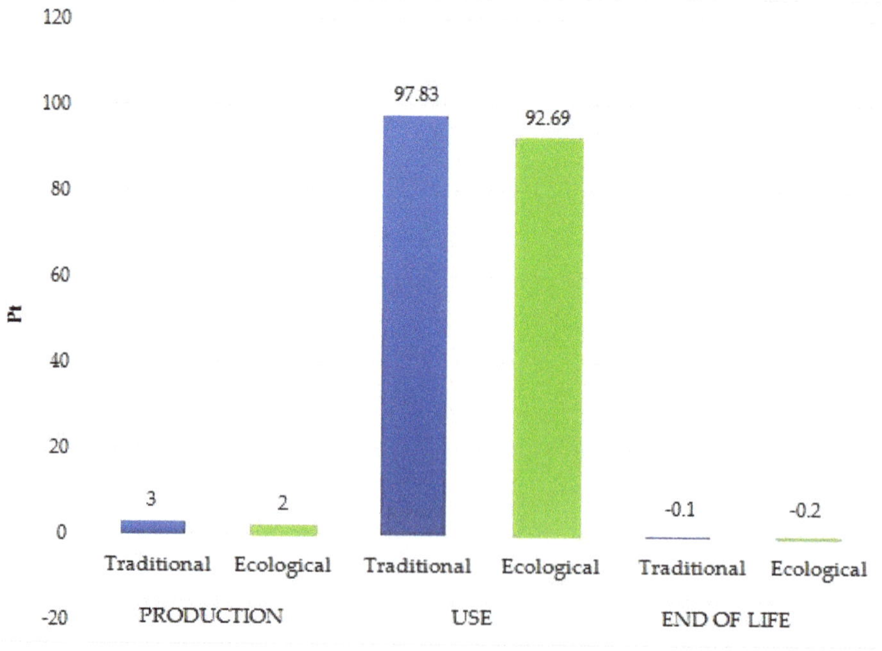

Figure 10. Results of grouping and weighting of environmental damage involved in the extraction of fossils at individual stages of the traditional and ecological tire life cycle.

4.1. Eco-Indicator 99

Table 2 presents the results of the characterization of environmental consequences occurring in the material life cycle of a traditional and ecological car tire based on impact categories. Among substances detrimental to human health, these were the inorganic compounds causing respiratory diseases that were established to have the highest negative impact. Their maximum emissions levels were obtained in the use-stage of the traditional tire (1.09×10^{-3} DALY). The analyses indicate that the recycling processes would result in the reduction of their environmental impact over the entire life by a total of 8.54×10^{-7} DALY, and in the case of the ecological tire by a total of 1.49×10^{-6} DALY. Among the factors of a negative effect on animal health and life, ecotoxic substances were determined to be the leading cause of hazard. Their emissions reach the highest level in the production of the traditional tire (2.90×10 PAF $\times$ m^2 $\times$ year). Recycling at the end of the product use stage of the life cycle would minimized the negative impact by $2.87 \times 10^{-1} \times$ PAF $\times$ m^2 $\times$ year in the conventional tire and $4.66 \times 10^{-1} \times$ PAF $\times$ m^2 $\times$ year in the eco-tire. Considering the hazards to plant organisms, the strongest negative effect is concerned with acidification or eutrophication of the natural environment. The highest level of the compound emissions responsible for these effects relate to the use of the traditional tire (3.64×10 PDF $\times$ m^2 $\times$ year). Recycling would reduce the negative impact by a total of 4.45×10^{-2} PDF $\times$ m^2 $\times$ year (the traditional tire) and by $-7.83 \times 10^{-2} \times$ PDF $\times$ m^2 $\times$ year (the ecological tire). Considering elevated energy consumption on account of raw material extraction, the highest level of harmful impact in this respect was recorded in fossil fuels. As in the previous cases, the most resource-engaging phase was the use of the traditional tire (4.11×10^3 MJ). It was estimated that over the entire life cycle of the traditional tire, recycling would introduce total energy savings amounting to 5.95 MJ, and in the eco-tire 8.32 MJ.

One of the key purposes of LCM is to determine elements in the life cycle of a given product that exert the heaviest impact on the environment. In traditional and ecological tires, the impact categories that were shown to have the most substantial negative consequences for the environment are, in the order of magnitude, the extraction of fossil fuels (an aggregate score for the traditional tire was 100.87 Pt and 94.82 Pt for the eco-tire), emissions of respiratory inorganic substances (33.32 Pt—the traditional tire, 30.71 Pt—the eco-tire), emissions of compounds causing climate change (14.19 Pt—the traditional tire, 13.14—the eco-tire), and acidification/eutrophication emissions (3.21 Pt—the traditional tire, 2.97 Pt—the eco-tire). For both tire technologies, the highest negative emission rates occurred in the use stage (Figure 5).

Table 3 shows the grouping and weighting results concerning the environmental impact at particular stages of the traditional and ecological tire life cycle, divided into impact categories. The negative environmental impact was the highest in the fossil fuel extraction. The total ecological burden of this activity was estimated in the traditional tire at 97.8 Pt, and 92.7 Pt in the case of the eco-tire technology. Recycling is likely to cause a 0.14 Pt reduction in the environmental impact of the traditional tire life cycle, and a 0.2 Pt decrease in the ecological tire. Considering the secondary factors, it was found that inorganic substances linked to respiratory diseases and compounds that cause climate change also exhibit a particularly adverse impact on the environment. For both categories, the strongest negative effect rates were calculated in the tire use stage. On the other hand, the category that displayed the lowest ecological threat concerned compounds contributing to the ozone layer depletion in the traditional and ecological tire life cycle.

The grouping and weighting of consequences to the environment for carcinogens with respect to the products life cycles are shown in Table 4 and Figure 6. Unlike in the previously reported results, the highest hazard on the part of carcinogens was determined for the production stage of the tires' life cycle: 1.51 Pt for the traditional tire and 1.17 Pt for the ecological tire. Considering particular chemicals, arsenic and cadmium ion emissions were found to reach particularly high levels, specifically, for As ions: 1.35 Pt (the traditional tire) and 1.05 Pt (the eco-tire), and for Cd ions: 0.12 Pt (the traditional tire) and 0.11 Pt (the ecological tire). The emissions levels could be improved by recycling, which would drive the decrease in the ecological burden of the entire life cycle of traditional tires by as much as

4.62×10^{-3} Pt (As ions) and 1.35×10^{-4} Pt (Cd ions). Exposure to arsenic compounds results in gastrointestinal irritation, vomiting and diarrhea, as well as skin lesions. In extreme cases, it can lead to death. Long-term exposure most often causes vascular diseases and the development of lung, skin, liver, kidney, and bladder cancer. Arsenic is usually present in the presence of other heavy metals, which increases the risk of interactions between them [50]. The toxic effect of cadmium is mainly related to its presence in the body in the form of free cadmium ions. It causes chronic poisoning, affects the metabolism of calcium and phosphorus compounds, and damages the proper mineralization of bones, increasing their fragility. Its harmful influence mainly affects the excretory and respiratory systems and the circulatory system [53,54].

Several observations emerge from the results of grouping and weighting performed to determine the environmental consequences resulting from the emissions of respiratory inorganic compounds throughout the traditional and ecological tire life cycle (Table 5 and Figure 7). The emission hazard was the highest in the tire use stage, amounting to 28.38 Pt in the case of the traditional tire and 26.89 Pt for the ecological tire. Sulphur oxide and nitrogen dioxide emissions were shown to constitute the greatest threat to health. For the former, they were 15.8 Pt (the traditional tire) and 15.0 Pt (the ecological tire), and for the latter 10.0 Pt (the traditional tire) and 9.49 Pt (the ecological tire), respectively. In both tires, recycling in the post-consumer waste management phase could bring in the drop in the harmful effect of Sulphur oxide by 4.07×10^{-3} Pt in relation to the life cycle of the traditional tire and by 3.60×10^{-3} Pt—the ecological tire. Sulphur dioxide is one of the most dangerous air pollutants. It is a colorless, highly toxic gas with a suffocating odor. Its high concentration in the air can damage the respiratory tract, while in plants it causes the death of leaves, reacting with water vapor or water droplets in clouds creates sulfuric acid, and forms sulphates with dust. Sulphur oxides also contribute to the occurrence of acid rain. The influence of nitrogen oxides also significantly affects the quality of the atmospheric air. Nitrogen dioxide (NO_2) is commonly found in work and municipal environments, resulting from the combustion of nitrogen-containing organic substances. Acute nitrogen dioxide poisoning causes pulmonary edema which can be fatal. Moreover, this compound may act clastogenic and contribute to the development of neoplasms. Sulphur compounds and NOx are the main components of London-type smog [55].

Table 6 and Figure 8 contain the results of grouping and weighting the environmental burden concerned with the emission of hazardous compounds—potential climate change factors. In the category of impact in question, the source of the most critical environmental impact should be traced to the stage of tire use: 12.39 Pt in the traditional tire technology and 11.74 Pt in the eco-tire life cycle. The highest emission levels were displayed for carbon dioxide. Recycling would curb the detrimental effect of carbon dioxide on the environment by 1.03×10^{-2} Pt in the life cycle of the traditional tire and by 1.24×10^{-3} Pt for the ecological tire. There are many risks to human health in a car tire plant. The increase in the amount of this poisonous gas in the room negatively affects the well-being of employees. Carbon dioxide is a colorless, odorless, and non-flammable gas. It is heavier than air and has a choking effect on humans, but in low concentration it is not toxic to the human body. Therefore, it should be ensured that the chemical composition of the air in the production plant is slightly different from the natural (fresh) air. The increase in carbon dioxide content in the room causes the worker to have an adverse effect of hyperventilation, i.e., abnormal and rapid breathing. At higher concentrations, it may lead to unconsciousness and even loss of life [54,55].

Table 7 and Figure 9 report the results of grouping and weighting the environmental effect of acidification/eutrophication agents linked to the traditional and ecological car tire life cycle. The direst consequences were linked to the tire use-stage, which was where the highest values of the eco-indicator points were estimated: 2.84 Pt for the conventional tire manufacturing technology and 2.69 Pt for the ecological tire. Among the compounds promoting acidification/eutrophication, nitrogen dioxide (1.76×10^{-3} Pt traditional tire, 2.72×10^{-3} ecological tire) and Sulphur oxide (7.61×10^{-4} Pt traditional tire) displayed particularly high ecological damage rates. If implemented, the post-consumer waste management of car tires could help reduce the detrimental impact of Sulphur oxide emissions over the entire life cycle of the traditional tire by 2.32×10^{-4} Pt and the ecological tire by 2.06×10^{-4} Pt.

Grouping and weighting results of the environmental consequences of fossil fuel mining that occur in the traditional and ecological car tire life cycle are presented in Table 8 and Figure 10. The highest negative environmental impact was observed in the use-stage of both products' life cycles: 97.83 Pt for the traditional tire design and 92.69 Pt for the ecological tire. Among the investigated processes, natural gas production exhibited a particularly high detrimental effect on the environment. The quantified effect of this factor was estimated at 97.69 Pt in the traditional tire manufacturing and 92.55 Pt in the ecological one.

4.2. CED

The highest energy consumption was recorded at the use-stage of the car tire life cycle. The most likely explanation for this observation is that setting the car in motion requires burning, in aggregate, extensive amounts of non-renewable fossil fuels. By reducing the fuel consumption, eco-friendly tires improve the average mileage, which translates into energy savings of 1500 MJ within a single life cycle. A similar mechanism is observed during the tire manufacturing stage—the technological process of one ecological tire production consumes 995 MJ energy less compared to the traditional tire manufacturing product (Figure 11).

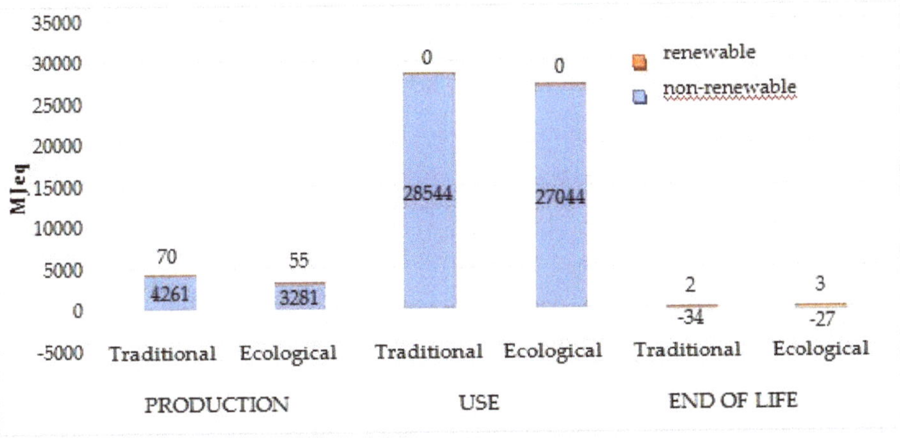

Figure 11. Cumulative Energy Demand (CED) assessment of energy consumption at individual stages of the traditional and ecological tire life cycle, including the type of raw material.

4.3. IPCC

The consumption of non-renewable fuels throughout the car tire use period is inherent in the emission of greenhouse gases in the atmosphere. In fact, any action to reduce the pollution rates will be considered positive from the perspective of climate protection. Fitting cars with environmentally friendly tires could potentially reduce harmful emissions by 119 kg CO_2eq compared to traditional tires (Figure 12). However, it would be the replacement of conventional fuels that power cars with renewable energy sources that would be sure to reduce the extent of the environmental burden throughout the life cycle of both conventional and green tires.

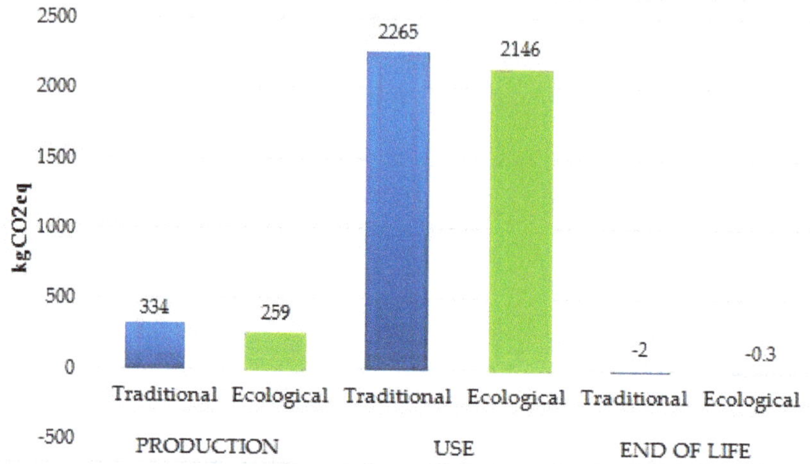

Figure 12. Intergovernmental Panel on Climate Change (IPCC) assessment of greenhouse gas (GHG) emissions at individual stages of the traditional and ecological tire life cycle.

5. Summary and Conclusions

Tire manufacturing is a complex process with a multiple-stage execution. Manufacturers follow in-house production standards and procedures, which explicitly specify the entire process from the acquisition of raw resources to quality management. Faulty quality control could contribute to the emergence of hidden defects in both material quality and dimensional and shape accuracy. While the defects may be visible in a tire at purchase, they could also emerge in use, thereby contributing to the deterioration of operating conditions and travel comfort [52,53].

Manufacturing activities have a significant (usually negative) impact on the environment, including due to the consumption of natural resources, emissions of harmful substances, and the generation of waste. The level of air quality is significantly lower, generating many threats to human health. Air pollution associated with the production of car tires is a serious health risk because it shortens life expectancy and contributes to the occurrence of many diseases, including heart disease, respiratory disease, and cancer. Inhaling air with a high content of harmful substances not only causes health problems of varying severity but can even cause death in the long term.

The aim of the work was achieved owing to the use of LCA as an effective tool for the environmental production of car tires made in a traditional and ecological manner. The conducted research shows that a tire made in accordance with the traditional technology generates more negative consequences for the environment compared to an ecological tire. This is due to a larger (by about 20%) amount of materials used in its production. The most harmful environmental impact at the manufacturing stage is due to the high level of consumption of the constantly depleting natural resources that are necessary in the manufacturing process of car tires. The extraction of fossil fuels is an important cause of environmental degradation and deterioration of human health.

As it results from the analyses, the key factor generating the most negative environmental consequences (apart from manufacturing) is the operation stage. The use of unconventional, renewable energy sources to power a passenger car could significantly contribute to reducing the negative impact on the environment (e.g., by reducing the impact of fossil fuel extraction and the processes of obtaining energy from them).

Considering the economies of scale, i.e., the number of passenger cars registered in Europe, the use of environmentally friendly (lighter weight) tires would save a large amount of material resources needed at the manufacturing stage. This would translate, in a broader perspective, into the improvement of the environment. The possibility of obtaining conventional energy resources is becoming more difficult and more limited. The extraction of coal and crude oil and their combustion, in addition to the negative impact on the greenhouse effect and air purity, also causes irreversible damage to the landscape, land, and disturbances in the management of groundwater and ecosystems.

The results of the conducted research show that the processes related to the extraction of fossil fuels and emissions of inorganic compounds causing respiratory diseases, but also emissions of compounds causing climate change, compounds causing acidification/eutrophication, and carcinogens cause the most negative environmental consequences.

However, it is possible to control the production process of car tires in such a way as to limit its negative impact on the environment. For this purpose, it is possible to:

- increase the importance of renewable raw materials in production processes, e.g., by obtaining rubber from plants such as Guayule instead of only traditional rubber tree, as well as replacing fossil raw materials with renewable materials, e.g., synthetic rubber from petroleum can be made from butadiene produced on the basis of bioethanol, fillers currently produced from crude oil or coal can be produced from vegetable fats and oils or obtaining silica from inedible rice husk, which is now intended for burning;
- create a tire design that allows for easier separation of individual materials that will be easy to identify during development after use;
- minimize the energy consumption and material consumption of the manufacturing, use, and post-use disposal of tires;
- precisely verify the life cycle scenarios of various types of car tires in real conditions, taking into account various models of post-use management.

Author Contributions: Conceptualization, P.K. and P.I.; methodology, P.K., P.I., B.-W.P., K.W. and T.A.; software, P.K., P.I., B.-W.P. and K.W.; validation, P.I. and B.-W.P.; formal analysis, P.K., B.-W.P. and K.W.; investigation P.I., B.-W.P., K.W. and T.A.; resources, P.K. and P.I.; data curation, P.I., B.-W.P. and T.A.; writing—original draft preparation, P.K., P.I.; writing—review and editing, P.K., B.-W.P., and T.A.; visualization, P.I., B.-W.P. and K.W.; supervision, P.K., P.I., and T.A.; project administration, P.K. and P.I. All authors have read and agreed to the published version of the manuscript.

Funding: This research received no external funding.

Conflicts of Interest: The authors declare no conflict of interest.

References

1. Kłos, Z. Introduction of environmental aspects in designing of machines. In *MATEC Web of Conferences*; EDP Sciences: Paris, France, 2019; Volume 254.
2. Guinée, J.; Heijungs, R.; Huppes, G.; Zamagni, A.; Masoni, P.; Buonamici, R.; Rydberg, T. Life cycle assessment: Past, present, and future. *Environ. Sci. Technol.* **2011**, *1*, 90–96. [CrossRef] [PubMed]
3. Ram, A.; Sharma, P. *A Study on Life Cycle Assessment, ICNASET, February 24–25*; Madhav University: Sirohi (Rajasthan), India, 2017.
4. Bałdowska-Witos, P.; Kruszelnicka, W.; Kasner, R.; Tomporowski, A.; Flizikowski, J.; Mrozinski, A. Impact of the plastic bottle production on the natural environment. Part 2. Analysis of data uncertainty in the assessment of the life cycle of plastic beverage bottles using the Monte Carlo technique. *Przem. Chem.* **2019**, *98*, 1668–1672.
5. International Organization for Standardization. *ISO 4223-1:2017—Definitions of Some Terms Used in the Tire industry—Part. 1: Pneumatic Tires*; International Organization for Standardization: Geneva, Switzerland, 2017.
6. Dell'Acqua, G.; De Luca, M.; Prato, C.; Prentkovskis, O.; Junevicius, R. The impact of vehicle movement on exploitation parameters of roads and runways: A short review of the special issue. *Transport* **2016**, *31*, 127–132. [CrossRef]
7. Kaka, S.; Himran, S.; Renreng, I.; Sutresman, O. The pneumatic actuators as vertical dynamic load simulators on medium weighted wheel suspension mechanism. *J. Phys. Conf. Ser.* **2018**, *962*, 1–9. [CrossRef]
8. Voiku, I.P.; Komissarov, I.A. Double-sided summer-winter vehicle tire: Advantages and prospects of industrial production. *IOP Conf. Ser. Mater. Sci. Eng.* **2020**, *709*, 3–5. [CrossRef]

9. Arachchige, U.; Sithari, M.; Rangana, H.; Tharakie, G.M.; Tharuka, H. Environmental Pollution by Tire Manufacturing Industry. *Int. J. Sci. Technol. Res.* **2018**, *8*, 80–81.

10. Lebreton, B.; Tuma, A. A quantitative approach to assessing the profitability of car and truck tire manufacturing. *Int. J. Prod. Econ.* **2006**, *2*, 639–652. [CrossRef]

11. Gratkowski, M.T. Burning characteristics of automotive tires. *Fire Technol.* **2014**, *50*, 379–391. [CrossRef]

12. Lopera Perez, J.C.; Kwok, C.Y.; Senetakis, K. Effect of rubber size on the behavior of sand-rubber mixtures: A numerical investigation. *Comput. Geotech.* **2016**, *80*, 199–214. [CrossRef]

13. Imbernon, L.; Norvez, S. From landfilling to vitrimer chemistry in rubber life cycle. *Europ. Pol. J.* **2016**, *82*, 347–376. [CrossRef]

14. Bokhari, A.; Chuaha, L.F.; Yusup, S.; Klemeš, J.J.; Akbar, M.M.; Kamil, R.N.M. Cleaner production of rubber seed oil methyl ester using a hydrodynamic cavitation: Optimization and parametric study. *J. Clean. Prod.* **2016**, *136*, 31–41. [CrossRef]

15. Azimov, Y.; Gilmanshin, I.; Gilmanshina, S. Modern technologies of waste utilization from industrial tire production. In Proceedings of the IOP Conference Series: Materials Science and Engineering, Kazan, Russia, 2–3 December 2015.

16. Sobaszek, Ł.; Gola, A.; Kozłowski, E. Prediction of variable technological operation times in production jobs scheduling. *IFAC Pap.* **2019**, *52*, 1301–1306. [CrossRef]

17. Ivan, G. Some aspects of tire impact on environment: Production, service and recycling. *Elastomery* **2002**, *6*, 24–38.

18. Kannan, D.; Diabat, A.; Shankar, K.M. Analyzing the drivers of end-of-life tire management using interpretive structural modeling (ISM). *The Intern. J. Adv. Manufact. Tech.* **2014**, *72*, 1603–1614. [CrossRef]

19. Rzeźnik, C.; Grześ, Z. Structural materials in agricultural machines. *J. Res. Appl. Agric. Eng.* **2017**, *62*, 100–103.

20. Berge, T.; Mioduszewski, P.; Ejsmont, J.; Świeczko-Żurek, B. Reduction of road traffic noise by source measures—Present and future strategies. *Noise Control. Eng. J.* **2017**, *65*, 549–559. [CrossRef]

21. Kole, P.J.; Löhr, A.J.; Van Belleghem, F.G.A.J.; Ragas, A.M.J. Wear and tear of tires: A stealthy sourceof microplastics in the environment. *Int. J. Environ. Res. Public Health* **2017**, *14*, 1265. [CrossRef]

22. Bulei, C.; Todor, M.P.; Heput, T.; Kiss, I. Directions for material recovery of used tires and their use in the production of new products intended for theindustry of civil construction and pavements. In Proceedings of the International Conference on Applied Sciences (ICAS2017), Hunedoara, Romania, 10–12 May 2017.

23. Uson, A.A.; German, F.F.; Zabalza, I.; Zambrana-Vasquez, D. Study of the environmental performance of end-of-life tire recycling through a simplified mathematical approach. *Therm. Sci.* **2012**, *16*, 889–899.

24. Forrest, M.J. *Recycling and Re-Use of Waste Rubber*; Smithers Rapra Technology Ltd.: Shawbury, UK, 2014.

25. Myhre, M.; Saiwari, S.; Dierkes, W.; Noordermeer, J. Rubber recycling: Chemistry, processing, and applications. *Rubber. Chem. Technol.* **2012**, *85*, 408–449. [CrossRef]

26. Martínez, J.; Puy, N.; Murillo, R.; Garcia, T.; Navarro, M.; Mastral, A. Waste tire pyrolysis—A review. *Renew. Sustain. Energy Rev.* **2013**, *23*, 179–213. [CrossRef]

27. Ramos, G.; Alguacil, F.; López, F. The recycling of end-of-life tires. Technological review. *Rev. Metal.* **2011**, *47*, 273–284.

28. Piotrowska, K.; Kruszelnicka, W.; Bałdowska-Witos, P.; Kasner, R.; Rudnicki, J.; Tomporowski, A.; Flizikowski, J.; Opielak, M. Assessment of the environmental impact of a car tire throughout its lifecycle using the lca method. *Materials* **2019**, *12*, 4177. [CrossRef] [PubMed]

29. Hunkeler, D.; Saur, K.; Stranddorf, H.; Rebitzer, G.; Finkbeiner, M.; Schmidt, W.P.; Jensen, A.A.; Christiansen, K. *Life Cycle Management*; SETAC Press: Pensacola, FL, USA, 2003; pp. 20–24.

30. Piasecka, I.; Bałdowska-Witos, P.; Piotrowska, K.; Tomporowski, A. Eco-energetical life cycle assessment of materials and components of photovoltaic power plant. *Energies* **2020**, *13*, 1385. [CrossRef]

31. Bałdowska-Witos, P.; Kruszelnicka, W.; Kasner, R.; Rudnicki, J.; Tomporowski, A.; Flizikowski, J. Impact of the plastic bottle production on the natural environment. Part. 1. Application of the ReCiPe 2016 assessment method to identify environmental problems. *Przem. Chem.* **2019**, *10*, 1662–1667. [CrossRef]

32. Kirchain, R.E., Jr.; Gregory, J.R.; Olivetti, E.A. Environmental life-cycle assessment. *Nat. Mater.* **2017**, *16*, 693–697. [CrossRef] [PubMed]

33. Mroziński, A.; Piasecka, I. Selected aspects of building, operation and environmental impact of offshore wind power electric plants. *Pol. Marit. Res.* **2015**, *22*, 86–92. [CrossRef]

34. Kłos, Z. Ecobalancial assessment of chosen packaging processes in food industry. *Int. J. Life Cycle Assess.* **2002**, *7*, 309. [CrossRef]
35. Jasiulewicz-Kaczmarek, M.; Gola, A. Maintenance 4.0 technologies for 2014 sustainable manufacturing—An Overview. *IFAC Pap.* **2019**, *52*, 91–96. [CrossRef]
36. Piasecka, I.; Tomporowski, A. Analysis of environmental and energetical possibilities of sustainable development of wind and photovoltaic power plants. *Probl. Sustain. Dev.* **2018**, *13*, 125–130.
37. Simla, T.; Stanek, W. Reducing the impact of wind farms on the electric power system by the use of energy storage. *Renew. Energy* **2020**, *145*, 772–782. [CrossRef]
38. Wolfson, A.; Dominguez-Ramos, A.; Irabien, A. From goods to services: The life cycle assessment perspective. *J. Serv. Sci. Res.* **2019**, *11*, 17–45. [CrossRef]
39. Guineē, J.B. Handbook on life cycle assessment operational guide to the ISO standards. *Int. J. Life Cycle Assess.* **2002**, *7*, 311–313. [CrossRef]
40. Bałdowska-Witos, P.; Piotrowska, K.; Kruszelnicka, W.; Błaszczak, M.; Tomporowski, A.; Opielak, M.; Kasner, R.; Flizikowski, J. Managing the uncertainty and accuracy of life cycle assessment results for the process of beverage bottle moulding. *Polymers* **2020**, *12*, 1320. [CrossRef] [PubMed]
41. Tomporowski, A.; Flizikowski, J.; Wełnowski, J.; Najzarek, Z.; Topoliński, T.; Kruszelnicka, W.; Piasecka, I.; Śmigiel, S. Regeneration of rubber waste using an intelligent grinding system. *Przem. Chem.* **2018**, *10*, 1659–1665.
42. Lewandowska, A. Environmental life cycle assessment as a tool for identification and assessment of environmental aspects in environmental management systems (EMS) part 1: Methodology. *Int. J. Life Cycle Assess.* **2011**, *16*, 178–186. [CrossRef]
43. Shvetsova, O.A.; Lee, J.H. Minimizing the environmental impact of industrial production: Evidence from south korean waste treatment investment projects. *Appl. Sci.* **2020**, *10*, 3489. [CrossRef]
44. Rebitzer, G.; Loerincik, Y.; Jolliet, O. Input-output life cycle assessment: From theory to applications. *Int. J. Life Cycle Assess.* **2002**, *7*, 174–176. [CrossRef]
45. Imran, U.; Ahmad, A.; Othman, M. Life cycle assessment of simulated hydrogen production by methane steam reforming. *Aust. J. Basic Appl. Sci.* **2017**, *11*, 43–50.
46. Olsen, S.; Christensen, F.; Hauschild, M.; Pedersen, F.; Larsen, H.; Torslov, J. Life cycle impact assessment and risk assessment of chemicals—A methodological comparison. *Environ. Impact Assess. Rev.* **2001**, *21*, 385–404. [CrossRef]
47. Piasecka, I.; Tomporowski, A.; Piotrowska, K. Environmental analysis of post-use management of car tires. *Przem. Chem.* **2018**, *97*, 1649–1653.
48. Tomporowski, A.; Piasecka, I.; Flizikowski, J.; Kasner, R.; Kruszelnicka, W.; Mroziński, A.; Bieliński, K. Comparision analysis of blade life cycles of land-based and offshore wind power plants. *Pol. Marit. Res.* **2018**, *25*, 225–233. [CrossRef]
49. Goedkoop, M.; Spriensma, R. *The Eco-Indicator 99—A Damage Oriented Method for Life Cycle Assessment Methodology Report*; Pré Consultants: Amersfoort, The Netherlands, 2000; pp. 100–101.
50. Dreyer, L.C.; Niemann, A.L.; Hauschild, M.Z. Comparison of three different LCIA methods: EDIP97, 521 CML2001 and Eco-indicator 99. *Int. J. Life Cycle Assess.* **2003**, *8*, 191–200. [CrossRef]
51. Chłopek, Z.; Lasocki, J. Application of the life cycle assessment method to the study on the ecological properties of the car. *Zesz. Nauk. Inst. Pojazdów* **2013**, *1*, 57–66.
52. Bovea, M.; Powell, J. Alternative scenarios to meet the demands of sustainable waste management. *J. Environ. Manag.* **2006**, *79*, 115–132. [CrossRef]
53. Tomporowski, A.; Opielak, M. Structural features versus multi-hole grinding efficiency. *Maint. Reliab.* **2012**, *14*, 222–227.
54. Kostrz, M.; Satora, P. The compounds responsible for air pollution. *Inżynieria Ekol.* **2017**, *18*, 89–95. [CrossRef]
55. Ociepa-Kubicka, A.; Ociepa, E. Toxic effects of heavy metals on plants, animals and humans. *Eng. Prot. Environ.* **2012**, *15*, 169–180.

Article

Game-Theoretic Analysis to Examine How Government Subsidy Policies Affect a Closed-Loop Supply Chain Decision

Izabela Ewa Nielsen [1,†], **Sani Majumder** [2,†] **and Subrata Saha** [1,*,†]

1   Department of Materials and Production, Aalborg University, DK 9220 Aalborg East, Denmark; izabela@mp.aau.dk

2   Betberia High School, Nadia 741123, West Bengal, India; sani81live@gmail.com

*   Correspondence: subrata.scm@gmail.com; Tel.: +45-91939202

†   These authors contributed equally to this work.

Received: 21 November 2019; Accepted: 19 December 2019; Published: 23 December 2019

Abstract: The pros and cons of government subsidy policies in a closed-loop supply chain (CLSC) setting on optimal pricing, investment decisions in improving product quality, and used product collection under social welfare (SW) optimization goal have not been examined comprehensively. This study compares the outcomes of three government policies under manufacturer-Stackelberg (MS) and retailer-Stackelberg (RS), namely (i) direct subsidy to the consumer, (ii) subsidy to the manufacturer to stimulate used product collection, and (iii) subsidy to the manufacturer to improve product quality. Results demonstrate that the greening level, used product collection, and SW are always higher under the RS game, but the rate of a subsidy granted by the government is always higher under the MS game. Profits for the CLSC members and SW are always higher if the government provides a subsidy directly to the consumer, but productivity of investment in the perspective of the manufacturer or government are less. In a second policy, the government organizations grant a subsidy to the manufacturer to stimulate used product collection, but it does not necessarily yield the desired outcome compared to others. In a third policy, the manufacturer receives a subsidy on a research and development (R&D) investment, but it yields a sub-optimal greening level. This study reveals that the outcomes of subsidy policies can bring benefit to consumers and add a degree of complication for CLSC members; government organizations need to inspect carefully among attributes, mainly product type, power of CLSC members, and investment efficiency for the manufacturer, before implementing any subsidy policies so that it can lead to an environmentally and economically viable outcome.

Keywords: production planning optimization; closed-loop green supply chain; government subsidy; stackelberg game; re-manufacturing

---

1. Introduction

In last two decades, a closed-loop supply chain (CLSC) is gaining increasing attention from both industry practitioners and academics due to alleged benefits for sustainable development and growing environmental awareness among consumers and environmental regulations [1–7]. Government organizations play an impressive role in the development of sustainable product manufacturing and re-manufacturing decisions; they can enforce strict legislation, as well as offer support through various subsidy policies. In Japan, the Ministry of Environment approved 5 billion yen in 2019 as a subsidy for the manufacturer to cover 33% to 50% of their equipment price to produce products with biodegradable bio-plastics (https://bioplasticsnews.com/2018/08/27/japan-government-bioplastics/). The government of China provides a subsidy in different ways. For example, a manufacturer in Hunan province receives a one-time subsidy to improve re-manufacturing activities, whereas in Hubei province, Sevalo Construction Machinery Re-manufacturing Co. Ltd. receives 1 million RMB as a research and development (R&D) subsidy to improve re-manufacturing activities [8]. In 2016, the Chinese government introduced the "Guidance on Promoting Green Consumption" program to achieve the long-term goal of stimulating green product consumption. Through the Technology & Quality Up-gradation Support scheme for MSMEs (TEQUP), the government of India provides a subsidy up to 25% of the project cost for implementation of energy efficient technology. The support became one of the key factors influencing growth for the companies like Banyan Nation, Karma Recycling; in fact, the former company received the Dell People's choice award for circular economy entrepreneur at the world economic forum in Davos (www.standupmitra.in/ Home/SubsidySchemesForAll). To encourage the consumer to procure an energy-efficient green vehicle, the United States government provides subsidies up to $7500 for the purchase of a plug-in electric vehicle [9]. Government organizations in the USA, such as The Ohio Environmental Protection Agency (EPA), has awarded a total of $1.24 million in recycling market development grant money to upgrade and install new equipment to increase the amount of recyclables (www.recyclingtoday.com/ article/ohio-epa-recycling-grants/). The European Union also put significant efforts to promote green product manufacturing and expedite product reuse, as well as introduce various financial packages to encourage the circular economy (European Commission, 2015). Recently, an innovative and flexible pan-European network of research funding organizations, supported by EU Horizon 2020, proposed a funding of 14.530 million euros (www.era-min.eu/sites/default/files/docs/call_text_2018_0.pdf). The above evidences explain that government subsidy policies are made in different ways.

Therefore, it is imperative to conduct comparative analysis for highlighting the pros and cons among those policies. Despite the necessity to explore the effect of different government subsidy policies in CLSCs under government social welfare (SW) maximization objective, comparative study is relatively sparse. This study considers omnipotence of three subsidy policies under the manufacturer-Stackelberg (MS) and retailer-Stackelberg (RS) games to pinpoint their effects and explores the answers to the following research issues:

1. How does the government social welfare optimization goal affect the optimal decision of a CLSC members?
2. Do power structures and product type affect the used product collection, pricing, and investment decision of the product?
3. Which policy stimulates the manufacturer to escalate investment in R&D and product collection activity?
4. What are the main barriers associated with each subsidy policy that overturn the government, as well as CLSC members' sustainability target?

In an attempt to answer the above questions and provide insights, we examined outcomes of eight scenarios and compared corresponding optimal decisions. For tractability, and in line with the CLSC configuration considered by previous studies [10], we mainly focused on single period optimal decision. However, product collection and network design [11] is an important aspect, we limited this study on the manufacturer collection mode only to keep our focus on the assessment of three subsidy policies under a three-stage game framework. In Policy C, the government provides a subsidy directly to the consumer [12] to stimulate a green product purchase. In Policy RE, the government shares a fraction of manufacturer investment effort to encourage used product return. In Policy T, the manufacturer receives a fraction of R&D investment from the government as a subsidy to improve the greening level (GL) [13]. To compare outcomes of subsidy policies, we derived an optimal decision under no subsidy as the benchmark, called Policy N. To explore the influence of a powerful retailer, models were formulated under both the MS and RS game frameworks; and results are compared. This study contributes to the present literature as follows: First, comparative analysis will help practitioners to understand the behavior of pricing, investment decision for the manufacturer in used product return, and R&D to improve GL in a CLSC. Second, the government sets the SW optimization goal and decides the amount of the subsidy. Therefore, results can provide a guideline for them before implementing subsidy policies. Third, according to the investment decision, green products can be categorized as development intensive green products (DIGPs); marginal cost-intensive green products (MIGPs); and marginal-development cost-intensive green products (MDIGPs) [14,15]. Examples belonging to the first categories are developing LED bulbs; integrating an adaptive product business model, energy star home appliances, technology for product-life extension, and high-speed electric cars. All of these require a substantial amount of R&D investments. On the other hand, there is installing lithium-ion car batteries or emission control devices, using biodegradable plastics for FMCG packaging, and the manufacturing cost increasing with per unit product, all of which belong to second category. To the best of the authors' knowledge, the effect of MDIGPs on CLSCs has not been explored yet. Therefore, this study provides a complete overview for the manufacturer on the investment decision to produce MDIGPs. Finally, comparative study in the perspective of participating members can assist to formulate a framework to design a subsidy policy for green product manufacturing and re-manufacturing.

This study is organized as follows. The following subsection a brief literature review is reported to highlight the position of our research in the literature. Assumptions and background of the models are presented in Section 2. CLSC decisions under subsidy policies are discussed in Section 3. In Section 4, managerial insights are drawn with numerical illustration. Finally, conclusions, limitations, and future research are presented in Section 5.

1.1. Literature Review

It is imperative for CLSC members trading with green product to consider the evolving behavior of consumers when making important strategic and operational decisions. Possibly, Ottman [16] first reported the opportunities and the pitfalls in green marketing. The author noted double benefit for the consumer, i.e., 'personal benefit' and 'environmental benefit', in green product procurement. Since then, numerous studies related to green supply chain (GSC) management explored the properties of optimal decision under price-GL sensitive demand in different perspectives [17–19]. We will discuss some recent works focusing on variation of optimal decisions in different games under price-GL sensitive demand. Ghosh and Shah [20] compared optimal decisions obtained for different games and stated that the GL increased in the MS game, but the consumers needs to pay more. Liu and Yi [21] established that pricing and GL changes considerably under various power structures when the manufacturer also invest in knowing consumer preference information in the big data environment, and the manufacturer needs to set the lowest wholesale price under the RS game. Yang and Xiao [22] explored optimal decisions for a GSC under governmental interventions and used triangular fuzzy numbers to describe the imprecise information. The authors found that RS game scenario is the best for all players if governmental interventions are strong enough. Nielsen et al. [23] explored characteristics of the three-level GSC in a two-period setting. The authors found that the manufacturer needs to trade with the product at lower GL if the distributor dominates the GSC. Chen et al. [24] examined pricing, along with the investment effort of both the manufacturer and retailer in a GSC. The authors found that the total GSC profits increased if members share the R&D expenditure but not individual of the manufacturer or retailer simultaneously. Dey et al. [14] found that the manufacturer's decision to produce MIGPs and DIGPs is highly sensitive to the power structure. The authors found that a powerful retailer might want to trade with MIGPs, which leads to less amount of profit for the manufacturer. In this direction, the recent works of Huang et al. [25] and Ranjan and Jha [26] are worthy of mention. The findings of the above cited articles support that the game structures always made an impact on the optimal decisions. However, the above studies explored the characteristics of a forward SC. We extended this stream of research and studied the properties of CLSC under price-GL sensitive demand. CLSC is one of the great interests in both business and academic researchers due to growing consumer awareness on environmental issues and regulations. In existing literature, CLSC models are studied to explore various perspectives. For example, Hong and Yeh [27], Ma et al. [28], and Saha et al. [29] compared optimal decisions in a CLSC under different collection mode. The authors formulated their models mainly under the MS game framework and explored the consequences where the manufacturer, retailer, or a third party, individually or jointly collects used products. On the other hand, CLSC coordination issues were comprehensively studied by Zhang and Ren [30], He et al. [31], and others. For example, Hong et al. [32], Johari and Motlagh [33], and He et al. [31] discussed effect of two-part tariff contract; Zhao et al. [34] used a commission fee contract, while a revenue-and-expense sharing contract is used by Xie et al. [35], and spanning revenue-cost sharing is used by Choi et al. [36]. On the other hand, the optimal decision under different game structures is discussed by Wang et al. [37], Gao et al. [38], Zheng et al. [39], and others. In those studies, the authors made an effort to highlight how the optimal decision changes according to game structure. We refer to recent review articles [2,40–42] on CLSC for detailed discussion. The environmental and operational measures of a CLSC network is another important aspect, where how to reduce number of vehicles to be used and the resulting carbon emissions, as well as the impact of re-manufacturing on environment, are studied extensively under an integer programming framework. We refer to the work by [43–45] for the detailed discussion in this direction. However, as mentioned earlier, it is difficult to ignore the influence of government organizations on the optimal decision in a CLSC, but literature is scanty in that direction. Zhang et al. [46] measured the supply-chain green efficiency (SCGE) of thirty-seven different industrial sectors in China and found that environmental policy management and innovation capacity of the manufacturer are important factors affecting SCGE. Researchers mostly explored the influence of a government subsidy in a forward SC. For example, Hafezalkotob [47], in addition to Sinayi and Rasti-Barzoki [12], explored the optimal decision where the consumer receives a subsidy directly from the government; Chen et al. [48] explored characteristics of optimal decisions when both the manufacturer and retailer receive a subsidy on per unit product; Safarzadeh and Rasti-Barzoki [49] discussed the optimal decision for a two-echelon SC when the manufacturer receives a subsidy on the R&D investment. To establish the position of the present study, we outline existing work on CLSCs under influence of a subsidy in Table 1.

Table 1. Comparison of existing studies with the present study. SW = social welfare; GL = greening level; MS = manufacturer-Stackelberg; RS = retailer-Stackelberg.

Study	Games	Effect of GL	Nature of Subsidy	SW Maximization
Mitra and Webster [50]	MS	No	To the manufacturer to increase re-manufacturing activity	No
Ma et al. [51]	MS	No	To consumers to procure new products	No
Shu et al. [52]	MS	No	To the manufacturer for re-manufacturing	No
Xiao et al. [53]	MS	No	To the manufacturer and consumer jointly	Yes
Heydari et al. [54]	MS	No	To the manufacturer and retailer for re-manufacturing	No
Jena et al. [55]	MS	No	Replacement subsidy to the customer and manufacturer	Yes
Jena et al. [56]	MS	No	To the manufacturer for re-manufacturing	No
Guo et al. [3]	MS	No	To the manufacturer for re-manufacturing	No
Wan and Hong [57]	MS	No	To the manufacturer for re-manufacturing and retailer for recycling	No
Saha et al. [58]	MS	No	To the manufacturer and to consumer based on the greening level	No
He et al. [59]	MS	No	Directly to consumers	Yes
Present study	MS and RS	Yes	Directly to consumers, to the manufacturer for improving quality and to the manufacturer for re-manufacturing	Yes

Table 1 demonstrates that most of the articles focused on the behavior of participants under a single subsidy policy, mostly under the MS game setting. The effect of joint influence of price-GL is also ignored. With growing awareness about green products, the influence of GL needs to be considered to obtain a pragmatic CLSC decision. Comparative study to explore preferences of the CLSC members, consumers, and government organizations are not examined in the previous literature.

In this study, the investment and pricing decisions of a CLSC members are explored by correlating the optimal decision of the government organization. This study can help practitioners to understand the pricing and investment patterns for the manufacturer under the MS and RS games in a CLSC setting. Comparative analysis conducted in this study on the efficiency of investment and consumer preference can help government organizations to cultivate a pragmatic subsidy policy.

2. Prerequisites and Assumptions

We considered eight different scenarios, namely Scenarios ij, $i \in \{m, r\}$, which signifies MS and RS games; $j \in \{C, RE, T, N\}$, which refers to the models where consumers receive a subsidy (C), the manufacturer receives a subsidy on the investment effort on recycling (RE), and the manufacturer receives a subsidy on the total R&D investment (T); and the benchmark decision model where the government organizations do not provide a subsidy (N), respectively. Therefore, the first index represents the game structure, and the second one represents the subsidy policy. The following notations presented in Table 2 are used to differentiate the decision and auxiliary variables in different scenarios:

Table 2. Decision and auxiliary variables.

w_i^j	wholesale price of the new/re-manufacturing product
p_i^j	market price
τ_i^j	collection rate of used products
θ_i^j	greening level
ρ_i^j	per unit subsidy received by the consumer from the government, $(\rho_i^j < p_i^j)$
η_i^j	subsidy received by the manufacturer on investment to improve recycling, $(0 \leq \eta_i^j \leq 1)$
μ_i^j	fraction of subsidy received by the manufacturer on investment to improve GL, $(0 \leq \mu_i^j \leq 1)$
π_{ki}^j	member k's profit, $k \in \{m, r\}$
π_{gi}^j	SW of government
Q_i^j	sales volume

The following assumptions are made to establish proposed models:

1. Similar to Nielsen et al. [60]; Dey et al. [14]; Dey and Saha [61], the market demand is linearly dependent on the retail price and GL, and its functional form is $D_i^j = a - p_i^j + \beta\theta_i^j$, where a represents potential intrinsic demand, and $\beta(> 0)$ represents GL sensitivity. Therefore, higher value for a means the consumer has the better perception about the product. For analytical simplicity, the coefficient of price sensitivity is normalized with the unit [15].

2. It is assumed that re-manufacturing cost is less compared to manufacturing cost, i.e., $c_r < c_m$ [6,57]. The manufacturer invests $CL(\tau_i^j, \alpha) = \alpha\tau_i^j D_i^j + \kappa\tau_i^{j^2}$ to collect used products, where $\kappa > 0$ represents the scaling parameter, and $\$\alpha > 0$ represents the monetary benefit's for the consumers for returning used products. If $\alpha = 0$, this assumption is similar to Ma et al. [28], Xiao et al. [53], and Wan and Hong [57]. τ_i^j represents the collection rate $(0 \leq \tau_i^j \leq 1)$. While optimizing objective functions, all members have access to the same information [51,59]. A portion δ $(0 \leq \delta \leq 1)$ of the collected used products converts into new one [57].

3. The manufacturer bears extra cost for green technology innovation. In a recent study by Zhang et al. [62], it was found that technological innovation have significant effects on regional industrial eco-efficiency. In this study, we assume that the manufacturer produces MDIGPs, and corresponding per unit and total R&D investment costs are considered as $\lambda_1\theta$ and $\lambda_2\theta^2$, respectively. Therefore, λ_1 and λ_2 represents the efficiency of the manufacturer on per unit investment and investment in R&D, respectively. If $\lambda_1 = 0$, then the manufacturer invests in producing DIGPs [18,19]. If $\lambda_2 = 0$, then the product converts to MIGPs [63]. The fixed cost for the retailer and manufacturer are normalized to zero [14,15]. The government organizations provide

a R&D subsidy on the total investment. As noted by Dey et al. [14], for MIGPs, the variable cost is directly proportional with the product quality, and it might not possible to recover the cost for the manufacturers. For example, installing emission reduction devices or packaging material are directly proportional to the unit product, but it is difficult for the manufacturer to recover the cost of those in the re-manufacturing process.

4. The influence of three subsidy policies was analyzed. In Policy C, the government provides a subsidy ρ_i^j on per unit product directly to consumers. Therefore, the consumers need to pay $p_i^j - \rho_i^j$ [12,64] for per unit purchase. In Policy RE, the manufacturer receives a subsidy η_i^j, $(0 < \eta_i^j \leq 1)$ on the investment effort on used product collection. In Policy T, the manufacturer receives a subsidy μ_i^j, $(0 < \mu_i^j \leq 1)$ on the R&D investment [13,65] to improve GL.

5. We find optimal decisions in a three-stage game to study the influence of government decision. Under the MS and RS games, the decision sequence is defined as follows:

- Step 1: The government decides the subsidy rate (ρ_i^C or η_i^{RE} or μ_i^T) by maximizing social welfare;
- Step 2: In the MS game, the manufacturer decides w_m^j, θ_m^j, and τ_m^j. In the RS game, the retailer decides the profit margin $m_r^j = p_r^j - w_r^j$;
- Step 3: In the MS game, the retailer decides the retail price p_m^j. In the RS game, the manufacturer decides w_r^j, θ_r^j, and τ_r^j.

Therefore, the government takes the responses of the CLSC members into consideration, while deciding the subsidy rate [48,59].

3. The Models

In this section, we formulate mathematical models to examine the characteristics of optimal pricing, re-manufacturing, and investment in R&D for CLSC members and the subsidy rate of government organizations. The scenarios wherein the consumer subsidy is provided are explored in Section 3.1, and then the scenarios where the manufacturer receives a subsidy on the investment effort for improving used product collection are discussed in Section 3.2, and finally, the equilibrium results are derived in Section 3.3, where the manufacturer receives a subsidy on the R&D investment. Optimal decisions between the MS and RS games are compared to explore characteristics of GL, collection rate, SW, and the government subsidy rate.

3.1. Optimal Decisions in Policy C

The manufacturer produces MDIGPs and sells to the retailer at wholesale price w_i^C. The retailer sells those to the customers at a price of p_i^C. The demand function in Policy C is $D_i^C = a - (p_i^C - \rho_i^C) + \beta\theta_i^C$. The manufacturer collects used product directly from the consumers for re-manufacturing. The government organization decides the subsidy rate by maximizing SW. A consumer subsidy on electronic vehicles is common in countries like China, Canada, Germany, Japan, etc. [66]. The profit functions for the retailer and manufacturer, and the SW for the government in Scenarios MC and RC are obtained as follows:

$$\pi_{ri}^C(p_i^C) = (p_i^C - w_i^C)D_i^C, \tag{1}$$

$$\pi_{mi}^C(w_i^C, \theta_i^C, \tau_i^C) = (w_i^C - \lambda_1\theta_i^C)D_i^C + (c_m\delta - \alpha - c_r)\tau_i^C D_i^C - \kappa\tau_i^{C2} - \lambda_2\theta_i^{C2}, \tag{2}$$

$$\pi_{gi}^C(\rho_i^C) = \pi_{ri}^C + \pi_{mi}^C + \frac{D_i^{C2}}{2} - \rho_i^C D_i^C. \tag{3}$$

The government's objective function includes the influence of profits for each member, the social aspect, i.e., consumer surplus (CS), and the total amount of the subsidy provided by the government organization [67,68]. The CS is the area of the demand curve below a given price, which can be expressed as $\frac{D_i^{C2}}{2}$. Optimal decisions in Policy C under the MS and RS games are presented in

Lemma 1,2, respectively. The detail derivations of optimal decisions are presented in Appendix A,B, respectively. Additional notations used throughout the study are presented at the end of Appendix A.

Lemma 1. *Optimal decision in Scenario MC are obtained as follows:*

$$\rho_m^C = \frac{6(a-c_m)\kappa\lambda_2}{\Delta_1}; \quad w_m^C = \frac{(aN_2-2c_m\kappa)\lambda_2-YZ\kappa}{\Delta_1}; \quad p_m^C = \frac{(aN_3-4c_m\kappa)\lambda_2-YZ\kappa}{\Delta_1}; \quad \theta_m^C = \frac{(a-c_m)\kappa Z}{\Delta_1}; \quad \tau_m^C = \frac{(a-c_m)X\lambda_2}{\Delta_1};$$

$$\pi_{mm}^C = \frac{(a-c_m)^2\kappa\lambda_2(M_1\kappa+N_3\lambda_2)}{\Delta_1^2}; \quad \pi_{rm}^C = \frac{4(a-c_m)^2\kappa^2\lambda_2^2}{\Delta_1^2}; \quad \pi_{gm}^C = \frac{(a-c_m)^2\kappa\lambda_2}{\Delta_1}; \quad Q_m^T = \frac{2(a-c_m)\kappa\lambda_2}{\Delta_1}, \text{ where } \Delta_1 = M_1\kappa - X^2\lambda_2.$$

Lemma 2. *Optimal decision in Scenario RC are obtained as follows:*

$$\rho_r^C = \frac{(a-c_m)\kappa(M_1\kappa+N_2\lambda_2)}{\Delta_1}; \quad w_r^C = \frac{aN_1\lambda_2+YZ\kappa}{\Delta_1}; \quad p_r^C = \frac{(a-c_m)M_1\kappa-c_mN_1\lambda_2+YZ\kappa}{\Delta_1}; \quad \theta_r^C = \frac{(a-c_m)Z\kappa}{\Delta_1}; \quad \tau_r^C =$$

$$\frac{(a-c_m)X\lambda_2}{\Delta_1}; \quad \pi_{mr}^C = \frac{(a-c_m)^2\kappa\lambda_2(2\kappa\lambda_2+\Delta_1)}{\Delta_1^2}; \quad \pi_{rr}^C = \frac{2(a-c_m)^2\kappa\lambda_2(2\kappa\lambda_2+\Delta_1)}{\Delta_1^2}; \quad \pi_{gr}^C = \frac{(a-c_m)^2\kappa\lambda_2}{\Delta_1}; \quad Q_m^C = \frac{2(a-c_m)\kappa\lambda_2}{\Delta_1}.$$

The concavity of profit functions for CLSC members and SW for the government in Scenarios MC and RC is ensured by condition $\Delta_1 > 0$ and $4\kappa > X^2$, respectively. It is found that feasible values of GLs (θ_m^C and θ_r^C) of the product exists if $\beta > \lambda_1$. Therefore, if per unit investment efficiency for the manufacturer is too low, but consumer sensitivity with green product is less, then the manufacturer cannot produce MDIGPs. The unit cost of the product increased with λ_1; in this circumstance, the manufacturer cannot compensate increasing cost. Therefore, results make sense. By comparing optimal decisions between the MS and RS games, the following theorem is proposed.

Theorem 1. *In Policy C*

1. *The greening levels, collection rates, and social welfare are identical under both games; and the amount of the subsidy on per unit product is higher under the MS game.*
2. *The greening levels, collection rates, subsidy rates, and social welfare decrease with respect to λ_1 and λ_2.*

We refer to Appendix C for the details of Theorem 1. Lemma 1,2 indicate that if the manufacturer is not efficient enough, then it is difficult to produce greener product. Per unit subsidy and the collection rate decreases with respect to λ_1 and λ_2, and as a result, SW decreased. Additionally, the manufacturer needs to reduce the total investment effort in improving used product collection if the manufacturing cost is high. Therefore, results are sensible. Except profits of CLSC members, game structures do no have any effect on the optimal decision. The manufacturer charges a higher wholesale price, and the retailer sets a higher retail price in the MS game because $w_m^C - w_r^C = \frac{2(a-c_m)\kappa\lambda_2}{\Delta_1} > 0$ and $p_m^C - p_r^C = \frac{(a-c_m)(\kappa(\beta-\lambda_1)^2+X^2\lambda_2)}{\Delta_1} > 0$, respectively. Due to more bargaining power, the subsidy rate is always higher in the MS game, but the consumer needs to pay the same price under the MS and RS games because $(p_m^C - \rho_m^C) - (p_r^C - \rho_r^C) = 0$. Overall, consumers remain unaffected, and the government needs to provide a higher per unit subsidy under the MS game, which does not ensure higher quality.

3.2. Optimal Decisions in Policy RE

The profit structure of CLSC members and the SW of government remain similar to the previous subsection, but the demand function converts to $D_i^{RE} = a - p_i^{RE} + \beta\theta_i^{RE}$. For example, in China, manufacturers and government organizations, such as China's National Development and Reform Commission in 2010, collaborated to promote the program "Comments on Boosting the Re-manufacturing Development" to encourage product reuse [52]. As discussed in Section 3.1, the profit functions for the retailer and manufacturer, and SW in Scenarios MRE and RRE are obtained as follows:

$$\pi_{ri}^{RE}(p_i^{RE}) = (p_i^{RE} - w_i^{RE})D_i^{RE}, \tag{4}$$

$$\pi_{mi}^{RE}(w_i^{RE}, \theta_i^{RE}, \tau_i^{RE}) = (w_i^{RE} - \lambda_1\theta_i^{RE} + X\tau_i^{RE})D_i^{RE} - (1 - \eta_i^{RE})\kappa\tau_i^{RE2} - \lambda\theta_i^{RE2}, \tag{5}$$

$$\pi_{gi}^{RE}(\eta_i^{RE}) = \pi_{ri}^{RE} + \pi_{mi}^{RE} + \frac{D_i^{RE^2}}{2} - \kappa \eta_i^{RE} \tau_i^{RE^2}. \tag{6}$$

Derivations of optimal decisions are similar to Policy C, hence omitted. The concavity of profit functions for CLSC members and SW for the government in Scenarios MRE and RRE is ensured by condition $\Delta_{2m} > 0$ and $\Delta_{2r} > 0$, respectively. Optimal decisions under the MS and RS games are presented in Lemma 3,4, respectively.

Lemma 3. *Optimal decision in Scenario MRE are obtained as follows:*

$$\eta_m^{RE} = \frac{6\lambda_2}{M_4}; \; w_m^{RE} = \frac{(a(4M_3\kappa - M_4 X^2) + 4c_m M_3 \kappa)\lambda_2}{\Delta_{2m}}; \; p_m^{RE} = \frac{4(8(3a+c_m)\kappa - 7aX^2)\lambda_2^2 - 2Z^2(a(6\kappa - X^2) + 2c_m\kappa)\lambda_2 + 2M_3 YZ\kappa}{\Delta_{2m}};$$

$$\theta_m^{RE} = \frac{(a-c_m)M_3 Z\kappa}{\Delta_{2m}}; \; \tau_m^{RE} = \frac{(a-c_m)M_4 X\lambda_2}{\Delta_{2m}}; \; \pi_{mm}^{RE} = \frac{(a-c_m)^2 M_3 \kappa \lambda_2}{\Delta_{2m}}; \; \pi_{rm}^{RE} = \frac{4(a-c_m)^2 M_3 \kappa^2 \lambda_2^2}{\Delta_{2m}^2}; \; \pi_{gm}^{RE} =$$

$$\frac{(a-c_m)^2 M_4 \kappa \lambda_2}{\Delta_{2m}}; \; Q_m^{RE} = \frac{2(a-c_m)M_4 \kappa \lambda_2}{\Delta_{2m}}, \; where \; \Delta_{2m} = M_3{}^2\kappa - M_4 X^2 \lambda_2.$$

Lemma 4. *Optimal decision in Scenario RRE are obtained as follows:*

$$\eta_r^{RE} = \frac{N_3\lambda_2 - Z^2\kappa}{2\kappa(M_2 + \lambda_2)}; \; w_r^{RE} = \frac{M_2(2(a+c_m)(\kappa - X^2)\lambda_2 + (c_m M_2 + YZ)\kappa) + X^2(c_m M_2 + YZ)\lambda_2}{2\Delta_{2r}}; \; p_r^{RE} =$$

$$\frac{\kappa Z^3((a-c_m)\beta - 2Y) + Z((a-c_m)(2\kappa Z + (4\kappa + X^2)\beta) - 2aN_4 Z)\lambda_2 + 2(a(7N_1 - 2X^2) + c_m(4\kappa + X^2))\lambda_2^2}{2\Delta_{2r}}; \; \theta_r^{RE} = \frac{(a-c_m)Z(M_2\kappa + X^2\lambda_2)}{2\Delta_{2r}};$$

$$\tau_r^{RE} = \frac{(a-c_m)X\lambda_2(M_2 + \lambda_2)}{\Delta_{2r}}; \; \pi_{mr}^{RE} = \frac{(a-c_m)^2\lambda_2(M_2\kappa + X^2\lambda_2)}{4\Delta_{2r}}; \; \pi_{rr}^{RE} = \frac{(a-c_m)^2\lambda_2(M_2\kappa + X^2\lambda_2))}{2\Delta_{2r}}; \; \pi_{gr}^{RE} =$$

$$\frac{(a-c_m)^2\lambda_2((M_1 + M_3 + M_2)\kappa + X^2\lambda_2)}{4\Delta_{2r}}; \; Q_r^{RE} = \frac{(a-c_m)\lambda_2(M_2\kappa + X^2\lambda_2)}{\Delta_{2r}}, \; where \; \Delta_{2r} = M_2{}^2\kappa - X^2(M_2 + 2\lambda_2)\lambda_2.$$

Recall that optimal subsidy rates in Policy C are directly proportional with market potential, and different results are obtained in Policy RE. Although the demand increases with market potential, it does not directly affect the subsidy rate. However, the government may have to spend more because the collection rates τ_m^{RE} and τ_r^{RE}, or overall demand Q_m^C and Q_r^C, increase with market potential. The following theorem highlights the characteristics of the optimal decision.

Theorem 2. *In Policy RE,*

1. *The greening levels, collection rate, and social welfare are always higher under the RS game: however, the subsidy rate is higher under the MS game*
2. *The greening levels, collection rates, subsidy rates, and social welfare decrease with respect to λ_1 and λ_2.*

We refer to Appendix D for the details of Theorem 2. The outcome of Theorem 2 differs from the previous one. A powerful retailer can enforce that the manufacturer produce and trade with greener product. The product collection rate is also higher under the RS game. Similar to Policy C, the subsidy rate is higher under the MS game. Therefore, one can find an indication that the sustainability goal can be achieved under the RS game in the presence of a subsidy.

3.3. Optimal Decisions in Policy T

In this subsection, models are formulated and the optimal decision is derived in Policy T. A similar type of subsidy policy is discussed in the forward SC setting by several researchers [59]. For example, to strengthen sustainable innovation, $400 million was allotted to fund the R&D of energy technologies as a part of the American Recovery and Reinvestment Act of 2009 [48]. The demand function in this policy becomes $D_i^T = a - p_i^T + \beta\theta_i^T$, and the corresponding profit functions of the manufacturer and retailer; and SW remains similar with previous subsections, and they are obtained as follows:

$$\pi_{ri}^T(p_i^T) = (p_i^T - w_i^T)D_i^T, \tag{7}$$

$$\pi_{mi}^T(w_i^T, \theta_i^T, \tau_i^T) = (w_i^T - \lambda_1\theta_i^T)D_i^T + X\tau_i^T D_i^T - \kappa\tau_i^{T^2} - (1 - \mu_i^T)\lambda_2\theta_i^{T^2}, \tag{8}$$

$$\pi_{gi}^{T}(\mu_{i}^{T}) = \pi_{ri}^{T} + \pi_{mi}^{T} + \frac{D_{i}^{T2}}{2} - \mu_{i}^{T}\lambda_{2}\theta_{i}^{T2}. \tag{9}$$

Note that the manufacturer receives a subsidy on the total R&D investment, not on per unit product $\lambda_{1}\theta_{i}^{T}, i = m, r$. If $\lambda_{1} = 0$ and $\mu_{i}^{T} = 0$, the profit functions become similar to Ghosh and Saha [18], as well as Song and Gao [19], where the authors examined the optimal decision for a forward SC setting where the government does not provide any subsidy. Similar to previous subsections, we derive the optimal decision for Policy T and omit the detailed derivation. Lemma 5,6 characterize the optimal decisions under the MS and RS games, respectively.

Lemma 5. *Optimal decision in Scenario MT are obtained as follows:*

$$\mu_{m}^{T} = \frac{6\kappa}{N_{5}}; \quad w_{m}^{T} = \frac{N_{5}YZ\kappa + N_{4}(aN_{2}+4c_{m}\kappa)\lambda_{2}}{\Delta_{3m}}; \quad p_{m}^{T} = \frac{N_{5}YZ\kappa + N_{4}(aN_{3}+2c_{m}\kappa)\lambda_{2}}{\Delta_{3m}}; \quad \theta_{m}^{T} = \frac{(a-c_{m})N_{5}Z\kappa}{\Delta_{3m}}; \quad \tau_{m}^{T} = $$
$$\frac{(a-c_{m})N_{4}X\lambda_{2}}{\Delta_{3m}}; \quad \pi_{mm}^{T} = \frac{(a-c_{m})^{2}N_{4}\kappa\lambda_{2}}{\Delta_{3m}}; \quad \pi_{rm}^{T} = \frac{4(a-c_{m})^{2}N_{4}^{2}\kappa^{2}\lambda_{2}^{2}}{\Delta_{3m}}; \quad \pi_{gm}^{T} = \frac{(a-c_{m})^{2}N_{5}\kappa\lambda_{2}}{\Delta_{3m}}; \quad Q_{m}^{T} = \frac{2(a-c_{m})N_{4}\kappa\lambda_{2}}{\Delta_{3m}},$$

where $\Delta_{3m} = N_{4}^{2}\lambda_{2} - N_{5}Z\kappa$.

Lemma 6. *Optimal decision in Scenario RT are obtained as follows:*

$$\mu_{r}^{T} = \frac{N_{3}\lambda_{2}-Z^{2}\kappa}{2(N_{2}+\kappa)\lambda_{2}}; \quad w_{r}^{T} = \frac{N_{2}(aN_{1}+c_{m}N_{3})\lambda_{2}+Z\kappa((aN_{1}-c_{m}(3N_{2}+2\kappa))\beta+(aN_{4}+c_{m}N_{2})\lambda_{1})}{2\Delta_{3r}}; \quad p_{r}^{T} = $$
$$\frac{N_{2}((a+c_{m})\kappa+aN_{2})\lambda_{2}-Z\kappa(a((Z+\beta)\kappa-N_{3}\lambda_{1})+c_{m}(N_{2}\beta+\kappa\lambda_{1}))}{\Delta_{3r}}; \quad \theta_{r}^{T} = \frac{(a-c_{m})Z\kappa(N_{2}+\kappa)}{\Delta_{3r}}; \quad \tau_{r}^{T} = \frac{(a-c_{m})X(N_{2}\lambda_{2}+Z^{2}\kappa)}{2\Delta_{3r}};$$
$$\pi_{mr}^{T} = \frac{(a-c_{m})^{2}\kappa(N_{2}\lambda_{2}+Z^{2}\kappa)}{4\Delta_{3r}}; \quad \pi_{rr}^{T} = \frac{(a-c_{m})^{2}\kappa(N_{2}\lambda_{2}+Z^{2}\kappa)}{2\Delta_{3r}}; \quad \pi_{gr}^{T} = \frac{(a-c_{m})^{2}\kappa((3N_{2}+2\kappa)\lambda_{2}+Z^{2}\kappa)}{4\Delta_{3r}};$$
$$Q_{r}^{T} = \frac{(a-c_{m})\kappa(N_{2}\lambda_{2}+Z^{2}\kappa)}{\Delta_{3r}}, \text{ where } \Delta_{3r} = N_{2}^{2}\lambda_{2} - N_{3}Z^{2}\kappa.$$

The concavity of profit functions for CLSC members and SW in Scenarios MT and RT is ensured by condition $\Delta_{3m} > 0$ and $\Delta_{3r} > 0$, respectively. Similar to Policy RE, optimal decisions differ according to the power of CLSC members, and increasing market potential does not effect the subsidy rates. By comparing optimal decisions, the following theorem is proposed.

Theorem 3. *In Policy T*

1. *The greening level, collection rate, and social welfare are always higher under the RS game; however, the subsidy rate is higher under the MS game.*
2. *The greening levels, collection rates, subsidy rates, and social welfare decreases with respect to λ_{1} and λ_{2}.*

We refer to Appendix E for the details of Theorem 3. Results of Theorem 2,3 are similar. Overall, SW and GL are higher under the RS game under Policy RE and T. In all three policies, the subsidy rates are higher under the MS game, but it does not guarantee the higher SW and GL. Therefore, a shift of power from the manufacturer to the retailer in a CLSC always encourages in the perspective of achieving the sustainability goal. By combining concavity conditions, one can conclude that the profits of CLSC members and SW are concave in three subsidy policies under the MS and RS games if $M_{1} > 0, N_{1} > 0$ and $\Delta_{1} > 0$. To obtain an overview of increment or decrements in GL, used product collection rate, profits of CLSC members, and SW, we derive the optimal decision in the absence of a subsidy.

3.4. Optimal Decisions in Absence of Subsidy

By substituting $\rho_{i}^{j} = 0$, or $\eta_{i}^{j} = 0$, or $\mu_{i}^{j} = 0$ in Equation (1),(2); (4),(5); (7),(8), respectively, one can obtain the profit functions of CLSC members in the absence of a subsidy. The corresponding optimal decisions under the MS and RS games are presented in Lemma 7,8, respectively.

Lemma 7. *Optimal decision in the absence of a subsidy under the MS game is as follows:*

$$w_{m}^{N} = \frac{aN_{1}\lambda_{2}+(YZ+4c_{m}\lambda_{2})\kappa}{\Delta_{4m}}; \quad p_{m}^{N} = \frac{aN_{3}\lambda_{2}+(YZ+2c_{m}\lambda_{2})\kappa}{\Delta_{4m}}; \quad \theta_{m}^{N} = \frac{(a-c_{m})Z\kappa}{\Delta_{4m}}; \quad \tau_{m}^{N} = \frac{(a-c_{m})X\lambda_{2}}{\Delta_{4m}}; \quad \pi_{mm}^{N} = \frac{(a-c_{m})^{2}\kappa\lambda_{2}}{\Delta_{4m}};$$
$$\pi_{rm}^{N} = \frac{4(a-c_{m})^{2}\kappa^{2}\lambda_{2}^{2}}{\Delta_{4m}^{2}}; \quad Q_{m}^{N} = \frac{2(a-c_{m})\kappa\lambda_{2}}{\Delta_{4m}}, \text{ where } \Delta_{4m} = N_{4}\lambda_{2} - \kappa Z^{2}.$$

Lemma 8. *Optimal decision in the absence of a subsidy under the RS game is as follows:*

$$w_r^N = \frac{(a+c_m)N_1\lambda_2+(YZ+c_mM_2)\kappa}{2\Delta_{4r}}; \; p_r^N = \frac{2aN_1\lambda_2+(aM_1+YZ+2c_m\lambda_2)\kappa}{2\Delta_{4r}}; \; \theta_r^N = \frac{(a-c_m)Z\kappa}{2\Delta_{4r}}; \; \tau_r^N = \frac{(a-c_m)X\lambda_2}{2\Delta_{4r}}; \; \pi_{mr}^N = \frac{(a-c_m)^2\kappa\lambda_2}{4\Delta_{4r}}; \; \pi_{rr}^N = \frac{(a-c_m)^2\kappa\lambda_2}{2\Delta_{4m}^2}; \; Q_r^N = \frac{(a-c_m)\kappa\lambda_2}{\Delta_{4r}}, \text{ where } \Delta_{4r} = N_2\lambda_2 - \kappa Z^2.$$

Unlike optimal decisions in Policy C, the outcomes differ between the MS and RS game in the absence of a subsidy. In this regards, one can conclude that government can weaken the effect of power of CLSc members by implementing Policy C. We use results of above two lemmas as the benchmark to compare outcomes.

4. Analysis and Discussions

In the previous section, results were compared to highlight the behavior of optimal decisions between the two games. In the following subsections, we evaluate gains and losses in the perspective of consumer, CLSC members, and the government organization.

4.1. Consumer's Perspective

The following theorem highlights consumers preference among three subsidy policies.

Theorem 4. *Irrespective of game structures, the greening level and sales volume are higher in Policy C.*

We refer to Appendix F for the details of Theorem 4. Theorem 4 demonstrates that consumers always receive product at a higher GL in Policy C, where the government can penetrate consumers directly. Therefore, direct monetary gains stimulate consumers to buy more product. If sales volumes increase, the manufacturer can compensate investment cost, and GL is also consequently increased. Recall that GL, collection rate, and retail price are identical under both games in Policy C. Therefore, Policy C outperforms others in the perspective of consumer benefit and green product consumption. Graphical representation of GLs, sales volumes, effective prices consumer needs to pay, and ratios of GLs with effective retail price in six scenarios is presented in Figure 1a–d. Parameter values are used for numerical examples as follows: $a = 500$, $\beta = 0.6$, $c_m = 50(\$/\text{unit})$, $c_r = 20(\$/\text{unit})$, $\alpha = 10(\$/\text{unit})$, $\delta = 0.7$, $\kappa = 1500$, $\lambda_1 = 0.3$, and $\lambda_2 = 1$. Note that technical restrictions on parameters values are considered to ensure optimal conditions.

Figure 1a,b justify the statement of Theorem 4. The effective retail price is less in Policy C compared to others, and GL is always higher in the RS game in all three policies. If the consumer perceives the retail price in their mind, then Policy C outperforms others because the ratio of GLs with retail price is maximum under Policy C. By comparing the ratios, consumers can figure out how much they need to pay to procure the product. Figure 1c. demonstrates that the consumer needs to pay a lesser price under Policy C. One can observe that the GLs are lower in Policy RE compared to others, and the consumer needs to pay a higher price in Policy T.

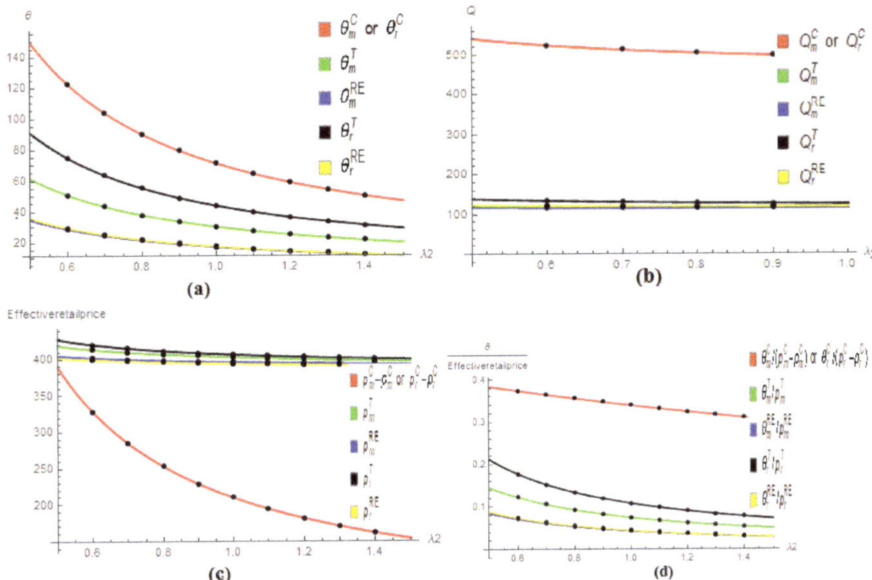

Figure 1. Graphical representation of (**a**) greening levels, (**b**) sales volumes, (**c**) effective retail prices, and (**d**) ratio of greening levels with effective retail prices in Scenarios MC, RC, MRE, RRE, MT, and RT.

4.2. Retailer and Manufacturer Perspectives

The following theorem is proposed to highlight the pros and cons for three policies in the perspective of CLSC members.

Theorem 5. *Under both games, the collection rate of used product and profits for each member are always higher in Policy C.*

We refer to Appendix G for the details of Theorem 5. The outcomes of Theorem 5 are consistent with Theorem 4. GL and sales volume are both higher in Policy C; consequently, CLSC members receive a higher profit in a green-sensitive market. Flexibility of investment for the manufacturer in improving GL and used product return is increased with market demand. In such a scenario, the retailer can also get benefited. The results demonstrate that fact. Graphical representation of the profits for the retailer and manufacturer, collection rate, total investment for the manufacturer to produce product (MI) ($MI_i^C = \lambda_1 \theta_i^C D_i^C + \lambda_2 \theta_i^{C2}$; $MI_i^{RE} = \lambda_1 \theta_i^R ED_i^{RE} + \lambda_2 \theta_i^{RE2}$; and $MI_i^T = \lambda_1 \theta_i^T D_i^T + (1 - \mu_i^T)\lambda_2 \theta_i^{T2}$), investment for the manufacturer in encouraging used product return (RI) ($RI_i^C = \alpha \tau_i^C D_i^C + \kappa \tau_i^{C2}$; $RI_i^{RE} = \alpha \tau_i^{RE} D_i^{RE} + (1 - \eta_i^{RE})\kappa \tau_i^{RE2}$; and $RI_i^T = \alpha \tau_i^T D_i^T + \kappa \tau_i^{T2}$), ratios of relative change of GL with investment to produce products ($\Delta \theta M_i^j = \frac{\theta_i^j - \theta_i^N}{MI_i^j - MI_i^N}$), and ratios of relative change of the collection rate with the investment effort to stimulate used product return ($\Delta \tau R_i^j = \frac{\tau_i^j - \tau_i^N}{RI_i^j - RI_i^N}$) in six scenarios is presented in Figure 2a–g.

Figure 2a–g support the statement of Theorem 5. In Policy RE, the manufacturer receives a subsidy to encourage the product collection, but it does not yield a higher return compared to Policy C. Note that the used product collection rates are less in Policy T compared to other two. Figure 2e demonstrates that the investment effort to stimulate used product return for the manufacturer is less in Policy RE; however, due to the government support, collection rate improved. In Policy T, the manufacturer has more flexibility in the R&D investment until the manufacturer can invest more in

Policy C. However, Figure 2f exhibits some doubts about the efficiency of Policy C. In Policy C, the total amount of investment for the manufacturer to improve GL is maximum, but the ratio of relative change in GL improvement is less. Therefore, higher investment does not ensure higher GL, especially in Policy C. Figure 2g demonstrates a noteworthy outcome in the perspective of designing subsidy policy. It demonstrates that the power of CLSC members should be considered before implementation of the subsidy policy. Interestingly, the manufacturer reduces the investment effort considerably under the RS game, but a reverse trend is observed under the MS game. Overall, investment efficiency in producing greener product and used product return reduced in Policy C.

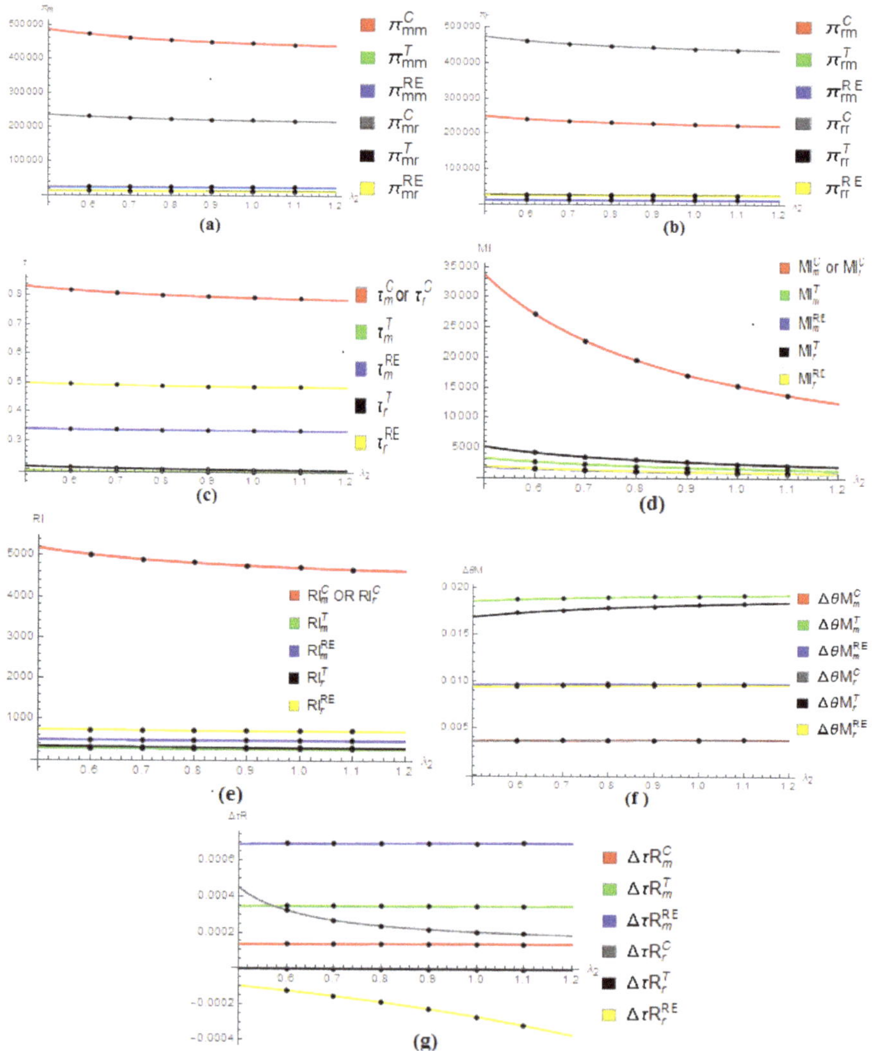

Figure 2. Graphical representation of (**a**) profit of manufacturer, (**b**) profit of retailer, (**c**) used product return, (**d**) manufacturer's R&D investment to produce product, (**e**) investment effort in encouraging used product collection, (**f**) ratios of relative change of greening levels with investment to produce greener product, and (**g**) ratios of relative change of return rate with investment effort to stimulate product return for the manufacturer in Scenarios MC, RC, MRE, RRE, MT, and RT.

4.3. Government Perspective

In this subsection, we compare SWs and the amount of government subsidy to explore consequence in the perspective of government organizations.

Theorem 6. *The social welfare and the amount of government subsidy is higher in Policy C in both the games.*

We refer Appendix H for the proof of Theorem 6. Theorem 6 demonstrates that the government expenditure and SW are always higher in Policy C. Therefore, the outcomes of Theorem 4–6 are very much alike. Higher subsidy cause higher profits, as well as GL, and SW consequently increased. Graphical representation of SW, the amount of government subsidy (GI) ($GI_i^C = \rho_i^C D_i^C$; $GI_i^{RE} = \eta_i^{RE} \kappa \tau_i^{RE2}$; and $GI_i^T = \mu_i^T \lambda_2 \theta_i^{T2}$), and the ratios of relative change of GL with total amount of government subsidy ($\Delta\theta G_i^j = \frac{\theta_i^j - \theta_i^N}{GI_i^j}$) in six scenarios is depicted in Figure 3a–c.

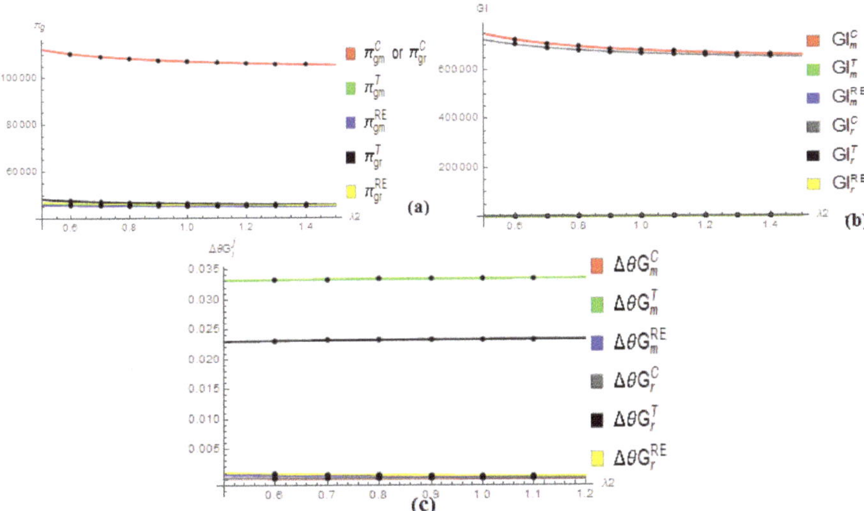

Figure 3. Graphical representation of (**a**) social welfare, (**b**) total amount of government subsidy in each scenario, and (**c**) ratios of relative change of greening levels with total amount of government subsidy in Scenarios MC, RC, MRE, RRE, MT, and RT.

The above figures support the statement of Theorem 6. However, if we investigate at the macro level, then one cannot draw a straightforward conclusion in favor of Policy C. By correlating Figure 2e with Figure 3c, the ratio of investment efficiency reflects the different consequence. In the perspective of the manufacturer and government organization, Policy C may lead to an inadequate investment decision. However, one cannot ignore the influence of consumers until they enjoy the higher privilege in Policy C because government support directly passes to the consumer. The total amount of expenditure for the government is too high in Policy C, which does not yield a higher relative improvement in GL.

4.4. When Manufacturer Produces Only DIGPs

In this study, it is assumed that the manufacturer produces MDIGPs but does not recover the cost for used product. Therefore, we conduct numerical experiment where the manufacturer produces DIGPs($\lambda_1 = 0$), which is more predominant in existing literature. The following figures represent the relative change of profits of the SC members ($\Delta\pi_{ki}^{j} = \frac{\pi_{ki}^{j}|_{\lambda_1=0} - \pi_{ki}^{j}}{\pi_{ki}^{j}}, k = m, r$); collection rates ($\Delta\tau_{i}^{j} = \frac{\tau_{i}^{j}|_{\lambda_1=0} - \tau_{i}^{j}}{\tau_{i}^{j}}$); GLs ($\Delta\theta_{i}^{j} = \frac{\theta_{i}^{j}|_{\lambda_1=0} - \theta_{i}^{j}}{\theta_{i}^{j}}$); total amount of government subsidies ($\Delta GI_{i}^{j} = \frac{GI_{i}^{j}|_{\lambda_1=0} - GI_{i}^{j}}{GI_{i}^{j}}$); and SWs ($\Delta\pi_{gi}^{j} = \frac{\pi_{gi}^{j}|_{\lambda_1=0} - \pi_{gi}^{j}}{\pi_{gi}^{j}}$).

It is expected that the CLSC members receive higher profits if unit production cost decreased. Figure 4a,b reflect that nature changes profits for CLSC members, SW, and GL, which also supports the expectation. CLSC members always receive a higher profit in Policy C, and GL is always maximum. However, the nature of used product collection and the amount of government subsidy changes significantly. Increment in the used the product collection rate and the amount of government subsidy are maximum in the RS game and in Policy T. Therefore, the government needs to examine the product type to frame an effective policy.

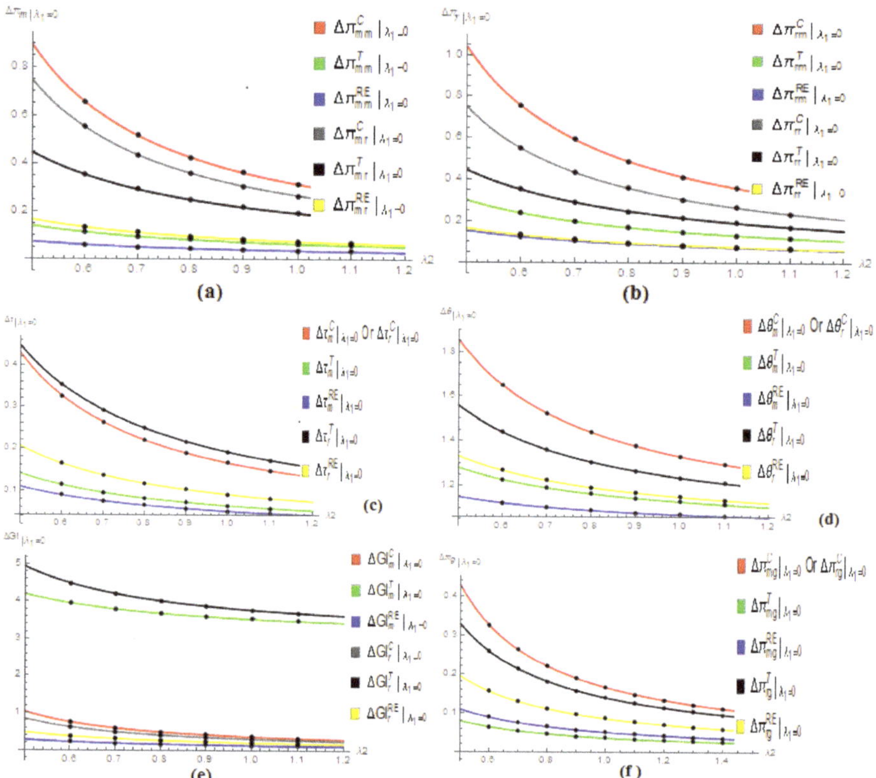

Figure 4. Graphical representation of (**a**) change in profits for the manufacturer, (**b**) change in profits for the retailer, (**c**) change in product collection rates, (**d**) change in greening levels ($\lambda_1 = 0$), (**e**) change in amount of government subsidies, and (**f**) change in social welfare in Scenarios MC, RC, MRE, RRE, MT, and RT ($\lambda_1 = 0$).

4.5. Overall Implications

The preceding discussion offers a rich amount of contextual detail based on the analytical and numerical evaluation. Subsidies make sense to encourage R&D activities in areas that would benefit society, stimulate greener product consumption with a society's environmental objectives, such as less contamination, cleaner air, etc., provide much-needed help to innovative startups, or support a manufacturer in surviving financial losses due to high R&D investment. However, there has been little discussion on comparative analysis among outcomes under the government SW optimization goal.

The present study discloses some eye opening issues. It has always been a topic of interest to consider which subsidy policy can lead to a pragmatic CLSC business model, or which is both environmentally and economically worthwhile for participating member and government organization. Based on the discussion, one can articulate that the optimal decision, preference, and implications of subsidy policies significantly change between MS and RS games. To maintain goodwill and dominate a green-sensitive market, it is always imperative for the retailer to sell greener product. However, the manufacturer receives a higher subsidy in the MS game; yet, GL and SW is higher in the RS game. Therefore, the power of a CLSC member adds a degree of conflict, and government organizations needs figure out the dynamics of power before implementing subsidy policies. Overall, Policy C under the RS game can drive toward encouraging outcomes in the perspective of consumers, retailer, and government organization.

It is commonly believed that a subsidy assists manufactures to produce greener products and trade them at low price to the consumers. To some extent, the results of the present study support the convention, but in the presence of government subsidy policy, CLSC members need to be prepared for sudden operational changes. In practice, government organizations sometimes commit to environmental policies for several years but afterward renege on their commitments. For example, the government of China recently recommended to withdraw a subsidy from the EV battery industry (https://chinaeconomicreview.com/subsidy-withdrawal-to-decimate-chinas-ev-industry/), and the government of India recently revised the amount of subsidy for the scheme "Faster Adoption and Manufacturing of Hybrid and Electric Vehicles" (https://energy.economictimes.indiatimes.com/news/power/govt-withdraws-sops-to-conventional-battery-vehicles-under-fame/65990495). If the government suddenly revises a subsidy amount due to a sudden fall of market demand, the manufacturer needs to adjust its production rate and to be prepared for adjustment of the entire operations and marketing activities. For example, when the Indian government revised the scheme and reduced the direct subsidy to consumers, car manufacturers faced market fall. A similar situation also reported in the UK is that "Subsidy cuts blamed for fall in UK sales of electrified vehicles" (www.theguardian.com/business/2019/jul/04/subsidy-cuts-blamed-for-fall-in-uk-sales-of-electrified-vehicles). Examples are similar with Policy C because, in the car industry, consumers directly receive a subsidy from government. Results indicate that Policy C clearly becomes a financial liability for government. Due to direct cash-transfer in Policy C, all the consumers enjoy a subsidy irrespective of income groups. Therefore, there is a possibility that government resources might become a drain, especially if high-income consumers take the subsidy. As observed earlier, GL and GIs are maximum in Policy C; therefore, this is where to find answer of the question how much additional amount needs to be paid for improving GL? Figure 5a,b, representing the ratio of GLs and GIs under Policy C, are drawn to obtain an overview in this direction.

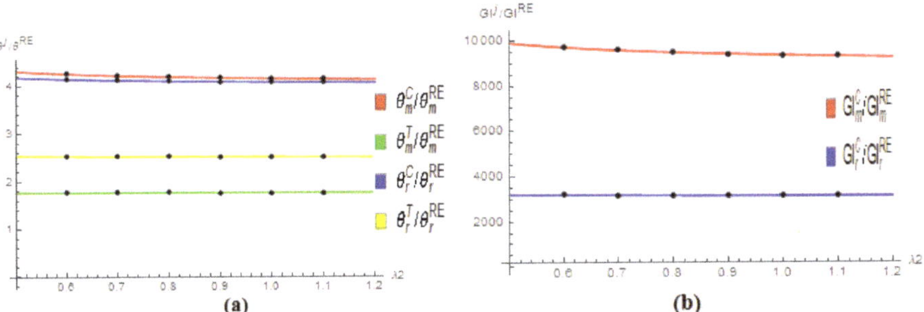

(a)   (b)

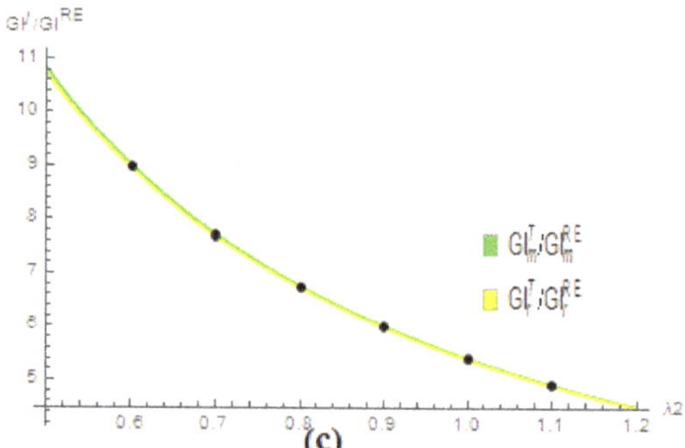

Figure 5. (**a**) Ratio greening levels in different subsidy policies. (**b**) Ratios of amount of government subsidy in Policy C and RE. (**c**) Ratios of amount of government subsidy in Policy T and RE.

By observing the vertical axis of Figure 5a–c, one can recognize the additional financial burden associated with Policy C, especially under the MS game. It is also found that in Policy RE, the government needs to spend less and consequently also lessen GL. In practice, there is a possibility that the manufacturer can strategically reduce investment effort in presence of a subsidy. Our study also supports this fact because the manufacturer clearly reduce its investment effort, as depicted in Figure 2g. Figure 2g also demonstrates that if the intention of the government is to improve used product collection, then Policy T or RE under the MS game can drive to the desirable outcome. Finally, in Policy T, the government provides a subsidy and anticipates that the manufacturer can produce greener product, while consumers benefit from low prices. However, Policy T also becomes pricier in the perspective of consumers, but the improvement of GL is maximum under this policy.

5. Conclusions

The formation of a sustainable CLSC in the presence of a government subsidy is one of the key issues because it does not make sense to pollute the world for higher profits. One the other hand, it is infeasible in the perspective of a government organization to spend large amounts that fail to create value. Therefore, it is always challenging to design a subsidy policy that can lead to pragmatic outcomes. In literature, comparative studies on optimal outcomes in the presence of government SW optimization goal are scanty.

Motivated by emerging practice, we formulated eight CLSC models to compare outcomes of three subsidy policies. The central result emerging from the analysis reveals that in Policy C, CLSC members receive higher profits, SW of the government organization higher, and the consumers receive products at a higher GL. Characteristics of the optimal decision under the MS and RS games are not concurrent; GL and used product collection are always higher in the RS game, and the government subsidy rate is always higher in the MS game. Whatever the nature of game structures, the consumer always receives product at lower price in Policy C. However, Policy C still has shortcomings. GL does improve as the R&D investment or amount of the government subsidy increases, but the rate of change is lowest. It is found that Policy C can be a substantial financial burden without too much improvement in GL and used product collection. Because the amount of the subsidy is maximum in Policy C, our study contradicts conventional beliefs that a higher subsidy level always improves the performance of the CLSC members. The present study discloses that any straightforward conclusion on the optimal preferences in the perspective of CLSC members, consumers, and government organization is challenging to be made. If the government wants to improve green product consumption among its community, then the government can implement Policy C. However, expenditure as a subsidy will increase considerably. If the government aims to utilize a subsidy expenditure in an effective way, then Policy T can lead to a decent outcome. However, SW will be less and the consumer needs to pay more. In Policy RE, a strategic manufacturer can reduce the investment effort in the presence of a subsidy. If the intention of government is to improve used product collection, then Policy T or RE under the MS game can drive to a desirable outcome. It is also observed that the manufacturer's decision to produce DIGPs or MDGIPs can also affect the outcomes of a subsidy policy. A retailer-dominated CLSC is always advantageous for the government; in that scenario, the government can reduce the amount of a subsidy, maximize SW, and the consumer receives greener products.

Therefore, this study can be extended in several directions. In practice, a retailer or third party is also involved in used product collection. Sometimes manufacturer and retailer can both be involved in collections. Therefore, one can examine optimal decisions in the different modes of collections i.e., manufacturer, retailer, third party, or their combined collection mode. We assume that the consumer cannot distinguish the difference between the new and re-manufactured products. However, consumers often value the re-manufactured product less than the new product [69]. Therefore, one can analyze the influence of subsidy policies where the CLSC members need to set different prices for new and re-manufactured product. We restricted our analysis under single period formulation; therefore, one can extend this analysis under two-period setting. Furthermore, it will be interesting to examine the behavior of a CLSC decision if the members agree to cooperate with each other through coordination contract mechanisms [70,71] under the influence of the government subsidy.

Author Contributions: S.S. and I.N. developed the concept. S.S. formulated the models. S.M. conducted all the numerical experiments. All the authors are equally contributed in manuscript writing. All authors have read and agreed to the published version of the manuscript.

Funding: This research received no external funding.

Acknowledgments: The authors deeply appreciate the valuable comments of three anonymous reviewers and the Associate Editor to improve this study.

Conflicts of Interest: The authors declare no conflict of interest.

Appendix A. Optimal Decision in Scenario MC

To obtain optimal response for the retailer in Equation (1), one needs to solve $\frac{d\pi_{rm}^C}{dp_m^C} = 0$. On simplification, $p_m^C(w_m^C, \theta_m^C, \tau_m^C, \rho_m^C) = \frac{a + w_m^C + \theta_m^C \beta + \rho_m^C}{2}$. Because $\frac{d^2\pi_{rm}^C}{dp_m^{C2}} = -2 < 0$, the profit function for the retailer is concave.

Therefore, substituting p_m^C in Equation (2), the profit function for the manufacturer is obtained as:

$$\pi_{mm}^C(w_m^C, \theta_m^C, \tau_m^C, \rho_m^C) = \frac{(w_m^C - \theta_m^C\lambda_1 + X\tau_m^C - c_m)(a - w_m^C + \beta\theta_m^C + \rho_m^C) - 2\kappa\tau_m^{C2} - 2\lambda_2\theta_m^{C2}}{2}$$

To obtain optimal response for the manufacturer on wholesale price and investment efforts, we need to solve $\frac{d\pi_{mm}^C}{dw_m^C} = 0$, $\frac{d\pi_{mm}^C}{d\tau_m^C} = 0$, and $\frac{d\pi_{mm}^C}{d\theta_m^C} = 0$, simultaneously. After simplification, the following response is obtained:

$$w_m^C = \frac{\kappa(a + c_m + \rho_m^C)(4\lambda_2 - \beta(\beta - \lambda_1)) + (\kappa(\beta^2 - \lambda_1^2) - X^2\lambda_2)(a + \rho_m^C)}{\kappa(8\lambda_2 - Z^2) - X^2\lambda_2}$$

$$\tau_m^C = \frac{(a - c_m + \rho_m^C)X\lambda_2}{\kappa(8\lambda_2 - Z^2) - X^2\lambda_2} \text{ and } \theta_m^C = \frac{\kappa Z(a - c_m + \rho_m^C)}{\kappa(8\lambda_2 - Z^2) - X^2\lambda_2}$$

Because, the manufacturer's profit function is a function of three variables, we compute the Hessian matrix(H_m^C) to verify concavity as follows:

$$H_m^C = \begin{bmatrix} \frac{\partial^2\pi_{mm}^C}{\partial w_m^{C2}} & \frac{\partial^2\pi_{mm}^C}{\partial w_m^C\partial\tau_m^C} & \frac{\partial^2\pi_{mm}^C}{\partial w_m^C\partial\theta_m^C} \\ \frac{\partial^2\pi_{mm}^C}{\partial w_m^C\partial\tau_m^C} & \frac{\partial^2\pi_{mm}^C}{\partial\tau_m^{C2}} & \frac{\partial^2\pi_{mm}^C}{\partial\tau_m^C\partial\theta_m^C} \\ \frac{\partial^2\pi_{mm}^C}{\partial w_m^C\partial\theta_m^C} & \frac{\partial^2\pi_{mm}^C}{\partial\tau_m^C\partial\theta_m^C} & \frac{\partial^2\pi_{mm}^C}{\partial\theta_m^{C2}} \end{bmatrix} = \begin{bmatrix} -1 & \frac{-X}{2} & \frac{\beta+\lambda_1}{2} \\ \frac{-X}{2} & -2\kappa & \frac{\beta X}{2} \\ \frac{\beta+\lambda_1}{2} & \frac{\beta X}{2} & -2\lambda_2 - \lambda_1\beta \end{bmatrix}$$

The values of first, second, and third principal minors are obtained as $H^C_{m1} = -1 < 0$; $H^C_{m2} = \frac{8\kappa - X^2}{4}$; and $H^C_{m3} = -\frac{(8\kappa - X^2)\lambda_2 - \kappa(\beta - \lambda_1)^2}{2}$, respectively. Therefore, profit function for the manufacturer is concave if $8\kappa > X^2$ and $(8\kappa - X^2)\lambda_2 - \kappa(\beta - \lambda_1)^2 > 0$.

Substituting optimal responses in Equation (3), the SW function for the government organization is obtained as $\pi^C_{gm}(\rho^C_m) = \frac{\kappa\lambda_2(a - c_m + \rho^C_m)((a - c_m)((14\kappa - X^2)\lambda_2 - \kappa Z^2) + \kappa Z^2 \rho^C_m - (2\kappa - X^2)\lambda_2 \rho^C_m)}{((8\kappa - X^2)\lambda_2 - \kappa(\beta - \lambda_1)^2)^2}$. Therefore, one can

obtain optimal subsidy rate by solving $\frac{d\pi^C_{gr}}{d\rho^C_m} = 0$. On Simplification, optimal subsidy rate is obtained as

$\rho^C_m = \frac{6(a - c_m)\kappa\lambda_2}{\kappa(2\lambda_2 - Z^2) - X^2\lambda_2}$. Note that $\frac{d^2\pi^C_{rm}}{d\rho^{C2}_m} = -\frac{2\kappa\lambda_2\Delta_1}{(\kappa(8\lambda_2 - (\beta - \lambda_1)^2) - X^2\lambda_2)^2} < 0$, i.e., π^C_{gm} is concave with respect

to subsidy rate if $\Delta_1 = \kappa(2\lambda_2 - (\beta - \lambda_1)^2) - X^2\lambda_2$. By using back substitution, we obtain optimal decision as presented in Lemma 1.

The following additional notations are used throughout the article for simplicity:
$M_1 = 2\lambda_2 - Z^2$, $M_2 = 4\lambda_2 - Z^2$, $M_3 = 8\lambda_2 - Z^2$, $M_4 = 14\lambda_2 - Z^2$, $N_1 = 2\kappa - X^2$, $N_2 = 4\kappa - X^2$, $N_3 = 6\kappa - X^2$, $N_4 = 8\kappa - X^2$, $N_5 = 14\kappa - X^2$, $X = c_m\delta - c_r - \alpha$, $Y = a\lambda_1 - c_m\beta$, $Z = \beta - \lambda_1$.

Appendix B. Optimal Decision in Scenario RC

First, we substitute $m^C_r = p^C_r - w^C_r$ in Equation (1)–(3) to obtain optimal decision in the RS game. To obtain the manufacturer response first, one needs to solve $\frac{d\pi^C_{mr}}{dw^C_r} = 0$, $\frac{d\pi^C_{mr}}{d\tau^C_r} = 0$, and $\frac{d\pi^C_{mr}}{d\theta^C_r} = 0$ simultaneously. Therefore, the manufacturer's response is obtained as follows:

$$w^C_r = \frac{(\kappa(2\lambda_2 + Z\lambda_1) - X^2\lambda_2)(a - m^C_r + \rho^C_r) + c_m\kappa(2\lambda_2 - \beta(\beta - \lambda_1))}{\kappa(4\lambda_2 - Z^2) - X^2\lambda_2}$$

$$\tau = \frac{(a - c_m - m^C_r + \rho^C_r)X\lambda_2}{\kappa(4\lambda_2 - Z^2) - X^2\lambda_2} \text{ and } \theta = \frac{(a - c_m - m^C_r + \rho^C_r)\kappa Z}{\kappa(4\lambda_2 - (\beta - \lambda_1)^2) - X^2\lambda_2}$$

Because, the profit function for the manufacturer is a function of three variables, therefore, we compute corresponding Hessian matrix (H^C_r) for the manufacturer profit function is as follows:

$$H^C_r = \begin{bmatrix} \frac{\partial^2\pi^C_{mr}}{\partial w^{C2}_r} & \frac{\partial^2\pi^C_{mr}}{\partial w^C_r \partial \tau^C_r} & \frac{\partial^2\pi^C_{mr}}{\partial w^C_r \partial \theta^C_r} \\ \frac{\partial^2\pi^C_{mr}}{\partial w^C_r \partial \tau^C_r} & \frac{\partial^2\pi^C_{mr}}{\partial \tau^{C2}_r} & \frac{\partial^2\pi^C_{mr}}{\partial \tau^C_r \partial \theta^C_r} \\ \frac{\partial^2\pi^C_{mr}}{\partial w^C_r \partial \theta^C_r} & \frac{\partial^2\pi^C_{mr}}{\partial \tau^C_r \partial \theta^C_r} & \frac{\partial^2\pi^C_{mr}}{\partial \theta^{C2}_r} \end{bmatrix} = \begin{vmatrix} -2 & -X & \beta + \lambda_1 \\ -X & -2\kappa & \beta X \\ \beta + \lambda_1 & \beta X & -2(\lambda_2 + \beta\lambda_1) \end{vmatrix}$$

The principal minors of above Hessian matrix are $H^C_{r1} = -2 < 0$; $H^T_{r2} = 4\kappa - X^2 > 0$; and $H^T_{r3} = -2(\kappa(4\lambda_2 + Z^2) - X^2\lambda_2)$, respectively. Consequently, the profit function will be concave if $4\kappa > X^2$ and $\kappa(4\lambda_2 + Z^2) > X^2\lambda_2$.

Substituting optimal response for the manufacturer in Equation (1), profit function for the retailer is obtained as $\pi^C_{rr}(m^C_r) = \frac{2m^C_r \kappa\lambda_2(a - c_m - m^C_r + \rho^C_r)}{\kappa(4\lambda_2 - Z^2) - X^2\lambda_2}$. Therefore, the optimal response for the retailer is

obtained by solving $\frac{d\pi^C_{rr}}{dm^C_r} = 0$. After simplification, $m^C_r = \frac{a - c_m + \rho^C_r}{2}$. The profit function of the retailer

is also concave because $\frac{d^2\pi^C_{rr}}{dm^{C2}_r} = \frac{-4\kappa\lambda_2}{\kappa(4\lambda_2 - Z^2) - X^2\lambda_2}$. Substituting optimal responses in Equation (3), the simplified value of the SW in Scenario RC is obtained as follows:

$$\pi^C_{gr}(\rho^C_r) = \frac{\kappa\lambda_2(a - c_m + \rho^C_r)((a - c_m)(14\kappa\lambda_2 - 3\kappa Z^2 - 3X^2\lambda_2) + \kappa Z^2 \rho^C_r - (2\kappa - X^2)\lambda_2 \rho^C_r)}{4(\kappa(4\lambda_2 - Z^2) - X^2\lambda_2)^2}$$

Therefore, the optimal subsidy rate will be obtained by solving $\frac{d\pi^C_{gr}}{d\rho^C_r} = 0$. On simplification,

we obtain the value of ρ^C_r as presented in Proposition (2). The SW under the RS game is concave

because, $\frac{d^2\pi^C_{rm}}{d\rho^{C2}_r} = -\frac{\kappa\lambda_2\Delta_1}{2(\kappa(4\lambda_2 - (\beta - \lambda_1)^2) - X^2\lambda_2)^2} < 0$ where $\Delta_1 = \kappa(2\lambda_2 - (\beta - \lambda_1)^2) - X^2\lambda_2$. By using back

substitution, we obtain optimal decision as presented in Lemma 2.

Appendix C. Proof of Theorem 1

The following inequalities ensure the proof of first part of Theorem 1:

$$\theta_m^C - \theta_r^C = 0;\ \tau_m^C - \tau_r^C = 0;\ \pi_{gm}^C - \pi_{gr}^C = 0;\ \rho_{0m}^C - \rho_{0r}^C = \frac{(a-c_m)(\kappa Z^2 + X^2\lambda_2)}{\Delta_1} > 0;$$

Differentiating optimal decisions in Lemma 1,2, with respect to λ_1 and λ_2, the following relations are obtained:

$$\frac{d\tau_m^C}{d\lambda_1} = \frac{d\tau_r^C}{d\lambda_1} = \frac{(a-c_m)XZ\kappa\lambda_2}{\Delta_1^2} < 0;\ \frac{d\tau_m^C}{d\lambda_2} = \frac{d\tau_r^C}{d\lambda_2} = \frac{-(a-c_m)XZ^2\kappa}{\Delta_1^2} < 0;\ \frac{d\theta_m^C}{d\lambda_1} = \frac{d\theta_r^C}{d\lambda_1} = \frac{-(a-c_m)\kappa(2Z^2\kappa+\Delta_1)}{\Delta_1^2} < 0;$$

$$\frac{d\theta_m^C}{d\lambda_2} = \frac{d\theta_r^C}{d\lambda_2} = \frac{-(a-c_m)N_1 Z\kappa}{\Delta_1^2} < 0;\ \frac{d\pi_{gm}^C}{d\lambda_1} = \frac{d\pi_{gr}^C}{d\lambda_1} = \frac{-2(a-cm)^2 Z\kappa^2\lambda_2}{\Delta_1^2} < 0;\ \frac{d\pi_{gm}^C}{d\lambda_2} = \frac{d\pi_{gr}^C}{d\lambda_2} = \frac{-(a-c_m)^2 Z^2\kappa^2}{\Delta_1^2} < 0;$$

$$\frac{d\rho_m^C}{d\lambda_1} = \frac{-12(a-c_m)^2 Z\kappa^2\lambda_2}{\Delta_1^2} < 0;\ \frac{d\rho_m^C}{d\lambda_2} = \frac{-6(a-c_m)^2 Z^2\kappa^2}{\Delta_1^2} < 0;\ \frac{d\rho_r^C}{d\lambda_1} = \frac{-8(a-c_m)^2 Z\kappa^2\lambda_2}{\Delta_1^2} < 0;\ \frac{d\rho_r^C}{d\lambda_2} = \frac{-4(a-c_m)^2 Z^2\kappa^2}{\Delta_1^2} <$$

$0.$ The above inequalities supports the claim in Theorem 1.

Appendix D. Proof of Theorem 2

The following inequalities ensure the proof of first part of Theorem 2:

$$\theta_r^{RE} - \theta_m^{RE} = \frac{(a-c_m)(Z^2(X^4+16N_1\kappa)\lambda_2^2 + Z^4\kappa(Z^2\kappa-2N_3\lambda_2) + 2X^2\lambda_2^2(5M_1\kappa+7N_3\lambda_2))}{2\Delta_{2m}\Delta_{2r}} > 0;$$

$$\tau_r^{RE} - \tau_m^{RE} = \frac{(a-c_m)X\lambda_2^2(24M_2\kappa\lambda_2 + M_3 Z^2\kappa + M_4 X^2\lambda_2)}{\Delta_{2m}\Delta_{2r}} > 0;\ \eta_r^{RE} - \eta_m^{RE} = \frac{2(6N_1-X^2)\lambda_2^2 - N_4 Z^2\lambda_2 + \kappa Z^4}{2\kappa(5\lambda_2-Z^2)(14\lambda_2-Z^2)} > 0;$$

$$\pi_{gr}^{RE} - \pi_{gm}^{RE} = \frac{(a-c_m)^2\lambda_2(2X^2(102\kappa-7X^2)\lambda_2^3 + (96\kappa^2-40X^2\kappa+X^4)Z^2\lambda_2^2 - 2(13\kappa-X^2)\kappa Z^4\lambda_2 + \kappa^2 Z^6)}{4\Delta_{2m}\Delta_{2r}} > 0.$$

Differentiating optimal decisions in Lemma 3,4, with respect to λ_1 and λ_2, the following relations are obtained:

$$\frac{d\tau_m^{RE}}{d\lambda_1} = \frac{-2(a-c_m)(M_4+6\lambda_2)M_3 XZ\kappa\lambda_2}{\Delta_{2m}^2} < 0;\ \frac{d\tau_m^{RE}}{d\lambda_2} = \frac{-(a-c_m)\kappa XZ^2(48\lambda_2^2 + 28M_2\lambda_2 + Z^4)}{\Delta_{2m}^2} < 0;$$

$$\frac{d\theta_m^{RE}}{d\lambda_1} = \frac{-(a-c_m)\kappa(2(M_4+6\lambda_2)Z^2 M_3^2\kappa + (2(17M_1+22\lambda_2)\lambda_2 + Z^4)\Delta_{2m})}{M_4\Delta_{2m}^2} < 0;$$

$$\frac{d\theta_m^{RE}}{d\lambda_2} = \frac{-(a-c_m)\kappa Z(M_4 X^2 Z^2 + 2M_3(4M_2\kappa+7N_1\lambda_2+2\kappa\lambda_2))}{M_4\Delta_{2m}^2} < 0;\ \frac{d\pi_{gm}^{RE}}{d\lambda_1} = \frac{-2(a-c_m)^2\kappa^2 Z\lambda_2}{\Delta_{2m}^2} < 0;$$

$$\frac{d\pi_{gm}^{RE}}{d\lambda_2} = \frac{-(a-c_m)^2 Z^2\kappa^2(48\lambda_2^2 + 28M_2\lambda_2 + Z^4)}{\Delta_{2m}^2} < 0;\ \frac{d\tau_r^{RE}}{d\lambda_1} = \frac{-2(a-c_m)XZ\lambda_2((4\kappa+X^2)\lambda_2^2 + 10M_1\kappa\lambda_2 + \kappa Z^4)}{\Delta_{2r}^2} < 0;$$

$$\frac{d\tau_r^{RE}}{d\lambda_2} = \frac{-(a-c_m)XZ^2((4\kappa+X^2)\lambda_2^2 + M_1\kappa\lambda_2 + \kappa Z^4)}{\Delta_{2r}^2} < 0;\ \frac{d\theta_r^{RE}}{d\lambda_2} = \frac{-(a-c_m)Z(2\kappa\Delta_{2r} + X^2(M_2 Z^2\kappa + (3N_1+2\kappa)\lambda_2^2))}{2\Delta_{2r}^2} < 0;$$

$$\frac{d\theta_r^{RE}}{d\lambda_1} = \frac{-(a-c_m)(N_1\lambda_2^2(16\kappa\lambda_2 + X^2(Z^2+\lambda_2)) + 4(\kappa^2-X^4)\lambda_2^3 + N_3 X^2\lambda_2^3 + 4\kappa^2\lambda_2^2(7\lambda_2-4Z^2) + Z^2\kappa(Z^4\kappa - 4Z^2\kappa\lambda_2 + 4X^2\lambda_2(M_2+\lambda_2)))}{2\Delta_{2r}^2} <$$

$0;$

$$\frac{d\pi_{gr}^{RE}}{d\lambda_1} = \frac{-(a-c_m)^2 Z^2(3M_1\kappa + N_2\lambda_2)(M_1\kappa + X^2\lambda_2)}{\Delta_{2r}^2} < 0;\ \frac{d\pi_{gr}^{RE}}{d\lambda_2} = \frac{-(a-c_m)^2 Z\lambda_2(3M_1\kappa + N_2\lambda_2)(M_1\kappa + X^2\lambda_2)}{\Delta_{2r}^2} < 0.$$

The above inequalities supports the claim in Theorem 2.

Appendix E. Proof of Theorem 3

The following inequalities ensure the proof of first part of Theorem 3:

$$\theta_r^T - \theta_m^T = \frac{(a-c_m)Z\kappa^2(N_5 Z^2\kappa + (2(X^2+4\kappa)N_2 + N_4^2)\lambda_2)}{\Delta_{3m}\Delta_{3r}} > 0;\ \mu_r^T - \mu_m^T = \frac{N_2^2\lambda_2 + \kappa(8\kappa\lambda_2 - Z^2 N_5)}{2(N_2+\kappa)N_5\lambda_2} > 0$$

$$\tau_r^T - \tau_m^T = \frac{(a-c_m)X(N_5 Z^2\kappa(M_2\kappa + 2N_1\lambda_2) + N_2\lambda_2(2\kappa(25M_1\kappa - 2X^2 M_2 + (7N_1-2X^2)\lambda_2) + 3X^4\lambda_2))}{2\Delta_{3m}\Delta_{3r}} > 0;$$

$$\pi_{gr}^T - \pi_{gm}^T = \frac{(a-c_m)^2\kappa(X^6\lambda_2^2 + X^2\kappa^2(Z^4 + 44\lambda_2^2) + 2Z^2\kappa(N_4(N_4+4\kappa)\lambda_2 + (M_3-6Z^2)\kappa^2))}{4\Delta_{3m}\Delta_{3r}} > 0.$$

Differentiating optimal decisions in Propositions 5 and 6, with respect to λ_1 and λ_2, the following relations are obtained:

$$\frac{d\tau_m^T}{d\lambda_1} = \frac{-2(a-c_m)N_4 N_5 XZ\kappa\lambda_2}{\Delta_{3m}^2} < 0;\ \frac{d\tau_m^T}{d\lambda_2} = \frac{-(a-c_m)N_4 N_5 XZ^2\kappa}{\Delta_{3m}^2} < 0;\ \frac{d\theta_m^T}{d\lambda_1} = \frac{-(a-c_m)N_2\kappa(2N_2 Z^2\kappa + \Delta_{3m})}{\Delta_{3m}^2} < 0;$$

$$\frac{d\theta_m^T}{d\lambda_2} = \frac{-(a-c_m)N_4^2 N_5 Z\kappa}{\Delta_{3m}^2} < 0;\ \frac{d\pi_{gm}^T}{d\lambda_1} = \frac{-2(a-c_m)^2 N_5^2 Z\kappa^2\lambda_2}{\Delta_{3m}^2} < 0;\ \frac{d\pi_{gm}^T}{d\lambda_2} = \frac{-(a-c_m)^2 N_5^2 Z^2\kappa^2}{\Delta_{3m}^2} < 0;$$

$$\frac{d\tau_r^T}{d\lambda_1} = \frac{2(a-c_m)X\kappa(2\kappa^2 + 9N_1\kappa + X^4)Z\lambda_2}{\Delta_{3r}^2} < 0;\ \frac{d\tau_r^T}{d\lambda_2} = \frac{-(a-c_m)(N_2+\kappa)N_2 Z^2 X\kappa}{\Delta_{3r}^2} < 0;$$

$$\frac{d\theta_r^T}{d\lambda_1} = \frac{-(a-c_m)\kappa(N_2^2\lambda_2 + N_3 Z^2\kappa)(N_2+\kappa)}{\Delta_{3r}^2} < 0;\ \frac{d\theta_r^T}{d\lambda_2} = \frac{-(a-c_m)(N_2+\kappa)N_2^2 Z\kappa}{\Delta_{3r}^2} < 0;$$

$$\frac{d\pi_{gr}^T}{d\lambda_1} = \frac{-2(a-c_m)^2(N_2+\kappa)^2 Z\kappa^2\lambda_2}{\Delta_{3r}^2} < 0;\ \frac{d\pi_{gr}^T}{d\lambda_2} = \frac{-(a-c_m)^2(N_2+\kappa)^2 Z^2\kappa^2}{\Delta_{3r}^2} < 0.$$ The above inequalities supports the claim in Theorem 3.

Appendix F. Proof of Theorem 4

Differences among GLs under the MS game are found as follows:

$$\theta_m^C - \theta_m^{RE} = \frac{6(a-c_m)Z\kappa\lambda_2(6\kappa\lambda_2+\Delta_1)}{\Delta_1\Delta_{2m}} > 0; \quad \theta_m^C - \theta_m^T = \frac{36(a-c_m)\kappa^3 Z\lambda_2}{\Delta_1\Delta_{3m}} > 0.$$

Similarly, differences among GLs under the RS game are found as follows:

$$\theta_r^C - \theta_r^{RE} = \frac{(a-c_m)Z(4\kappa\lambda_2+\Delta_1)(2\kappa\lambda_2+\Delta_1)}{2\Delta_1\Delta_{2r}} > 0; \quad \theta_r^C - \theta_r^T = \frac{(a-c_m)\kappa^2 Z(4\kappa\lambda_2+\Delta_1)}{\Delta_1\Delta_{3r}} > 0.$$

Differences among sales volumes under the MS game are found as follows:

$$Q_m^C - Q_m^{RE} = \frac{12(a-cm)\kappa\lambda_2^2(6\kappa\lambda_2+\Delta_1)}{2\Delta_1\Delta_{2m}} > 0; \quad Q_m^C - Q_m^T = \frac{12(a-cm)\kappa^2\lambda_2(6\kappa\lambda_2+\Delta_1)}{2\Delta_1\Delta_{2m}} > 0.$$

Differences among sales volumes under the RS game are found as follows:

$$Q_r^C - Q_r^{RE} = \frac{(a-cm)\lambda_2(4\kappa-\lambda_2+\Delta_1)(2\kappa-\lambda_2+\Delta_1)}{\Delta_1\Delta_{2r}} > 0; \quad Q_r^C - Q_r^T = \frac{(a-cm)\kappa(4\kappa-\lambda_2+\Delta_1)(2\kappa-\lambda_2+\Delta_1)}{\Delta_1\Delta_{3r}} > 0.$$

The above relations ensure the claim in Theorem 4.

Appendix G. Proof of Theorem 5

The profit differences for the manufacturer in three subsidy policies are obtained as follows:

$$\pi_{mm}^C - \pi_{mm}^{RE} = \frac{6(a-c_m)^2\kappa\lambda_2^2(2M_4\kappa\Delta_1+\lambda_2\Delta_{3m})}{\Delta_1^2\Delta_{2m}} > 0; \quad \pi_{mm}^C - \pi_{mm}^T = \frac{6(a-c_m)^2\kappa^2\lambda_2(6\kappa\lambda_2+\Delta_1)\Delta_1+\lambda_2\Delta_{3m})}{\Delta_1^2\Delta_{3m}} > 0;$$

$$\pi_{mr}^C - \pi_{mr}^{RE} = \frac{(a-c_m)^2\lambda_2(4\kappa\lambda_2+\Delta_1)(3Z^4\kappa^2+(18\kappa N_1+X^4)\lambda_2^2+2\kappa\lambda_2(M_1\kappa-2Z^2(N_2+\kappa)))}{4\Delta_1^2\Delta_{2r}} > 0;$$

$$\pi_{mr}^C - \pi_{mr}^T = \frac{(a-c_m)^2\kappa(4\kappa\lambda_2+\Delta_1)(\Delta_1^2+2\Delta_1(N_3+\kappa)\lambda_2+8\kappa^2\lambda_2^2)}{4\Delta_1^2\Delta_{3r}} > 0.$$

Similarly, the profit differences for the retailer in three subsidy policies are obtained as follows:

$$\pi_{rm}^C - \pi_{rm}^{RE} = \frac{48(a-cm)^2\kappa^2\lambda_2^3(\Delta_1+6\kappa\lambda_2)(M_1^2\kappa+\lambda_2(11\Delta_1+Z^2(X^2+2\kappa)+14\kappa\lambda_2))}{\Delta_1^2\Delta_{2m}^2} > 0;$$

$$\pi_{rm}^C - \pi_{rm}^T = \frac{48(a-c_m)^2\kappa^3\lambda_2^2(N_4^2\lambda_2((\kappa+X^2)\lambda_2+2\Delta_1)+(N_4+3\kappa)\kappa^2 Z^2)}{\Delta_1^2\Delta_{3m}^2} > 0;$$

$$\pi_{rr}^C - \pi_{rr}^{RE} = \frac{(a-c_m)^2\lambda_2(\Delta_1+4\kappa\lambda_2)(\Delta_1^2-2Z^2\kappa\Delta_1+2\Delta_1(N_3+X^2+\kappa)\lambda_2+8\kappa^2\lambda_2^2)}{\Delta_1^2\Delta_{2r}} > 0;$$

$$\pi_{rr}^C - \pi_{rr}^T = \frac{(a-c_m)^2\kappa(4\kappa\lambda_2+\Delta_1)(\Delta_1^2+2\Delta_1(N_3+\kappa)\lambda_2+8\kappa^2\lambda_2^2)}{2\Delta_1^2\Delta_{3r}} > 0;$$

Finally, the differences among used product return rates are obtained as follows:

$$\tau_m^C - \tau_m^{RE} = \frac{36(a-c_m)X\kappa\lambda_2^3}{\Delta_1\Delta_{2m}} > 0; \quad \tau_m^C - \tau_m^T = \frac{6(a-c_m)X\kappa\lambda_2(6\kappa\lambda_2+\Delta_1)}{\Delta_1\Delta_{3m}} > 0;$$

$$\tau_r^C - \tau_r^{RE} = \frac{(a-c_m)X\lambda_2^2(4\kappa\lambda_2+\Delta_1)}{\Delta_1\Delta_{2r}} > 0; \quad \tau_r^C - \tau_r^T = \frac{(a-c_m)X(4\kappa\lambda_2+\Delta_1)}{2\Delta_1\Delta_{3r}} > 0.$$

Therefore, the theorem is proved.

Appendix H. Proof of Theorem 6

The differences among SWs measures in the MS and RS games are computed as follows:

$$\pi_{gm}^C - \pi_{gm}^{RE} = \frac{36(a-c_m)^2\kappa^2\lambda_2^3}{\Delta_1\Delta_{2m}} > 0; \quad \pi_{gm}^C - \pi_{gm}^T = \frac{36(a-c_m)\kappa^3\lambda_2^2}{\Delta_1\Delta_{3m}} > 0;$$

$$\pi_{gr}^C - \pi_{gr}^{RE} = \frac{(a-c_m)^2\lambda_2(2\kappa\lambda_2+\Delta_1)^2}{4\Delta_1\Delta_{2r}} > 0; \quad \pi_{gr}^C - \pi_{gr}^T = \frac{(a-c_m)^2\kappa(4\kappa\lambda_2+\Delta_1)^2}{4\Delta_1\Delta_{3r}} > 0.$$

Similarly, the differences among total amount of subsidies in the MS and RS games are computed as follows:

$$GI_m^C - GI_m^{RE} = \frac{6(a-c_m)^2\kappa\lambda_2^2(2\kappa\Delta_{2m}^2-M_4 X^2\lambda_2\Delta_1^2)}{\Delta_1^2\Delta_{2m}^2} > 0; \quad GI_m^C - GI_m^T = \frac{6(a-c_m)^2\kappa^2\lambda_2(2\lambda_2\Delta_{3m}^2-N_5 Z^2\kappa\Delta_1^2)}{\Delta_1^2\Delta_{3m}^2} > 0;$$

$$GI_r^C - GI_r^{RE} = \frac{(a-c_m)^2\lambda_2(N_3\lambda_2-Z^2\kappa)(4\kappa\Delta_{2r}^2-(M_2+\lambda_2)X^2\lambda_2\Delta_1^2)}{\Delta_1^2\Delta_{2r}^2} > 0; \quad GI_r^C - GI_r^T =$$

$$\frac{(a-c_m)^2\kappa(N_3\lambda_2-Z^2\kappa)(4\lambda_2\Delta_{3r}^2-(N_2+\kappa)Z^2\kappa\Delta_1^2)}{\Delta_1^2\Delta_{3r}^2} > 0$$

Therefore, the theorem is proved.

References

1. Abdallah, T.; Diabat, A.; Simchi-Levi, D. Sustainable supply chain design: A closed-loop formulation and sensitivity analysis. *Prod. Plan. Control* **2011**, *23*, 120–133. [CrossRef]

2. Govindan, K.; Soleimani, H.; Kannan, D. Reverse logistics and closed-loop supply chain: A comprehensive review to explore the future. *Eur. J. Oper .* **2015**, *240*, 603–626. [CrossRef]

3. Guo, J.; He, L.; Gen, M. Optimal strategies for the closed-loop supply chain with the consideration of supply disruption and subsidy policy. *Comput. Ind. Eng.* **2019**, *128*, 886–893. [CrossRef]

4. Iris, Ç.; Lam, J.S.L. A review of energy efficiency in ports: Operational strategies, technologies and energy management systems. *Renew. Sustain. Energy Rev.* **2019**, *112*, 170–182. [CrossRef]

5. Mishra, J.L.; Hopkinson, P.G.; Tidridge, G. Value creation from circular economy-led closed loop supply chains: A case study of fast-moving consumer goods. *Prod. Plan. Control.* **2018**, *29*, 509–521. [CrossRef]

6. Savaskan, R.C.; Bhattacharya, S.; Van Wassenhove, L.N. Closed-loop supply chain models with product Re-manufacturing. *Manag. Sci.* **2004**, *50*, 239–252. [CrossRef]

7. Shaharudin, M.R.; Tan, K.C.; Kannan, V.; Zailani, S. The mediating effects of product returns on the relationship between green capabilities and closed-loop supply chain adoption. *J. Clean. Prod.* **2019**, *211*, 233–246. [CrossRef]

8. Wang, Y.; Chang, X.; Chen, Z.; Zhong, Y.; Fan, T. Impact of subsidy policies on recycling and Re-manufacturing using system dynamics methodology: A case of auto parts in China. *J. Clean. Prod.* **2014**, *74*, 161–171. [CrossRef]

9. Helveston, J.P.; Liu, Y.; Feit, E.M.; Fuchsa, E.; Klampfl, E.; Michalekad, J.J. Will subsidies drive electric vehicle adoption? Measuring consumer preferences in the U.S. and China. *Transp. Res. Part A Policy Pract.* **2015**, *73*, 96–112. [CrossRef]

10. Han, X.; Feng, B.; Pu, X. Modelling decision behaviours in pricing game of closed-loop supply chains. *J. Oper. Soc.* **2015**, *66*, 1052–1060. [CrossRef]

11. Iris, Ç.; Asan, S.S. A review of genetic algorithm applications in supply chain network design. In *Computational Intelligence Systems in Industrial Engineering*; Atlantis Press: Paris, France, 2012; pp. 203–230.

12. Sinayi, M.; Rasti-Barzoki, M. A game theoretic approach for pricing, greening, and social welfare policies in a supply chain with government intervention. *J. Clean. Prod.* **2018**, *196*, 1443–1458. [CrossRef]

13. Liu C.; Xia G. Research on the dynamic interrelationship among R&D investment, technological innovation and economic growth in China. *Sustainability* **2018**, *10*, 4260.

14. Dey, K.; Roy, S.; Saha, S. The impact of strategic inventory and procurement strategies on green product design in a two-period supply chain. *Int. J. Prod. Res.* **2019**, *57*, 1915–1948. [CrossRef]

15. Zhu, W.; He, Y. Green pro duct design in supply chains under competition. *Eur. J. Oper. Res.* **2017**, *258*, 165–180. [CrossRef]

16. Ottman, J.A. *Green Marketing: Challenges & Opportunities for the New Marketing Age*; NTC Publishing Group: Lincolnwood, IL, USA, 1995.

17. Basiri, Z.; Heydari, J. A mathematical model for green supply chain coordination with substitutable products. *J. Clean. Prod.* **2017**, *145*, 232–249. [CrossRef]

18. Ghosh, D.; Shah, J. Supply chain analysis under green sensitive consumer demand and cost sharing contract. *Int. J. Prod. Econ.* **2015**, *164*, 319–329. [CrossRef]

19. Song, H.; Gao, X. Green supply chain game model and analysis under revenue-sharing contract. *J. Clean. Prod.* **2018**, *170*, 183–192. [CrossRef]

20. Ghosh, D.; Shah, J. A comparative analysis of greening policies across supply chain structures. *Int. J. Prod. Econ.* **2012**, *135*, 568–583. [CrossRef]

21. Liu, P.; Yi, S.P. Pricing policies of green supply chain considering targeted advertising and product green degree in the Big Data environment. *J. Clean. Prod.* **2017**, *164*, 1614–1622. [CrossRef]

22. Yang, D.; Xiao, T. Pricing and green level decisions of a green supply chain with governmental interventions under fuzzy uncertainties. *J. Clean. Prod.* **2017**, *149*, 1174–1187. [CrossRef]

23. Nielsen, I.E.; Majumder, S.; Saha, S. Exploring the intervention of intermediary in a green supply chain. *J. Clean. Prod.* **2019**, *233*, 1525–1544 [CrossRef]

24. Chen, X.; Wang, X.; Zhou, M. Firms' green R&D cooperation behavior in a supply chain: Technological spillover, power and coordination. *Int. J. Prod. Econ.* **2019**, *218*, 118–134.

25. Huang, Y.; Wang, K.; Zhang, T.; Pang, C. Green supply chain coordination with greenhouse gases emissions management: A game-theoretic approach. *J. Clean. Prod.* **2016**, *112*, 2004–2014. [CrossRef]

26. Ranjan, A.; Jha, J.K. Pricing and coordination strategies of a dual-channel supply chain considering green quality and sales effort. *J. Clean. Prod.* **2019**, *218*, 409–424. [CrossRef]

27. Hong, I.; Yeh, J.S. Modeling closed-loop supply chains in the electronics industry: A retailer collection application. *Transp. Res. Part E Logist. Transp. Rev.* **2012**, *48*, 817–829. [CrossRef]

28. Ma, Z.J.; Zhang, N.; Dai, Y.; Hu, S. Managing channel profits of different cooperative models in closed-loop supply chains. *Omega* **2016**, *59*, 251–262.

29. Saha, S.; Sarmah, S.P.; Moon, I. Dual channel closed-loop supply chain coordination with a reward-driven Re-manufacturing policy. *Int. J. Prod. Res.* **2016**, *54*, 1503–1517. [CrossRef]

30. Zhang, C.T.; Ren, M.L. Closed-loop supply chain coordination strategy for the remanufacture of patented products under competitive demand. *Appl. Math.* **2016**, *40*, 6243–6255. [CrossRef]

31. He, Q.; Wang, N.; Yang, Z.; He, Z.; Jiang, B. Competitive collection under channel inconvenience in closed-loop supply chain. *Eur. J. Oper. Res.* **2019**, *155*, 155–166. [CrossRef]

32. Hong, X.; Xu, L.; Du, P.; Wang, W. Joint advertising, pricing and collection decisions in a closed-loop supply chain. *Int. J. Prod. Econ.* **2015**, *167*, 12–22. [CrossRef]

33. Johari, M.; Motlagh, S.H. Coordination of social welfare, collecting, recycling and pricing decisions in a competitive sustainable closed-loop supply chain: A case for lead-acid battery. *Ann. Oper. Res.* **2019**. [CrossRef]

34. Zhao, J.; Wei, J.; Sun, X. Coordination of fuzzy closed-loop supply chain with price dependent demand under symmetric and asymmetric information conditions. *Ann. Oper. Res.* **2017**, *257*, 469–489. [CrossRef]

35. Xie, J.P.; Liang, L.; Liu, L.; Ieromonachou, P. Coordination contracts of dual-channel with cooperation advertising in closed-loop supply chains. *Int. J. Prod. Econ.* **2017**, *183*, 528–538. [CrossRef]

36. Choi, T.M.; Li, Y.; Xu, L. Channel leadership, performance and coordination in closed loop supply chains. *Int. J. Prod. Econ.* **2013**, *146*, 371–380. [CrossRef]

37. Wang, W.; Zhang, Y.; Zhang, K.; Bai, T.; Shang, J. Reward-penalty mechanism for closed-loop supply chains under responsibility-sharing and different power structures. *Int. J. Prod. Econ.* **2015**, *170*, 178–190. [CrossRef]

38. Gao, J.; Han, H.; Hou, L.; Wang, H. Pricing and effort decisions in a closed-loop supply chain under different channel power structures. *J. Clean. Prod.* **2016**, *112*, 2043–2057. [CrossRef]

39. Zheng, B.; Yang, C.; Yang, J.; Zhang, M. Dual channel closed loop supply chains: Forward channel competition, power structures and coordination. *Int. J. Prod. Res.* **2017**, *55*, 3510–3527. [CrossRef]

40. Bressanelli, G.; Perona, M.; Saccani, N. Challenges in supply chain redesign for the circular economy: A literature review and a multiple case study. *Int. J. Prod. Res.* **2008**. [CrossRef]

41. Coenen, J.; Van der Heijden, R.E.; Van Riel, A.C. Understanding approaches to complexity and uncertainty in closed-loop supply chain management: Past findings and future directions. *J. Clean. Prod.* **2018**, *201*, 1–13. [CrossRef]

42. Diallo, C.; Venkatadri, U.; Khatab, A.; Bhakthavatchalam, S. State of the art review of quality, reliability and maintenance issues in closed-loop supply chains with Re-manufacturing. *Int. J. Prod. Res.* **2017**, *55*, 1277–1296. [CrossRef]

43. Bottani, E.; Casella, G. Minimization of the environmental emissions of closed-loop supply chains: A case study of returnable transport assets management. *Sustainability* **2018**, *10*, 329. [CrossRef]

44. Garg, K.; Kannan, D.; Diabat, A.; Jha, P.C. A multi-criteria optimization approach to manage environmental issues in closed loop supply chain network design, *J. Clean. Prod.* **2015**, *100*, 297–314. [CrossRef]

45. Govindan, K.; Darbari, J.D.; Agarwal, V.; Jha, P.C. Fuzzy multi-objective approach for optimal selection of suppliers and transportation decisions in an eco-efficient closed loop supply chain network. *J. Clean. Prod.* **2017**, *165*, 1598–1619. [CrossRef]

46. Zhang, J.; Chang, Y.; Wang, C.; Zhang, L. The green efficiency of industrial sectors in China: A comparative analysis based on sectoral and supply-chain quantifications. *Resour. Conserv. Recycl.* **2018**, *132*, 269–277. [CrossRef]

47. Hafezalkotob, A. Modeling intervention policies of government in price-energy saving competition of green supply chains. *Comput. Ind. Eng.* **2018**, *119*, 247–261. [CrossRef]

48. Chen, J.Y.; Dimitrov, S.; Pun, H. The impact of government subsidy on supply chains' sustainability innovation. *Omega* **2019**, *86*, 42–58. [CrossRef]

49. Safarzadeh, S.; Rasti-Barzoki, M. A game theoretic approach for pricing policies in a duopolistic supply chain considering energy productivity, industrial rebound effect, and government policies. *Energy* **2019**, *167*, 92–105. [CrossRef]

50. Mitra, S.; Webster, S. Competition in Re-manufacturing and the effects of government subsidies. *Int. J. Prod. Econ.* **2008**, *111*, 287–298. [CrossRef]

51. Ma, W.; Zhao, Z.; Ke, H. Dual-channel closed-loop supply chain with government consumption-subsidy. *Eur. J. Oper. Res.* **2013**, *226*, 221–227. [CrossRef]

52. Shu, T.; Peng, Z.; Chen, S.; Wang, S.; Lai, K.K.; Yang, H. Government subsidy for Re-manufacturing or carbon tax rebate: Which is better for firms and a low-carbon economy. *Sustainability* **2017**, *9*, 156. [CrossRef]

53. Xiao, L.; Wang, X.; Chin, K.S.; Qin, Y. Competitive strategy in Re-manufacturing and the effects of government subsidy. *J. Syst. Sci. Syst. Eng.* **2017**, *26*, 417–432. [CrossRef]

54. Heydari, J.; Govindan, K.; Jafari, A. Reverse and closed loop supply chain coordination by considering government role. *Transp. Res. Part D Transp. Environ.* **2017**, *52*, 379–398. [CrossRef]

55. Jena, S.K.; Ghadge, A.; Sarmah, S.P. Managing channel profit and total surplus in a closed-loop supply chain network. *J. Oper. Soc.* **2018**, *69*, 1345–1356. [CrossRef]

56. Jena, S.K.; Sarmah, S.P.; Padhi, S.S. Impact of government incentive on price competition of closed-loop supply chain systems. *INFOR Inf. Syst. Oper. Res.* **2018**, *56*, 192–224. [CrossRef]

57. Wan, N.; Hong, D. The impacts of subsidy policies and transfer pricing policies on the closed-loop supply chain with dual collection channels. *J. Clean. Prod.* **2019**, *224*, 881–891. [CrossRef]

58. Saha, S.; Majumder, S.; Nielsen, I.E. Dilemma in two game structures for a closed-loop supply chain under the influence of government incentives. *J. Ind. Eng. Int.* **2019**. [CrossRef]

59. He, P.; He, Y.; Xu, H. Channel structure and pricing in a dual-channel closed-loop supply chain with government subsidy. *Int. J. Prod. Econ.* **2019**, *213*, 108–123. [CrossRef]

60. Nielsen, I.E.; Majumder, S.; Sana, S.; Saha, S. Comparative analysis of government incentives and game structures on single and two-period green supply chain. *J. Clean. Prod.* **2019**, *235*, 1371–1398. [CrossRef]

61. Dey, K.; Saha, S. Influence of procurement decisions in two-period green supply chain. *J. Clean. Prod.* **2018**, *190*, 388–402. [CrossRef]

62. Zhang, J.; Liu, Y.; Chang, Y.; Zhang, L. Industrial eco-efficiency in China: A provincial quantification using three-stage data envelopment analysis. *J. Clean. Prod.* **2017**, *143*, 238–249. [CrossRef]

63. Mussa, M.; Rosen, S. Monopoly and product quality. *J. Econ. Theory* **1978**, *18*, 301–317. [CrossRef]

64. Shu, T.; Huang, C.; Chen. S, Wang, S.; Lai, K.K. Trade-Old-for-Remanufactured Closed-Loop Supply Chains with Carbon Tax and Government Subsidies. *Sustainability* **2018**, *10*, 3935. [CrossRef]

65. Saha, S.; Majumder, S.; Nielsen, I.E. Is it a strategic move to subsidized consumers instead of the manufacturer? *IEEE Access* **2019**, doi:10.1109/ACCESS.2019.2954376. [CrossRef]

66. Huang, J.; Leng, M.; Liang, L.; Liu, J. Promoting electric automobiles: Supply chain analysis under a government's subsidy incentive scheme. *IIE Trans.* **2013**, *45*, 826–844. [CrossRef]

67. Fu, J.; Chen, X.; Hu, Q. Subsidizing strategies in a sustainable supply chain. *J. Oper. Soc.* **2017**, *69*, 283–295. [CrossRef]

68. Liu, Y.; Quan, B., Xu, Q.; Forrest, J.Y. Corporate social responsibility and decision analysis in a supply chain through government subsidy. *J. Clean. Prod.* **2019**, *2098*, 436–447. [CrossRef]

69. Ferrer, G.; Swaminathan, J.M. Managing new and differentiated remanufactured products. *Eur. J. Oper. Res.* **2010**, *203*, 370–379. [CrossRef]

70. Saha, S. Supply chain coordination through rebate induced contracts. *Transp. Res. E Logist Transp. Rev.* **2013**, *50*, 120–137. [CrossRef]

71. Saha, S.; Goyal, S.K. Supply chain coordination contracts with inventory level and retail price dependent demand. *Int. J. Prod Econ.* **2015**, *161*, 140–152. [CrossRef]

MDPI
St. Alban-Anlage 66
4052 Basel
Switzerland
Tel. +41 61 683 77 34
Fax +41 61 302 89 18
www.mdpi.com

Applied Sciences Editorial Office
E-mail: applsci@mdpi.com
www.mdpi.com/journal/applsci